AF356540

# Statistical Inference from High Dimensional Data

# Statistical Inference from High Dimensional Data

Editor

**Carlos Fernandez-Lozano**

MDPI • Basel • Beijing • Wuhan • Barcelona • Belgrade • Manchester • Tokyo • Cluj • Tianjin

*Editor*
Carlos Fernandez-Lozano
Department of Computer Science,
Faculty of Computer Science,
University of A Coruña
Spain

*Editorial Office*
MDPI
St. Alban-Anlage 66
4052 Basel, Switzerland

This is a reprint of articles from the Special Issue published online in the open access journal *Entropy* (ISSN 1099-4300) (available at: https://www.mdpi.com/journal/entropy/special_issues/high_dimensional).

For citation purposes, cite each article independently as indicated on the article page online and as indicated below:

LastName, A.A.; LastName, B.B.; LastName, C.C. Article Title. *Journal Name* **Year**, *Volume Number*, Page Range.

**ISBN 978-3-0365-0944-0 (Hbk)**
**ISBN 978-3-0365-0945-7 (PDF)**

# Contents

# About the Editor

**Carlos Fernandez-Lozano** (Assistant Professor at the University of A Coruña) is a biomedical data scientist with deep interest in discovering the complex relationships between different biological levels. His research track is multidisciplinary as he was trained in computer science, machine learning, bioinformatics, and biostatistics. His research is focused on how biological interactions are manifested at the disease level through the use, development and application of kernel-based computational approaches that integrate different levels of biological data on the microorganism, gene, protein, and medical imaging axis.

# Preface to "Statistical Inference from High Dimensional Data"

The contributions to this Special Issue "Statistical Inference from High Dimensional Data" show that high-dimensional data exist in real-world data. Moreover, complex, heterogeneous, and multifactorial diseases such as cancer are studied from a multi-dimensional point of view: gene expression data, copy number alteration, or metabolites data, for example.

In [1], a novel integrative analysis of CNA and GE based on differentially co-expressed gene sets is proposed to explore the impact on cancer progression and prognosis by using set-wise interaction distances with Renyi's relative entropy. Using lower-grade glioma data from the TCGA project, this model revealed pathways that are consistent with the molecular pathophysiology of the glioma. Using renal cell carcinoma data from the same multidimensional genome atlas in [2], the authors explain common problems that may occur during metabolomic and transcriptomic data integration finding, at the same time, poor-prognosis clusters of patients that decrease the performance of the models in renal cell carcinoma.

Omic data analysis must deal with some difficulties due to the high number of features according to the low number of cases. Feature selection approaches such as those proposed in [3] rank genes according to their relevance with the variable of interest. In this case, this approach evaluates a maximum relevance minimum redundancy approach over gene expression data. Feature selection is a well-known approach in machine learning, in [4] a review of feature selection approaches using data integration of different omic platforms is presented. Integrative approaches add domain knowledge from external biological resources during the feature selection process, adapting techniques to the biological problem. In [5], the authors discuss how to initially model gene data and the impact of this process on the later integrative analysis. Moreover, in [6], the authors reviewed gene set analysis approaches used for microarrays, transcriptome data, and genome-wide association data analysis. In fact, it is necessary to focus on the initial step of data analysis and how a researcher can move from a sequence of letters to a complete genome. For example, with some organisms with short genomes such as HIV-1, a novel approach is proposed where the process is carried out with the complete genome sequence [7], conventional approaches for virus subtype classification are done using multi-sequence alignments.

Thus, novel methodological approaches are necessary with the high dimensional data and quantity of research that is being completed. For example, a novel multi-objective feature selection algorithm for inference from high-dimensional data is proposed [8] using entropy-based measures and evaluated using the earth observation data set for land-cover classification. Feature selection approaches using mutual information or Pearson correlation coefficients to reduce the number of single nucleotide polymorphisms before genome-wide association studies [9]. Problems of higher complexity tend to incorporate variables ranging to thousands, not all of them relevant or informative, following an approach to explore local networks of score-based approaches in [10], the authors investigated different levels of candidate parent sets pruning. In [11], an extension of a parsimonious topic model is proposed for text corpora, increasing the model sparsity and optimizing the analysis of the same word across different topics. Moreover, a novel semi-supervised method for dimensionality reduction extending anchor graph regularization is proposed in [12] using kernel-based approaches. In some cases, the difficulty lies in modeling relationships such as the structure and the function of biological molecules such as proteins [13], the authors calculate residue cluster classes from 3D structures to effectively learn the structural and functional classifications of proteins using machine learning.

At present, it is increasingly likely to have more and more data available, mainly due to the ability to generate such data using multiple and different sensors. Stellar spectra analysis is proposed in [14] using artificial neural networks applied to the Morgan–Keenan system for the classification of stars. A three-year dataset from the traffic police department was used to develop a complex risk model of e-bike riders in [15] using gradient boosted decision trees. At times, new data came from image-generating sensors, for example, in [16] the authors developed a machine learning model in image-based authentication using mouse-click data. Finally, a kernel-based approach for contrast-enhanced spectral mammography is proposed as a tool for breast cancer diagnosis using textural features with a feature importance analysis using an additional tree-based approach [17].

In summary, this Special Issue provides insight into statistical inference from high dimensional data analysis. This type of data is very common in the biological domain although it is increasingly possible to generate and analyze data that meet high dimensionality requirements offering an interesting overview of novel methods to deal with dimensionality reduction, feature selection, and data integration approaches suggesting the need for further research.

Acknowledgments

We express our thanks to the authors of the contributions of this Special Issue, and the journal referees for their valuable comments and suggestions. Finally, special thanks to the journal Entropy for its support during the development of this Special Issue.

Conflicts of Interest

The authors declare no conflict of interest.

**References**

1. Cho, S.B. Set-Wise Differential Interaction between Copy Number Alterations and Gene Expressions of Lower-Grade Glioma Reveals Prognosis-Associated Pathways. *Entropy* **2020**, *22*, 1434. https://doi.org/10.3390/e22121434.

2. Gogolewski, K.; Kostecki, M.; Gambin, A. Low Entropy Sub-Networks Prevent the Integration of Metabolomic and Transcriptomic Data. *Entropy* **2020**, *22*, 1238. https://doi.org/10.3390/e22111238.

3. Das, S.; Rai, S.N. Statistical Approach for Biologically Relevant Gene Selection from High-Throughput Gene Expression Data. *Entropy* **2020**, *22*, 1205. https://doi.org/10.3390/e22111205.

4. Yousef, M.; Kumar, A.; Bakir-Gungor, B. Application of Biological Domain Knowledge Based Feature Selection on Gene Expression Data. *Entropy* **2021**, *23*, 2. https://doi.org/10.3390/e23010002.

5. Huminiecki, Ł. Models of the Gene Must Inform Data-Mining Strategies in Genomics. *Entropy* **2020**, *22*, 942. https://doi.org/10.3390/e22090942.

6. Das, S.; McClain, C.J.; Rai, S.N. Fifteen Years of Gene Set Analysis for High-Throughput Genomic Data: A Review of Statistical Approaches and Future Challenges. *Entropy* **2020**, *22*, 427. https://doi.org/10.3390/e22040427.

7. Ma, Y.; Yu, Z.; Tang, R.; Xie, X.; Han, G.; Anh, V.V. Phylogenetic Analysis of HIV-1 Genomes Based on the Position-Weighted K-mers Method. *Entropy* **2020**, *22*, 255. https://doi.org/10.3390/e22020255.

8. Koprivec, F.; Kenda, K.; Šircelj, B. FASTENER Feature Selection for Inference from Earth Observation Data. *Entropy* **2020**, *22*, 1198. https://doi.org/10.3390/e22111198.

9. Guo, H.; Yu, Z.; An, J.; Han, G.; Ma, Y.; Tang, R. A Two-Stage Mutual Information Based Bayesian Lasso Algorithm for Multi-Locus Genome-Wide Association Studies. *Entropy* **2020**, *22*, 329.

https://doi.org/10.3390/e22030329.

10. Guo, Z.; Constantinou, A.C. Approximate Learning of High Dimensional Bayesian Network Structures via Pruning of Candidate Parent Sets. *Entropy* 2020, *22*, 1142. https://doi.org/10.3390/e22101142.

11. Wang, H.; Miller, D. Improved Parsimonious Topic Modeling Based on the Bayesian Information Criterion. *Entropy* **2020**, *22*, 326. https://doi.org/10.3390/e22030326.

12. Liu, J.; Zhao, M.; Kong, W. Sub-Graph Regularization on Kernel Regression for Robust Semi-Supervised Dimensionality Reduction. *Entropy* **2019**, *21*, 1125. https://doi.org/10.3390/e21111125.

13. Fontove, F.; Del Rio, G. Residue Cluster Classes: A Unified Protein Representation for Efficient Structural and Functional Classification. *Entropy* **2020**, *22*, 472. https://doi.org/10.3390/e22040472.

14. Dafonte, C.; Rodríguez, A.; Manteiga, M.; Gómez, Á.; Arcay, B. A Blended Artificial Intelligence Approach for Spectral Classification of Stars in Massive Astronomical Surveys. *Entropy* **2020**, *22*, 518. https://doi.org/10.3390/e22050518.

15. Wang, C.; Kou, S.; Song, Y. Identify Risk Pattern of E-Bike Riders in China Based on Machine Learning Framework. *Entropy* **2019**, *21*, 1084. https://doi.org/10.3390/e21111084.

16. Lee, K.; Lee, S.-Y. Improved Practical Vulnerability Analysis of Mouse Data According to Offensive Security based on Machine Learning in Image-Based User Authentication. *Entropy* **2020**, *22*, 355. https://doi.org/10.3390/e22030355.

17. Losurdo, L.; Fanizzi, A.; Basile, T.M.A.; Bellotti, R.; Bottigli, U.; Dentamaro, R.; Didonna, V.; Lorusso, V.; Massafra, R.; Tamborra, P.; Tagliafico, A.; Tangaro, S.; La Forgia, D. Radiomics Analysis on Contrast-Enhanced Spectral Mammography Images for Breast Cancer Diagnosis: A Pilot Study. *Entropy* **2019**, *21*, 1110. https://doi.org/10.3390/e21111110.

**Carlos Fernandez-Lozano**
*Editor*

*Article*

# Set-Wise Differential Interaction between Copy Number Alterations and Gene Expressions of Lower-Grade Glioma Reveals Prognosis-Associated Pathways

**Seong Beom Cho**

Department of Biomedical Informatics, College of Medicine, Gachon University, Seongnam-Daero 1342, Korea; sbcho1749@gmail.com

Received: 22 October 2020; Accepted: 16 December 2020; Published: 18 December 2020

**Abstract:** The integrative analysis of copy number alteration (CNA) and gene expression (GE) is an essential part of cancer research considering the impact of CNAs on cancer progression and prognosis. In this research, an integrative analysis was performed with generalized differentially coexpressed gene sets (gdCoxS), which is a modification of dCoxS. In gdCoxS, set-wise interaction is measured using the correlation of sample-wise distances with Renyi's relative entropy, which requires an estimation of sample density based on omics profiles. To capture correlations between the variables, multivariate density estimation with covariance was applied. In the simulation study, the power of gdCoxS outperformed dCoxS that did not use the correlations in the density estimation explicitly. In the analysis of the lower-grade glioma of the cancer genome atlas program (TCGA-LGG) data, the gdCoxS identified 577 pathway CNAs and GEs pairs that showed significant changes of interaction between the survival and non-survival group, while other benchmark methods detected lower numbers of such pathways. The biological implications of the significant pathways were well consistent with previous reports of the TCGA-LGG. Taken together, the gdCoxS is a useful method for an integrative analysis of CNAs and GEs.

**Keywords:** copy number alteration; gene expression; integrative analysis; Renyi's relative entropy; the cancer gene atlas project; lower-grade glioma

---

## 1. Introduction

Copy number alteration (CNA) is a cytogenetic hallmark of cancer pathophysiology [1]. Due to the aberrant behavior of cancer cell proliferation and differentiation, genomic sequences can be amplified or deleted in cancer cells. The CNA can cause the abnormal expression of oncogenes or tumor suppressor genes. These abnormal expressions are related to cancer progression or poor prognosis [2–6]. For this reason, the identification of the copy number aberration has been a key issue in cancer research [7–9].

The array comparative genomic hybridization (aCGH) facilitated the discovery of the CNAs in cancer [7]. The paradigm of high-throughput technology, which is a massive parallelization of single experiments, was directly applied to the aCGH method. Consequently, researchers can obtain information about copy numbers on a genome-wide scale using the aCGH platform. Studies on many types of cancers revealed copy number anomalies in various genomic regions with the aCGH technology [8–12]. Recently, researchers have used a single nucleotide polymorphism (SNP) microarray platform for the detection of CNAs [13]. For the detection of CNAs, specific probes are inserted in the microarray platform. Several algorithms had been developed for analysis of the CNAs using the SNP microarray platform [14–17].

Although the microarray platforms enable the efficient screening of the CNAs, they give no information about gene expression (GE). For the identification of their impact on GE, they should be validated at the transcription level because the GEs of CNA loci can show no significant change [18]. To this end, the GE microarray or RNA sequencing platform can be used concurrently on the same samples that are applied to the CNA-detecting platform for accurate detection of the CNAs having an effect on transcription. The underlying assumption of the integrative analysis of the CNA and GE is straightforward: if the CNAs of genomic loci co-vary with the expression level of genes, it indicates that the genomic loci are likely to influence the GE.

The integrative analysis of the CNV and GE datasets has been focused on single gene-wise correlations or regression-based approaches that found significant relationships between CNA and GE, which are focused on identifying the coordinated variation between CNA and GE. To capture the variation, several computational methods were applied [19,20]. Lathi et al. reviewed and classified such methods into four categories, including two-step-, regression- and correlation-based approaches, and latent variable models [20]. The two-step approach consists of detecting CNA lesions and testing the association of the lesions and differential gene expressions. Regression- and correlation-based approaches are dependent on the corresponding statistical models that have been widely used in the data analysis, and some modifications of the original models are applied. Latent variable models are used to model the shared and independent signals between CNA and GE. This approach has an advantage in that it directly models the signal and noise, but has the disadvantage of high computation time.

In addition to the single gene-wise method, gene set approaches were also applied to the integrative analysis of CNA and GE. Menezes et al. used the global test to identify the relationship between single copy number alteration and corresponding gene set expression profiles [21]. By mapping neighbor expression probes to a single aCGH probe, they identified the CNAs that influenced the gene set expression profiles using the global test. The other gene set approach identified relationships between sets of CNAs and sets of expression values using canonical correlation analysis. Peng et al. applied the multivariate regression method for the set-wise analysis of CNAs and GEs [22]. To deal with the high dimensionality of genomic data, they used a regularization process. The canonical correlation analysis is a multivariate analysis method for detecting similarity between two variable sets. Lahti et al. used the canonical correlation method to determine a regional set of copy numbers and gene expression changes [23], which includes a probabilistic approach that is robust to small sample sizes. In another research, the elastic net approach was adopted to reduce the number of variables in the genomic data [24]. Similarly, selecting sparse subsets of variables of CCA instead of considering all combinations of genomic variables is proposed to consider high dimensional variables of genomic data [25].

In this research, the integrative analysis of CNA and gene expression is performed in terms of the gene set approach. The rationale for the set-wise analysis was to identify biological findings that were not detected by the single gene-wise analysis. Moreover, conditional changes in the similarity between CNAs and gene expressions are explicitly tested to identify whether a pair of CNAs and GEs is associated with the condition, which indicates that the CNAs and GEs are likely to be involved in the biology of the condition. For this purpose, the dCoxS method is modified to capture the variation between heterogenous omics data, especially for CNAs. The dCoxS was originally designed to detect interaction between a pair of GEs [26]. The interaction implies similarity between GEs, which is measured by the correlation between sample-wise distances in the GE matrices. For the identification of interactions between CNAs and GEs, dCoxS is able to be applied directly. However, if the CNAs data are in a segmented form, the dCoxS may not identify the combination effect of CNA loci in the determination of interaction because the dCoxS uses productive kernels for the estimation of sample-wise distances. Since the productive kernel computes the bandwidth parameters of the variables from the standard deviation of each variable, which can show monotonic variations in the segmented values of CNAs that represent only three statuses of gain, loss, and normal, the productive kernel may not be appropriate for the segmented CNA data. In this research, multivariate normal density estimation was applied, which integrates the correlation structure of the CNAs explicitly. Here, the modified method is named generalized dCoxS (gdCoxS), and it can analyze heterogenous omics datasets. The performance of the gdCoxS is tested using simulation data and lower-grade glioma of the cancer genome atlas program data.

## 2. Materials and Methods

### 2.1. Identification of Conditional Change of Interactions between Set-Wise CNAs and GEs

The overview of analysis is illustrated in Figure 1. The dCoxS method was originally developed for detecting significant changes in the interaction of a pair of gene expression matrices between different conditions. In the dCoxS, conditional similarity between two gene set expression profiles was determined by the correlation of sample-wise distances in the expression profiles, which was defined as the interaction score (IAS).

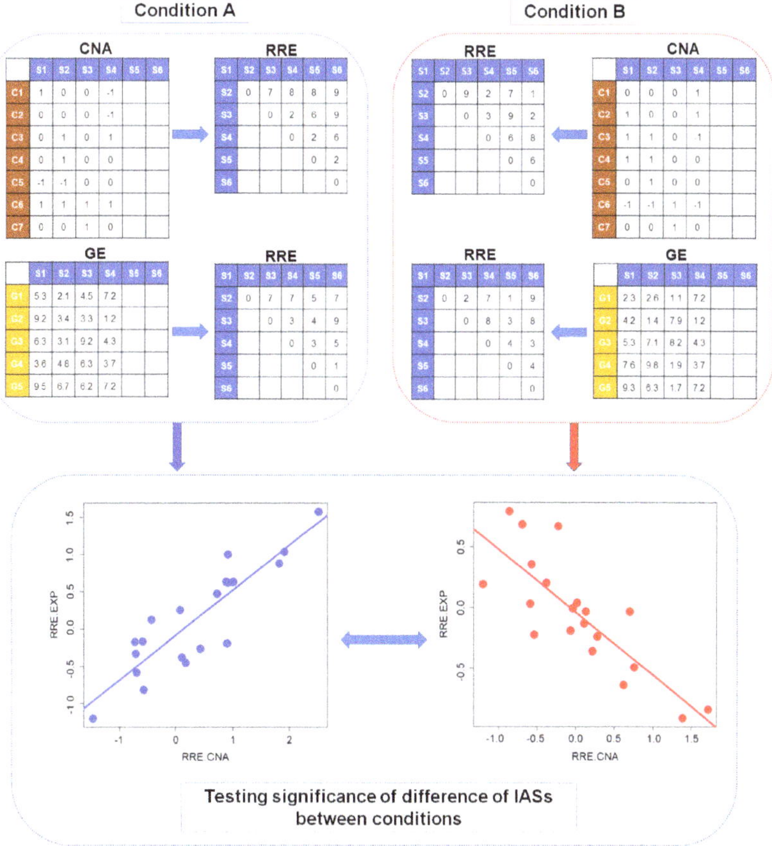


**Figure 1.** Overall analysis flow of generalized differentially coexpressed gene sets (gdCoxS). In each condition, copy number and gene expression matrices are converted to matrices of sample-wise distances that are measured by Renyi's relative entropies. Then, interactions are determined by the computation of correlation coefficients of sample-wise distances from the copy number and gene expression matrix. CNAs: copy number alterations; GEs: gene expressions; IAS: interaction score; RREs: sample-wise distances with Renyi's relative entropies.

For the estimation of sample-wise distances, Renyi's relative entropy is estimated by the ratio of densities from two different samples. The densities were computed using the multivariate productive kernel that multiplies the single density values and bandwidth parameters obtained from standard deviations of the variables. The dCoxS performs well in the estimation of differential interaction between a set of gene expressions. However, when the method is applied to CNAs and gene expressions, the productive kernel may not represent the dynamics of CNA changes because it integrates no explicit correlation structure into the density estimation. The CNA status includes only three possible values, which are loss, neutral and gain, and these are frequently coded as $-1, 0$ and $1$, respectively. Since the CNAs occur in a small portion of samples, it is likely that the density of the CNA matrix had small variations because combinations of the CNA status are not considered explicitly in the dCoxS. Thus, in this analysis, a multivariate normal density estimation that uses a covariance matrix representing combinations of the CNA status is adopted. The multivariate density function is:

$$f(x) = \frac{1}{\sqrt{(2\pi)^{p/2}|\Sigma|^{1/2}}} e^{-(x-\mu)'\Sigma^{-1}(x-\mu)/2} \tag{1}$$

where $n$ and $p$ represent the number of samples and variables. The $\mu$ is a mean vector of CNA or GE profiles, and $\hat{\Sigma}^{1/2}$ is the square root of the estimated covariance matrix. In practice, $n$ was the number of samples and $d$ was the number of CNAs or GEs in a pathway. The corpcor R package was used for the shrinkage estimation of the covariance matrix and its inverse form [27] to handle the computation of high-dimensional matrices that are frequently possible with various types of genomics data ($n < p$).

For each corresponding copy number and expression matrix, sample-wise distances were measured with Renyi's quadratic divergence.

$$D_2(P\|Q) = \log \frac{\hat{f}_h(S_i)}{\hat{f}_h(S_j)} \tag{2}$$

In Equation (2), $D_2(P\|Q)$ represents Renyi's quadratic diversity [26]. The $S_i$ and $S_j$ indicate different samples. The $\hat{f}_h(S_i)$ and $\hat{f}_h(S_j)$ are the probabilistic densities of the samples $S_i$ and $S_j$. Therefore, the higher divergence implies that two samples are more distant from each other.

Using the Renyi's diversity, set-wise CNA and expression matrices were transformed to sample-wise distance matrices. The upper trigonal members of the sample-wise distance matrices were used for the computation of the IAS. The IAS was obtained through the correlation coefficient between the upper trigonal members of the sample-wise distance matrices.

$$IAS = \frac{\sum_{i<j}(RE^C - \overline{RE^C})(RE^G - \overline{RE^G})}{\sqrt{\sum_{i<j}(RE^C - \overline{RE^C})^2}\sqrt{\sum_{i<j}(RE^G - \overline{RE^G})^2}} \tag{3}$$

In Equation (3), $RE^C$ and $RE^G$ are the sample-wise distance (relative entropy) matrices of the CNAs and GEs, respectively. After the IASs were determined in each condition, the significance of the IAS and the differences in the IAS between conditions were tested non-parametrically (Supplementary Methods).

*2.2. Simulation Analysis*

Since the IAS is used for determining the similarity between set-wise CNA and gene expression matrices, unlike the original application, a simulation study tests whether the IAS reflects the similarity between CNAs and GEs.

First, a CNA matrix was generated using binomial distribution. In general, CNA occurs in a small proportion of samples. Neutral status was therefore set to the predefined proportion of total samples. Then, gain (+1) or loss (−1) status was assigned to the rest of the samples using binomial distribution with number of trials = 1 and probability = 0.5. The rbinom R function generates a 0 or 1 status according to the predefined probability, and 0 is assigned to the −1. The proportion of samples having CNAs in the total sample was selected among the predefined values (0.1, 0.2, 0.3, 0.4 and 0.5) for each simulated CNA.

After the generation of CNAs, the GEs matrix with similarity with the CNA matrix was simulated. The random values from the normal distribution with different standard deviation (SD) values were added to a simulated CNA for the generation of GEs having various similarities according to the SD values. To simulate a GE matrix having less similarity with the CNA matrix, a greater SD value was applied in the generation of random numbers.

Power analysis was also performed with the simulation data. First, two random CNAs–GEs pairs were generated. The CNA matrices were generated by the same method used in similarity analysis. Then, a random expression matrix was generated and the same matrix was used as an expression matrix in both conditions. The random expression matrix was generated by random numbers from standard normal distribution. Since the CNA matrices were different and the expression matrices were the same between conditions, this generated the true differential interaction of CNAs and GEs between conditions. Simulation data were generated with different parameters, including the number of samples and genes.

### 2.3. Analysis of TCGA-Multiomics Data

In addition to the simulation study, to test whether the current approach identifies valid biological phenomena, TCGA-LGG data were analyzed. The data were downloaded from the genomic data commons (GDC) portal (https://portal.gdc.cancer.gov/), and clinical information was also obtained from the portal. For the detection of CNAs and GEs, Affymetrix 6.0 SNP microarray and Illumina Hiseq 2500 sequencing platform were used, respectively.

For the set-wise CNA expression interaction analysis of the TCGA-LGG data, biological pathway information was used. The current analysis framework can be applied straightforwardly to gene sets that are constructed with the other biological knowledge, such as gene ontology. The pathway information, which is mainly compiled from the Bio-Carta (www.biocarta.com), KEGG (www.genome.jp/kegg) and the Reactome (www.reactome.org) websites, was downloaded from MSigDB of the Broad Institute (https://www.gsea-msigdb.org/gsea/msigdb/index.jsp).

### 2.4. Comparison with Single Gene-Wise CNA Expression Analysis

One of the strengths of the gene set-wise analysis was that it could identify slight changes in genomic signals [28]. Maybe the strength came from the modeling of the interaction between the elements of the gene sets. To find out whether the current set-wise approach had the same advantage, the detection of significant changes in the CNAs and gene expression profiles was performed single gene-wisely. However, previous methods are not implemented to model the difference in interaction between conditions. Therefore, applicable methods for testing the differential change in the interaction of CNAs and GEs between conditions were applied. First, correlation-based single CNA and GE analysis was performed (See Supplementary Methods), and Mantel statistics with different distance measures, including Euclidean, Manhattan and Mahalanobis distances, that were available to the differential interaction analysis, were applied for comparison with Renyi's relative entropy and Mantel statistics in the analysis of gdCoxS.

## 3. Results

### 3.1. Simulation Analysis Results

To generate simulation data for testing whether IAS represents similarity between CNAs and GEs, CNA matrices that have 20, 50, and 100 variables, and 100 samples, were generated. For each simulated CNA, the proportion of the CNA in the total samples was randomly selected from among the predefined frequencies as described in the methods. When a CNA matrix was generated, random values from the normal distribution with SD = 0.01 were added, which resulted in high IAS between the CNA and GE matrices. The second GE matrix was generated by adding random values from normal distribution with SD = 0.1 to the previously generated GE matrix. Likewise, the $i$-th GE matrix was generated by adding random values from normal distribution with SD = $(i - 1) \times 0.1$ to the $(i - 1)$-th GE matrix. This generated GE matrices that were less similar to the simulated CNA matrix compared with the previously generated matrix. For each simulated CNA matrix, five GE matrices were generated in total, and this process was iterated 1000 times. When the number of variables in a GE matrix was 100, the same CNA vectors were repeatedly sampled and used for the generation of the GE matrix. Figure 2 shows that the IAS represents the similarity between CNAs and GEs. Each point indicates the mean IAS between the CNA matrices and the simulated expression matrices, with corresponding SD values. In general, the mean IASs were highest when SD was 0.01, and they became lower with increasing SD. The mean IAS was lowest with SD = 0.4 in all simulations. Besides mean values, the paired t tests of the IASs were highly significant between IASs from different SDs ($p < 2.2 \times 10^{-16}$). These indicate that the IAS represents similarity between CNAs and GEs. Since the CNA and GE matrices are different types of data, the simulated matrices should have different distributions. While it was obvious that the simulated CNA matrices have binomial distributions, it was not clear that the simulated GE matrices have multivariate normal distributions that are frequently used in the simulation of a gene expression matrix, because they were generated by adding numbers from binomial and normal distributions. Therefore, normality tests were applied to the GE matrices and the result showed that the matrices had multivariate normal distributions with Bonferroni's multiple testing correction (data not shown).

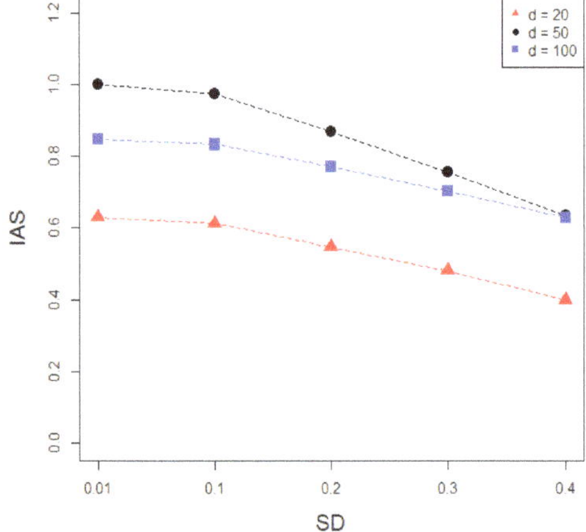


**Figure 2.** Results of simulation study for measuring similarity between copy number alterations (CNAs) and gene expressions (GEs). The red, black and blue dots and lines indicate the numbers of variables in the gene expression matrix, as 20, 50, and 100, respectively. As standard deviations (SDs) increase, the mean IASs decline.

Power analysis was performed with changes in the number of samples and number of elements in the simulation data. The number of samples included {100, 200, 400}, and the number of variables in the set were set within {10, 20, 30}. The number of permutations was set to 100. Figure 3 shows the results of the power analysis. There was an obvious trend whereby the power of gdCoxS and dCoxS increased as the number of samples was elevated. However, dCoxS had a decreasing power as the number of elements in the gene sets increased, regardless of the number of samples, while dCoxS showed the best performance with the smallest number of elements ($n = 10$). Since the gdCoxS used a covariance matrix for estimating the relationship between variables, gdCoxS captured the difference in CNA matrices more efficiently than the dCoxS, which adopted the productive kernel in estimating density without the use of such a covariance matrix, which was more evident in the higher number of elements in the gene set. Considering the high-dimensional characteristics of functional genomics data, the gdCoxS is a more efficient and robust method, which can detect the dynamics between matrices from two different sets of genomic data.

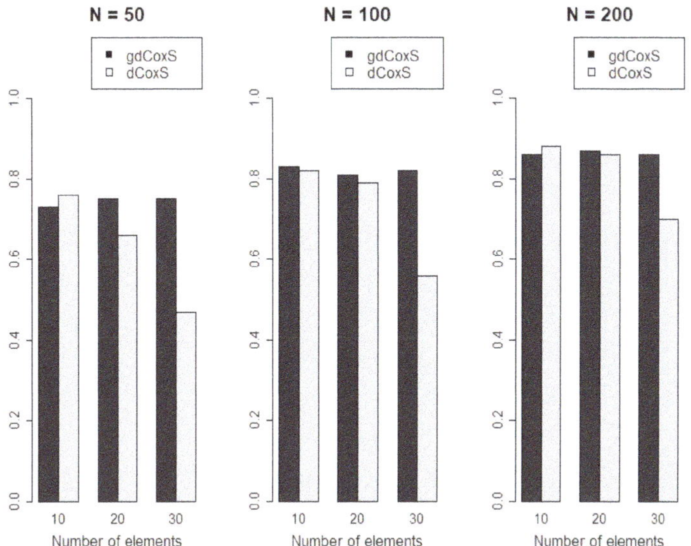


**Figure 3.** Results of power analysis. The number of x axis is the number of variables in the gene sets, and the y axis represents power. When the number of samples is higher, the overall powers of gdCoxS are higher than the powers with lower number of samples, regardless of the number of variables. The dCoxS shows, however, an obvious trend of decreasing power with elevating numbers of elements of gene sets. N; number of simulation samples in each class.

## 3.2. Real Data Analysis

In the TCGA-LGG, genes of CNA and expression data were mapped to the ensemble identifier system. Since the pathway information used gene symbols, the mapping table of the HUGO Gene Nomenclature Committee (HGNC) for gene symbols and ensemble identifiers was used for mapping gene symbols to ensemble identifier (https://www.genenames.org/download/cus-tom/). The CNA data had 533 samples and the RNA sequencing data had 530 samples. Of the samples, 507 samples with CNA, gene expression and survival information were used in the analysis. In the MSigDB, there were 1335 canonical pathways from the open databases including the KEGG, BioCarta and Reactome. The class was labeled into two groups according to the survival status (death = 98, survival more than 5 years = 409). In the analysis of the TCGA-LGG dataset, the GDC provided CNA information that had been computed using the Genomic Identification of Significant Targets in Cancer (GISITC) algorithm [17]. The CNA information of 12,117 ensemble genes, that were matched to the genes of the 1335 items of MSigDB pathway information, were applied in this analysis. The RNA sequencing

(RNA-seq) data has 60,483 transcripts in total, and 13,339 transcripts were mapped to the ensemble identifiers of all the pathway information in the 1335 pathways. Since the RNA-seq data had different batches, a batch effect adjustment was performed with Combat-seq program [29]. After the adjustment, the RNA-seq data were normalized using the quantile normalization method. First, zero values were treated as missing values and they were imputed using the impute R package with default parameters [30]. The data were then log-transformed and the quantile normalization was applied. For the quantile normalization, the normalize.quantiles function of the preprocessCore R package was used [31]. In the real data analysis, pathway gene sets having more than 10 elements were arbitrarily selected for analysis. In total, 1282 pathways were applied for this analysis. The numbers of CNA ensemble identifiers of each pathway ranged from 10 to 933 (median = 23). Those of the pathway expression matrices lay between 10 and 941 (median = 23).

For each pathway, CNA and expression matrices with elements of the pathway were generated, and the differential interaction of two matrices between the survival and death group was computed. To test the significance of the difference of IASs between conditions, a permutation test was applied with 26,000 repeats of permutation.

In the gdCoxS analysis, Bonferroni's multiple testing correction was applied (adjusted $p$ value = $3.9 \times 10^{-5}$). With the threshold, 577 pathways were found to exhibit significantly different interactions of CNAs and expressions of the pathways between the survival and death groups of TCGA-LGG patients (Table 1 and Supplementary Table S1).

**Table 1.** Pathways showing upper and lower top 5 significant results in gdCoxS analysis. The total results are listed in Supplementary Table S1.

| Pathway Database | Pathways | $N_{\text{CNA}}$ [1] | $N_{\text{EXP}}$ [2] | IAS.S [3] | IAS.NS [4] | diffIAS [5] |
|---|---|---|---|---|---|---|
| PID | IL3_PATHWAY | 10 | 10 | 0.023 | 0.407 | −52.037 |
| REACTOME | PROTEIN_METHYLATION | 14 | 14 | 0.197 | 0.524 | −48.578 |
| REACTOME | DUAL_INCISION_IN_GG_NER | 14 | 14 | 0.081 | 0.430 | −48.002 |
| BIOCARTA | FORMATION_OF_INCISION_COMPLEX_IN_GG_NER | 26 | 26 | 0.176 | 0.501 | −47.231 |
| REACTOME | MICRORNA_MIRNA_BIOGENESIS | 10 | 10 | 0.146 | 0.476 | −47.219 |
| REACTOME | TRIGLYCERIDE_CATABOLISM | 15 | 11 | 0.148 | −0.179 | 41.893 |
| REACTOME | DEGRADATION_OF_CYSTEINE_AND_HOMOCYSTEINE | 11 | 10 | 0.168 | −0.159 | 41.940 |
| BIOCARTA | EGF_PATHWAY | 14 | 14 | 0.200 | −0.145 | 44.284 |
| KEGG | CYTOSOLIC_DNA_SENSING_PATHWAY | 16 | 15 | 0.240 | −0.127 | 47.355 |
| REACTOME | GLYCOSPHINGOLIPID_METABOLISM | 31 | 28 | 0.256 | −0.136 | 50.660 |

[1] $N_{\text{CNA}}$: number of variables in copy number matrix; [2] $N_{\text{CNA}}$: number of variables in gene expression matrix; [3] IAS.S: interaction score in survival group; [4] IAS.NS: interaction score in non-survival group; [5] diffIAS: difference of interaction scores; PID: pathway interaction database; KEGG: Kyoto Encyclopedia of Genes and Genomes.

In the result, 274 pathways showed increased interactions of CNAs and GEs in the non-survival group, which indicated that variations in CNAs and GEs were more harmonized. On the other hand, 303 pathways had decreased interactions in the non-survival group. The IAS of the IL3_PATHWAY from the pathway interaction database (PID) increased from 0.023 in the survival group to 0.407 in the non-survival group, which was the greatest absolute diffIAS among the results (Figure 4). The 'GLYCOSPHINGOLIPID_METABOLISM' pathway from the REACTOME database had the greatest positive diffIAS (= 50.66), which implied that the coordination of the CNAs and GEs of the pathway in the survival group was disrupted in the non-survival group. While the IAS of the pathway CNAs and GEs was 0.256 in the survival group, it decreased (−0.136) in the non-survival group (Figure 4).

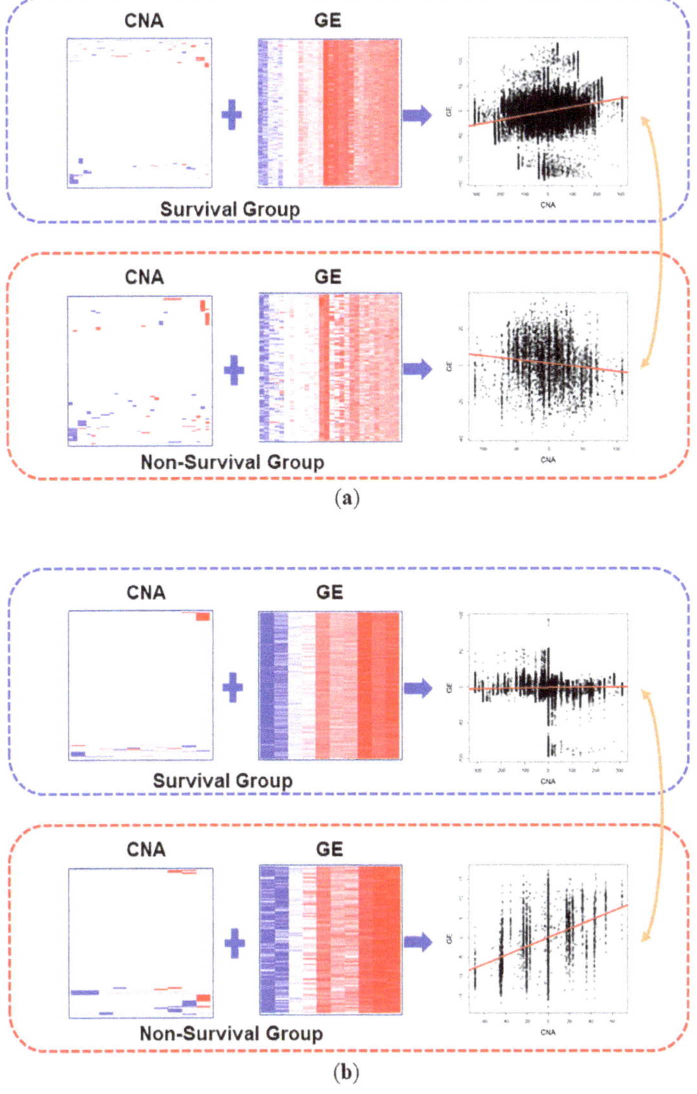

**Figure 4.** Heatmap and scatter plot of sample-wise distances of pathway copy number alteration and gene expression matrices of the significant results. Note that pathway CNA matrices contain substantial portions of neutral status. The orders of genes in the CNA and GE matrices are set to the same in the survival and non-survival groups. (**a**) Results of 'GLYCOSPHINGOLIPID_METABOLISM'. (**b**) Results of 'IL3_PATHWAY' pathway gene set. The scatter plots are made up of plotting sample-wise distances from CNA and GE matrices. The slopes of red lines in the scatter plots indicate interaction scores of each condition.

For the benchmark analysis of gdCoxS, differential co-expression analysis and Mantel statistics were applied. The differential coexpression analysis includes an estimation of the correlation coefficient between a CNA and GE in each condition, and a test of the significance of the difference in correlations between conditions (See Supplementary Methods). In the single gene-wise differential coexpression analysis, cis and trans regulation were considered, and only the CNAs and GEs that were used in the pathway analysis were included to avoid the loss of power that resulted from a large number of statistical tests. First, 6202 CNAs and 6233 GEs were selected and correlations between the CNAs and GEs were computed in each condition, and the differences in the correlations were tested (Supplementary Methods). After the Bonferroni's multiple testing correction, there was no significant result from the Bonferroni's multiple testing correction (adjusted $p < 1.29 \times 10^{-9}$).

In the benchmark analysis, the Mantel statistics were also applied to compare the performance of gdCoxS when different similarity measures other than Renyi's relative entropy were applied (Supplementary Methods). Different statistics, including Euclidean, Manhattan and Mahalanobis, which could compute interactions between CNAs and GEs, were applied. Although the Mantel test with different measures showed substantial numbers of significant results, the numbers were far less than those of the gdCoxS analysis (Supplementary Tables S2–S4, respectively). In the result, the Mantel statistics with the Mahalanobis distance using the covariance matrix showed the largest number of significant results ($n = 171$).

## 4. Discussion

In this research, the gdCoxS performs an integrative analysis of CNAs and GEs. In the simulation analysis, the gdCoxS shows an improvement in the performance in terms of power, especially with larger numbers of gene set elements. In the real data analysis, the gdCoxS detected 577 significant results, while the single gene-wise differential coexpression analysis gave no significant result, and the set-wise analysis with Mantel statistics identified fewer significant pathways than gdCoxS. These results seem to indicate that the gdCoxS outperforms the other benchmark methods.

When the single gene differential coexpression analysis was applied, no significant results could be found in the result of the single gene-wise analysis. However, gene set methods including gdCoxS and Mantel tests identified a lot of significant pathway CNA–GE set pairs. These findings clearly indicate the benefit of gene set-wise analysis, which has more power to detect significant interactions between CNAs and GEs. In the benchmark study using Mantel statistics, the results with Mahalanobis distance showed a far better performance than the other measures. This seems to result from the fact that the Mahalanobis distance uses a covariance matrix that can capture the relationship between elements of gene sets. This finding supports the validity of the concept in gdCoxS, which is an application of the multivariate density function with covariance information to capture the relationship between CNAs explicitly. The dCoxS method was not compared in the real data analysis because variations in sample-wise distances in CNA matrices tended to be zero, which made the computation of IAS intractable. Among the pathways, more than a thousand of pathway CNA matrices showed such variations. This finding strongly indicates that the productive kernel of the dCoxS was not suitable for detecting combinatorial variations in CNAs. In the benchmark analysis, the set-wise methods, such as modified canonical correlation analysis (CCA), that were presented in the introduction could be applied. However, the methods can estimate the similarity between CNA and GE matrices only, and the differences in the similarities between conditions were not considered. Moreover, the methods provided no statistical testing for the estimation of $P$ values. Therefore, the comparison between the gdCoxS and the modified CCA was not possible.

In the result, many pathways were related to the glioma pathophysiology in previous studies. For example, 10 pathways were related to p53, which has impacts on the glioma pathophysiology (Supplementary Table S1). The mutation and inactivation of p53 is related to the proliferation and progression of glioma, invasion, and anti-apoptotic activity [32–35]. It is possible that copy number alterations in p53-related pathways disrupt the CNAs–GE relationship in the favorable group of LGG. The significant change in IASs between the CNAs and GEs of the p53-related pathways in the non-survival group seems to implicate a disrupted regulatory relationship between CNAs and GEs. Considering the role of p53 in the prognosis of many types of cancers [35], these results indicate the validity of gdCoxS analysis. Among the p53-related pathways, the "53 regulates transcription of caspase activators and caspases" pathway is interesting because the result indicated that the differential interaction of CNAs and GEs in the pathway was associated with the apoptosis that is critical to the survival of cancer genes. There are supportive results to this finding. In the pathway, p53 regulates caspase 10, which is associated with apoptotic signaling in glioblastoma [36], and capase 10 induced cellular death in response to the chemotherapeutic agent, which has a possibility of prolonged survival [37]. In the mouse experiment, the ATM gene was involved in the suppression of glioblastoma by the down-regulation of glioblastoma-associated genes such as the PDGFRA gene [38]. P63, which is another member of the pathway, was revealed to suppress tumor growth by up-regulating caspase 1 expression [39]. These seem to be consistent with the results of the significant differential interaction of CNAs and GEs between survival and non-survival groups.

The EGF pathway also indicated the validity of the analysis result (Table 1). The EGF receptor (EGFR) and its downstream signaling is frequently aberrant in cancers, especially in glioma [40]. EFGR gene amplification and overexpression can be observed in approximately 40% of glioblastoma [41]. Since the EGFR signaling is associated with the apoptosis, proliferation and invasion of cancer cells [42], the EGFR was investigated as a therapeutic target in previous studies [43]. The significant change in the interaction of CNAs and GEs in the EGF pathway between the survival and non-survival groups seems to be further supportive evidence of the fact that the EGF and its receptor have a therapeutic potential. It is notable that the homocysteine pathway ('DEGRADATION OF CYSTEINE AND HOMOCYSTEINE' from REACTOME database) was highly ranked in the significant results. It is well known that the homocysteine metabolism is aberrant in cancers, including glioma [43], and the homocysteine level is associated with the death of a human glioblastoma cell line [44]. Moreover, the variant of the methylenetetrahydrofolate reductase was shown to be significantly associated with patient survival [45,46]. Considering these, the interaction between CNAs and GEs in the homocysteine pathway seems to be related to the pathophysiology of the lower-grade glioma.

In conclusion, the set-wise identification of the interaction between CNAs and GEs revealed pathways that are consistent with the molecular pathophysiology of lower-grade glioma, which was not found in single-variable analysis. This gene set method for performing the integrative analysis of multi-omics data will promote the discovery of hidden biologic mechanisms.

**Supplementary Materials:** The Supplementary Materials are available online at http://www.mdpi.com/1099-4300/22/12/1434/s1.

**Funding:** This research was funded by the Gil Medical Center (grant number: FRD2020-08).

**Conflicts of Interest:** The author declares no conflict of interest.

## References

1. Beroukhim, R.; Mermel, C.H.; Porter, D.; Wei, G.; Raychaudhuri, S.; Donovan, J.; Barretina, J.; Boehm, J.S.; Dobson, J.; Urashima, M.; et al. The landscape of somatic copy-number alteration across human cancers. *Nature* **2010**, *463*, 899–905. [CrossRef]
2. Hirsch, F.R.; Varella-Garcia, M.; Cappuzzo, F. Predictive value of EGFR and HER2 overexpression in advanced non-small-cell lung cancer. *Oncogene* **2009**, *28*, S32–S37. [CrossRef]
3. Ono, M.; Kuwano, M. Molecular mechanisms of epidermal growth factor receptor (EGFR) activation and response to gefitinib and other EGFR-targeting drugs. *Clin. Cancer Res.* **2006**, *12*, 7242–7251. [CrossRef]
4. Yang, L.; Li, Y.; Wei, Z.; Chang, X. Coexpression network analysis identifies transcriptional modules associated with genomic alterations in neuroblastoma. *Biochim. Biophys. Acta Mol. Basis Dis.* **2018**, *1864*, 2341–2348. [CrossRef]
5. Chang, X.; Zhao, Y.; Hou, C.; Glessner, J.; McDaniel, L.; Diamond, M.A.; Thomas, K.; Li, J.; Wei, Z.; Liu, Y.; et al. Common variants in MMP20 at 11q22.2 predispose to 11q deletion and neuroblastoma risk. *Nat. Commun.* **2017**, *8*, 569. [CrossRef]
6. Lopez, G.; Conkrite, K.L.; Doepner, M.; Rathi, K.S.; Modi, A.; Vaksman, Z.; Farra, L.M.; Hyson, E.; Noureddine, M.; Wei, J.S.; et al. Somatic structural variation targets neurodevelopmental genes and identifies SHANK2 as a tumor suppressor in neuroblastoma. *Genome Res.* **2020**, *30*, 1228–1242. [CrossRef] [PubMed]

7.  Pinkel, D.; Segraves, R.; Sudar, D.; Clark, S.; Poole, I.; Kowbel, D.; Collins, C.; Kuo, W.L.; Chen, C.; Zhai, Y.; et al. High resolution analysis of DNA copy number variation using comparative genomic hybridization to microarrays. *Nat. Genet.* **1998**, *20*, 207–211. [CrossRef] [PubMed]

8.  Kaur, S.; Vauhkonen, H.; Bohling, T.; Mertens, F.; Mandahl, N.; Knuutila, S. Gene copy number changes in dermatofibrosarcoma protuberans—A fine-resolution study using array comparative genomic hybridization. *Cytogenet. Genome Res.* **2006**, *115*, 283–288. [CrossRef] [PubMed]

9.  Kim, M.Y.; Yim, S.H.; Kwon, M.S.; Kim, T.M.; Shin, S.H.; Kang, H.M.; Lee, C.; Chung, Y.J. Recurrent genomic alterations with impact on survival in colorectal cancer identified by genome-wide array comparative genomic hybridization. *Gastroenterology* **2006**, *131*, 1913–1924. [CrossRef]

10. Stransky, N.; Vallot, C.; Reyal, F.; Bernard-Pierrot, I.; de Medina, S.G.; Segraves, R.; de Rycke, Y.; Elvin, P.; Cassidy, A.; Spraggon, C.; et al. Regional copy number-independent deregulation of transcription in cancer. *Nat. Genet.* **2006**, *38*, 1386–1396. [CrossRef] [PubMed]

11. Staaf, J.; Torngren, T.; Rambech, E.; Johansson, U.; Persson, C.; Sellberg, G.; Tellhed, L.; Nilbert, M.; Borg, A. Detection and precise mapping of germline rearrangements in BRCA1, BRCA2, MSH2, and MLH1 using zoom-in array comparative genomic hybridization (aCGH). *Hum. Mutat.* **2008**, *29*, 555–564. [CrossRef] [PubMed]

12. Yi, Y.; Nowak, N.J.; Pacchia, A.L.; Morrison, C. Chromosome 11 genomic changes in parathyroid adenoma and hyperplasia: Array CGH, FISH, and tissue microarrays. *Genes Chromosomes Cancer* **2008**, *47*, 639–648. [CrossRef] [PubMed]

13. Pitea, A.; Kondofersky, I.; Sass, S.; Theis, F.J.; Mueller, N.S.; Unger, K. Copy number aberrations from Affymetrix SNP 6.0 genotyping data-how accurate are commonly used prediction approaches? *Brief. Bioinform.* **2018**, *21*, 272–281. [CrossRef] [PubMed]

14. Yau, C.; Mouradov, D.; Jorissen, R.N.; Colella, S.; Mirza, G.; Steers, G.; Harris, A.; Ragoussis, J.; Sieber, O.; Holmes, C.C. A statistical approach for detecting genomic aberrations in heterogeneous tumor samples from single nucleotide polymorphism genotyping data. *Genome Biol.* **2010**, *11*, R92. [CrossRef]

15. Sun, W.; Wright, F.A.; Tang, Z.; Nordgard, S.H.; Van Loo, P.; Yu, T.; Kristensen, V.N.; Perou, C.M. Integrated study of copy number states and genotype calls using high-density SNP arrays. *Nucleic Acids Res.* **2009**, *37*, 5365–5377. [CrossRef]

16. Van Loo, P.; Nordgard, S.H.; Lingjærde, O.C.; Russnes, H.G.; Rye, I.H.; Sun, W.; Weigman, V.J.; Marynen, P.; Zetterberg, A.; Naume, B.; et al. Allele-specific copy number analysis of tumors. *Proc. Natl. Acad. Sci. USA* **2010**, *107*, 16910–16915. [CrossRef]

17. Mermel, C.H.; Schumacher, S.E.; Hill, B.; Meyerson, M.L.; Beroukhim, R.; Getz, G. GISTIC 2.0 facilitates sensitive and confident localization of the targets of focal somatic copy-number alteration in human cancers. *Genome Biol.* **2011**, *12*, R41. [CrossRef]

18. De Tayrac, M.; Etcheverry, A.; Aubry, M.; Saikali, S.; Hamlat, A.; Quillien, V.; Le Treut, A.; Galibert, M.D.; Mosser, J. Integrative genome-wide analysis reveals a robust genomic glioblastoma signature associated with copy number driving changes in gene expression. *Genes Chromosomes Cancer* **2009**, *48*, 55–68. [CrossRef]

19. Louhimo, R.; Lepikhova, T.; Monni, O.; Hautaniemi, S. Comparative analysis of algorithms for integration of copy number and expression data. *Nat. Methods* **2012**, *9*, 351–355. [CrossRef]

20. Lahti, L.; Schäfer, M.; Klein, H.U.; Bicciato, S.; Dugas, M. Cancer gene prioritization by integrative analysis of mRNA expression and DNA copy number data: A comparative review. *Brief. Bioinform.* **2013**, *14*, 27–35. [CrossRef]

21. Menezes, R.X.; Boetzer, M.; Sieswerda, M.; van Ommen, G.J.; Boer, J.M. Integrated analysis of DNA copy number and gene expression microarray data using gene sets. *BMC Bioinform.* **2009**, *10*, 203. [CrossRef] [PubMed]

22. Peng, J.; Zhu, J.; Bergamaschi, A.; Han, W.; Noh, D.Y.; Pollack, J.R.; Wang, P. Regularized multivariate regression for identifying master predictors with application to integrative genomics study of breast cancer. *Ann. Appl. Stat.* **2010**, *4*, 53–77. [CrossRef] [PubMed]

23. Lahti, L.; Myllykangas, S.; Knuutila, S.; Kaski, S. Dependency detection with similarity constraints. In Proceedings of the 2009 IEEE International Workshop on Machine Learning for Signal Processing, Grenoble, France, 1–4 September 2009; IEEE: Piscataway, NJ, USA, 2009; pp. 89–94.

24. Waaijenborg, S.; Zwinderman, A.H. Penalized canonical correlation analysis to quantify the association between gene expression and DNA markers. *BMC Proc.* **2007**, *1*, S122. [CrossRef] [PubMed]

25. Parkhomenko, E.; Tritchler, D.; Beyene, J. Genome-wide sparse canonical correlation of gene expression with genotypes. *BMC Proc.* **2007**, *1*, S119. [CrossRef]

26. Cho, S.B.; Kim, J.; Kim, J.H. Identifying set-wise differential co-expression in gene expression microarray data. *BMC Bioinform.* **2009**, *10*, 109. [CrossRef]

27. Schäfer, J.; Strimmer, K. A Shrinkage Approach to Large-Scale Covariance Matrix Estimation and Implications for Functional Genomics. *Stat. Appl. Genet. Mol. Biol.* **2005**, *4*, 32. [CrossRef]

28. Segal, E.; Friedman, N.; Kaminski, N.; Regev, A.; Koller, D. From signatures to models: Understanding cancer using microarrays. *Nat. Genet.* **2005**, *37*, S38–S45. [CrossRef]

29. Zhang, Y.; Parmigiani, G.; Johnson, W. ComBat-seq: Batch effect adjustment for RNA-seq count data. *NAR Genom. Bioinform.* **2020**. [CrossRef]

30. Hastie, T.; Tibshirani, R.; Narasimhan, B.; Chu, G. *Impute: Imputation for Microarray Data*; R Package Version 1.62.0; GitHub, Inc.: San Francisco, CA, USA, 2020.

31. Bolstad, B. *preprocessCore: A Collection of Pre-Processing Functions*, R Package Version 1.50.0; Available online: https://github.com/bmbolstad/preprocessCore (accessed on 14 July 2020).

32. England, B.; Huang, T.; Karsy, M. Current understanding of the role and targeting of tumor suppressor p53 in glioblastoma multiforme. *Tumour. Biol.* **2013**, *34*, 2063–2074. [CrossRef]

33. Krex, D.; Mohr, B.; Appelt, H.; Schackert, H.K.; Schackert, G. Genetic analysis of a multifocal glioblastoma multiforme: A suitable tool to gain new aspects in glioma development. *Neurosurgery* **2003**, *53*, 1377–1384. [CrossRef]

34. Djuzenova, C.S.; Fiedler, V.; Memmel, S.; Katzer, A.; Hartmann, S.; Krohne, G.; Zimmermann, H.; Scholz, C.J.; Polat, B.; Flentje, M.; et al. Actin cytoskeleton organization, cell surface modification and invasion rate of 5 glioblastoma cell lines differing in PTEN and p53 status. *Exp. Cell Res.* **2015**, *330*, 346–357. [CrossRef] [PubMed]

35. Park, C.M.; Park, M.J.; Kwak, H.J.; Moon, S.I.; Yoo, D.H.; Lee, H.C.; Park, I.C.; Rhee, C.H.; Hong, S.I. Induction of p53-mediated apoptosis and recovery of chemosensitivity through p53 transduction in human glioblastoma cells by cisplatin. *Int. J. Oncol.* **2006**, *28*, 119–125. [CrossRef] [PubMed]

36. Kandoth, C.; McLellan, M.D.; Vandin, F.; Ye, K.; Niu, B.; Lu, C.; Xie, M.; Zhang, Q.; McMichael, J.F.; Wyczalkowski, M.A.; et al. Mutational landscape and significance across 12 major cancer types. *Nature* **2013**, *502*, 333–339. [CrossRef] [PubMed]

37. Valdés-Rives, S.A.; Casique-Aguirre, D.; Germán-Castelán, L.; Velasco-Velázquez, M.A.; González-Arenas, A. Apoptotic Signaling Pathways in Glioblastoma and Therapeutic Implications. *Biomed. Res. Int.* **2017**, *2017*, 7403747. [CrossRef]

38. Mohr, A.; Deedigan, L.; Jencz, S.; Mehrabadi, Y.; Houlden, L.; Albarenque, S.M.; Zwacka, R.M. Caspase-10: A molecular switch from cell-autonomous apoptosis to communal cell death in response to chemotherapeutic drug treatment. *Cell Death Differ.* **2018**, *25*, 340–352. [CrossRef] [PubMed]

39. Blake, S.M.; Stricker, S.H.; Halavach, H.; Poetsch, A.R.; Cresswell, G.; Kelly, G.; Kanu, N.; Marino, S.; Luscombe, N.M.; Pollard, S.M.; et al. Inactivation of the ATMIN/ATM pathway protects against glioblastoma formation. *Elife* **2016**, *5*, e08711. [CrossRef]

40. Celardo, I.; Grespi, F.; Antonov, A.; Bernassola, F.; Garabadgiu, A.V.; Melino, G.; Amelio, I. Caspase-1 is a novel target of p63 in tumor suppression. *Cell Death Dis.* **2013**, *4*, e645. [CrossRef]

41. Xu, H.; Zong, H.; Ma, C.; Ming, X.; Shang, M.; Li, K.; He, X.; Du, H.; Cao, L. Epidermal growth factor receptor in glioblastoma. *Oncol. Lett.* **2017**, *14*, 512–516. [CrossRef]

42. Hatanpaa, K.J.; Burma, S.; Zhao, D.; Habib, A.A. Epidermal growth factor receptor in glioma: Signal transduction, neuropathology, imaging, and radioresistance. *Neoplasia* **2010**, *12*, 675–684. [CrossRef]

43. Manfred, W.M.; Maire, C.L.; Lamszus, K. EGFR as a Target for Glioblastoma Treatment: An Unfulfilled Promise. *CNS Drugs* **2017**, *31*, 723–735.

44. Škovierová, H.; Vidomanová, E.; Mahmood, S.; Sopková, J.; Drgová, A.; Červeňová, T.; Halašová, E.; Lehotský, J. The Molecular and Cellular Effect of Homocysteine Metabolism Imbalance on Human Health. *Int. J. Mol. Sci.* **2016**, *17*, 1733. [CrossRef] [PubMed]

45. Hasan, T.; Arora, R.; Bansal, A.K.; Bhattacharya, R.; Sharma, G.S.; Singh, L.R. Disturbed homocysteine metabolism is associated with cancer. *Exp. Mol. Med.* **2019**, *51*, 1–13. [CrossRef] [PubMed]
46. Linnebank, M.; Semmler, A.; Moskau, S.; Smulders, Y.; Blom, H.; Simon, M. The methylenetetrahydrofolate reductase (MTHFR) variant c.677C>T (A222V) influences overall survival of patients with glioblastoma multiforme. *Neuro Oncol.* **2008**, *10*, 548–552. [CrossRef] [PubMed]

**Publisher's Note:** MDPI stays neutral with regard to jurisdictional claims in published maps and institutional affiliations.

*Article*

# Low Entropy Sub-Networks Prevent the Integration of Metabolomic and Transcriptomic Data

**Krzysztof Gogolewski *, Marcin Kostecki and Anna Gambin ***

Institute of Informatics, University of Warsaw, 02-097 Warsaw, Poland; mrckostecki@gmail.com
* Correspondence: k.gogolewski@mimuw.edu.pl (K.G.); aniag@mimuw.edu.pl (A.G.)

Received: 15 September 2020; Accepted: 27 October 2020; Published: 31 October 2020

**Abstract:** The constantly and rapidly increasing amount of the biological data gained from many different high-throughput experiments opens up new possibilities for data- and model-driven inference. Yet, alongside, emerges a problem of risks related to data integration techniques. The latter are not so widely taken account of. Especially, the approaches based on the flux balance analysis (FBA) are sensitive to the structure of a metabolic network for which the low-entropy clusters can prevent the inference from the activity of the metabolic reactions. In the following article, we set forth problems that may arise during the integration of metabolomic data with gene expression datasets. We analyze common pitfalls, provide their possible solutions, and exemplify them by a case study of the renal cell carcinoma (RCC). Using the proposed approach we provide a metabolic description of the known morphological RCC subtypes and suggest a possible existence of the poor-prognosis cluster of patients, which are commonly characterized by the low activity of the drug transporting enzymes crucial in the chemotherapy. This discovery suits and extends the already known poor-prognosis characteristics of RCC. Finally, the goal of this work is also to point out the problem that arises from the integration of high-throughput data with the inherently nonuniform, manually curated low-throughput data. In such cases, the over-represented information may potentially overshadow the non-trivial discoveries.

**Keywords:** genome-scale metabolic networks; information redundancy; metabolic landscapes analysis; graph entropy; renal cell carcinoma; transcriptomics

---

## 1. Introduction

Observing the technological progress of the recent decades one can notice that it facilitated an access to vast biomedical data resources describing different molecular levels (so-called -omics data). Consequently, our comprehension of biological processes becomes more profound as well as reliable. These facts open up a broad field of data integration, that aims to infer from various data collection platforms taking into account known biological dependencies between them.

**Motivations.** In the literature, it is broadly emphasized that the integration of *-omics* data improves understanding of various phenomena. The modern studies and corresponding literature emphasize the significant role of the integration of different *-omics* data types for better comprehension of a phenomenon of interest [1]. However, these integration procedures, thanks to a wider perspective, are not only meant to provide a new insight into a specific phenomena, but above all should constitute a deeper understanding of the general genotype-phenotype gap and the relationship between these two layers [2]. Importantly, each unique set of -omics data can be integrated using various statistical methods, and thus may result in an unprecedented outcome. As a consequence, each year a number of reviews are published to track and summarize the current state of the art in the field of -omics integration, which one can check for more detailed discussions [3,4].

In this work, however, the aim is to focus on a particular problem of the transcriptomic and metabolomic data integration. We report an interesting phenomenon related to the analysis of individual metabolic networks in the context of transcriptomic data. The conducted study allowed to detect a common and robust artificial pattern identifiable for each data cohort from the Cancer Genome Atlas that splits cancer-specific patients into two to four clusters. Interestingly, each cluster is characterized by a peculiar activity pattern among reactions hubs. In the article, we prove that these observations result mostly from the uneven metabolic information distribution within the metabolic network. To the best of our knowledge, no one has yet commented on the problem that the structure of a metabolic network can introduce when used along with integrative methods, in particular when used in the context of the flux balance analysis (FBA) [5].

**Related research.** One of the first attempts to this type of integration was suggested by Covert and Palsson, where authors infer the binary enzymatic activity from the transcriptomic continuous signal [6]. Next, a cascade of methods was proposed to harness transcriptomic data for analysis of metabolic networks. Among the most discussed methods we can point out: E-Flux [7], GIMME [8], GIMMEp [9], iMAT [10], INIT [11], MADE [12], mCADRE [13], PROM [14], RELATCH [15]. Apart from some minor details, fundamentally these approaches differ at three basic levels: (i) the way of inference of the enzymatic activity from transcriptomic data; (ii) the determination of the flow capacity boundaries from transcriptomic data; (iii) number of samples needed to create a mathematical model. These methods along with their other features, were already discussed and compared in few reviews where detailed descriptions can be found [16–19]. Therefore, here, we focus on a summary of what type of biomedical outcomes they have provided so far.

Using this type of models, where a general metabolic network becomes context-specific through integration with a transcriptome of specific tissue or organism few groups of researchers have already reported some interesting discoveries. With nearly 2000 samples of breast tumor Leoncikas et al. discovered a novel poor prognosis cluster characterized by local production of serotonin along with active extracellular matrix and membrane remodeling reactions [20]. Li et al. by integrating transcriptomic knowledge with human metabolic network suggest a supervised method to predict novel drug-target interaction [21]. In their work they predict related metabolic reactions and enzyme targets for approved cancer drugs, and predict drug targets with statistically high confidence rate. Reconstruction of a genome-scale metabolic models for 126 human tissues and cell types including healthy and tumor type was derived using the Recon 1 human metabolic network accompanied by transcriptomic data [13]. Among all, the set of models includes 26 tumour-specific models accompanied by their normal counterparts, in particular 30 models of brain tissue subtypes were determined. Next, using the modified version of iMAT, differential fatty acid uptake into mitochondria along with arachidonic acid and eicosanoid metabolism were suggested to explain different proliferative rates and invasiveness between PC-3/M (highly proliferative, cancer stem cells) and PC-3/S (highly invasive, epithelial-mesenchymal-transition-like properties) subpopulations derived from prostate cancer cell line [22].

In summary, the general biomedical aim of this approach is to model the gene–protein–reaction interactions in the form of a metabolic network and infer multiple biological properties, e.g., post-transcriptional gene activities or intensity and activity of metabolic reactions, that can explain the nature of analyzed sample.

**Our results.** In this paper, we want to draw attention to a specific problem related to the task of the data integration and analysis based on the general metabolic networks along with transcriptomic data. First, we present how the specific structure of sub-networks can inflate the importance of specific groups of enzymatic reactions, which may lead to incomplete or questionable conclusions. Next, in order to cope with that obstacle, we suggest a possible routine that can track and eliminate the unwanted problem influencing the metabolic knowledge obtained from the integration procedure. Furthermore, we present our results from the analysis of the TCGA renal cell carcinoma (RCC) dataset, which point out reactions discriminating patients in the context of the observed clinical factors. Additionally, we report the discovery of a poor-prognosis cluster of patients along with its characterization. Finally, we point out and discuss some already existing methods of analysis of metabolic networks that are likely to be influenced by this subtle yet significant network structural property. The preliminary version of this study was published as an extended abstract in the proceedings of the 2018 IEEE International Conference on Bioinformatics and Biomedicine (BIBM), pp. 96–101; Madrid, Spain.

## 2. Materials and Methods

### 2.1. The Human Genome-Scale Metabolic Reconstruction

In general, a metabolic network is a set of reactions among elements of a given set of metabolites. Each reaction may be associated with a specific genetic rule that needs to be met in order for reaction to occur. If a reaction has no genetic rule assigned to it, it can be triggered whenever all substrates are available. A genetic rule usually describes which genes/proteins need to be active/present in the system to carry out the reaction. In particular, a rule may require simultaneous presence of several enzymes (e.g., the transfer of L-Oh-Proline by the Apical Imino Amino Acid Transporters in Kidney and Intestine requires both Transmembrane Protein 27 and Solute Carrier Family 6), or alternative enzymes that can catalyze the reaction (e.g., efflux of 2-hydroxy-atorvastatin-lactone into bile that is supported by ATP binding cassette subfamily C member 2 or subfamily B member 1). Finally, each reaction has also a lower and upper bound describing minimal and maximal flow of metabolites through this reaction.

On the other hand, formally, the network can be considered as a Petri net with conditional transitions of places, where by a conditional transition we mean additional Boolean formula that needs to be met in order for transition to occur (see Figure 1 for the example).

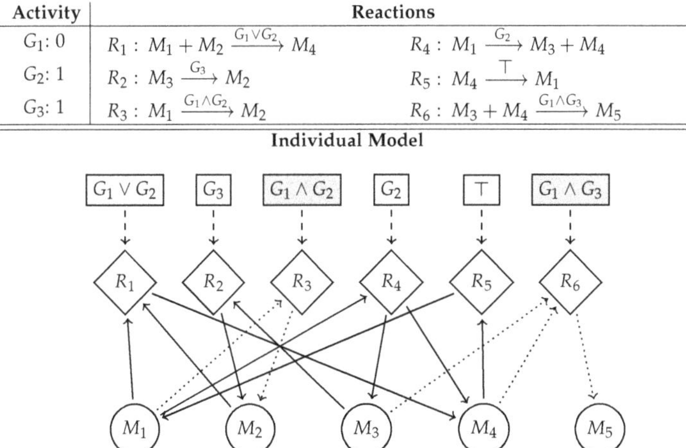



**Figure 1.** An example of metabolic network with genetic rules. The example is composed of six reactions $R_i$ among five metabolites $M_j$. Among these reactions there are five enzymatic reactions that are coordinated by the enzymes related to the three specific genes $G_k$. All arrows describe the flow of metabolites through reactions according to the above list of reactions. Using the gene activity pattern, a general network can be turned into transcriptom specific and represented as a Petri net with conditional transactions. Here we can see that inactivity of gene $G_1$ results in silencing (dotted line) of the reactions $R_3$ and $R_6$ which represses the production of the metabolite $M_2$ and eliminates the production of the $M_5$ in the system.

In our study we used the human genome-scale metabolic reconstruction model, RECON 2.2 that details known metabolic reactions occurring in humans, and thereby holds substantial promise for studying complex diseases and phenotypes [23]. The model described 7785 reactions (i.e., transitions of a Petri Net), between 2654 metabolites (i.e., places), out of which 4742 depended on a genetic rule (i.e., condition of a transition). We will refer to these reactions as enzymatic reactions. Each genetic rule was a Boolean formula in a disjunctive normal form (DNF), where each Boolean variable corresponds to the activity state (active or not) of one of 1670 unique genes. Additionally, all of the metabolites in the network were distributed among 10 compartments. In total, we considered 6048 compartment-specific metabolites. For more details see Table 1 and Figure 2.

### 2.2. Metabolic Landscapes

In order to construct a sample-specific metabolic model from a general metabolic network given a transcriptomic pattern the following procedure was applied. First, we specified the thermodynamic conditions describing the acceptable capacity of reactions fluxes, i.e., set the minimal and the maximal flow levels through each reaction. Next, we evaluated each genetic rule, based on the transcriptomic data to decide which reactions could occur in the initial model (see an example in Figure 1). Finally, the obtained, individualized model could be further studied taking into consideration the metabolic network structure. In our work, the linear programming problem was formulated and the simplex algorithm was run to find the steady-state flux distribution. This procedure provided two types of information. First, it suggested a new pattern of fulfillment among the genetic rules (binary-valued), where each change, in contrast to the original value, could be considered as a post-transcriptional change. Second, it outputted a vector of fluxes that met the criteria of the linear problem that was solved. These fluxes could provide information about the metabolic reactions or pathways that were mostly exploited given such expression patterns.

**Table 1.** The table provides a summary of the RECON 2.2 metabolic model.

| | | | |
|---|---|---|---|
| | | enzymatic | 4742 |
| | | transport | 3043 |
| all reactions | 7785 | exchange [a] | 701 |
| | | demand [b] | 44 |
| | | reverse | 3782 |
| | | directed | 4003 |
| | | unique substances | 2654 |
| all metabolites [c] | 6048 | boundary | 722 |
| | | compartments | 10 |
| | | simple rules | 2912 |
| all genetic rules | 4742 | complex rules | 1830 |
| | | unique rules | 1341 |
| | | unique genes | 1670 |

[a] Exchange reactions describe in- and outflow of metabolites through the system boundary. [b] Demand reactions are intra-network, unlimited sinks or sources of metabolites degradation or production. [c] An additional metabolite (in contrast to the summary reported in [23]) was required for the *R_biomass_other* reaction.

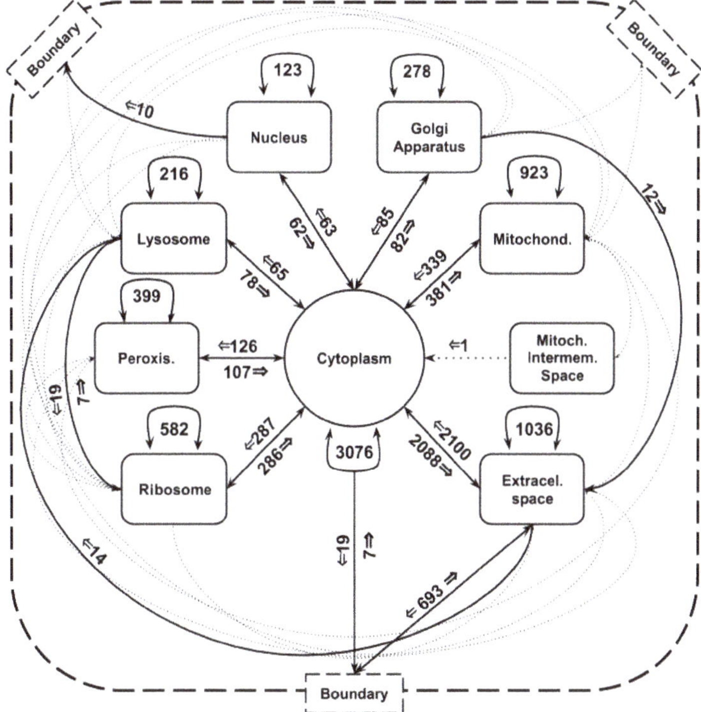



**Figure 2.** The figure presents an outline of the reactions distribution in the metabolic model of RECON 2.2. There are nine inter-cellular compartments depicted and the additional boundary component which represents external entry and exit of metabolites into the metabolism network model. Each line connecting two compartments represents all reactions that involve metabolites from these compartments. The number of these reactions and their direction is assigned to each line. To keep the figure readable, for less than 10 reactions we use a dotted, thinner line.

## 2.3. TCGA Transcriptomic Data

In this study we used 20 RNA-seq transcriptomic datasets provided by TCGA, each composed of cancerous and control samples and accompanied by clinical data. For the purpose of our work, each dataset was subjected to a standard preprocessing routine with recount R package [24]. The raw counts were scaled by the total coverage of the sample, that is, the area under the curve (AUC) of the coverage. Next, only genes that were composing genetic rules in the RECON 2.2 model were selected. Finally a number of 5 read counts was selected to be a threshold for a gene to be considered as active. After Leucken et al. the thresholding procedures were as permissive as possible, so that downstream analysis could be conducted and interpreted [25].

In the presented case study we used the kidney cancer dataset (renal cell carcinoma, RCC) composed of the 897 primary tumor and 129 normal tissue samples. Tumor tissues were also classified by their morphological type subtype into four groups: 527 clear cell RCC, 290 papillary RCC, 66 chromophobe RCC and 14 unclassified RCC samples. For 201 patients the survival data were also available. Additionally, we used the brain cancer dataset (707 samples) to showcase the artifacts of the clustering results obtained from the metabolic landscapes.

### 2.4. Steady-State Flux Distribution

In order to determine the activity state of each reaction in a personalized model we made use of the approach proposed by Shlomi et al. [26]. We formulated a mixed-integer linear programming (MILP) problem (see Equation (1)) and solve it using Gurobi solver [27]. The goal was to maximize an objective function (1a) with respect to stoichiometric (1b), thermodynamic (1c) and transcriptomic (1d–g) constraints. For detailed description see the Methods section in [26].

The solution of the described problem provided a binary vector of activity states for each reaction which we refer to as a metabolic landscape.

$$
\begin{aligned}
&\max_{v,y^+,y^-} \sum_{r \in \mathcal{A}} (y_r^+ + y_r^-) + \sum_{r \in \mathcal{I}} y_r^+ \quad \text{s.t.} \quad &\text{(a)}\\
&S \cdot v = 0 \quad &\text{(b)}\\
&v_{min} \leq v \leq v_{max} \quad &\text{(c)}\\
&v_{min,r}(1 - y_r^+) \leq v_r - y_r^+ \qquad r \in \mathcal{A} \quad &\text{(d)}\\
&v_{max,r}(1 - y_r^-) \geq v_r + y_r^- \qquad r \in \mathcal{A} \quad &\text{(e)}\\
&v_{min,r}(1 - y_r^+) \leq v_r \qquad r \in \mathcal{I} \quad &\text{(f)}\\
&v_{max,r}(1 - y_r^+) \geq v_r \qquad r \in \mathcal{I} \quad &\text{(g)}\\
&v \in \mathbb{R}^m, \ y_r^+, y_r^- \in \{0,1\}
\end{aligned}
\tag{1}
$$

In the described setting the FBA problem was formulated as a linear program and solved using the simplex algorithm. However, it should be emphasized that the structure of the underlying graph was crucial in context of the flux analysis and the existence of dense sub-graphs loosely communicating with neighbouring vertices strongly influenced the final outcome of the algorithm.

### 2.5. Graph Entropy

To detect the possible sources of the structural redundancy in a metabolic network we applied the measure of entropy on a graph structure. To account for the context of the metabolic fluxes we made use of the entropy based on the inner and outer neighbours of vertices within a given sub-graph [28].

Namely, for a graph $G = (V, E)$ and its sub-graph $G' = (V', E')$ where $V' \subset V$ and $E' \subset E$ contains all edges adjacent to vertices in $V'$ we define the probability of the inner ($\mathcal{P}_I$) and outer ($\mathcal{P}_O$) flow of a vertex $v \in V'$ as:

$$
\mathcal{P}_I(v) = \frac{I(v)}{N(v)}; \qquad \mathcal{P}_O(v) = \frac{O(v)}{N(v)} = 1 - \frac{I(v)}{N(v)}
$$

where $I(v), O(v)$ are the number of neighbours of $v$ that: belong to $V'$ and not belong to $V'$, respectively, while $N(v)$ are all neighbours of $v$. Based on that we define the entropy $H(v)$ of a vertex $v$ as:

$$
\mathcal{H}(v) = -\mathcal{P}_I(v) \log_2 P_I(v) - \mathcal{P}_O(v) \log_2 \mathcal{P}_O(v)
$$

and consequently, the entropy of a graph G induced by sub-graph G', $H_{G'}(G)$, is defined as:

$$
\mathcal{H}_{G'}(G) = \sum_{v \in V'} \mathcal{H}(v)
$$

Given the above definitions, we determine the entropy of a group of enzymatic reactions within the RECON 2.2 metabolic network. Additionally, we evaluate the entropy induced by groups of random reactions sampled from the entire network. These steps aim to depict the graph in terms of the entropy and thus connectivity and the flow of metabolic information, that is subject to the metabolic landscape analysis.

### 2.6. Binary Data Analysis

Since the final outcome of a linear programming problem is a set of binary vectors, it is necessary to introduce specific methods to analyze them. In the literature there are described various statistical methods that are specifically dedicated to the analysis of binary data, including: the non-negative matrix factorization [29], the sparse logistic PCA method proposed by Lee et al. [30], and the variational factorization method employing independent beta latent densities [31].

Nonetheless, in our case, in order to explore the data, track the potential latent variables and visualize results we use the standard principal component analysis (PCA), which provides numerically stable and robust results and is also used to define a function for separating variables selection.

#### PCA Loadings-Based Variables Selection

Let $M \in \{0,1\}^{m \times n}$ be a binary data matrix with $m$ observations and $n$ variables, $L = (L^{(1)}, \ldots, L^{(n)})$ be a loadings matrix and $P$ scores (or principal components) matrix, satisfying $P = ML$. By definition $P$ is a representation of $M$ in a new basis that is composed of vectors of $L$. For the purpose of binary landscapes analysis, we construct a set of highest valued coordinates from first $k$ directions of the rotation matrix. The general parametrized function $\mathcal{L}$ describes the selection procedure:

$$\mathcal{L}_{f,k}(L) = \bigcup_{i=1}^{k} \left\{ j : |L_k^{(i)}| \geq f\left(L^{(i)}\right) \right\}$$

where $f$ is the function determining the threshold for the value of a coordinate to be assumed significant and $k$ is the number of vectors from loading matrix that are considered. The function is used to select the groups of reactions with the highly correlated activity pattern between all samples. In parallel, the results from PCA Loadings-based variable selection were compared with the overall entropy analysis of the network structure considering the interaction between the groups of enzymatic reactions. It turned out that the groups requiring the adjustment belong to the group of clusters characterized by the low entropy and loosely connected to the other vertices in the network.

To determine differentiating variables in the studied datasets, two measures were used. First, the Jaccard Index is used to determine differentiating reactions between two groups of samples. For any $z \in \{0,1\}^n$ we define $z^1 = \sum_i z_i$ and $z^0 = \sum_i (1 - z_i)$. Given that the Jaccard Index of two binary vectors $\mathcal{J} : \{0,1\}^n \times \{0,1\}^m \to [0,1]$ is defined as:

$$\mathcal{J}(x,y) = \frac{\min\left(x^1, y^1\right) + \min\left(x^0, y^0\right)}{\max\left(x^1, y^1\right) + \max\left(x^0, y^0\right)}$$

The index is non-linear and depends on the lengths of vectors, thus evaluates to 0 when two vectors don't share any element, increases with an increasing number of shared elements between vectors. It reaches the maximal value of 1 when both vectors share the same elements and the same lengths. In the process of variables selection as a threshold value we set 0.25 of the maximal value that the index can reach (given the lengths of compared vectors).

Additionally, the Tanimoto similarity measure of two binary vectors $\mathcal{T} : \{0,1\}^n \times \{0,1\}^n \to [0,1]$ is defined as:

$$\mathcal{T}(x,y) = \frac{n - |\{i : x_i \neq y_i\}|}{n + |\{i : x_i \neq y_i\}|}$$

and is used as a distant measure between two binary metabolic landscapes.

## 3. Results

### 3.1. Metabolic Network Structure Problems

The first step was to apply the procedure suggested by Shlomi et al. to the cohort of the TCGA data in order to verify if there existed a common metabolic pattern among various types of tumors, or alternatively, if there existed any metabolic biomarkers which may be of diagnostic importance.

As mentioned in the introduction, our preliminary results suggested that, surprisingly, each data set was composed of two to four clearly separable clusters of patients. In the case of 13 out of 20 TCGA cancer datasets, the corresponding set of metabolic landscapes was well-separated into four clusters that, on average, were differentiated by the activity of 180 reactions. Additionally, for the remaining seven cancer-specific landscape datasets there were on average 95 reactions differentiating samples into two clusters (see Figure 3 for an example based on brain cancer dataset). Moreover, it turned out that almost all (>95%) discriminating (using $\mathcal{L}_{max,2}$ function) reactions in these clusters were coordinated by two main enzymes that were encoded by: *SLCO1A2* (solute carrier organic anion transporter family member 1A2; coordinating superfamily of 94 Amino Acid-Polyamine-Organocation reactions [32]) and *SLC7A9* (solute carrier family 7 member 9; coordinating superfamily of 79 Resistance-Nodulation-Cell Division reactions [33]) genes. These observations may lead to a conclusion that cancers in general have natural subfamilies that can be described by the activity of specific groups of metabolic reactions and thus also activity of particular enzyme encoding genes. Activation of *SLCO1A2* is related to the development and functioning of the immune system, organismal system for calibrated responses to potential internal or invasive threats and is highly over expressed in breast cancer tissues [34]. On the other hand, *SLC7A9* enables the transportation of substances (such as macromolecules, small molecules, ions) into, out of or within compartments of a cell, or between cells. Additionally, as a co-enzyme with *SLC3A1* coordinates group of five transportation/exchange reactions of L-Cystine, L-Alanine and L-Ornithine, that were also detected as differentiating the cancer data. This observation implies activation of *SLC3A1*, that was recently reported to promote breast cancer tumorigenesis [35].

Even though these literature reports may sound promising, we report another observation related to the analysis performed on a dataset with 500 randomly generated gene activities that were subjected to the metabolic analysis (see the right panel of the Figure 3). Despite the fact that the gene activity dataset did not include any relevant information, we were able to identify two groups of reactions differentiating samples into four well-separable clusters. These observations, undoubtedly, put in question all results related to the clustering of metabolic landscapes performed on all cancer datasets, since the analysis inferred about the knowledge that did not come from the data but from the topology and structure of metabolic network.

The analysis of the RECON 2.2 network structure revealed that there existed groups of reactions that were associated with the same genetic rule (top 10 most common genetic rules were related to over 900 reactions). Even though these reactions were biologically non-redundant (each of them was functional), they very often formed well-connected sub-networks that were loosely connected with the rest of the network. As a consequence they made it problematic to apply the FBA methods. As the way to identify such structures we proposed to apply the notion of entropy induced by sub-networks. It turned out that these sub-networks minimized the entropy.

In the Figure 4 one can notice the structural properties of the RECON 2.2 metabolic network. We discovered that among the groups of enzymatic reactions that follow the same genetic rule, there was a representation of these that consisted of several dozen of reactions and yet induced a very low entropy of the sub-network. The entropy- and structural-based analysis revealed that these reactions were very well connected within their group, however they were loosely connected to the neighbouring vertices outside their group. Such characteristics imposed the risk of the uneven flow of the metabolic information throughout the network and made it of low probability that initial transcriptomic conditions would be changed due to the influence of the other neighboring reaction states. In effect, this may result in the detection of network topology artifacts rather than new information on metabolic reactions.

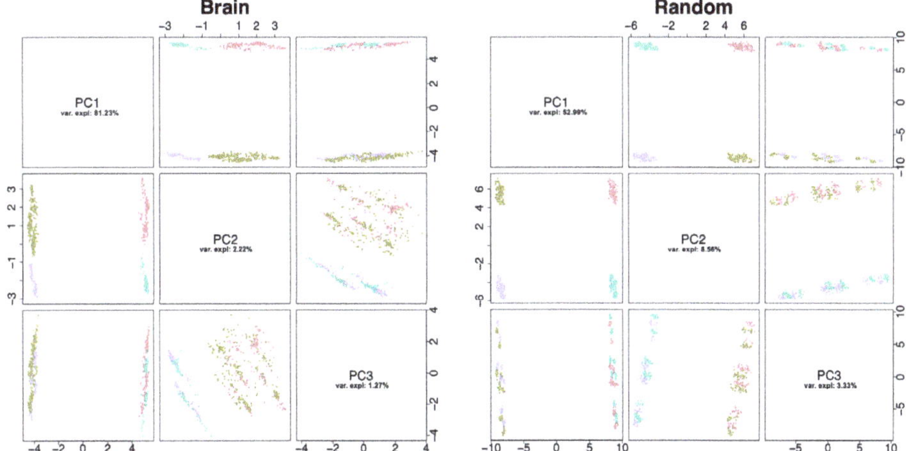

**Figure 3.** The comparison of the first three principal components of metabolic landscapes determined for brain cancer (**left**) and random (**right**) datasets. In both cases samples form well-separating clusters that can be identified by the activity pattern of overrepresented gene rules. For the brain dataset: *SLC7A9* and *SLC28A3*. For the random dataset: *SLC7A6* and *SLCO1B1*.

For this reason we proposed two adjustment methods that take into account this phenomena. The aim was to transform the data so that the statistical analysis did not detect artificially induced data separation, but rather may result in a discovery of subtle differences in metabolic activity between samples possibly related to novel metabolic biomarkers.

*3.2. Computational Workflow*

Due to the discovery of the artifacts in the results of the metabolic landscape analysis we propose the following pipeline of procedures, that aims to reduce the statistical redundancy immersed in the RECON 2.2 network structure.

As the prerequisite, the entropy levels of each group of enzymatic reactions are determined with the tools described in the Materials and Methods section. These are the base of the control for redundant activity patterns in the statistical inference from the reaction activities.

First, we processed the RNA-seq transcriptomic data and converted it to the gene activity matrix using the threshold of five counts. The matrix was then used to create a personalized metabolic network model. For each of these, a MILP problem was formulated and solved with the Gurobi solver based on the simplex algorithm. The solution was composed of the sample-specific metabolic landscapes, which were subjected to statistical analysis. The analysis of the data was supported by the two-fold verification if the redundancy among groups of reactions existed. On one hand, using the PCA loadings-based variables selection $\mathcal{L}$ function defined in the subsection Binary Data Analysis, on the other assessing the entropy of the detected group of reactions. Based on these two conditions, if needed, the adjustment of these reactions was performed.

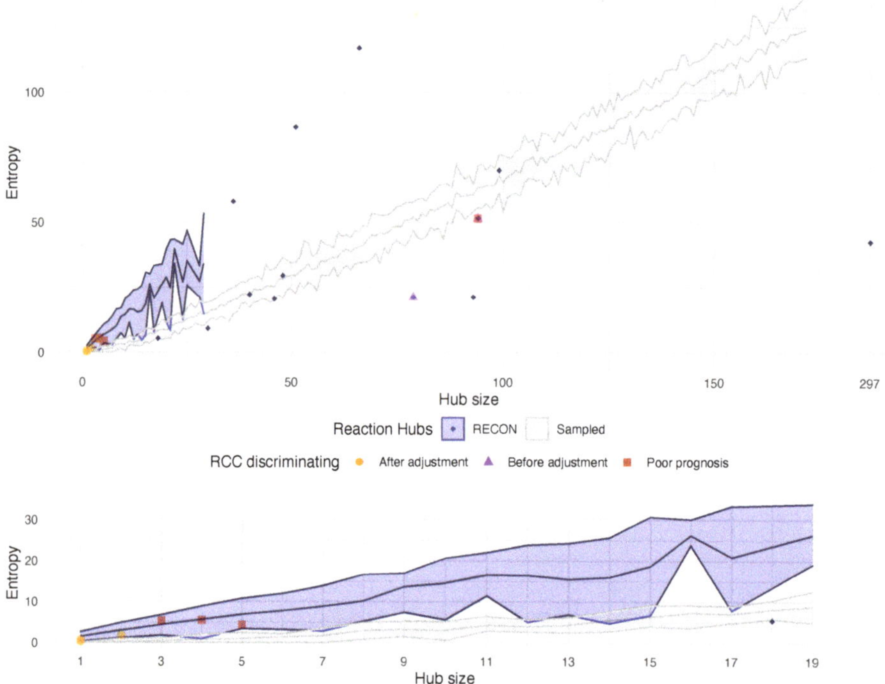

**Figure 4.** **The entropy levels induced by specific sub-networks in RECON 2.2 by their size.** The upper panel depicts the entropy level induced by each sub-network of enzymatic reactions following the same genetic rule (the blue area and outlying blue points). The grey area shows the entropy levels for random groups of reactions of a given size. The groups of reactions that characterize the RCC samples before and after the adjustment, as well as, the potential biomarker reactions for the discovered poor-prognosis cluster are depicted by corresponding points. The bottom panel is a close-up view of the low-number groups of reactions.

The data transformation was based on the aggregation of the activity states for groups of reactions with respect to compartments to which belong their substrates and products. Namely, each reaction was labeled with a name of form $s_1 \ldots s_k$-$p_1 \ldots p_j$, where $s_1, \ldots s_k$ and $p_1, \ldots, p_j$ are alphabetically ordered names of compartments that, respectively, substrates and products belonged to. Next, all reactions with the same genetic rule and the assigned label were aggregated into the one represent reaction with the level of activity equal to the average of activities in the group.

Finally, inference from the transformed data structure through the hierarchical clustering using the binary distance measures (e.g., the Tanimoto similarity), correlation with the clinical data, selection of discriminatory features and functional analysis of determined clusters was performed. The outline of the described workflow is also depicted in the Figure 5. Additionally, in the Github repository: https://github.com/storaged/metabolic-landscape the code and the scripts used in this workflow are available. For the purpose of this study the workflow was executed on the machine with Mac OS.

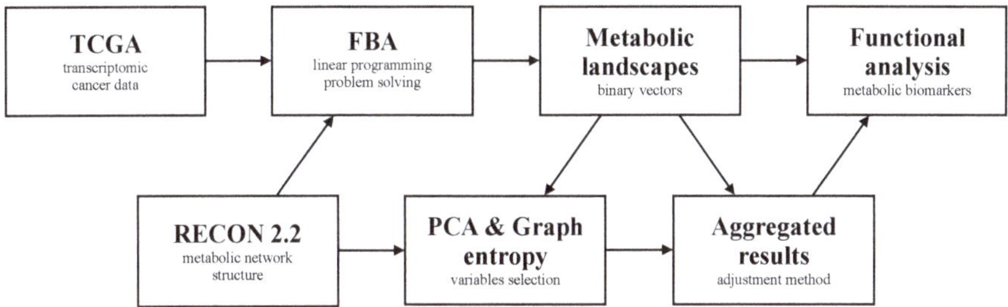

**Figure 5.** The workflow outline highlighting the main steps in the proposed data analysis method.

### 3.3. Metabolic Landscape Adjustment for the Renal Cell Carcinoma

In order to validate the proposed workflow we performed the metabolic landscape analysis on the TCGA dataset of renal cell carcinoma (RCC). After performing the adjustment methods on the metabolic landscapes we reported a significant improvement in samples clustering, both in the sense of unwanted network structure-dependent clusters composition and correlation with clinical data (see Figure 6). Additionally, we took advantage of the provided prerequisite and confirmed that all of the adjusted groups of reactions fell into the group of the low-entropy level (see Figure 4).

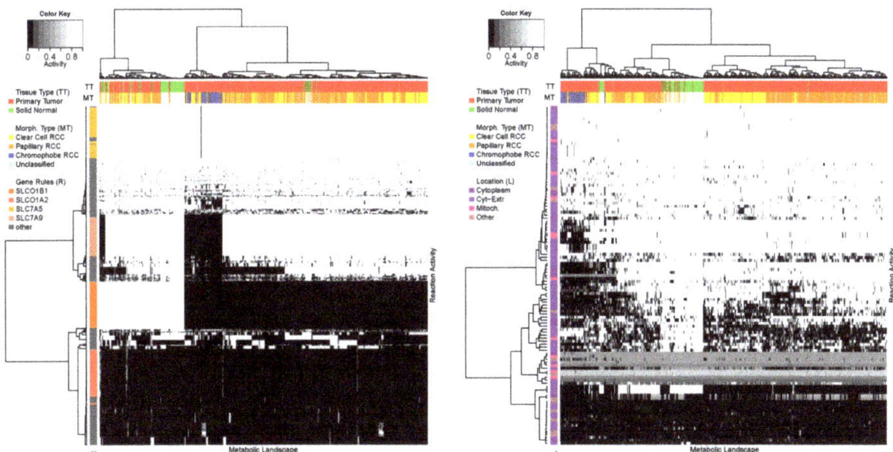

**Figure 6.** The comparison of reactions activity before (**left**) and after (**right**) the adjustment. On the left panel, the vertical strip marks reactions associated with the same genetic rule (the orange scale colors) noticeably determining the clustering of all landscapes. In both panels, horizontal stripes represent the Tissue Type (TT) and the Morphological Type (MT) of all samples. One can see, how the adjustment improves the correlation of the data with the clinical variables, especially the morphological type. Finally, the vertical strip on the right panel presents that correspondence related to the compartments (the purple scale colors) was introduced.

We removed the amplified activity pattern of reactions coordinated by the same genetic rule, that influenced the clustering of samples in an unwanted way. After the reduction there were no significant, discriminating reactions associated with the same genetic rule. This step also resulted in more reliable clustering of data according to clinical observations, e.g., normal or tumor tissue type or morphological type of a tumor sample.

The results of our analysis of the renal cell carcinoma TCGA dataset were consistent with latest reports, that indicated *SLC6A3* as a experimentally confirmed biomarker for RCC [36].

The transcriptomic signal of *SLC6A3* in our data was clearly discriminating biomarker of Clear Cell RCC subtype (6.89 logFC).

However, thanks to the analysis of metabolic landscapes we further suggest potential biomarkers that correspond to specific, known RCC subtypes [37]. The literature so far reports *CXCL16* gene as a significantly expressed in papillary RCC with others still waiting for their validation [38]. Nonetheless, the analysis of metabolic landscapes suggests two transport reactions that discriminate the Papillary RCC subtype. Both of them are supported alternatively by the already reported *SLC6A3* or *SLC6A2*, other member of the same Solute Carrier family. The transportation activity state of dopamine and norepinephrine via sodium symport between cytoplasm and extracellular space are well separating the Papillary RCC subtype from other samples. Namely, these reactions are predicted to be inactive in Papillary samples. Our results reported 247 out of 291 ($\approx$85%) papillary samples and 42 out of 737 ($\approx$6%) other samples characterized by inactivation of these reactions. This observation was also confirmed by the purely transcriptomic data, which suggested down regulation of both genes ($-3.62$ and $-2.2$ logFC of *SLC6A2* and *SLC6A3*, respectively) for papillary samples. This observation suggested a simultaneous drop of the activity of both *SLC6A2* and *SLC6A3* as a potential diagnostic biomarker.

In case of the Chromophobe RCC (ChRCC) subtype, before the adjustment of the metabolic landscapes two genetic rules were in fact separating this subtype from the other cancer samples: (i) the extracellular space and cytoplasm exchange reactions supported by the complex of *SLC3A1* and *SLC7A9* (five reactions involving L- Cystine, L-Alanine and L-Leucine) and (ii) reactions controlled by *SLC7A9* that was involved in 79 reactions. However, after the adjustment two potential biomarkers were found: inactivation of sodium-dependent transport of (i) phosphate, supported by *SLC17A1* and (ii) ascorbate supported by *SLC23A1*. Inactive phosphate and ascorbate transport characterized, respectively, 65 and 62 out of 66 ChRCC samples; 45 and 28 out of 774 other tumor samples. Even in the literature reports it is still not clear how the phosphate transportation or concentration level and absorption via *SLC23A1* of ascorbate influences the cancer cells [39]. We point to these factors, both reaction activity and transcriptomic/proteomic levels, as possible biomarkers of ChRCC.

*3.4. Poor Prognosis Cluster*

Finally, we reported a candidate for a new cancer subtype resulting from the statistical and functional analysis of metabolic landscape clusters. Based on the samples clustering after the adjustment (see Figure 6) we further studied the obtained six clusters of samples. Among them, there were four homogeneous clusters composed mainly of: the healthy tissues (Control), the Chromophobe RTCC subtype (Chr-basal), the Clear Cell RTCC (CC-basal) and the Papillary RTCC (Pap-basal). Even though, the pattern of the remaining two clusters was not related to their morphological type we noticed a statistically significant (*p*-value: 0.002) difference in the survival time in one of the clusters (see Figure 7), which we addressed as a poor prognosis cluster.

The first and natural step here is to compare our clustering with the previously published [40]. In particular, the membership of the CpG island methylator phenotype (CIMP) cluster is of high interest, because of its bad prognosis properties that were investigated in [41]. It turns out that the entire CIMP cluster was part of the much larger (10 vs. 102) poor prognosis cluster that was identified based on its metabolic profile. Additionally, our poor prognosis cluster also contained five out of six samples classified as metabolically divergent (MD) in the literature. While CIMP-associated tumors showed increased expression of key genes involved in glycolysis, our cluster metabolic pattern was also associated with deregulations in glycan biosynthesis. However, the important characteristic feature was the reduced expression of genes related to amino acid transport.

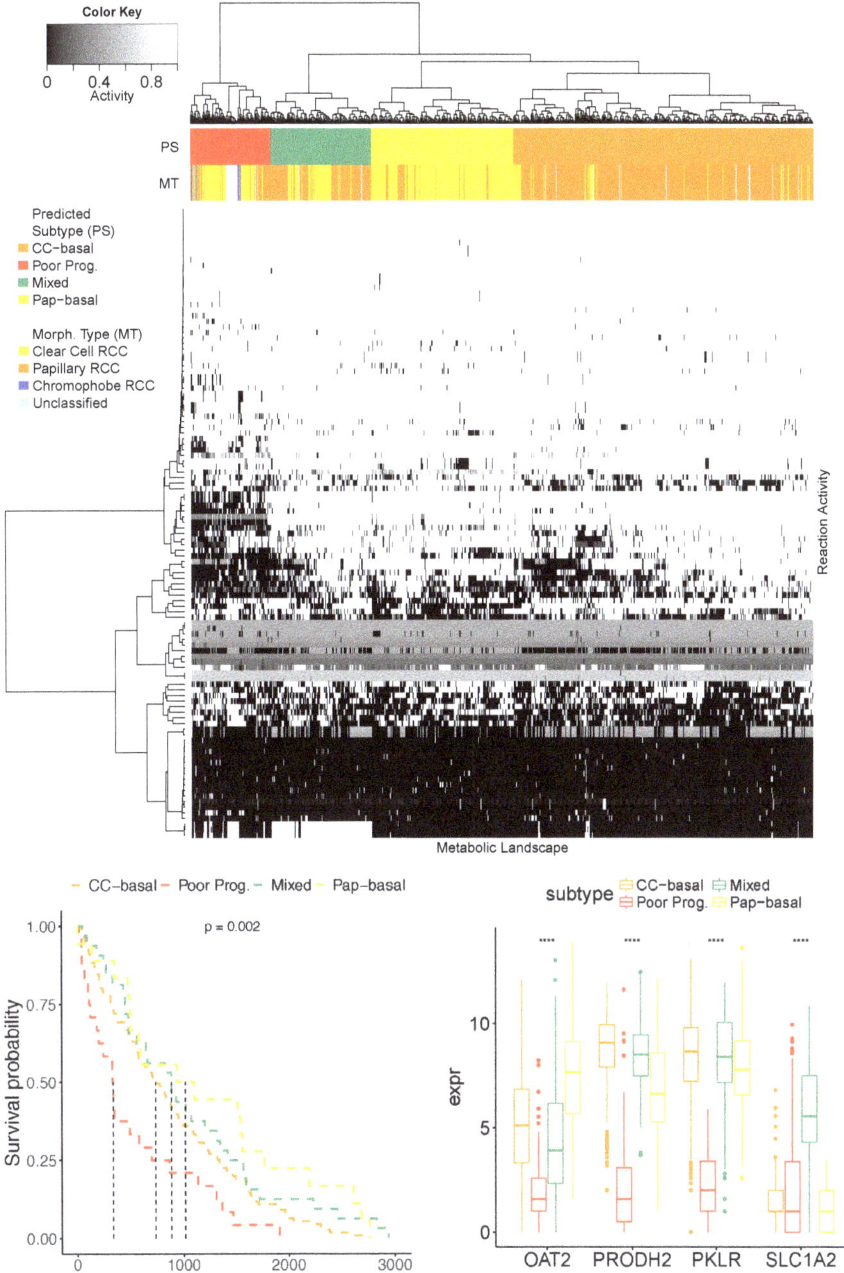



**Figure 7. The analysis of the poor prognosis cluster.** The heatmap presents the activity of reactions from the four clusters determined with the hierarchical clustering. The upper horizontal stripes compare the predicted subtypes (PS) labeling with the known morphological types (MT). The bottom-left panel presents the Kaplan–Meier survival curves, which present the significantly lower survival time of patients from the poor prognosis cancer (*p*-value: 0.002). The bottom-right panel compares the expression level of the genes characterizing the poor prognosis cluster.

To deepen the nature of these newly discovered clusters, we performed a differential analysis of four clusters (using also CC-basal and Pap-basal) in order to describe a metabolic as well as genetic nature of this cluster. We determined a set of 106 differentiating reactions coordinated by four genes: *OAT2* (3), *PRODH2* (4), *PKLR* (5) and *SLC1A2* (94), see Figure 4. Among these reactions we report orotate-glutamate antiport, uptake of allopurinol and oxypurinol by the hepatocytes, mitochondrial proline Oxidase (NAD) and dehydrogenase, and reactions involved in pyruvate metabolism. Genes coordinating these reactions reveal a specific expression pattern of the poor prognosis cluster, i.e., the low expression and the consequent inactivation of corresponding reactions. It is worth to emphasize, that the adjustment procedure provides a confirmation of the significance of the hub of reactions governed by the *SLC1A2* gene (see the corresponding datapoint in the Figure 4).

Additionally, we performed the functional analysis of top 100 differentiating genes, using the DAVID on-line tool. The analysis provided a consistent output indicating functions and keywords commonly related to modifications in transport and symport reactions highlighting transmembrane transport activity. Finally, pathway analysis performed with KEGG implied deregulations in the glycan biosynthesis and metabolism pathway.

The above observations can be preliminarily verified with literature reports. In a metabolic sense, abnormalities in glycan biosynthesis that we observe were reported as a significant factor of cancer cells phenotype and biology almost two decades ago [42]. It should be emphasized that recent studies stress the particular role of amino acid transporters in the pathogenesis of cancer. Specifically, the deregulation of these genes leads to metabolic reprogramming changing intracellular amino acid levels, which may underlie the molecular processes that explain poor prognosis in detected cluster [43]. Moreover, in our study, the poor prognosis properties can be influenced by the low activity of *SLC1A2* belonging to the organic-anion-transporting polypeptide (OATP) family, because it is responsible for transport of anticancer drugs (e.g., methotrexate used in chemotherapy) [44] and overall uptake of. Similarly, low expression of *OAT2* was reported to influence a poor response to antitumor UFT-based chemotherapy in colorectal cancer patients [45]. Loss of *PRODH2* was also reported in cancer [46], however no links with cancer prognostics were reported. The above summary may constitute an introductory justification for the possible existence of the poor prognosis cluster in RCC mainly conditioned by the chemoresistance dictated by the activity of its potential biomarker genes.

## 4. Discussion and Further Research

Concluding, in our article we recall an important yet barely discussed data analysis and integration problem of the two-fold nature. Using the TCGA datasets and the metabolic network model we presented how the bioinformatical and statistical data analysis may lead to outwardly interesting biological and medical observations, which may be justified by the literature reports. However, not only do we prove that these observations are the inevitable consequence of the model assumptions (as genetic rules induce clusters artificially), but also we show that these assumptions conceal the current state of knowledge about cell metabolism, as many discovery claims may testify.

We believe that the identified problem is important enough to look at alternative approaches and verify if they are not exposed to the similar obstacle. However, as it was already mentioned, there is a strong evidence that the problem highlighted in this work may be present in other researches that dealt with the analysis of the FBA outcome. The reason, even though is very subtle, is related to the linearity of the solving method applied to the non-linear topology of the problem support. In particular, the existence of internally strongly connected sub-graphs that are loosely connected with their neighbouring vertices influences the graph flow approaches. The metabolic information cannot be flawlessly passed on between reactions and thus interferenced by particular information-flow blockers. Here, we briefly discuss several alternative approaches as an extension to the literature overview presented in the Introduction.

To the scope of the traditional linear programming approach to solve the FBA problem we add what Lee et al. proposed in their work. The redefinition of the main objective function as well as an emphasis put on the more informative nature of absolute gene expression measurements provided by RNA-seq data was presented. Authors set an optimization problem to maximize the correlation between an observed transcriptomic profile and levels of reactions fluxes, that is solved using the MILP solver. Authors provide a case study based on estimation of exometabolic flux in *Saccharomyces cerevisiae* and show that their method outperforms the traditional approach upon maximisation of the rate of biomass production.

Additionally, there is also a bunch of approaches that extends the approach to the FBA problems onto the field of Bayesian statistics. One example of an extension of the FBA via Bayesian factor modeling was suggested by Angione et al. [47]. The extension bases on an assumption that high-dimensional data are generated from the hidden lower-dimensional factors that are shared across data samples. As a consequence, as a first step authors solve the bi-level FBA problem, which extends the standard problem by interdependencies among enzymes and genetic rules coordinating metabolic reactions. Next, Bayesian matrix factorization modeling with Gaussian Markov random field is incorporated to perform pathway analysis that takes into account reaction-pathway memberships as prior knowledge. Using an metabolic model of *Escherichia coli* authors present how their model tracks changes in pathway responsiveness for variuos experimental conditions.

Lately, in [48] authors presented an approach that results in flux posterior represented as an unimodal truncated multivariate normal (TMVN) distribution. Using a MCMC Gibbs sampler implemented in Matlab a group of in-silico experiments was performed to highlight the capabilities of the model. The presented approach allows to characterize the genome-scale flux covariances, reveal flux couplings, and determine genome-scale number of intracellular unobserved fluxes in *Clostridium acetobutylicum* from 13C data based on a small set of intracellular flux measurements.

Interestingly, the problem of flux estimation was also approached from the perspective of the thermodynamic analysis. Zhu et al. introduce a novel, two-step optimization method termed as thermodynamic optimum searching [49]. First, the original FBA is used to determine the maximum growth rate. Then, the most thermodynamically favored solution is acquired by solving a nonlinear optimization problem in which the growth rate is fixed to the maximum. The later aim is achieved by maximizing the entropy production rate while minimizing energy usage and deviation from the second law of thermodynamics related to the the minimum magnitude of the Gibbs free energy change and the maximum entropy production principle. The method is supported with five *E. Coli* case studies presenting the improved accuracy of predictions compared to the standard FBA.

What should be emphasized here is that none of the recalled articles mentioned any observation on the relationship between the structure of the metabolic network and their final outcome. Additionally, an alarming fact is the use of methods such as the categorization of the transcriptomic signal, linear (or semi-linear) methods of solving the flow problem, binarization of data and their interpretation based by clustering. All of these observations allow the reader to ask if any control for artifacts was performed. In order to compare the methods, approaches, objectives of all the mentioned works in this article a table summary is provided in the Appendix A.

To conclude, the future directions in which we would like to conduct this research include incorporating other metabolic networks for the landscape analysis, but also improved adjustment methods, that would define and take into account the confidence of the reaction. One idea in that direction, would be to incorporate some experimental validations from the literature reports. Ideally, we would like to reach the gold standard using measurements from the 13C-fluxomics technology, where metabolic precursors enriched with 13C and quantified by mass spectrometry or NMR [50,51], as well as from transcriptomic assays. However, such data has so far been collected for relatively small systems, e.g., single pathways [52] or other simple organisms [53], while our amendment concerns the whole human cell metabolism. Another option is to take into account the entropy-based approach and propose a sample-specific adjustment method. Finally, we also suggest defining a probability of reaction activity, that may be introduced instead of currently considered binary activity state. This would allow to make use of the continuous information included in the expression data, rather than reduced binary signal. We believe that these improvements may lead to better understanding of cancer biology and phenotype resulting from the observed differences in metabolic activity.

**Author Contributions:** The individual contributions of each author of this article presents as follows: conceptualization, K.G., M.K. and A.G.; methodology, K.G.; software, K.G. and M.K; validation, K.G. and M.K. formal analysis, K.G.; data curation, K.G.; writing—original draft preparation, K.G. and A.G.; visualization, K.G.; supervision, A.G.; project administration, A.G.; funding acquisition, A.G. All authors have read and agreed to the published version of the manuscript.

**Funding:** This research and APC was funded by the polish National Science Centre (NCN) grant no. 2014/12/W/ST5/00592.

**Acknowledgments:** Authors thank to the doctoral studies program funded by NCBR (The National Centre for Research and Development, Poland) project Descartes.

**Conflicts of Interest:** The authors declare no conflict of interest.

## Abbreviations

The following abbreviations are used in this manuscript:

| | |
|---|---|
| ATP | Adenosine Triphosphate |
| ChRCC | Chromophobe Renal Cell Carcinoma |
| DAVID | Database for Annotation, Visualization, and Integrated Discovery |
| DNF | Disjunctive Normal Form |
| FBA | Flux Balance Analysis |
| LP | Linear Programming |
| MCMC | Markov Chain Monte Carlo |
| MILP | Mixed-Integer Linear Programming |
| PCA | Principal Components Analysis |
| RCC | Renal Cell Carcinoma |
| RNA | Ribonucleic Acid |
| TCGA | The Cancer Genome Atlas |
| TMVN | Truncated Multivariate Normal |

## Appendix A

A literature search of articles was conducted, and Table A1 summarizes and compares the literature and research that was related to the FBA problem and integration of the transcriptomic and metabolic knowledge. Let us emphasize that nearly each objective in the presented works is related to detection of activation profiles. Yet, artifact detection is not considered in any of the articles.

**Table A1.** Summary of the literature related to the problem of integrating the transcriptomic and metabolomic in the context of the flux balance analysis (FBA) problem.

| Author | Approach | Software | Data Used | Objective | Problem Size |
| --- | --- | --- | --- | --- | --- |
| Leoncikas et al. (2016) [20] | MILP | R$^{(-)}$ | microarray detection *p*-value | activity of flux states in a network | 3311 reactions, 2766 metabolites, 2004 genes |
| Li et al. (2010) [21] | reduced MILP | Matlab$^{(-)}$ | gene expression of NCI-60 cancer cell lines | reaction flux and activity for all the reactions | 3742 reactions, 2766 metabolites, 1905 genes |
| Wang et al. (2012) [13] | Model Building Algorithm | Matlab$^{(+)}$ | gene expression or literature evidence | activity for all the reactions | 3311 reactions, 2766 metabolites, 2004 genes |
| Marín de Mas et al. (2018) [22] | MILP | FASIMU$^{(+)}$ | microarray transcriptomic data | activity for all the reactions | 3311 reactions, 2766 metabolites, 2004 genes |
| Mhamdi et al. (2020) [54] | Bayesian | - | RNA-seq data | prediction of the biomass production | 4008 reactions, 3292 metabolites |
| St. John P. C. et al. (2019) [55] | Bayesian | Python$^{(+)}$ | metabolite, boundary flux, enzyme measurements [56] | infer posterior distributions in kinetic parameters of a metabolic network | 203 metabolites, 240 reactions |
| Heinonen et al. (2019) [48] | Bayesian | Matlab$^{(+)}$ | metabolic networks | find posterior probabilities of fluxes states in a network | 4388 reactions, 2936 genes |
| Angione et al. (2015) [47] | semi- Bayesian | Matlab$^{(+)}$ | microarray expression profiles | detect pathway cross-correlations and predict metabolic pathway activation profiles | 2583 reactions, 466 genes |
| Heino et al. (2007) [57] | Bayesian | Matlab$^{(-)}$ | ischemic biopsy data [58] | find posterior probabilities of fluxes states in a network | 26 reactions, 53 metabolites |

## References


1. Huang, S.; Chaudhary, K.; Garmire, L.X. More Is Better: Recent Progress in Multi-Omics Data Integration Methods. *Front. Genet.* **2017**, *8*, 84. [CrossRef] [PubMed]
2. Gjuvsland, A.B.; Vik, J.O.; Beard, D.A.; Hunter, P.J.; Omholt, S.W. Bridging the genotype-phenotype gap: What does it take? *J. Physiol.* **2013**, *591*, 2055–2066. [CrossRef] [PubMed]
3. Wanichthanarak, K.; Fahrmann, J.F.; Grapov, D. Genomic, Proteomic, and Metabolomic Data Integration Strategies. *Biomark. Insights* **2015**, *10*, 1–6. [CrossRef] [PubMed]
4. Fondi, M.; Lio, P. Multi -omics and metabolic modelling pipelines: Challenges and tools for systems microbiology. *Microbiol. Res.* **2015**, *171*, 52–64. [CrossRef] [PubMed]
5. Orth, J.D.; Thiele, I.; Palsson, B. What is flux balance analysis? *Nat. Biotechnol.* **2010**, *28*, 245–248. [CrossRef] [PubMed]
6. Covert, M.W.; Palsson, B. Transcriptional regulation in constraints-based metabolic models of Escherichia coli. *J. Biol. Chem.* **2002**, *277*, 28058–28064. [CrossRef] [PubMed]
7. Colijn, C.; Brandes, A.; Zucker, J.; Lun, D.S.; Weiner, B.; Farhat, M.R.; Cheng, T.Y.; Moody, D.B.; Murray, M.; Galagan, J.E. Interpreting expression data with metabolic flux models: predicting Mycobacterium tuberculosis mycolic acid production. *PLoS Comput. Biol.* **2009**, *5*, e1000489. [CrossRef]
8. Becker, S.A.; Palsson, B.O. Context-specific metabolic networks are consistent with experiments. *PLoS Comput. Biol.* **2008**, *4*, e1000082. [CrossRef]
9. Bordbar, A.; Mo, M.L.; Nakayasu, E.S.; Schrimpe-Rutledge, A.C.; Kim, Y.M.; Metz, T.O.; Jones, M.B.; Frank, B.C.; Smith, R.D.; Peterson, S.N.; et al. Model-driven multi-omic data analysis elucidates metabolic immunomodulators of macrophage activation. *Mol. Syst. Biol.* **2012**, *8*, 558. [CrossRef] [PubMed]
10. Zur, H.; Ruppin, E.; Shlomi, T. iMAT: An integrative metabolic analysis tool. *Bioinformatics* **2010**, *26*, 3140–3142. [CrossRef]
11. Agren, R.; Bordel, S.; Mardinoglu, A.; Pornputtapong, N.; Nookaew, I.; Nielsen, J. Reconstruction of genome-scale active metabolic networks for 69 human cell types and 16 cancer types using INIT. *PLoS Comput. Biol.* **2012**, *8*, e1002518. [CrossRef]
12. Jensen, P.A.; Papin, J.A. Functional integration of a metabolic network model and expression data without arbitrary thresholding. *Bioinformatics* **2011**, *27*, 541–547. [CrossRef]
13. Wang, Y.; Eddy, J.A.; Price, N.D. Reconstruction of genome-scale metabolic models for 126 human tissues using mCADRE. *BMC Syst. Biol.* **2012**, *6*, 153. [CrossRef] [PubMed]
14. Chandrasekaran, S.; Price, N.D. Probabilistic integrative modeling of genome-scale metabolic and regulatory networks in Escherichia coli and Mycobacterium tuberculosis. *Proc. Natl. Acad. Sci. USA* **2010**, *107*, 17845–17850. [CrossRef]
15. Kim, J.; Reed, J.L. RELATCH: Relative optimality in metabolic networks explains robust metabolic and regulatory responses to perturbations. *Genome Biol.* **2012**, *13*, R78. [CrossRef] [PubMed]
16. Mardinoglu, A.; Nielsen, J. Systems medicine and metabolic modelling. *J. Intern. Med.* **2012**, *271*, 142–154. [CrossRef]
17. Masoudi-Nejad, A.; Asgari, Y. Metabolic cancer biology: Structural-based analysis of cancer as a metabolic disease, new sights and opportunities for disease treatment. *Semin. Cancer Biol.* **2015**, *30*, 21–29. [CrossRef]
18. Kim, M.K.; Lun, D.S. Methods for integration of transcriptomic data in genome-scale metabolic models. *Comput. Struct. Biotechnol. J.* **2014**, *11*, 59–65. [CrossRef]
19. Blazier, A.S.; Papin, J.A. Integration of expression data in genome-scale metabolic network reconstructions. *Front. Physiol.* **2012**, *3*, 299. [CrossRef]
20. Leoncikas, V.; Wu, H.; Ward, L.T.; Kierzek, A.M.; Plant, N.J. Generation of 2,000 breast cancer metabolic landscapes reveals a poor prognosis group with active serotonin production. *Sci. Rep.* **2016**, *6*, 19771. [CrossRef] [PubMed]
21. Li, L.; Zhou, X.; Ching, W.K.; Wang, P. Predicting enzyme targets for cancer drugs by profiling human metabolic reactions in NCI-60 cell lines. *BMC Bioinform.* **2010**, *11*, 501. [CrossRef]
22. Marin de Mas, I.; Aguilar, E.; Zodda, E.; Balcells, C.; Marin, S.; Dallmann, G.; Thomson, T.M.; Papp, B.; Cascante, M. Model-driven discovery of long-chain fatty acid metabolic reprogramming in heterogeneous prostate cancer cells. *PLoS Comput. Biol.* **2018**, *14*, e1005914. [CrossRef]

23. Swainston, N.; Smallbone, K.; Hefzi, H.; Dobson, P.D.; Brewer, J.; Hanscho, M.; Zielinski, D.C.; Ang, K.S.; Gardiner, N.J.; Gutierrez, J.M.; et al. Recon 2.2: From reconstruction to model of human metabolism. *Metabolomics* **2016**, *12*, 109. [CrossRef] [PubMed]

24. Collado-Torres, L.; Nellore, A.; Jaffe, A.E. recount workflow: Accessing over 70,000 human RNA-seq samples with Bioconductor. *F1000Res* **2017**, *6*, 1558. [CrossRef]

25. Luecken, M.D.; Theis, F.J. Current best practices in single-cell RNA-seq analysis: A tutorial. *Mol. Syst. Biol.* **2019**, *15*, e8746. [CrossRef] [PubMed]

26. Shlomi, T.; Cabili, M.N.; Herrgard, M.J.; Palsson, B.; Ruppin, E. Network-based prediction of human tissue-specific metabolism. *Nat. Biotechnol.* **2008**, *26*, 1003–1010. [CrossRef]

27. Gurobi Optimization, L. Gurobi Optimizer Reference Manual. 2018. Available online: https://www.gurobi.com/documentation/ (accessed on 26 October 2020)

28. Kenley, E.C.; Cho, Y. Entropy-Based Graph Clustering: Application to Biological and Social Networks. In Proceedings of the 2011 IEEE 11th International Conference on Data Mining, Vancouver, BC, Canada, 11–14 December 2011; pp. 1116–1121.

29. Zhang, Z.; Li, T.; Ding, C.; Zhang, X. Binary Matrix Factorization with Applications. In Proceedings of the Seventh IEEE International Conference on Data Mining (ICDM 2007), Omaha, NE, USA, 28–31 October 2007; pp. 391–400. [CrossRef]

30. Lee, S.; Huang, J.Z.; Hu, J. Sparse logistic principal components analysis for binary data. *Ann. Appl. Stat.* **2010**, *4*, 1579–1601. [CrossRef] [PubMed]

31. Li, T. A General Model for Clustering Binary Data. In Proceedings of the Eleventh ACM SIGKDD International Conference on Knowledge Discovery in Data Mining, Chicago, IL, USA, 21–24 August 2005; ACM: New York, NY, USA, 2005; pp. 188–197. [CrossRef]

32. Jack, D.L.; Paulsen, I.T.; Saier, M.H. The amino acid/polyamine/organocation (APC) superfamily of transporters specific for amino acids, polyamines and organocations. *Microbiology* **2000**, *146 Pt 8*, 1797–1814. [CrossRef]

33. Tseng, T.T.; Gratwick, K.S.; Kollman, J.; Park, D.; Nies, D.H.; Goffeau, A.; Saier, M.H. The RND permease superfamily: An ancient, ubiquitous and diverse family that includes human disease and development proteins. *J. Mol. Microbiol. Biotechnol.* **1999**, *1*, 107–125. [PubMed]

34. Zhou, Y.; Yuan, J.; Li, Z.; Wang, Z.; Cheng, D.; Du, Y.; Li, W.; Kan, Q.; Zhang, W. Genetic polymorphisms and function of the organic anion-transporting polypeptide 1A2 and its clinical relevance in drug disposition. *Pharmacology* **2015**, *95*, 201–208. [CrossRef]

35. Jiang, Y.; Cao, Y.; Wang, Y.; Li, W.; Liu, X.; Lv, Y.; Li, X.; Mi, J. Cysteine transporter SLC3A1 promotes breast cancer tumorigenesis. *Theranostics* **2017**, *7*, 1036–1046. [CrossRef] [PubMed]

36. Hansson, J.; Lindgren, D.; Nilsson, H.; Johansson, E.; Johansson, M.; Gustavsson, L.; Axelson, H. Overexpression of Functional SLC6A3 in Clear Cell Renal Cell Carcinoma. *Clin. Cancer Res.* **2017**, *23*, 2105–2115. [CrossRef]

37. Muglia, V.F.; Prando, A. Renal cell carcinoma: Histological classification and correlation with imaging findings. *Radiol. Bras.* **2015**, *48*, 166–174. [CrossRef]

38. McGuire, B.B.; Fitzpatrick, J.M. Biomarkers in renal cell carcinoma. *Curr. Opin. Urol.* **2009**, *19*, 441–446. [CrossRef]

39. Wohlrab, C.; Phillips, E.; Dachs, G.U. Vitamin C Transporters in Cancer: Current Understanding and Gaps in Knowledge. *Front. Oncol.* **2017**, *7*, 74. [CrossRef]

40. Linehan, W.M.; Linehan, W.M.; Spellman, T.; Ricketts, J.; Creighton, J.; Fei, S.; Wheeler, A.D.; Murray, A.; Schmidt, L.; Vocke, D.; et al. Comprehensive Molecular Characterization of Papillary Renal-Cell Carcinoma. *N. Engl. J. Med.* **2016**, *374*, 135–145. [PubMed]

41. Ricketts, C.J.; De Cubas, A.A.; Fan, H.; Smith, C.C.; Lang, M.; Reznik, E.; Bowlby, R.; Gibb, E.A.; Akbani, R.; Beroukhim, R.; et al. The Cancer Genome Atlas Comprehensive Molecular Characterization of Renal Cell Carcinoma. *Cell. Rep.* **2018**, *23*, 313–326. [CrossRef]

42. Brockhausen, I. Pathways of O-glycan biosynthesis in cancer cells. *Biochim. Biophys. Acta* **1999**, *1473*, 67–95. [CrossRef]

43. Kandasamy, P.; Gyimesi, G.; Kanai, Y.; Hediger, M.A. Amino acid transporters revisited: New views in health and disease. *Trends. Biochem. Sci.* **2018**, *43*, 752–789. [CrossRef]

44. Thakkar, N.; Lockhart, A.C.; Lee, W. Role of Organic Anion-Transporting Polypeptides (OATPs) in Cancer Therapy. *AAPS J.* **2015**, *17*, 535–545. [CrossRef]

45. Nishino, S.; Itoh, A.; Matsuoka, H.; Maeda, K.; Kamoshida, S. Immunohistochemical analysis of organic anion transporter 2 and reduced folate carrier 1 in colorectal cancer: Significance as a predictor of response to oral uracil/ftorafur plus leucovorin chemotherapy. *Mol. Clin. Oncol.* **2013**, *1*, 661–667. [PubMed]
46. Loayza-Puch, F.; Agami, R. Monitoring amino acid deficiencies in cancer. *Cell Cycle* **2016**, *15*, 2229–2230. [CrossRef] [PubMed]
47. Angione, C.; Pratanwanich, N.; Lio, P. A Hybrid of Metabolic Flux Analysis and Bayesian Factor Modeling for Multiomic Temporal Pathway Activation. *ACS Synth. Biol.* **2015**, *4*, 880–889. [CrossRef] [PubMed]
48. Heinonen, M.; Osmala, M.; Mannerstrom, H.; Wallenius, J.; Kaski, S.; Rousu, J.; Lahdesmaki, H. Bayesian metabolic flux analysis reveals intracellular flux couplings. *Bioinformatics* **2019**, *35*, i548–i557. [CrossRef] [PubMed]
49. Zhu, Y.; Song, J.; Xu, Z.; Sun, J.; Zhang, Y.; Li, Y.; Ma, Y. Development of thermodynamic optimum searching (TOS) to improve the prediction accuracy of flux balance analysis. *Biotechnol. Bioeng.* **2013**, *110*, 914–923. [CrossRef] [PubMed]
50. Niittylae, T.; Chaudhuri, B.; Sauer, U.; Frommer, W.B. Comparison of quantitative metabolite imaging tools and carbon-13 techniques for fluxomics. *Methods Mol. Biol.* **2009**, *553*, 355–372.
51. Long, C.P.; Antoniewicz, M.R. High-resolution 13C metabolic flux analysis. *Nat. Protoc.* **2019**, *14*, 2856–2877. [CrossRef]
52. Hanke, T.; Noh, K.; Noack, S.; Polen, T.; Bringer, S.; Sahm, H.; Wiechert, W.; Bott, M. Combined fluxomics and transcriptomics analysis of glucose catabolism via a partially cyclic pentose phosphate pathway in Gluconobacter oxydans 621H. *Appl. Environ. Microbiol.* **2013**, *79*, 2336–2348. [CrossRef] [PubMed]
53. Daniels, W.; Bouvin, J.; Busche, T.; Ruckert, C.; Simoens, K.; Karamanou, S.; Van Mellaert, L.; Friojonsson, O.H.; Nicolai, B.; Economou, A.; et al. Transcriptomic and fluxomic changes in Streptomyces lividans producing heterologous protein. *Microb. Cell. Fact.* **2018**, *17*, 198. [CrossRef]
54. Mhamdi, H.; Bourdon, J.; Larhlimi, A.; Elloumi, M. Bayesian Integrative Modeling of Genome-Scale Metabolic and Regulatory Networks. *Informatics* **2020**, *7*, 1. [CrossRef]
55. St. John, P.C.; Strutz, J.; Broadbelt, L.J.; Tyo, K.E.J.; Bomble, Y.J. Bayesian inference of metabolic kinetics from genome-scale multiomics data. *PLOS Comput. Biol.* **2019**, *15*, 1–23. [CrossRef] [PubMed]
56. Hackett, S.R.; Zanotelli, V.R.T.; Xu, W.; Goya, J.; Park, J.O.; Perlman, D.H.; Gibney, P.A.; Botstein, D.; Storey, J.D.; Rabinowitz, J.D. Systems-level analysis of mechanisms regulating yeast metabolic flux. *Science* **2016**, *354*. [CrossRef]
57. Jenni Heino, J.; Tunyan, K.; Calvetti, D.; Somersaloa, E. Bayesian flux balance analysis applied to a skeletal muscle metabolic model. *J. Theor. Biol.* **2007**, *248*, 91–110. [CrossRef] [PubMed]
58. Katz, A. G-1,6-P2, Glycolysis, and Energy Metabolism During Circulatory Occlusion in Human Skeletal Muscle. *Am. J. Physiol.* **1988**, *255*, 140–144. [CrossRef] [PubMed]

**Publisher's Note:** MDPI stays neutral with regard to jurisdictional claims in published maps and institutional affiliations.

*Article*

# Statistical Approach for Biologically Relevant Gene Selection from High-Throughput Gene Expression Data

**Samarendra Das** [1,2,3,4] **and Shesh N. Rai** [3,4,5,6,7,8,*]

[1]  Division of Statistical Genetics, Indian Council of Agricultural Research (ICAR)-Indian Agricultural Statistics Research Institute, PUSA, New Delhi 110012, India; samarendra.das@louisville.edu

[2]  Netaji Subhas-Indian Council of Agricultural Research (ICAR) International Fellow, Indian Council of Agricultural Research, Krishi Bhawan, New Delhi 110001, India

[3]  Biostatistics and Bioinformatics Facility, JG Brown Cancer Center, University of Louisville, Louisville, KY 40292, USA

[4]  School of Interdisciplinary and Graduate Studies, University of Louisville, Louisville, KY 40292, USA

[5]  Alcohol Research Center, University of Louisville, Louisville, KY 40292, USA

[6]  Department of Hepatobiology and Toxicology, University of Louisville, Louisville, KY 40292, USA

[7]  Department of Bioinformatics and Biostatistics, University of Louisville, Louisville, KY 40292, USA

[8]  Wendell Cherry Chair in Clinical Trial Research, University of Louisville, Louisville, KY 40292, USA

[*]  Correspondence: shesh.rai@louisville.edu; Tel.: +1-502-426-0016

Received: 8 September 2020; Accepted: 21 October 2020; Published: 25 October 2020

**Abstract:** Selection of biologically relevant genes from high-dimensional expression data is a key research problem in gene expression genomics. Most of the available gene selection methods are either based on relevancy or redundancy measure, which are usually adjudged through post selection classification accuracy. Through these methods the ranking of genes was conducted on a single high-dimensional expression data, which led to the selection of spuriously associated and redundant genes. Hence, we developed a statistical approach through combining a support vector machine with Maximum Relevance and Minimum Redundancy under a sound statistical setup for the selection of biologically relevant genes. Here, the genes were selected through statistical significance values and computed using a nonparametric test statistic under a bootstrap-based subject sampling model. Further, a systematic and rigorous evaluation of the proposed approach with nine existing competitive methods was carried on six different real crop gene expression datasets. This performance analysis was carried out under three comparison settings, i.e., subject classification, biological relevant criteria based on quantitative trait loci and gene ontology. Our analytical results showed that the proposed approach selects genes which are more biologically relevant as compared to the existing methods. Moreover, the proposed approach was also found to be better with respect to the competitive existing methods. The proposed statistical approach provides a framework for combining filter and wrapper methods of gene selection.

**Keywords:** SVM; MRMR; bootstrap; gene expression; biological relevance; subject classification

---

## 1. Background

The emergence of high-throughput sequencing technologies exponentially increase the size of output data in genome sciences with respect to a number of features [1]. For example, gene expression (GE) studies generate the expression measurements of several thousand(s) of genes for tissue samples over two contrasting conditions in a single study [2,3]. These huge amounts of expression data are being generated for complex traits, and are deposited in public domain databases, such as NCBI GEO, ArrayExpress, etc., over the years by researchers across the globe [4,5]. Further, these publicly available high-throughput data need to be analyzed in order to gain valid biological insights. One such aspect of this research is to select genes, which are highly relevant to the phenotype/trait under study, out of several thousands of genes in the data. This is called feature selection in machine learning in general and gene selection in genomics [5–7]. Gene selection has been the focused area of functional genomics research, and thus several statistical and machine learning approaches have been developed for this purpose [8,9]. Here, the main aim is to select relevant genes which are highly informative for the condition/trait (i.e., reduce the curse of high-dimensionality in GE data [5,6,10,11]), and use them as predictors for diagnosing a disease [7,8,12,13] or to understand the stress response mechanisms in plants [6,10]. Further, the selected genes can also be used as predictors for other predictive analysis, i.e., subjects classification [7,8,11], gene regulation modeling [14], gene network analysis [5,6], etc., which enhances the stability, power and feasibility of the developed models [15].

Gene selection methods can be grouped into: (i) filter; and (ii) wrapper methods [9,16]. Filter methods select individual genes or gene subset based on a performance measure computed from the data with respect to class variables regardless of the predictive modeling algorithm [17]. These methods include univariate approaches such as *t*-test [18,19], Fold change [19], F-score [20,21], Volcano plot [18], Wilcoxon's statistic (Wilcox) [22,23], information gain (IG) [9,24], gain ratio (GR) [9,24], symmetric uncertainty [19], etc. These methods select genes by only considering their relevance within a level of the experimental condition/trait. However, these approaches may not be sufficient to discover some complex relationships among genes (i.e., gene-gene interactions) for certain conditions/traits, under which the data is generated [10]. Therefore multivariate filter approaches, such as Pearson's Correlation (PCR), Spearman's rank correlation [9,24], Maximum Relevance and Minimum Redundancy (MRMR) [20,25,26], etc. have been developed to select genes from GE data [9,16]. Recently, MRMR method was applied to single-cell transcriptomics data for selection of relevant transcripts responsible for colorectal cancer [27].

Wrapper methods select gene subsets by assessing the performance of the predictive modelling algorithm [28]. In other words, this class of gene selection methods are embedded in the classification process. For instance, a wrapper method evaluates the gene subsets based on the classifiers' performance on GE data and selects the most relevant gene subset. However, the Wrapper methods have better performance over filter methods [9,16], but are more complex, and computationally expensive [28]. This class includes support vector machine-recursive feature elimination (SVM-RFE) [8,29], multiple SVM-RFE (MSVM-RFE) [30], Monte Carlo feature selection algorithm (with SVM classifier) [31] and random forest (RF) [11] to name a few. Further, hybrids of filter and wrapper methods are also reported in literature (known as embedded methods [9]) such as combination of SVM-RFE with MRMR weights (SVM-MRMR) [13], SVM with F-score and other methods [21] to select relevant genes from GE data. Moreover, the MRMR method [20] in conjunction with incremental feature selection and Dagging algorithms [32] were used for gene selection through integrating cross platforms data such as expression quantitative trait loci and genome-wide association study [33].

Besides hybrid gene selection methods through combining ReliefF with ant colony optimization [34] and particle swarm optimization [35], algorithms are also developed to select cancer-responsible genes from GE data. Moreover, the existing methods select genes through the weights (i.e., gene ranking criteria) computed from single high-dimensional GE data, which leads to the selection of spuriously associated and redundant genes (i.e., genes may not be informative but are correlated with other relevant genes) [5,6]. Therefore, the permutation procedures are used to compute statistical significance values for genes [6]. However, it has some serious limitations, such as being highly sensitive to a small permutation of experimental conditions (i.e., class labels) [5,6], computationally slow [36,37], cannot possibly give any significant *p*-values after multiple testing adjustments [37,38] and large permutations are required to get a significant *p*-value [37]. To address such issues, bootstrap procedures are used in gene selection which ably remove the spurious associations of genes with the classes and other genes [5,6,39].

Gene selection methods are mostly used to select cancer-responsible genes from GE datasets, and subsequently used for patient classification (e.g., with and without cancer) [6–8,13,15,34–40]. There are limited studies available in literature to systematically explore the performance of gene selection methods on crop GE datasets as there are typically limited experimental data available. Further, the performance of the existing methods were usually assessed through computation of post selection classification accuracy (CA) on cancer GE datasets [7,8,13,15,39,40]. In other words, these techniques are adjudged based on their ability to discriminate the GE samples between case and control groups through training classifiers like SVM [31]. Here, it is worthy to note, this traditional criterion is statistically sound but may not be biologically relevant for performance evaluation of gene selection methods [39,41]. For instance, a gene selection technique identified a set of genes which accurately predicted the class of GE samples for a salinity vs. control GE study in rice, but it fails to tell whether these selected genes are biologically relevant or not to the salinity stress. Hence, it is pertinent to evaluate the gene selection methods with respect to biology-based criteria. For this purpose, data related to traits, such as quantitative trait locus (QTLs) and gene ontology (GO) for model crop plants may be used, which are hugely available in public domains.

We, therefore, propose an improved statistical approach (BSM=Bootstrap-SVM-MRMR) that combines MRMR filter with SVM wrapper method to minimize the redundancy among genes and improve the relevancy of genes with the traits/phenotype under a sound statistical setup. Through this, relevant genes are selected from a high-dimensional GE data through the statistical significance values computed using a nonparametric (NP) test statistic under a bootstrap-based subject sampling model. Further, the comparative performance analysis of the proposed BSM approach is carried out with nine existing competitive methods (i.e., IG [9,24], GR [9,24], *t*-test [18,19], F-score [20,21], MRMR [12,20], SVM-RFE [8,29], SVM-MRMR [13], PCR [9,24] and Wilcox [22,23]). The comparative performance measures include CA along with its standard error computed through varying sliding windows size technique, and three biological criteria based on QTL [42] and GO [43] terms. We demonstrate these procedures on six publicly available, independent crop GE datasets, and find that the BSM approach outperforms in terms of classification and biological relevance criteria compared to the existing methods.

## 2. Materials and Methods

### 2.1. Motivation

The GE datasets, from various experiments conducted to understand the behavior of biological mechanisms, are hugely available in public domain databases. For example, GE datasets generated for 125,376 experiments over 19,893 Microarray platforms consisting of data on 3,406,218 samples are available in NCBI GEO database until the current date [4]. Usually, researchers use data from single experiments to test their methodology or select genes for further study. For instance, Wang et al. (2013) used the salinity stress GE samples from GSE14403 to test their methodology and select salinity responsive genes to understand salinity tolerance mechanism in rice [6]. Such a study is important but may not be enough to test the hypothesis of salinity tolerance in rice due to limited sample size. Hence, the real challenge is to integrate or combine the GE datasets generated for same or cross platforms over different experimental conditions and test the methodology(s) on the meta-data. Moreover, meta-analysis of data generated by GE experiments for the same or related stress(es) is essential to enhance the sensitivity of the hypothesis under consideration for drawing valid biological conclusions. Therefore, we performed meta-analysis on GE datasets corresponding to different stresses from multiple experiments and tested the performance of methods on these metadata, as shown in Table 1. The outlines of meta-analysis are given in Figure 1A.

**Table 1.** Rice gene expression datasets used in the study.

| Sl. No. | Descriptions | #Series | Series ID | #Genes | #Samples | Stress Type |
|---------|--------------|---------|-----------|--------|----------|-------------|
| 1. | Salinity stress | 3. | GSE14403, GSE16108, GSE6901. | 6637 | 45 (23, 22) | Abiotic |
| 2. | Cold stress | 4. | GSE31077, GSE33204. GSE37940, GSE6901. | 8840 | 28 (15, 13) | Abiotic |
| 3. | Drought stress | 5. | GSE6901, GSE26280. GSE21651, GSE23211. GSE24048. | 9078 | 70 (35, 35) | Abiotic |
| 4. | Bacterial (xanthomonas) stress | 3. | GSE19239, GSE36093. GSE36272. | 8356 | 74 (37, 37) | Biotic |
| 5. | Fungal (blast) stress | 2. | GSE41798, GSE7256. | 7072 | 26 (13, 13) | Biotic |
| 6. | Insect (brown plant hopper) stress | 1. | GSE29967. | 7241 | 18 (12, 6) | Biotic |

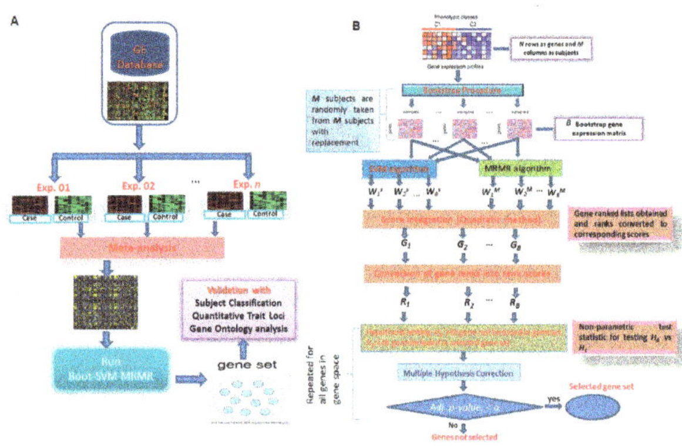

**Figure 1.** Operational procedure for data integration and the use of proposed BSM approach. (**A**) Outlines for the data integration used in this study for the application of BSM approach. The first step indicates the integration and meta-analysis of GE datasets obtained from various GE studies. Then gene selection methods are applied on the meta GE data. (**B**) Flowchart depicting the implemented algorithm of BSM approach. $W_i^{(S)}$'s and $W_i^{(M)}$'s are the $N$-dimensional vectors of weights computed through SVM and MRMR approach, respectively. $G_i$'s and $R_i$'s are the $N$-dimensional vectors of gene lists and corresponding gene rank scores. SVM and MRMR stand for Maximum Relevance and Minimum Redundancy and support vector machine algorithms. $p_i$-value is statistical significance value for $i^{th}$ gene. $\alpha$ is the desired level of statistical significance.

## 2.2. Data Source

Rice GE experimental datasets were collected from the Gene Expression Omnibus database (GEO) of NCBI for platforms GPL2025 (www.ncbi.nlm.nih.gov/geo/query/acc.cgi?acc=GPL2025) [4]. Here, we used the rice data, as it is a model crop plant, has a large amount of GE, other related biological (QTL and GO) datasets are available publicly, and its genome is well annotated. The selected GE datasets were generated under biotic (bacterial (*Xanthomonas*), fungal (Blast), insect (Brown plant hopper) and abiotic (salinity, cold and drought) stresses in rice. The summary and details of these datasets are given in Table 1 and Supplementary Table S1, respectively. Initially, the raw CEL files of the collected samples were processed using Robust Multichip Average algorithm available in *affy* Bioconductor package of R [44]. This procedure involves background correction, quantile normalization and summarization by median polish approach. Further, the log2 scale transformed expression data for the collected experimental samples were used for meta-analysis to remove the outlier samples (Supplementary Document S1). The GE samples from 3, 4, 5, 3 and 2 independent studies for salinity, cold, drought, bacterial and fungal stresses, respectively, were integrated (Table 1) through the meta-analysis (under the parameters settings in Supplementary Table S2) to obtain the meta-data. For instance, the salinity stress dataset, originating from 3 independent studies, are available in GEO database under the accession numbers GSE14403, GSE16108 and GSE6901 and consist of expression measurements for over 45 samples. Then, these meta-datasets for the respective stresses were further used to remove the control and irrelevant features through the preliminary genes selection to reduce the computational complexity and dimensions of the datasets. For instance, out of 57,381 genes in drought stress, the control (123) and irrelevant (48180) genes were filtered out by setting the fold change and *p*-value (from *t*-test) parameters as 1 and 0.05, respectively, through the preliminary gene selection. The detail process of data collection, meta-analysis and preliminary gene selection for the datasets are given in Supplementary Document S1. Then, the processed datasets (Table 1) were used for further data analysis. Further, the QTL datasets for the stresses in rice, viz. salinity, drought, cold, insect, fungal and bacterial, were collected from the Gramene QTL database (http://www.gramene.org/qtl/) [45]. The lists of the respective stress responsive QTLs along with their mapped positions on the genome are given in Supplementary Document S2. The GO annotations data of the rice genome used in this study were collected from *AgriGO* database [46].

*2.3. Methods*

### 2.3.1. Notations

Let $X_{N \times M} = [x_{im}]$ be the GE data matrix, where $x_{im}$ represents the expression of $i^{th}$ ($i = 1, 2, \ldots, N$) gene in $m^{th}$ ($m = 1, 2, \ldots, M$) sample/subject; $x_m$ be the $N$-dimensional vector of expression values of genes for $m^{th}$ sample; $y_m$ be the outcome variable for target class label of $m^{th}$ sample and take values $\{+1, -1\}$ for case and control conditions, respectively; $M_1$ and $M_2$ be the number of GE samples in case and control classes, respectively, $(M_1 + M_2 = M)$; $(\overline{x}_{i1}, S_{i1}^2)$ and $(\overline{x}_{i2}, S_{i2}^2)$ be the mean and variance of $i^{th}$ gene for case and control classes, respectively; $\overline{x}_i$ be the mean of $i^{th}$ gene across all $M$ samples; $S_{ij}$ be the covariance between $i^{th}$ and $j^{th}$ genes.

### 2.3.2. Maximum Relevance and Minimum Redundancy (MRMR) Filter

MRMR method aims at selecting maximally relevant and minimally redundant sets of genes for discriminating the tissue samples (e.g., case vs. control). This method is extensively used for selection of cancer-responsible genes from high-dimensional GE data for patient classification (i.e., with and without cancer) [12,20,26]. For continuous GE data (e.g., Microarrays), the relevance measure for $i^{th}$ gene over the given classes (i.e., case and control) is computed through F-statistic [12] and is expressed as:

$$F(i) = \frac{M_1 (\overline{x}_{i1} - \overline{x}_i)^2 + M_2 (\overline{x}_{i2} - \overline{x}_i)^2}{\left\{ (M_1 - 1) S_{i1}^2 + (M_2 - 1) S_{i2}^2 \right\} / (M - 2)} \tag{1}$$

Further, the redundancy measure in MRMR method is computed through Pearson's correlation (ignoring the class information) for continuous GE data [12] and is given as

$$R(i, j) = Corr\left(x_i, x_j\right) = \frac{S_{i,j}}{S_i S_j} = \frac{\sum_{m=1}^{M} (x_{im} - \overline{x}_i)\left(x_{jm} - \overline{x}_j\right)}{\sqrt{\sum_{m=1}^{M} (x_{im} - \overline{x}_i)^2} \sqrt{\sum_{m=1}^{M} \left(x_{jm} - \overline{x}_j\right)^2}} \tag{2}$$

In MRMR method, genes are ranked by the combination of relevance, and redundancy measures under F-score with correlation quotient scheme for continuous GE data [12,20,26]. The weights computed through MRMR method for gene ranking can be expressed in terms of Equations (1) and (2) and is given as:

$$w_i = F(i) / \left\{ \frac{1}{N-1} \sum_{\substack{j=1 \\ j \neq i}}^{N} |R(i,j)| \right\} \qquad \forall \ i = 1, 2, \ldots, N \tag{3}$$

where $w_i$ ($\geq 0$) is the weight associated with $i^{th}$ gene. The functions $F(i)$ and $R(i,j)$ in Equation (3) represent the F-statistic for $i^{th}$ gene and Pearson's correlation coefficient between $i^{th}$ and $j^{th}$ genes. In other words, $i^{th}$ gene weight is F-statistic adjusted with average absolute correlation of $i^{th}$ gene with the remaining genes.

### 2.3.3. Support Vector Machine (SVM)

SVM method is used for selection of genes (in a 2 group case) from high-dimensional GE data [29]. Let $\{x_m, y_m\} \in R^N \times \{-1, 1\}$ be the input given to SVM. Here, we wish to find out a hyperplane that divides the GE samples/subjects for case $(y_m = 1)\bar{x}_i$ from that of control class $(y_m = -1)$ in such a way that the distance between the hyperplane and the point, $\bar{x}_i \, x_m$, is maximum. Then the hyperplane can be written as:

$$\sum_{i=1}^{N} k_i x_{im} + b = 0 \qquad \forall \ m = 1, 2, \ldots, M \tag{4}$$

where $k_i$ and $b$ are the weight of $i^{th}$ gene and bias, respectively. Here, we assume that the GE samples for 2 classes are linearly separable. In other words, we can select 2 parallel hyperplanes that separate the case and control classes in such a way that the distance between them is maximum.

For case class, the hyperplane becomes:

$$\sum_{i=1}^{N} k_i x_{ip} + b = 1 \qquad \text{for any } p = 1, 2, \ldots, M_1 \tag{5}$$

For control class, the hyperplane becomes:

$$\sum_{i=1}^{N} k_i x_{iq} + b = -1 \qquad \text{for any } q = 1, 2, \ldots, M_2 \tag{6}$$

The expressions in Equations (5) and (6) can be combined as:

$$y_m \left( \sum_{i=1}^{N} k_i x_{im} + b \right) = 1 \qquad \forall \ m = 1, 2, \ldots, M \tag{7}$$

Here, we wish to maximize the distance between the case, and control hyperplanes in Equation (5) and Equation (6), respectively, under the constraint that there will be no GE samples between these 2 hyperplanes given in Equation (7). Mathematically, it can be written as:

$$\sum_{i=1}^{N} \frac{k_i}{(\sum k_i)^2} |x_{ip} - x_{iq}| = \frac{2}{(\sum k_i)^2} \tag{8}$$

So, to maximize the distance between the planes in Equation (8), we need to minimize $\frac{(\sum_i k_i)^2}{2}$ under the constraint of Equation (7). Mathematically, it can be written as:

$$L_p = \min_{k_i} \frac{(\sum_i k_i)^2}{2} + \sum_{m=1}^{M} \varphi_m \left\{ 1 - y_m \left( \sum_{i=1}^{N} k_i x_{im} + b \right) \right\} \ \forall \ m = 1, 2, \ldots, M \tag{9}$$

where $\varphi_m$ ($\geq 0$): Lagrange multiplier. Here, $k_i$'s are obtained by minimizing the objective function in Equation (9). Through the principle of maxima-minima, we have:

$$\frac{\partial L_p}{\partial k_i} = \sum_i k_i - \sum_i \left(\sum_{m=1}^{M} \varphi_m y_m x_{im}\right) = 0 \text{ and } \frac{\partial L_p}{\partial b} = \sum_{m=1}^{M} \varphi_m y_m = 0 \tag{10}$$

The value of $k_i$ can be obtained through solving the system of linear equations given in Equation (10) and is expressed as:

$$k_i = \sum_{m=1}^{M} \varphi_m y_m x_{im} \text{ with } \sum_{m=1}^{M} \varphi_m y_m = 0 \text{ and } \varphi_m \geq 0 \tag{11}$$

Here, $|k_i|$ ($\geq 0$) in Equation (11) is used as a metric for the ranking of genes in the GE data [29]. Alternatively, $k_i^2$ as a gene ranking metric can also be derived by using Taylor series approximation [47], which is given in Supplementary Document S3.

2.3.4. Proposed Hybrid Approach of Gene Selection

MRMR method may not yield optimal CA because it performs independently of the classifier and is only involved in selection of genes [13]. On the contrary, SVM method of gene selection does not consider the redundancy among genes (i.e., gene-gene correlations) while selecting genes [13]. Hence, Mundra and Rajapakse (2010) have developed a gene selection method by taking linear combination of weights computed through MRMR and SVM methods [13], and is given as:

$$SL_i = \delta w_i + (1 - \delta)|k_i| \tag{12}$$

where parameter $\delta \in [0, 1]$ decides the tradeoff between SVM and MRMR weights. The $SL_i$ in Equation (12) is highly dependent on the value of $\delta$. In other words, the choice of $\delta$ may alter the order of genes by MRMR ($w_i$) or by SVM ($k_i$), especially when $w_i$ and $k_i$ are negatively correlated. Hence, we propose a statistical approach by combining SVM and MRMR weights under sound statistical framework, where genes are selected through $p$-values computed using the NP test statistic, which is described as follows.

First, we normalized the $w_i$ and $k_i$'s through minimax normalization. Then $w_i$ and $k_i$ were ranked based on the ascending order of their magnitudes and assigned ranks $\gamma_i^{MR}$ and $\gamma_i^{SV}$ for $i^{th}$ gene, respectively. Then, we developed a technique, i.e., quadratic integration, for integrating the gene scores based on ranks, which automatically assigned more weights to the higher value of $w_i$ and $k_i$. Now, the quadratic integration score can be expressed as:

$$SD_i = \frac{\beta \gamma_i^{MR} w_i^{norm} + (1 - \beta)\gamma_i^{SV} |k_i|^{norm}}{\beta \gamma_i^{MR} + (1 - \beta)\gamma_i^{SV}} \tag{13}$$

where $w_i^{norm}$ and $|k_i|^{norm}$ are the normalized values, expressed in Equation (14) and Equation (15), respectively.

$$w_i^{norm} = \left(w_i - \min_i w_i\right) / \left(\max_i w_i - \min_i w_i\right) \tag{14}$$

$$|k_i|^{norm} = \left(|k_i| - \min_i |k_i|\right) / \left(\max_i |k_i| - \min_i |k_i|\right) \tag{15}$$

Further, $\beta (\in (0, 1))$ in Equation (13) is determined empirically from the data through a 5-fold cross validation technique. The detail procedure for determining the optimum value of $\beta$ is given in Supplementary Document S4. If $SD_i$ in Equation (13) is used alone for ranking of genes, it will become a filter approach and lead to selection of spuriously associated genes. Hence, we used a bootstrap procedure under a subject sampling model setup to obtain the empirical distribution of $SD_i$

for computation of statistical significance value for $i^{th}$ ($i = 1, 2, \ldots, N$) gene. Here, the used bootstrap procedure is described below.

The $M$ samples (as columns) in the GE data matrix, either belonging to case or control, can be considered as subjects/units in a population model, as shown in Equation (16).

$$(x_1, y_1), \ (x_2, y_2), \ldots, \ (x_m, y_m), \ldots, \ (x_{M-1}, y_{M-1}), (x_M, y_M) \tag{16}$$

Here, we assume that the subjects are independent and identically distributed, but the genes within each subject may be correlated. In the bootstrap procedure, $M$ units are randomly drawn from $M$ population units in Equation (16) with a replacement to constitute a bootstrap GE data matrix, i.e., $X_{NXM}^{(b)}$ ($M$ units serve as $M$ columns of $X$). This process is repeated $B$ times to get $B$ bootstrap GE data matrices, i.e., $X_{NXM}^{(1)}, \ X_{NXM}^{(2)}, \ldots, \ X_{NXM}^{(b)}, \ldots, \ X_{NXM}^{(B)}$. Here, $B$ (i.e., number of bootstrap samples) depends on several factors, such as number of units in the population model in Equation (16) and must be sufficiently large. So, we set $B = 200$ as several empirical studies showed that the number of bootstrap samples required for an estimation procedure is ~200 [6,48].

Now, the $B$ bootstrap GE data matrices are given as the input to Equations (3), (11) and (13) to compute the $SD$ scores, and subsequently gene ranking was performed on each of the $B$ bootstrap GE data matrices.

Let $P_{ib}$, be a random variable ($rv$) that shows the position of $i^{th}$ gene in $b^{th}$ bootstrap GE matrix. Then, another $rv$ can be defined based on $P_{ib}$ (without loss of generality), given as:

$$R_{ib} = \frac{N + 1 - P_{ib}}{N}; \ 0 \leq R_{ib} \leq 1 \tag{17}$$

where $R_{ib}$ in Equation (17) is the rank score of $i^{th}$ ($i = 1, 2, \ldots, N$) gene in $b^{th}$ ($b = 1, 2, \ldots, B$) bootstrap GE matrix. Here, it may be noted that the distribution of the rank scores of genes, computed from a bootstrap GE data matrix, is symmetric around the median value (as rank scores are a function of ranks). The values of the median and the third quartile ($Q_3$) are given as 0.5 and 0.75, respectively.

To decide whether $i^{th}$ gene is biologically relevant or not to the condition/trait under study, the following null hypothesis can be tested.

$$H_0 : R_i \leq Q_3 \ (i - th \text{ gene is not so relevant to the trait})$$

$$H_1 : R_i > Q_3 \ (i - th \text{ gene is relevant to the trait})$$

where $R_i$ is the rank score for $i^{th}$ gene over all possible bootstrap samples.

To obtain the distribution of test statistic under $H_0$, we define another $rv$ $Z_{ib}$, as:

$$Z_{ib} = \begin{cases} 1 & |R_{ib} - Q_3| > 0 \\ 0 & |R_{ib} - Q_3| < 0 \end{cases} \tag{18}$$

Let $r_{ib}$ be another $rv$ represents the rank assigned to $(R_{ib} - Q_3)$ (after arranging in ascending order of their magnitudes). To test $H_0$ vs. $H_1$ the test statistic for $i^{th}$ gene, $W_i$, was developed, and is given as:

$$W_i = \sum_{b=1}^{B} Z_{ib} r_{ib} = \sum_{b=1}^{B} U_{ib} \ (say) \tag{19}$$

In other words, $W_i$ in Equation (19) is the sum of the ranks of positive signed scores for $i^{th}$ gene over $B$ bootstrap samples. Further, $U_{ib}$ in Equation (19) is a Bernoulli $rv$, and its probability mass function can be given as:

$$P[U_{ib} = u_{ib}] = \begin{cases} \frac{3}{4} & if \ u_{ib} = 0 \\ \frac{1}{4} & if \ u_{ib} = 1 \end{cases} \tag{20}$$

Here, the expected value and variance of $W_i$ in Equation (19) under $H_0$ can be obtained as:

$$E(W_i) = \sum_{b=1}^{B} E(U_{ib}) = \sum_{b=1}^{B} \left(0.\frac{3}{4} + b.\frac{1}{4}\right) = \frac{1}{4}\sum_{b=1}^{B} b = \frac{B(B+1)}{8} \tag{21}$$

The variance becomes:

$$\begin{aligned} V(W_i) &= E\left(W_i^2\right) - [E(W_i)]^2 \\ &= \sum_{b=1}^{B} E\left(U_{ib}^2\right) - \sum_{b=1}^{B} E(U_{ib})^2 \\ &= \sum_{b=1}^{B} \left(\frac{b^2}{4} - \frac{b}{16}^2\right) = \frac{B(B+1)(2B+1)}{32} \end{aligned} \tag{22}$$

As $B$ is sufficiently large, then under central limit theorem, the distribution of $W_i$ in Equation (19) becomes:

$$Z_i = \frac{W_i - E(W_i)}{\sqrt{V(W_i)}} \xrightarrow{d} N(0, 1) \tag{23}$$

Through Equation (23), the *p*-value for $i^{th}$ ($i = 1, 2, \dots, N$) gene is computed and similarly this testing procedure is repeated for the remaining $N - 1$ genes. Let $p_1, p_2, \dots, p_N$ be the corresponding *p*-values for all the genes in GE data, and $\alpha$ be the level of significance. Here, we assume that all genes in the GE data are equally important for the trait development, hence, we employed Hochberg procedure [49] for correcting the multiple testing, and to compute the adjusted (*adj.*) *p*-values for genes. It is worthy to note that Hochberg's procedure is computationally simple, quite popular in genomic data analysis [50] and more powerful than Holm's procedure [51]. The algorithm for Hochberg's procedure [49] is as follows.

Step 1. If $p_{(l)} > \alpha$, then retain corresponding null hypothesis ($H_{(l)}$) and go to the next step. Otherwise, reject it and stop.

Step $i = 2, 3, \dots, N - 1$. If $p_{(N-i+1)} > \alpha/i$, then retain $H_{(N-i+1)}$ and go to the next step. Otherwise, reject all remaining hypotheses and stop.

Step $N$. If $p_{(1)} > \alpha/N$, then retain ($H_{(1)}$). Else reject it.

Now, the *adj. p-values* are given recursively beginning with the largest *p*-value [49]:

$$\tilde{p}_{(i)} = \begin{cases} p_{(i)} \quad if \; i=N \\ \min(\tilde{p}\,(i+1),\ (N-i+1)p_{(i+1)}) \end{cases} \quad if \; i = N-1, \dots, 1 \tag{24}$$

Further, based on the computed *adj. p*-values, the relevant genes are selected from the high dimensional GE data. In other words, lesser value of *adj. p*-value may indicate more relevance of the gene for the target trait and vice-versa. The outlines and key analytical steps of the proposed BSM approach are shown in Figure 1B.

## 2.4. Comparative Performance Analysis of the Proposed Approach

The comparative performance analysis of the proposed BSM approach with respect to 9 competitive gene selection methods (Supplementary Document S5) was carried out on 6 different rice GE datasets (Table 1). For this purpose, different gene sets (*G*) of various sizes given in Supplementary Table S10 were selected through the 10 gene selection methods including the proposed BSM approach. Then, these gene sets were validated with respect to subject classification, QTL testing and GO analysis.

### 2.4.1. Performance Analysis with Subject Classification

Under this comparison setting, the performance of the gene selection methods (Supplementary Document S5) including the proposed approach were assessed in terms of subject classification using mean CA and standard error (SE) in CA computed through a varying sliding window size technique [5,39]. Here, we used the varying window size technique to study the impact of gene ranking on classification of subjects. In other words, genes in *G* were validated with their ability to discriminate the class labels of subjects/samples between case ($+1$), and control ($-1$). Further, the gene set selected through a method which provides maximum discrimination between the subjects of 2 groups (i.e., case vs. control) through CA will be considered as highly relevant gene sets. The expressions for mean CA and SE in CA computed through varying window size technique are given in Equations (25) and (26).

Let *n* be the size of *G*, *S* be the size of the windows (i.e., size refers to number of ranked genes) and *L* be the sliding length. Then, the total number of windows becomes $K = (n - S)/L$. The genes in *G*, arranged in different windows along with their expression values, were then used in SVM classifiers with 4 basis-functions, i.e., linear (SVM-LBF), radial (SVM-RBF), polynomial (SVM-PBF) and Sigmoidal (SVM-SBF) to compute CA over a 5-fold cross validation. Let, $CA_1, CA_2, \ldots, CA_K$ be the CA's for each sliding windows, then the mean CA and SE in CA can be defined as:

$$\mu_{CA}^G = \frac{\left(\sum_{k=1}^{K} CA_k\right)}{K} \tag{25}$$

$$SE_{CA}^G = \sqrt{\frac{\sum_{k=1}^{K} \left(CA_k - \mu_{CA}^G\right)^2}{K}} \tag{26}$$

Here, we took different combinations of *n*, *S* and *L*, as given in Supplementary Table S10, to compute $\mu_{CA}^G$ and $SE_{CA}^G$ for the comparative performance analysis of the gene selection methods (Supplementary Document S5). The higher value of $\mu_{CA}^G$ and a lower value of $SE_{CA}^G$ indicates the better performance of the gene selection method, and vice-versa.

### 2.4.2. Performance Analysis with QTL Testing

The comparative criteria based on subject classification are popularly used for assessing the performance of gene selection methods [7,8,12,13,15,39,40]. However, these criteria fail to tell the biological relevancy of the genes selected through the gene selection methods [41]. Hence, under this comparative setting we assessed the performance of the proposed and existing methods through their ability to select genes which are associated with QTL regions. For this purpose, the criteria given in Equations (27) and (29) are developed.

$$Qstat = \sum_{t=1}^{|Q|} \sum_{i=1}^{n} I_{q_t}(g_i) \tag{27}$$

where *G*: gene set selected by a method, *Qstat*: *rv* whose values represent the number of genes covered by QTLs, *Q*: set of associated QTLs, and the indicator function present in Equation (27) is represented in Equation (28).

$$I_{q_t}(g_i) = \begin{cases} 1 & if \; g_i^c[a,] \geq q_t^c[d,] \; and \; g_i^c[,b] \leq q_t^c[,e] \\ 0 & else \end{cases} \tag{28}$$

where, $g_i^c \; [a, b] \in G$ (*a* and *b* represent start and stop positions in terms of bp of the gene $g_i$ on chromosome *c*) and $q_t^c \; [d, e] \in Q$ (*d* and *e* represents the start and stop positions of the QTL $q_t$ on chromosome *c*).

Here, the *Qstat rv* follows a hyper-geometric distribution and its distribution function is given in Equation (29).

$$P[Qstat = v] = 1 - \binom{M}{v}\binom{N-V}{n-v} \Big/ \binom{N}{n} \tag{29}$$

where *V*: total number of genes covered by the QTLs in the whole GE data and *v*: number of genes in *G* that are covered by QTLs. Further, using the Equation (29), the statistical significance value (*p*-value) associated with the *G* can be computed. In other words, this *p*-value reveals the enrichment significance of $G$ with trait specific QTLs. Here, the higher values of $Qstat$ and $-log_{10}(p\text{-value})$ indicate the better performance of the gene selection method, and vice-versa.

### 2.4.3. Performance Analysis with GO Enrichment

GO analysis involves the annotation of gene functions under 3 taxonomic categories, i.e., molecular function (MF), biological process (BP) and cellular component (CC) [43]. This analysis helps in evaluating the functional similarities among the genes in $G$ [52], as there exists a direct relationship between semantic similarity of gene pairs with their structural (sequence) similarity [53,54]. Under this comparison setting, we assessed the performance of 10 gene selection methods including the proposed method using GO based biologically relevant criterion. In other words, first different gene sets were selected through these methods, then GO based criterion was computed for each selected gene set. For this purpose, we developed a GO based semantic distance measure to assess the GO based biologically relevancy of $G$ selected thorough the proposed and existing gene selection methods. The GO based semantic distance measure $(d_{ij})$ between $i^{th}$ and $j^{th}$ genes can be expressed in Equation (30), as:

$$d_{ij\,(i \neq j)}^{GO} = 1 - \frac{\left| GO_i \cap GO_j \right|}{\left| GO_i \cup GO_j \right|} \qquad \forall\, i,\, j = 1,\, 2,\, \ldots,\, n \qquad (30)$$

where $GO_i = \{go_{i1}, go_{i2}, \ldots, go_{il}\}$ and $GO_j = \{go_{j1}, go_{j2}, \ldots, go_{jJ}\}$ are the 2 sets of GO terms that annotate $i^{th}$ and $j^{th}$ genes in $G$, respectively. Further, the GO based average biologically relevant score for $G$ (for a gene selection method) can be developed based on Equation (30) and is shown in Equation (31).

$$D_G^{avg} = \frac{2}{n(n-1)} \sum_{\substack{i,\,j=1 \\ i \neq j}}^{n} d_{ij}^{GO} \qquad (31)$$

where $D_G^{avg}$ in Equation (31) represents the average biologically relevant score for $G$ based on GO annotations. Using Equation (31), the $D_G^{avg}$ scores under MF, BP and CC taxonomies were computed for each of the gene sets selected through different methods. A lower value of $D_G^{avg}$ indicates better performance of the gene selection method and vice-versa.

## 3. Results and Discussion

### 3.1. Computation of Genes Selection Criteria through Proposed Approach

The distributions of weights computed from SVM-MRMR method [13] and adj. $p$-values for genes computed from the proposed BSM approach for abiotic and biotic stresses in rice are shown in Figure 2 and Figure S3, respectively. The distributions of SVM-MRMR weights of genes for salinity stress data contained values, which were not so clearly separated (i.e., higher values from lower values) (Figure 2A). In other words, the genes relevant to salinity stress condition were not well visualized from Figure 2A. However, from the distribution of adj. $p$-values computed through the proposed approach, it was observed that the relevant genes were well separated from the irrelevant genes, and a small number of genes found to be statistically significant (i.e., relevant to salinity stress) (Figure 2(A1)). In other words, for salinity stress data, the separation between relevant and irrelevant genes can be well visualized from Figure 2(A1) as compared to Figure 2A. Similar interpretations can be observed for other stress datasets, viz. cold, drought, bacterial, fungal and insect (Figure 2 and Figure S3). Hence, the comparative graphical analysis showed a clear distinction between relevant and irrelevant genes through the proposed BSM approach as compared to the existing SVM-MRMR approach. In other words, this comparative analysis showed the improvement of BSM approach over the SVM-MRMR method (Figure 2 and Figure S3), at least in terms TABLE of visualization. Further, the relevant genes selection using adj. $p$-values computed through the NP test statistic was more statistically sound as it is independent from the distribution of GE data, and corrected over multiple hypothesis testing. These metrics (values between 0 and 1) are scientifically well defined and statistically calculated biologically interpretable values to genome researchers and experimental biologists, as compared to gene ranks/weights. In BSM approach, a significant $p$-value gives confidence that the given gene is relevant for the target condition/trait.

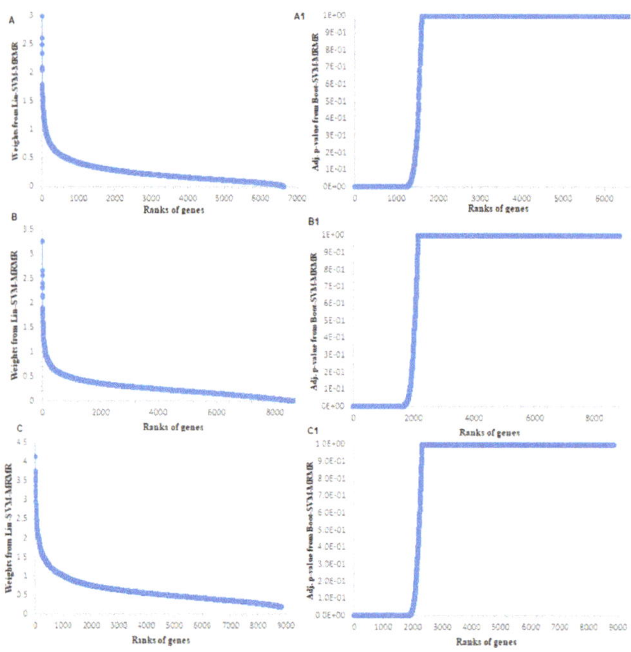

**Figure 2.** Graphical analysis of the proposed BSM approach with SVM-MRMR approach for abiotic stress datasets. Distribution of gene weights computed from SVM-MRMR approach for the abiotic stresses. The distributions of gene weights from the SVM-MRMR are shown for (**A**) salinity; (**B**) cold; and (**C**) drought stress datasets in rice. Distribution of adjusted $p$-values computed from the proposed BSM approach for the abiotic stresses. The distributions of the adjusted $p$-values are shown for (**A1**) salinity; (**B1**) cold; and (**C1**) drought stress datasets.

### 3.2. Comparative Performance Analysis Based on Subject Classification

We used $\mu_{CA}^{G}$ and $SE_{CA}^{G}$ computed through the varying sliding window size technique as statistically necessary criteria for performance analysis of the proposed BSM approach on six different GE datasets. Here, these measures were computed over five-fold cross validations through training the SVM-LBF, SVM-PBF, SVM-RBF and SVM-SBF classifiers. The results are shown in Figures 3 and 4 for abiotic stresses and in Supplementary Figure S4 for biotic stresses. The values of CA and SE in CA are also given in Supplementary Document S6. For cold stress data in rice, the $\mu_{CA}^{G}$ computed through SVM-LBF classifier for the proposed BSM approach was observed to be higher than other gene selection methods followed by SVM-RFE and SVM-MRMR over all selected gene sets Figure 3. This indicated the better performance of the BSM approach in terms of its ability to classify the subjects/samples through selecting relevant genes from cold stress GE data. Further, the values of $SE_{CA}^{G}$ from SVM-LBF classifier for the BSM approach was found to be much less (over all the gene sets) than that of nine existing gene selection methods considered in this study (Supplementary Document S6). This shows that the genes selected through the proposed BSM approach is much more relevant (informative), and robust than other methods. For instance, the gene set of size 50 (i.e., optimum gene set) provided satisfactory results in terms of higher $\mu_{CA}^{G}$ and lower $SE_{CA}^{G}$ irrespective of the gene selection methods used (Table S12 of Document S6). For cold stress data, similar interpretations can be made for SVM-PBF, SVM-RBF and SVM-SBF classifiers from Figures 3 and 4.

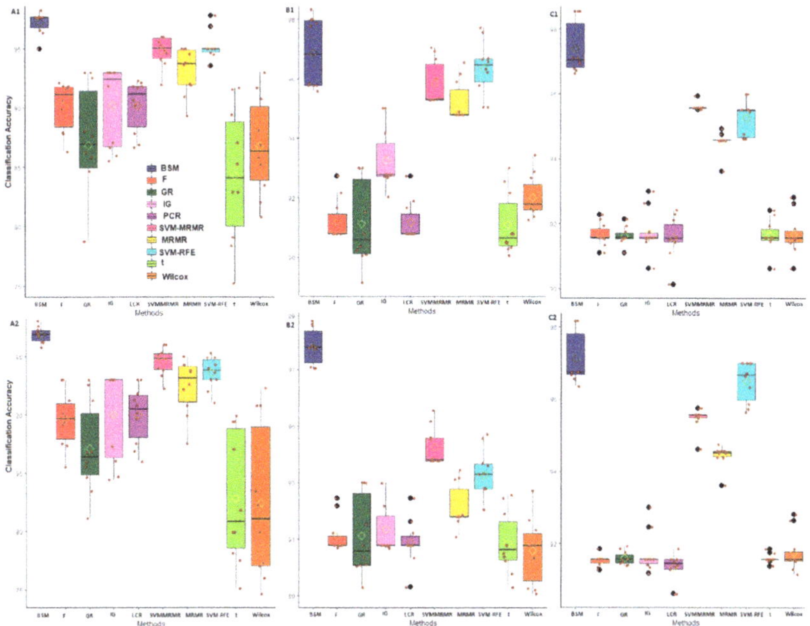

**Figure 3.** Classification-based comparative performance analysis of gene selection methods through SVM-LBF and SVM-PBF classifiers for abiotic stress datasets. The horizontal axis represents the gene selection methods. The vertical axis represents post selection classification accuracy obtained by using varying sliding window size technique. The classification accuracies over the window sizes are presented as boxes. The bars on the boxes represent the standard errors. The distributions of classification accuracies are shown for cold stress with SVM-LBF (**A1**), and SVM-PBF (**A2**) classifiers. The distributions of classification accuracies are shown for salinity stress with SVM-LBF (**B1**) and SVM-PBF (**B2**) classifiers. The distributions of classification accuracies are shown for drought stress with SVM-LBF (**C1**) and SVM-PBF (**C2**) classifiers.

For salinity stress data, the $\mu_{CA}^{G}$ (except gene sets of sizes 500, 1000 and 1500) computed for the proposed BSM approach through the SVM-LBF classifier were found to be higher than other methods followed by SVM-RFE and SVM-MRMR (Figure 3, Document S6). This indicated the proposed approach was fairly better, and competitive with the popular methods, i.e., SVM-RFE, SVM-MRMR. However, for SVM-PBF classifier, the $\mu_{CA}^{G}$ over all the gene sets for the BSM approach was higher than all other considered gene selection methods followed by SVM-RFE (Figure 3, Document S6). Furthermore, the $SE_{CA}^{G}$ computed through SVM-LBF and SVM-PBF classifiers for the BSM approach was found to be the least followed, by SVM-RFE (Document S6), indicating the selection of robust and relevant gene sets. Similar interpretation can be made for SVM-RBF and SVM-SBF classifiers from Figure 4. It was observed that the $\mu_{CA}^{G}$ from SVM-SBF classifier was found to be least (with high $SE_{CA}^{G}$) among the four classifiers for all the datasets (Figure 4 and Figure S4, Document S6). Here, it is pertinent to note that the sigmoid basis function may not be recommended to use in SVM training for real crop GE datasets. Furthermore, similar interpretations can be made for other abiotic (i.e., drought) and biotic (i.e., bacterial, fungal and insect) stress GE datasets (Figures 3 and 4, and Figure S4 and Document S6.

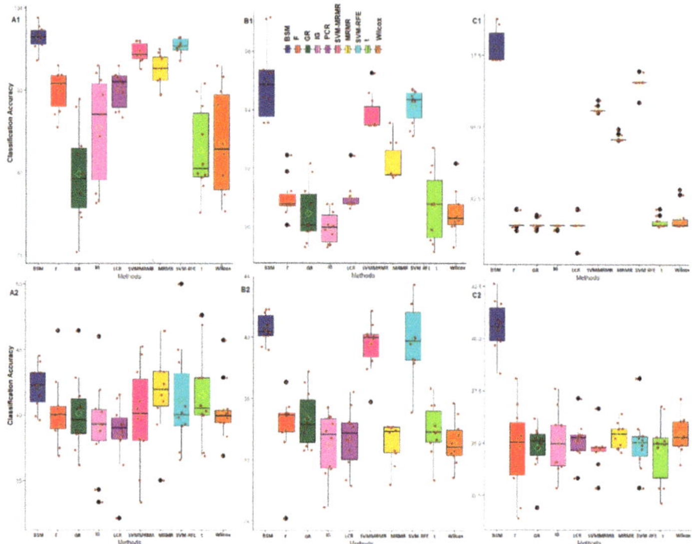

**Figure 4.** Classification-based comparative performance analysis of gene selection methods through SVM-RBF and SVM-SBF classifiers for abiotic stress datasets. The horizontal axis represents the gene selection methods. The vertical axis represents post selection classification accuracy obtained by using varying sliding window size technique. The classification accuracies over the window sizes are presented as boxes. The distributions of classification accuracies are shown for cold stress with SVM-RBF (**A1**) and SVM-SBF (**A2**) classifiers. The distributions of classification accuracies are shown for salinity stress with SVM-RBF (**B1**) and SVM-SBF (**B2**) classifiers. The distributions of classification accuracies are shown for drought stress with SVM-RBF (**C1**) and SVM-SBF (**C2**) classifiers.

The classification-based performance metrics can be considered as statistically necessary to check the informativeness and robustness of the selected genes. Through such analysis, it was found that the BSM approach performed better in terms of selecting informative and robust genes from the high-dimensional GE data as compared to other competitive methods such as SVM-RFE, MRMR, SVM-MRMR and the information theoretic measures. The reason may be attributed to the inclusion of bootstrap-based subject sampling model along with the self-contained statistical tests, which reduces the spurious association of genes with the target trait as well as with other genes. Further, the performance of BSM approach, in terms of the ability to classify the GE samples, was found to be better as compared to multivariate approaches, i.e., MRMR, SVM-MRMR, univariate approaches, i.e., *t*-test, F-score, Wilcox and informative theoretic measures, i.e., IG and GR. Here, it is worthy to note that the multivariate approaches performed better as compared to the univariate approaches when assessed under classification-based criteria, as the former considers gene-gene associations.

### 3.3. Comparative Performance Analysis Based on QTL Testing

We used publicly available QTL data to statistically measure the biological relevancy of the genes selected through the proposed and existing gene selection method(s). The main rationale behind such analysis is that the genes selected for a stress condition (through a gene selection method) are expected to be overlapped with the stress-specific QTL regions. Therefore, the QTLs and genes selected through these 10 gene selection methods, including the proposed BSM, were mapped to the whole rice genome using an MSU rice genome browser [55]. Further, the list of mapped QTLs for different abiotic (i.e., salinity, cold and drought), and biotic (i.e., bacterial, fungal and insect) stresses in rice along with their chromosomal positions in the genome are given in Supplementary Document S2.

The biological relevance of the selected genes through both proposed and existing gene selection methods were measured through two criteria, i.e., *Qstat* and $-log10(p\text{-value})$. The distributions of *Qstat* and $-log10(p\text{-value})$ over the selected genes for the six different datasets in rice are given in Figures 5 and 6, respectively. For salinity stress data, the values of *Qstat* over all the gene sets of sizes (<1000) selected through the proposed BSM approach were found to be higher than that of SVM-MRMR, SVM-RFE, MRMR, IG, F, Wilcox and PCR (Figure 5A). Further, it may be noted that the proposed approach was equally competitive with the univariate gene selection method such as a *t*-test, while they are assessed in terms of *Qstat* (Figure 5A). For higher gene set sizes (>1000), the value of *Qstat* for Wilcox method was found to be higher than that of proposed and existing approaches (Figure 5A) in the same data. This may be attributed to that the Wilcox method is nonparametric and does not depend on the distribution of GE data. For cold stress data, the values of *Qstat* statistic for all the selected gene sets through the BSM approach were higher than that of other existing methods (Figure 5B). This indicates that the performance of the proposed BSM approach is better in terms selecting cold stress related biologically relevant genes that are mostly overlapped with cold stress QTL regions in rice. Similar interpretations can be made for other abiotic (drought) and biotic (bacterial, fungal and insect) stress datasets in rice (Figure 5). Here, it is worthy to note that the *Qstat* is a linear function of the number of genes selected (through a gene selection approach), number of QTLs reported for that stress and length of the QTL regions (Figure 5).

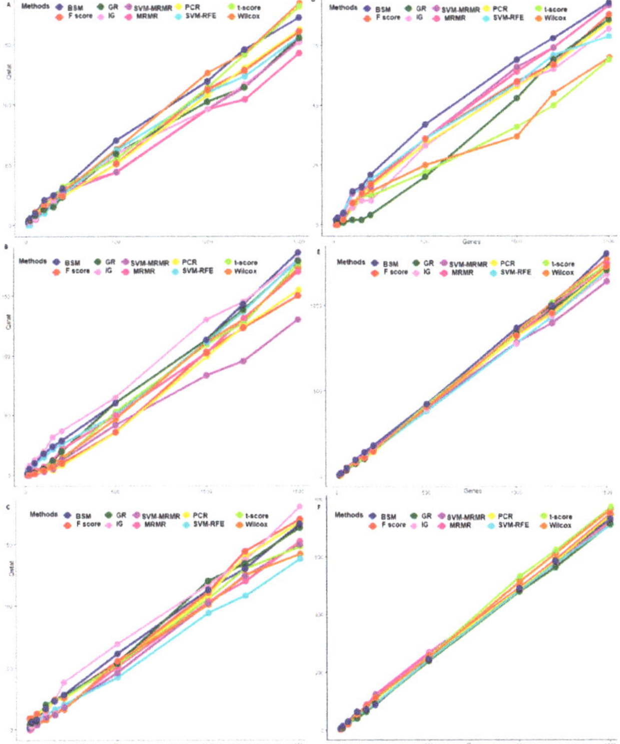

**Figure 5.** Comparative performance analysis of gene selection methods through distribution of *Qstat* statistic. The horizontal axis represents the informative gene sets obtained through gene selection methods. The vertical axis represents the value of *Qstat* statistic. The distribution of *Qstat* statistic are shown for (**A**) salinity; (**B**) cold; (**C**) drought; (**D**) bacterial; (**E**) fungal and (**F**) insect stress datasets in rice. The lines in different colors represent different gene selection methods.

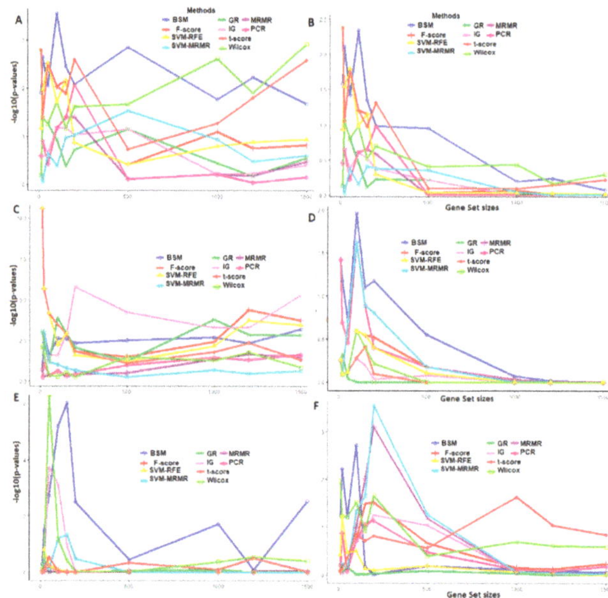

**Figure 6.** Comparative performance analysis of gene selection methods through distribution of *p*-values from QTL-hypergeometric test. The horizontal axis represents the gene sets obtained through gene selection methods. The vertical axis represents the value of −log10(*p*-value) from QTL-hypergeometric test. The distribution of −log10(*p*-value) are shown for (**A**) salinity; (**B**) cold; (**C**) drought; (**D**) bacterial; (**E**) fungal, and (**F**) insect stress datasets in rice. The lines in different colors represent different gene selection methods.

For salinity stress data, the $-log10$(*p*-value) from hypergeometric test over all the selected gene sets for the proposed BSM approach was observed to be higher than other existing gene selection methods (except t and GR) (Figure 6). In other words, genes selected by the BSM approach were more enriched with the underlying salinity responsive QTLs as compared to other existing methods. Similar interpretations can be made for other abiotic (i.e., cold and drought), and biotic (i.e., bacterial, fungal and insect) stress datasets in rice (Figure 6). Moreover, it is interesting to note that the values of *Qstat* and $-log10$(*p*-value) for (univariate) methods, such as t, F, PCR, Wilcox, IG and GR were found to be higher than that of the existing (multivariate) methods, such as MRMR, SVM-MRMR (Figures 5 and 6). This indicates the better and equally competitive performance of univariate over multivariate methods of gene selection when evaluated through QTL-based biological relevancy criteria. Such observations are not expected in statistics, but are well established in biology through previous studies [56].

Adjudging the performance of gene selection methods based on only classification might lead to the selection of biologically irrelevant genes. Therefore, we used criteria-based on QTLs to test the biological relevancy of the selected genes through proposed, and existing gene selection methods. Through this performance analysis, it was found that the BSM approach selects more biological relevant genes measured in terms of overlapping of the selected genes with given QTL regions as compared to multivariate approaches, i.e., MRMR, SVM-MRMR and machine learning approaches such as SVM-RFE. The proposed BSM approach was equally competitive (and better) with univariate approaches, i.e., *t*-test, F-score, Wilcox and information theoretic measures, i.e., IG and GR, when QTL-based criteria are considered. Through the QTLs-hypergeometric test analysis, it was evident that genes selected through the proposed BSM approach were more statistically enriched with the QTL regions.

### 3.4. Comparative Performance Analysis Based on GO Analysis

The comparative performance analysis of the proposed BSM approach with nine competitive gene selection methods (Document S5) was carried out through GO based distance analysis on six different rice GE datasets. Here, we set $n$ (i.e., number of selected genes) as 10, 20, 50, 100, 150, 200 and 500. Then, the selected genes, through the 10 gene selection methods, including the proposed BSM, were annotated with the GO terms under MF, BP and CC categories using *AgriGO* database [46]. The results from this analysis for abiotic stresses under MF, BP and CC GO categories are given in Tables 2–4 respectively and for biotic stresses in Supplementary Document S7. For salinity stress data, under the MF category, the values of GO-based average distance scores for the proposed BSM approach were found to be less than that of nine existing methods over all the selected gene sets (Table 2). This indicated that the proposed approach selected more (molecular) functionally similar genes which were responsible salinity tolerance in rice. Similar results can be found for BP and CC GO-based distance analysis for the same stress data (Table 2). In other words, the proposed BSM approach selects more biologically relevant genes attributed to each GO category for salinity stress as compared to the other nine competitive methods (Table 2). For bacterial stress, the values of GO-based average distance score under MF, BP and CC GO categories for the proposed BSM approach were found to be the least among other gene selection methods (Supplementary Document S7). Similar interpretations can be made for other abiotic (i.e., cold and drought) and biotic (i.e., fungal and insect) datasets in rice (Figures 2–4, Supplementary Document S7). Through this analysis, it was found that the proposed BSM approach performed better in terms of selecting more functionally relevant genes, which conferred biotic and abiotic stresses tolerance in rice.

**Table 2.** Comparative Performance analysis of the gene selection methods through MF GO-based biologically relevant score for abiotic stresses in rice.

| Methods | MRMR | SVM | SVM-MRMR | G | GR | Wilcox | t | PCR | F | BSM |
|---|---|---|---|---|---|---|---|---|---|---|
| | | | | Salt stress in rice | | | | | | |
| 10 | 0.98 | 0.95 | 0.97 | 0.92 | 0.89 | 0.93 | 0.93 | 0.96 | 0.96 | 0.88 |
| 20 | 0.97 | 0.89 | 0.93 | 0.92 | 0.86 | 0.89 | 0.89 | 0.91 | 0.91 | 0.86 |
| 50 | 0.92 | 0.91 | 0.92 | 0.90 | 0.90 | 0.87 | 0.87 | 0.92 | 0.92 | 0.85 |
| 100 | 0.92 | 0.90 | 0.89 | 0.90 | 0.88 | 0.87 | 0.88 | 0.92 | 0.91 | 0.83 |
| 150 | 0.90 | 0.89 | 0.90 | 0.89 | 0.88 | 0.87 | 0.87 | 0.90 | 0.91 | 0.83 |
| 200 | 0.90 | 0.89 | 0.88 | 0.89 | 0.87 | 0.88 | 0.88 | 0.90 | 0.90 | 0.84 |
| 500 | 0.90 | 0.90 | 0.89 | 0.90 | 0.90 | 0.89 | 0.90 | 0.89 | 0.89 | 0.83 |
| | | | | Cold stress in rice | | | | | | |
| 10 | 0.82 | 0.84 | 0.82 | 0.92 | 0.99 | 0.92 | 0.87 | 0.77 | 0.77 | 0.75 |
| 20 | 0.93 | 0.88 | 0.93 | 0.95 | 0.93 | 0.88 | 0.90 | 0.91 | 0.88 | 0.71 |
| 50 | 0.91 | 0.88 | 0.91 | 0.93 | 0.90 | 0.91 | 0.91 | 0.92 | 0.92 | 0.73 |
| 100 | 0.91 | 0.90 | 0.91 | 0.90 | 0.88 | 0.91 | 0.91 | 0.91 | 0.91 | 0.74 |
| 150 | 0.90 | 0.89 | 0.90 | 0.89 | 0.89 | 0.89 | 0.90 | 0.91 | 0.91 | 0.72 |
| 200 | 0.90 | 0.89 | 0.90 | 0.89 | 0.88 | 0.89 | 0.90 | 0.90 | 0.90 | 0.73 |
| 500 | 0.90 | 0.88 | 0.90 | 0.90 | 0.89 | 0.88 | 0.89 | 0.88 | 0.89 | 0.73 |
| | | | | Drought stress in rice | | | | | | |
| 10 | 0.82 | 0.86 | 0.81 | 0.90 | 0.93 | 0.65 | 0.76 | 0.76 | 0.76 | 0.71 |
| 20 | 0.79 | 0.86 | 0.78 | 0.91 | 0.90 | 0.80 | 0.81 | 0.81 | 0.81 | 0.75 |
| 50 | 0.88 | 0.84 | 0.87 | 0.88 | 0.90 | 0.84 | 0.88 | 0.89 | 0.89 | 0.75 |
| 100 | 0.89 | 0.89 | 0.88 | 0.89 | 0.89 | 0.88 | 0.88 | 0.88 | 0.88 | 0.76 |
| 150 | 0.88 | 0.88 | 0.87 | 0.89 | 0.88 | 0.88 | 0.88 | 0.88 | 0.88 | 0.76 |
| 200 | 0.88 | 0.88 | 0.87 | 0.88 | 0.89 | 0.89 | 0.88 | 0.88 | 0.88 | 0.74 |
| 500 | 0.88 | 0.88 | 0.87 | 0.88 | 0.88 | 0.89 | 0.88 | 0.87 | 0.87 | 0.73 |

Values in the last column represent dissimilarity scores obtained from proposed BSM approach.

**Table 3.** Comparative Performance analysis of the gene selection methods through BP GO-based biologically relevant score for abiotic stresses in rice.

| Methods | MRMR | SVM | SVM-MRMR | IG | GR | Wilcox | t | PCR | F | BSM |
|---|---|---|---|---|---|---|---|---|---|---|
| | | | | Salt stress in rice | | | | | | |
| 10 | 0.86 | 0.94 | 0.86 | 0.92 | 0.97 | 0.90 | 0.90 | 0.88 | 0.88 | 0.83 |
| 20 | 0.90 | 0.91 | 0.90 | 0.89 | 0.91 | 0.92 | 0.92 | 0.84 | 0.85 | 0.84 |
| 50 | 0.89 | 0.90 | 0.88 | 0.88 | 0.90 | 0.88 | 0.89 | 0.88 | 0.88 | 0.82 |
| 100 | 0.88 | 0.89 | 0.86 | 0.89 | 0.89 | 0.85 | 0.86 | 0.89 | 0.87 | 0.82 |
| 150 | 0.87 | 0.89 | 0.90 | 0.88 | 0.89 | 0.85 | 0.85 | 0.89 | 0.89 | 0.83 |
| 200 | 0.87 | 0.89 | 0.86 | 0.88 | 0.89 | 0.84 | 0.85 | 0.89 | 0.88 | 0.82 |
| 500 | 0.87 | 0.89 | 0.87 | 0.87 | 0.89 | 0.86 | 0.86 | 0.86 | 0.86 | 0.82 |
| | | | | Cold stress in rice | | | | | | |
| 10 | 0.79 | 0.82 | 0.79 | 0.86 | 0.94 | 0.91 | 0.90 | 0.79 | 0.79 | 0.79 |
| 20 | 0.93 | 0.89 | 0.93 | 0.90 | 0.88 | 0.86 | 0.88 | 0.90 | 0.86 | 0.82 |
| 50 | 0.88 | 0.89 | 0.88 | 0.90 | 0.88 | 0.88 | 0.87 | 0.89 | 0.90 | 0.71 |
| 100 | 0.88 | 0.89 | 0.88 | 0.89 | 0.87 | 0.90 | 0.88 | 0.89 | 0.89 | 0.74 |
| 150 | 0.89 | 0.88 | 0.89 | 0.88 | 0.88 | 0.88 | 0.87 | 0.88 | 0.88 | 0.73 |
| 200 | 0.89 | 0.87 | 0.89 | 0.87 | 0.87 | 0.87 | 0.87 | 0.88 | 0.84 | 0.73 |
| 500 | 0.88 | 0.86 | 0.88 | 0.86 | 0.86 | 0.84 | 0.86 | 0.87 | 0.83 | 0.71 |
| | | | | Drought stress in rice | | | | | | |
| 10 | 0.86 | 0.79 | 0.85 | 0.81 | 0.89 | 0.62 | 0.83 | 0.83 | 0.83 | 0.73 |
| 20 | 0.84 | 0.79 | 0.83 | 0.89 | 0.90 | 0.80 | 0.84 | 0.84 | 0.84 | 0.72 |
| 50 | 0.88 | 0.81 | 0.87 | 0.88 | 0.88 | 0.81 | 0.88 | 0.88 | 0.88 | 0.72 |
| 100 | 0.87 | 0.84 | 0.86 | 0.88 | 0.88 | 0.84 | 0.86 | 0.87 | 0.87 | 0.72 |
| 150 | 0.86 | 0.84 | 0.85 | 0.88 | 0.88 | 0.84 | 0.87 | 0.87 | 0.87 | 0.71 |
| 200 | 0.86 | 0.84 | 0.85 | 0.87 | 0.87 | 0.85 | 0.86 | 0.86 | 0.86 | 0.72 |
| 500 | 0.87 | 0.85 | 0.86 | 0.86 | 0.87 | 0.87 | 0.86 | 0.85 | 0.83 | 0.72 |

Values in the last column represent dissimilarity scores obtained from proposed BSM approach.

**Table 4.** Comparative Performance analysis of the gene selection methods through CC GO-based biologically relevant score for abiotic stresses in rice.

| | MRMR | SVM | SVM-MRMR | IG | GR | Wilcox | t | PCR | F | BSM |
|---|---|---|---|---|---|---|---|---|---|---|
| | | | | Salt stress in rice | | | | | | |
| 10 | 0.77 | 0.71 | 0.70 | 0.94 | 0.97 | 0.93 | 0.93 | 0.95 | 0.95 | 0.78 |
| 20 | 0.88 | 0.87 | 0.85 | 0.92 | 0.90 | 0.91 | 0.91 | 0.88 | 0.88 | 0.81 |
| 50 | 0.88 | 0.89 | 0.86 | 0.92 | 0.92 | 0.90 | 0.90 | 0.89 | 0.89 | 0.84 |
| 100 | 0.88 | 0.90 | 0.8 | 0.91 | 0.89 | 0.86 | 0.86 | 0.88 | 0.88 | 0.83 |
| 150 | 0.87 | 0.90 | 0.87 | 0.90 | 0.89 | 0.86 | 0.87 | 0.88 | 0.88 | 0.83 |
| 200 | 0.87 | 0.89 | 0.85 | 0.90 | 0.90 | 0.88 | 0.89 | 0.88 | 0.88 | 0.83 |
| 500 | 0.88 | 0.90 | 0.88 | 0.89 | 0.90 | 0.88 | 0.89 | 0.87 | 0.87 | 0.82 |
| | | | | Cold stress in rice | | | | | | |
| 10 | 0.78 | 0.80 | 0.78 | 0.96 | 0.81 | 0.87 | 0.86 | 0.70 | 0.70 | 0.70 |
| 20 | 0.88 | 0.86 | 0.88 | 0.96 | 0.87 | 0.87 | 0.89 | 0.81 | 0.83 | 0.71 |
| 50 | 0.86 | 0.89 | 0.86 | 0.90 | 0.85 | 0.84 | 0.85 | 0.89 | 0.90 | 0.73 |
| 100 | 0.88 | 0.90 | 0.88 | 0.90 | 0.81 | 0.83 | 0.84 | 0.87 | 0.87 | 0.74 |
| 150 | 0.88 | 0.89 | 0.88 | 0.90 | 0.82 | 0.82 | 0.86 | 0.87 | 0.88 | 0.74 |
| 200 | 0.87 | 0.90 | 0.87 | 0.90 | 0.84 | 0.85 | 0.86 | 0.87 | 0.85 | 0.73 |
| 500 | 0.88 | 0.89 | 0.88 | 0.89 | 0.86 | 0.97 | 0.86 | 0.88 | 0.87 | 0.73 |
| | | | | Drought stress in rice | | | | | | |
| 10 | 0.82 | 0.86 | 0.81 | 0.91 | 0.89 | 0.83 | 0.87 | 0.87 | 0.87 | 0.74 |
| 20 | 0.89 | 0.85 | 0.88 | 0.93 | 0.90 | 0.87 | 0.89 | 0.89 | 0.89 | 0.74 |
| 50 | 0.86 | 0.88 | 0.85 | 0.91 | 0.87 | 0.87 | 0.88 | 0.88 | 0.88 | 0.73 |
| 100 | 0.87 | 0.87 | 0.86 | 0.89 | 0.86 | 0.87 | 0.88 | 0.88 | 0.88 | 0.74 |
| 150 | 0.87 | 0.87 | 0.86 | 0.90 | 0.85 | 0.85 | 0.87 | 0.87 | 0.87 | 0.74 |
| 200 | 0.87 | 0.87 | 0.86 | 0.89 | 0.86 | 0.86 | 0.87 | 0.87 | 0.87 | 0.73 |
| 500 | 0.87 | 0.86 | 0.86 | 0.89 | 0.87 | 0.88 | 0.87 | 0.86 | 0.85 | 0.72 |

Values in the last column represent dissimilarity scores obtained from proposed BSM approach.

The GO-based distance analysis showed that higher functional similarities (which may have biological functions important to stress tolerance) exist among the genes selected by the BSM, as compared to existing methods. The performance of the BSM was found to be better and equally competitive with the univariate approaches, viz. t-score, F-score, Wilcox and correlation-based approaches in terms of selecting genes which are biologically relevant (in terms of GO annotations) for the target trait/condition. It is worthy to note that the univariate approaches performed better as compared to the multivariate approaches under the biology-based criteria, but the former performed poorer than the latter under classification-based criteria. This indicates the real biological complexity for assessing the performance of gene selection approaches on real data. Therefore, we used the comprehensive framework of performance analysis of the gene selection methods under both statistical necessary and biologically relevant criteria. The comparative performance analysis revealed that the proposed BSM approach is better as well as competitive under the classification, QTL and GO-based criteria.

*3.5. Comparative Performance Analysis Based on Runtime*

The recursive feature elimination algorithms-based gene selection methods such as SVM-RFE and SVM-MRMR are computationally intensive and time consuming. So, we used the runtime criterion to evaluate the performance of these gene selection methods. Here, the runtime refers to the amount of computational time required to analyze the GE data through running the codes of the respective methods in R software (v. 4.0.1). The detail results from the runtime-based evaluation of gene selection methods is given in Supplementary Document S8. For bacterial stress GE data (with 8356 genes over 74 samples), SVM-RFE and SVM-MRMR took ~90 and 80 min respectively to analyze on a 2-core DELL PC with 8 GB RAM with Intel(R) Core (TM) i3-6100U CPU at 2.30GHz. On the contrary, the BSM approach took ~25 min to analyze the same GE data to obtain biologically informative genes (Table S20). The BSM method required less computational time than popular methods of gene selection such as SVM-RFE and with much better performance in terms of obtaining biologically informative gene sets. Similar interpretations can be made for the gene selection methods based on the runtime criterion to analyze the remaining five datasets (Tables S21–S23).

## 4. Developed R Software Package

To popularize the use of the proposed gene selection approach among the users, we developed an R software package which includes BSM R package and accompanying documentation with examples. This package is supplied with the manuscript as supplementary information and also available in https://github.com/sam-uofl/BSM. The guidelines for the use of BSM R package is given in Supplementary Document S8. This software is capable of computing weights for gene selection through MRMR, SVM and SVM-MRMR methods, and also provide functions for computing *p*-values and adjusted *p*-values through a BSM approach for different parameter options. Further, it also allows different functions for selecting relevant gene sets through MRMR, SVM, SVM-MRMR and BSM gene selection approaches.

## 5. Conclusions

In GE genomics, the main aim is to select relevant genes which can be used as predictors for the development of statistical/classification models to handle high dimensionality in GE data. Therefore, we proposed an improved BSM statistical approach for gene selection from GE data, which was both effective in reducing redundancy among the genes and improves biological relevancy of genes with the target trait. Here, the genes were selected based on the assessment of the statistical significance of the self-contained null hypothesis under a sound computational framework. Usually, thousand(s) of null hypotheses are tested simultaneously in GE data analysis which increases the chance of selection of false positive genes. Hence, through the proposed BSM approach an adjusted *p*-value was assigned to each gene after multiple test adjustments, and relevant genes were selected based on the adjusted *p*-values. The BSM approach was based on the NP test statistic(s) which does not depend on the distribution of the GE data unlike *t*-test. Further, the bootstrap procedure in the BSM can minimize the redundancy among genes as well as reduce the spurious association of genes with traits during gene selection. The proposed approach was also less computationally expensive compared to SVM-RFE and SVM-MRMR and can be implemented on a personal or workstation computer for analyzing large GE datasets. Furthermore, we used a comprehensive framework of performance analysis of the gene selection methods under statistically necessary and biologically relevant criteria. More specifically, the tested gene selection methods included SVM-RFE from Wrapper, SVM-MRMR and proposed BSM from hybrid (embedded) and the remaining seven from the filter categories. The comparative analysis revealed the proposed approach has the features of an ideal technique of gene selection, as it performed better under both statistically necessary and biologically relevant criteria. Moreover, this study provided a systematic and rigorous evaluation of the gene selection methods under a multi-criteria decision setup on multiple real datasets. It also provided a framework to researchers to comparatively study the available methods, which will guide genome researchers and experimental biologists to select the best method(s) objectively. The proposed approach may provide a statistical template for combing other filter and wrapper gene selection methods under a sound and effective computational environment.

**Supplementary Materials:** The following are available online at http://www.mdpi.com/1099-4300/22/11/1205/s1.

**Author Contributions:** Conceived and designed the study, S.D.; developed the methodologies, S.D.; developed the R-code and R package, S.D.; contributed materials, S.D. and S.N.R.; drafted the manuscript, S.D.; corrected the manuscript, S.D. and S.N.R.; funding acquisition, S.N.R. All authors have read and agreed to the published version of the manuscript.

**Funding:** This study was fully supported by Netaji Subhas-ICAR International Fellowship, OM No. 18(02)/2016-EQR/Edn. (SD) of Indian Council of Agricultural Research (ICAR), New Delhi, India. It was supported in part by Wendell Cherry Chair in Clinical Trial Research Fund (SNR), multiple National Institutes of Health (NIH), USA grants (SNR) (5P20GM113226, PI: McClain; 1P42ES023716, PI: Srivastava; 5P30GM127607-02, PI: Jones; 1P20GM125504-01, PI: Lamont; 2U54HL120163, PI: Bhatnagar/Robertson; 1P20GM135004, PI: Yan; 1R35ES0238373-01, PI: Cave; 1R01ES029846, PI: Bhatnagar; 1R01ES027778-01A1, PI: States;), and Kentucky Council on Postsecondary Education grant (SNR) (PON2 415 1900002934, PI: Chesney). The content is solely the responsibility of the authors and does not necessarily represent the views of NIH or ICAR.

**Acknowledgments:** Authors duly acknowledge the support obtained from ICAR-Indian Agricultural Statistics Research Institute, New Delhi, India.

**Conflicts of Interest:** The authors declare no conflict of interest.

**Availability of Data and Material:** All the secondary data used in this study are available in the NCBI database. The proposed methods are implemented in the developed R package and R codes are freely available at http://github/sam-uofl/BSM.

## References

1. Reuter, J.A.; Spacek, D.V.; Snyder, M.P. High-Throughput Sequencing Technologies. *Mol. Cell* **2015**, *58*, 586–597. [CrossRef] [PubMed]
2. Trevino, V.; Falciani, F.; Barrera-Saldaña, H.A. DNA Microarrays: A Powerful Genomic Tool for Biomedical and Clinical Research. *Mol. Med.* **2007**, *13*, 527–541. [CrossRef] [PubMed]
3. Charpe, A.M. DNA Microarray. In *Advances in Biotechnology*; Springer: New Delhi, India, 2014; pp. 71–104. [CrossRef]
4. Barrett, T.; Wilhite, S.E.; Ledoux, P.; Evangelista, C.; Kim, I.F.; Tomashevsky, M. NCBI GEO: Archive for functional genomics data sets—Update. *Nucleic Acids Res.* **2012**, *41*, D991–D995. [CrossRef]
5. Das, S.; Meher, P.K.; Rai, A.; Bhar, L.M.; Mandal, B.N. Statistical approaches for gene selection, hub gene identification and module interaction in gene co-expression network analysis: An application to aluminum stress in soybean (*Glycine max* L.). *PLoS ONE* **2017**, *12*, e0169605. [CrossRef] [PubMed]
6. Wang, J.; Chen, L.; Wang, Y.; Zhang, J.; Liang, Y.; Xu, D. A Computational Systems Biology Study for Understanding Salt Tolerance Mechanism in Rice. *PLoS ONE* **2013**, *8*, e64929. [CrossRef] [PubMed]

7.   Golub, T.R.; Slonim, D.K.; Tamayo, P.; Huard, C.; Gaasenbeek, M.; Mesirov, J.P. Molecular Classification of Cancer: Class Discovery and Class Prediction by Gene Expression Monitoring. *Science* **1999**, *286*, 531–537. [CrossRef]

8.   Guyon, I.; Weston, J.; Barnhill, S.; Vapnik, V. Gene selection for cancer classification using support vector machines. *Mach. Learn.* **2002**. [CrossRef]

9.   Saeys, Y.; Inza, I.; Larranaga, P. A review of feature selection techniques in bioinformatics. *Bioinformatics* **2007**, *23*, 2507–2517. [CrossRef]

10.   Liang, Y.; Zhang, F.; Wang, J.; Joshi, T.; Wang, Y.; Xu, D. Prediction of Drought-Resistant Genes in Arabidopsis thaliana Using SVM-RFE. *PLoS ONE* **2011**, *6*, e21750. [CrossRef]

11.   Díaz-Uriarte, R.; Alvarez de Andrés, S. Gene selection and classification of microarray data using random forest. *BMC Bioinform.* **2006**, *7*, 3. [CrossRef]

12.   Peng, H.; Long, F.; Ding, C. Feature selection based on mutual information: Criteria of Max-Dependency, Max-Relevance, and Min-Redundancy. *IEEE Trans. Pattern Anal. Mach. Intell.* **2005**. [CrossRef]

13.   Mundra, P.A.; Rajapakse, J.C. SVM-RFE with MRMR Filter for Gene Selection. *IEEE Trans. Nanobioscience* **2010**, *9*, 31–37. [CrossRef] [PubMed]

14.   Das, S.; Pandey, P.; Rai, A.; Mohapatra, C. A computational system biology approach to construct gene regulatory networks for salinity response in rice (*Oryza sativa*). *Indian J. Agric. Sci.* **2015**, *85*, 1546–1552.

15.   Kursa, M.B. Robustness of Random Forest-based gene selection methods. *BMC Bioinform.* **2014**. [CrossRef] [PubMed]

16.   Inza, I.; Larrañaga, P.; Blanco, R.; Cerrolaza, A.J. Filter versus wrapper gene selection approaches in DNA microarray domains. *Artif. Intell. Med.* **2004**. [CrossRef] [PubMed]

17.   Lazar, C.; Taminau, J.; Meganck, S.; Steenhoff, D.; Coletta, A.; Molter, C. A survey on filter techniques for feature selection in gene expression microarray analysis. *IEEE/ACM Trans. Comput. Biol. Bioinform.* **2012**. [CrossRef] [PubMed]

18.   Cui, X.; Churchill, G.A. Statistical tests for differential expression in cDNA microarray experiments. *Genome Biol.* **2003**. [CrossRef]

19.   Das, S.; Meher, P.K.; Pradhan, U.K.; Paul, A.K. Inferring gene regulatory networks using Kendall's tau correlation coefficient and identification of salinity stress responsive genes in rice. *Curr. Sci.* **2017**, *112*. [CrossRef]

20.   Ding, C.; Peng, H. Minimum redundancy feature selection from microarray gene expression data. Computational Systems Bioinformatics CSB2003 Proceedings of the 2003 IEEE Bioinformatics Conference CSB2003. *IEEE Comput. Soc.* **2003**, 523–528. [CrossRef]

21.   Chen, Y.W.; Lin, C.J. Combining SVMs with various feature selection strategies. *Stud. Fuzziness Soft Comput.* **2006**. [CrossRef]

22.   Hossain, A.; Willan, A.R.; Beyene, J. An improved method on wilcoxon rank sum test for gene selection from microarray experiments. *Commun. Stat. Simul. Comput.* **2013**. [CrossRef]

23.   Troyanskaya, O.G.; Garber, M.E.; Brown, P.O.; Botstein, D.; Altman, R.B. Nonparametric methods for identifying differentially expressed genes in microarray data. *Bioinformatics* **2002**. [CrossRef] [PubMed]

24.   Cheng, T.; Wang, Y.; Bryant, S.H. F Selector: A Ruby gem for feature selection. *Bioinformatics* **2012**, *28*, 2851–2852. [CrossRef]

25.   Radovic, M.; Ghalwash, M.; Filipovic, N.; Obradovic, Z. Minimum redundancy maximum relevance feature selection approach for temporal gene expression data. *BMC Bioinform.* **2017**, *18*, 9. [CrossRef]

26.   Ding, C.; Peng, H. Minimum redundancy feature selection from microarray gene expression data. *J. Bioinform. Comput. Biol.* **2005**, *3*, 185–205. [CrossRef]

27.   Zhang, G.-L.; Pan, L.-L.; Huang, T.; Wang, J.-H. The transcriptome difference between colorectal tumor and normal tissues revealed by single-cell sequencing. *J. Cancer* **2019**, *10*, 5883–5890. [CrossRef]

28.   Kohavi, R.; John, G.H. Wrappers for feature subset selection. *Artif. Intell.* **1997**. [CrossRef]

29.   Hearst, M.A.; Dumais, S.T.; Osuna, E.; Platt, J.; Scholkopf, B. Support vector machines. *IEEE Intell. Syst.* **1998**, *13*, 18–28. [CrossRef]

30.   Duan, K.B.; Rajapakse, J.C.; Wang, H.; Azuaje, F. Multiple SVM-RFE for gene selection in cancer classification with expression data. *IEEE Trans. Nanobioscience* **2005**. [CrossRef]

31. Tao, X.; Wu, X.; Huang, T.; Mu, D. Identification and Analysis of Dysfunctional Genes and Pathways in CD8+ T Cells of Non-Small Cell Lung Cancer Based on RNA Sequencing. *Front. Genet.* **2020**. [CrossRef] [PubMed]

32. Ting, K.M.; Witten, I.H. Stacking bagged and dagged models. In *ICML '97: Proceedings of the Fourteenth International Conference on Machine Learning*; Douglas, H., Fisher, E.D., Eds.; Morgan Kaufmann Publishers Inc.: San Francisco, CA, USA, 1997; pp. 367–375.

33. Li, J.R.; Huang, T. Predicting and analyzing early wake-up associated gene expressions by integrating GWAS and eQTL studies. *Biochim. Biophys. Acta Mol. Basis Dis.* **2018**. [CrossRef] [PubMed]

34. Sun, L.; Kong, X.; Xu, J.; Xue, Z.; Zhai, R.; Zhang, S. A Hybrid Gene Selection Method Based on ReliefF and Ant Colony Optimization Algorithm for Tumor Classification. *Sci. Rep.* **2019**. [CrossRef]

35. Mahi, M.; Baykan, Ö.K.; Kodaz, H. A new hybrid method based on Particle Swarm Optimization, Ant Colony Optimization and 3-Opt algorithms for Traveling Salesman Problem. *Appl. Soft Comput.* **2015**, *30*, 484–490. [CrossRef]

36. Sohn, I.; Owzar, K.; George, S.L.; Kim, S.; Jung, S.H. A permutation-based multiple testing method for time-course microarray experiments. *BMC Bioinform.* **2009**. [CrossRef]

37. Ritchie, M.E.; Phipson, B.; Wu, D.; Hu, Y.; Law, C.W.; Shi, W. limma powers differential expression analyses for RNA-sequencing and microarray studies. *Nucleic Acids Res.* **2015**, *43*, e47. [CrossRef]

38. Knijnenburg, T.A.; Wessels, L.F.A.; Reinders, M.J.T.; Shmulevich, I. Fewer permutations, more accurate P-values. *Bioinformatics* **2009**. [CrossRef]

39. Das, S.; Rai, A.; Mishra, D.C.; Rai, S.N. Statistical approach for selection of biologically informative genes. *Gene* **2018**, *655*. [CrossRef]

40. Lai, C.; Reinders, M.J.T.; van't Veer, L.J.; Wessels, L.F.A. A comparison of univariate and multivariate gene selection techniques for classification of cancer datasets. *BMC Bioinform.* **2006**. [CrossRef]

41. Das, S.; Rai, A.; Mishra, D.C.; Rai, S.N. Statistical Approach for Gene Set Analysis with Trait Specific Quantitative Trait Loci. *Sci. Rep.* **2018**, *8*, 2391. [CrossRef]

42. Tiwari, S.; Kumar, V.; Singh, B.; Rao, A.; Mithra, S.V.A. Mapping QTLs for Salt Tolerance in Rice (Oryza sativa L) by Bulked Segregant Analysis of Recombinant Inbred Lines Using 50K SNP Chip. Yadav RS, editor. *PLoS ONE* **2016**, *11*, e0153610. [CrossRef]

43. Gene Ontology Consortium. The Gene Ontology (GO) database and informatics resource. *Nucleic Acids Res.* **2004**. [CrossRef]

44. Gautier, L.; Cope, L.; Bolstad, B.M.; Irizarry, R.A. Affy—Analysis of Affymetrix GeneChip data at the probe level. *Bioinformatics* **2004**. [CrossRef] [PubMed]

45. Ware, D. Gramene: A resource for comparative grass genomics. *Nucleic Acids Res.* **2002**. [CrossRef] [PubMed]

46. Tian, T.; Liu, Y.; Yan, H.; You, Q.; Yi, X.; Du, Z. AgriGO v2.0: A GO analysis toolkit for the agricultural community, 2017 update. *Nucleic Acids Res.* **2017**. [CrossRef]

47. Sahani, M.; Linden, J. *Advances in Neural Information Processing Systems, Processing Systems: Proceedings from the 2002, 2003*; MIT Press: Cambridge, MA, USA, 2003. [CrossRef]

48. Efron, B.; Tibshirani, R.J. *An Introduction to the Bootstrap*; Springer: Boston, MA, USA, 1993. [CrossRef]

49. Benjamini, Y.; Hochberg, Y. Multiple Hypotheses Testing with Weights. *Scand. J. Stat.* **1997**, *24*, 407–418. [CrossRef]

50. Li, Q.; Brown, J.B.; Huang, H.; Bickel, P.J. Measuring reproducibility of high-throughput experiments. *Ann. Appl. Stat.* **2011**, *5*, 1752–1779. [CrossRef]

51. Chen, S.-Y.; Feng, Z.; Yi, X. A general introduction to adjustment for multiple comparisons. *J. Thorac. Dis.* **2017**, *9*, 1725–1729. [CrossRef] [PubMed]

52. Mazandu, G.K.; Mulder, N.J. Information content-based gene ontology functional similarity measures: Which one to use for a given biological data type? *PLoS ONE* **2014**. [CrossRef]

53. Lord, P.W.; Stevens, R.D.; Brass, A.; Goble, C.A. Investigating semantic similarity measures across the gene ontology: The relationship between sequence and annotation. *Bioinformatics* **2003**. [CrossRef]

54. Wang, J.Z.; Du, Z.; Payattakool, R.; Yu, P.S.; Chen, C.F. A new method to measure the semantic similarity of GO terms. *Bioinformatics* **2007**. [CrossRef]

55. Ouyang, S.; Zhu, W.; Hamilton, J.; Lin, H.; Campbell, M.; Childs, K. The TIGR Rice Genome Annotation Resource: Improvements and new features. *Nucleic Acids Res.* **2007**. [CrossRef] [PubMed]
56. Glazko, G.V.; Emmert-Streib, F. Unite and conquer: Univariate and multivariate approaches for finding differentially expressed gene sets. *Bioinformatics* **2009**. [CrossRef] [PubMed]


**Publisher's Note:** MDPI stays neutral with regard to jurisdictional claims in published maps and institutional affiliations.

*Article*

# Residue Cluster Classes: A Unified Protein Representation for Efficient Structural and Functional Classification

**Fernando Fontove** [1] **and Gabriel Del Rio** [2,*]

[1]  C3 Consensus, Miguel Hidalgo, CDMX, Mexico City 11510, Mexico; fernando.fontove@c3consensus.com
[2]  Department of Biochemistry and Structural Biology, Instituto de Fisiología Celular, UNAM, Mexico City 04510, Mexico
*  Correspondence: gdelrio@ifc.unam.mx

Received: 1 March 2020; Accepted: 7 April 2020; Published: 20 April 2020

**Abstract:** Proteins are characterized by their structures and functions, and these two fundamental aspects of proteins are assumed to be related. To model such a relationship, a single representation to model both protein structure and function would be convenient, yet so far, the most effective models for protein structure or function classification do not rely on the same protein representation. Here we provide a computationally efficient implementation for large datasets to calculate residue cluster classes (RCCs) from protein three-dimensional structures and show that such representations enable a random forest algorithm to effectively learn the structural and functional classifications of proteins, according to the CATH and Gene Ontology criteria, respectively. RCCs are derived from residue contact maps built from different distance criteria, and we show that 7 or 8 Å with or without amino acid side-chain atoms rendered the best classification models. The potential use of a unified representation of proteins is discussed and possible future areas for improvement and exploration are presented.

**Keywords:** residue cluster class; structural classification; functional classification

---

## 1. Introduction

Proteins are molecules found in living organisms and participate in many diverse cellular and molecular functions. It is generally recognized that the three-dimensional (3D) structures of proteins are related to their functions, yet the relationship remains to be elucidated, since many attempts to predict the correct functions of proteins based on their 3D structures still result in false predictions [1].

Different approaches have been described to account for such a relationship, including those based on physical and chemical forces [2,3], protein sequence and phylogenies [4–6] and others. Machine-learning (ML) based models are currently the best models for predicting 3D protein structures [7] and protein functions [8]. ML models require object representations in the form of a set of features; these features are numeric values; hence, objects are represented by vectors. Yet, while protein structure and function are indeed related (both observations are derived from the same object, the protein, and hence are related), it is questionable whether at the model level these are actually related. For instance, the features used so far to predict the 3D structure are different from those used to predict protein function [9,10]; hence, the reliability of ML methods when predicting a protein's structure or function may be simply the consequence of having better representations for each aspect of the protein (i.e., 3D structure or function). We believe having a unique protein representation with which to efficiently predict 3D protein structure and function would provide a mathematical framework to explore the relationship between these two fundamental aspects of proteins, 3D structure and function.

We have previously described a representation of 3D structure that is learnable (that is, it displays a pattern that any ML heuristic model should be able to detect); such a representation allowed us to identify structural neighbors and classify the protein's 3D structure with the best performance and reliability reported so far [11]. The representation is based on counting the 26 different maximal clique classes that are derived from the 3D structure and protein sequence given a contact distance threshold of 5 Å, including atoms of the side chains; we referred to these maximal cliques as residue cluster classes or RCCs (see Figure 1 and Materials and Methods). In that previous work, we tested two ML algorithms (random forest and support vector machine) that were adequate for the structural classification of the data. In the present work, we developed a computationally efficient implementation for computing RCCs on large datasets of 3D protein structures. This allowed us to further explore the protein structure classification and distribute this implementation freely (see Supplementary Materials); this implementation incorporates some variations in the contact definition (different distances and the exclusion of the side-chain atoms). Here we also explored in a systematic way for dozens of ML models to identify the optimal model and corresponding hyper-parameters for such tasks, and corroborated that random forest rendered the best algorithm. Furthermore, we showed that ML models built from RCCs were able to efficiently learn protein functions, yet there is plenty of room for improvement. Hence, our results provide the first unified, high-dimensional description of proteins useful for learning both 3D structure classification and protein function using ML methods.

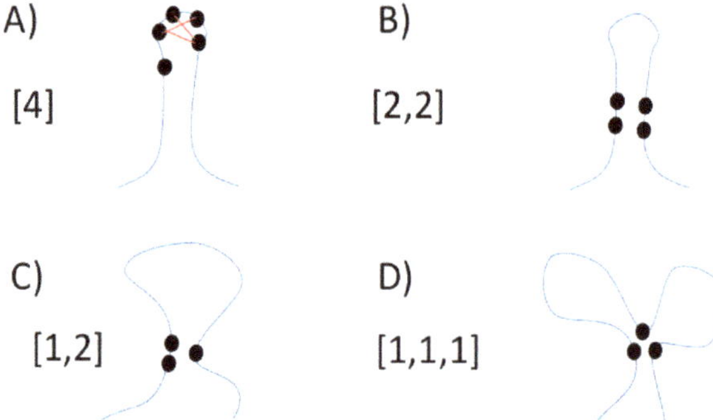

**Figure 1. Residue Cluster Classes**. Folded proteins (represented by blue lines) highlighting residues (black circles) in contact forming different classes of maximal cliques; a clique is a set of nodes (in the case of proteins, nodes are amino acid residues) wherein all nodes are in contact with each other; a maximal clique is that clique that is not part of another clique. (**A**) Maximal clique of class [4], where four sequence adjacent residues form a clique; these four residues are indicated on the top of the image connected by red lines (edges) and cannot be extended to form a clique of size five with another residue. (**B**) Maximal clique of class [2,2], where two sequence adjacent residues form a clique with other two adjacent residues; the four residues are not adjacent. (**C**) Maximal clique of class [1,2], where two sequence adjacent residues form a clique with a separated third residue. (**D**) Maximal clique of class [1,1,1] where three sequence non-adjacent residues form a clique. The regions separating these non-adjacent residues in the protein sequence could include different secondary structure elements (e.g., loops, helices).

## 2. Materials and Methods

### 2.1. RCC Calculation Implementation

The pseudo code to obtain the RCC from PDB entries is depicted in Algorithm 1:

---

**Algorithm 1: RCC calculation**

Input:     Cartesian coordinates of protein atoms
Output:    RCC coordinates

1:   Generate graph with the contacts of all residues. Two residues are in contact if they are within a distance threshold.
2:   Calculate maximal cliques using Tomita algorithm (see below)
3:   Calculate RCC from maximal cliques.

---

### 2.1.1. Contact Map Calculation

We accelerated the calculation of all contacts given a distance threshold $d$ with the use of several geometric restrictions:

(i) The hashing approach using a 3D grid over the space. We generated a grid over the 3D space of width $d$ that corresponded with the contact threshold distance; thus, given a residue we can identify the cube that contains it by using a hashing function; every cube in the grid has a list of all the residues that are inside of it. Given a residue $r$ and the cube in the grid where it is found, $G_r$, we know that any other residue inside $G_r$ must be in contact with $r$. Additionally, if a residue $s$ is in contact with $r$, it must be that the cubes $G_r$ and $G_s$ are neighbors in the grid (if they were not, then $r$ and $s$ must be at a distance greater than $d$). This way we reduced the search space dramatically (see Figure 2), which turns the complexity of the algorithm to $2{*}n$ ($n$ is the number of amino acids in a protein) in the linear portions of the protein and $16{*}n$ on average (given that at 5 Å a residue is in contact with four other residues on average).

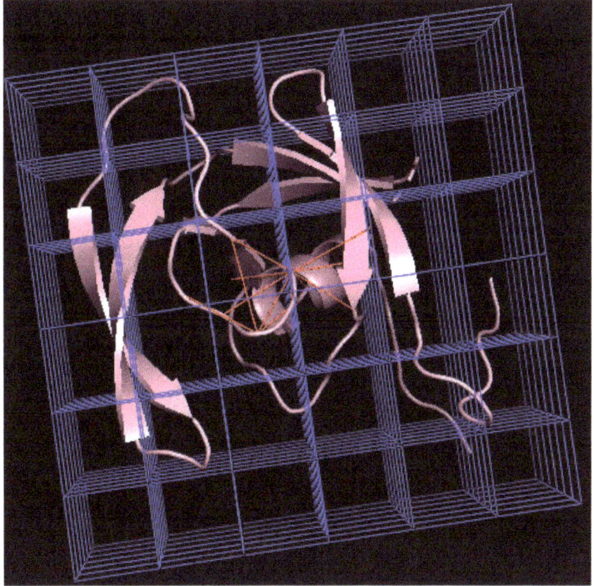

**Figure 2. Hash-like approach to build a contact map.** A grid is placed on top of the protein; its width is the contact distance. At the center, the residue of interest and the red lines connecting the residue of interest with the close by residues. Any residue in contact with the residue of interest must be in a neighboring cube; if it is inside the same cube, it must be in contact. The image was generated using CMView (version 1.1.1) [12], and the PyMol script DrawGridBox [13].

(ii) Rules for quick evaluation (see Figure 3). For a residue $r$ we enveloped it in a sphere of radius $r_d$ and center $r_c$. For two residues $r$ and $s$, let $D$ be the distance between $r_c$ and $s_{c'}$. They must be in contact if

$$D < d, \tag{1}$$

and cannot be in contact if

$$\text{minimum } \{D - r_d, D - s_d\} > d. \tag{2}$$

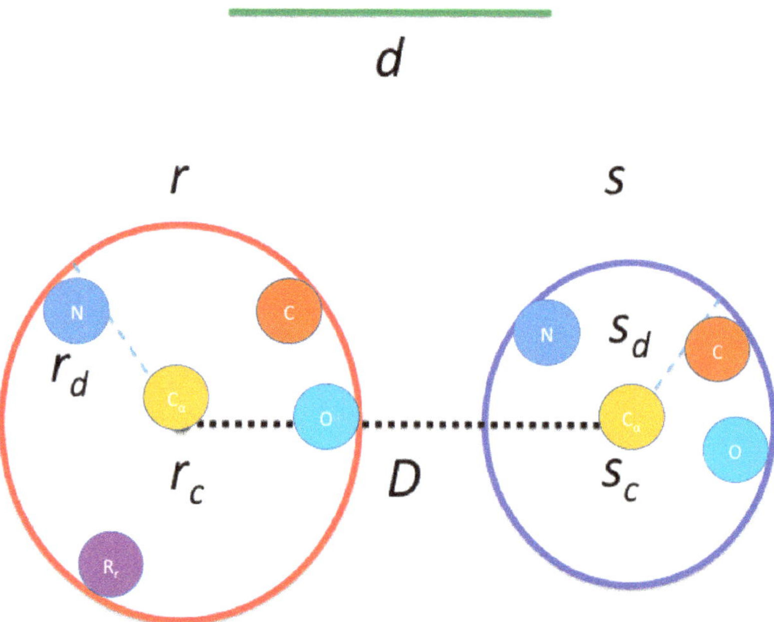


**Figure 3. Contact map algorithm.** $r$ and $s$ are two residues potentially in contact; each is enveloped in a sphere with center $r_c$ or $s_c$, and radius $r_d$ or $s_d$, respectively. The contact distance is $d$ and the distance between $r_c$ and $s_c$ is $D$. For each sphere, the center is calculated as the geometric center for all the atoms in the residue, including the side-chain atoms (the center does not necessarily overlap with any atom), and the radius is the distance between the center and the farthest atom, which may vary depending on the length and structure of the side chain. In the case in which the side chain is ignored, the spheres may still be not identical due to the elasticity of the bonds between the atoms (i.e., the distance between the nitrogen and carbon alpha atoms is not exactly constant, just like the internal angles of the backbone).

Finally, they must be in contact if a sphere of radius $d$ and center $r_c$ overlaps half the volume of the sphere enveloping $s$ or *vice versa* (though this last rule was turned off in the final implementation due to the overhead of the calculation being too similar to the time saved by it). Any pair of residues that do not fulfill one of these rules needs the explicit computation of the distance between its pair of atoms to be performed (with an early stop if a contact is found).

In the scenario in which the side chains are ignored, the spheres are calculated with the remaining atoms of the residue. The geometric restrictions work in the same way as if the spheres were calculated with all the atoms in the residue, and similarly, when a pair of residues does not meet the criteria for quick evaluation, the explicit computation of the distances between their pair of atoms is performed exclusively with the atoms in the backbone. The pseudo-code for this calculation is detailed in Algorithm 2:

---

**Algorithm 2: Contact map calculation**

---

    Input:   Cartesian coordinates of protein atoms
    Output:   Contact graph
    For each residue $r$ in the protein:
        Calculate its cube in the grid $G_r$ using the hash function.
        Get the list of residues $L$ inside the cubes neighboring $G_r$.
        For each residue $s$ in $L$:
            Calculate $D$ distance between $r_c$ and $s_c$.
            Quick inclusion if $D < d$ and continue.
            Quick exclusion if minimum $\{D{-}r_d, D{-}s_d\} > d$ and continue.
            Perform contact check for each pair of atoms from $r$ and $s$.

---

### 2.1.2. Maximal Cliques Calculation

Calculating the maximal cliques from a graph is a classical problem in computer science presented by Tomita [14], and the fastest implementation to our knowledge by Eppstein and Strash [15] was used in our method. To use said implementation, we converted the residue identifier to an index starting at 0; hence, the first residue in a 3D protein structure is labeled 0, the second 1 and so on.

### 2.1.3. RCC Calculation

We only considered maximal cliques containing at least three residues and at most six residues, which are grouped in 26 cluster classes. This grouping for a given protein produces different frequencies for each of these 26 clusters, wherein the first 3 frequencies correspond with maximal cliques with 3 residues ([1,1,1], [1,2] and [3]); the next 5 frequencies are maximal cliques with 4 residues ([1,1,1,1], [1,1,2], [2,2], [1,3] and [4]); the next 7 frequencies are maximal cliques containing 5 residues ([1,1,1,1,1], [1,1,1,2], [1,1,3], [1,4], [1,2,2], [2,3] and [5]); and the last 11 frequencies include maximal cliques with 6 residues ([1,1,1,1,1,1], [1,1,1,1,2], [1,1,1,3], [1,1,2,2], [1,1,4], [1,2,3], [1,5], [2,2,2], [2,4], [3,3] and [6]). The number of residues that are adjacent in the protein sequence define the class of a residue cluster. For instance, an RCC with 3 residues in which all residues are not adjacent in the protein sequence is referred to as [1,1,1]; an RCC with 4 residues in which 2 residues are adjacent in the protein sequence, and other 2 residues are also adjacent in the protein sequence (e.g., residues 45 and 46, and residues 101 and 102) is represented as [2,2] (see Figure 1). The regions separating these non-adjacent residues in the protein sequence could include different secondary structure elements (e.g., loops, helices). The maximum size of cliques is limited to 6 due to larger numbers being extremely rare at 5 Å and was kept as is in all the runs in order to keep the results comparable and avoid potential over-fitting.

### 2.2. RCC Database

For the 3D structural classification, we obtained the RCC for every protein domain reported in CATH (version 4.2.0) using 7 different distance cut-off values (5, 6, 7, 8, 9, 10 and 15 Å) and including or not the atoms of the amino acid side chain; hence, for each protein domain, 16 RCC representations were obtained. Our RCC dataset includes 354,079 different proteins (we noted that CATH 4.2 included 434,857 different domains, yet for several of these there was no PDB associated, so not all CATH domains are included in our dataset). A total of 2,831,584 different RCCs corresponding to those representations with amino acid side-chains or without side (see Supplementary Materials).

For the functional classification, we used the Gene Ontology (GO) database (version 2.0, generated on October 2017) and calculated the RCCs for every protein in the PDB database [16]. Similarly to the 3D structure, we used the 7 cut-off values and inclusion or exclusion of the side chains. By cross-referencing the 869,535 functions reported in GO (C: 176,437; F: 422,681; P: 270,416) with the 354,079 protein domains in the PDB, we obtained a database of 4,991,252 annotated proteins (C: 1,192,742; F: 2,2,85,509; P: 1,513,001) with their corresponding RCCs and known protein functions. The total number of annotated proteins is higher than the PDB protein domain because several proteins have multiple annotated functions, and a single protein sequence may have multiple chains in a single PDB file. To calculate the RCCs for all these PDB entries, we used our code and execute it on a 64-bit Intel-Xeon linux-based server with 24 cores and 256 GB of RAM.

*2.3. Model Training and Testing*

For identifying a model to classify protein structures, the full set of RCCs was labeled according to the annotated CATH classification in each of its 2 levels: 4 Classes (all alpha; all beta; a mixture of alpha and beta; or little secondary structure) and 41 Architectures. This full set was used to train models explored using the AutoWeka plugin [17], which performs an optimization over the ML models included in Weka (J48, DecisionTable, GaussianProcess, M5P, Kstar, LMT, PART, SMO, BayesNet, NaiveBayes, JRip, SimpleLogistic, LinearRegression, VotedPerceptron, SGD, Logistic, OneR, MultilayerPerceptron, REPTree, IBk, M5Rules, RandomForest, RandomTree and SMOreg; and the meta classifiers, which combine the previous models in different ways: Vote, Stacking, Bagging, RandomSubSpace, AttributeSelectedClassifier and RandomCommittee), their hyper-parameters and their select attributes (BestFirst, GreedyStepwise and CfsSubsetEval); AutoWeka was executed first for 20 minutes and identified as the best models those built using RCC at 7 or 8 Å with our without side chains. These same models were further analyzed under AutoWeka for 24 hours. In all these cases, random forest with hyper-parameters unlimited tree depth, no attribute selection and 100 iterations rendered the best results. The rest of the training datasets were run using this algorithm and hyper-parameters on a 64-bit Intel-Xeon linux-based server with 24 cores and 128 GB of RAM. Finally, 10-fold cross validation using the WEKA package was also performed to all these training sets [18].

The statistical parameters (accuracy, precision, correctly classified instances) reported by Weka and AutoWeka were used to evaluate the learning performance of the classifiers.

## 3. Results

The code implementing an efficient computation of residue cluster classes (RCCs) is freely (see Supplementary Materials); the computational efficiency of this implementation is reported in Table 1.

**Table 1.** Computer efficiency of our RCC implementation.

| PDB ID* | Length** | Reading*** (ms) | Graph Calculation (ms) | Maximal Cliques (ms) | Total Time (ms) |
|---------|----------|-----------------|------------------------|----------------------|-----------------|
| 1ORN (A) | 214 | 39 | 19 | 6 | 64 |
| 2HOX (A) | 425 | 110 | 37 | 6 | 153 |
| 3GVK (A) | 644 | 202 | 47 | 6 | 255 |
| 1F8N (A) | 818 | 329 | 67 | 6 | 402 |

* The letter in parentheses corresponds with the chain used for the listed PDB entries. ** Number of amino acid residues in the protein analyzed. *** Time taken in milliseconds (ms) reading the PDB file.

This computational efficiency allowed us to compute RCC values for all protein structures in the PDB database (354,079 protein domains [19]) in less than 1 hour (see Figure 4) on a 64-bit Intel-Xeon linux-based server with 24 cores and 128 GB of RAM. Considering that the PDB database includes about 10,000 new entries every year, which tend to include 200 residues each, this implementation would be able to compute these new entries in 11 minutes on average.

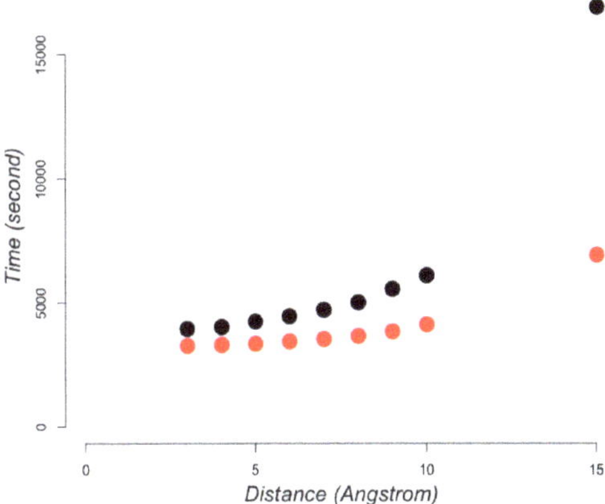


**Figure 4.** Time to compute RCC in PDB dataset. Computing 354,079 protein domains as annotated in CATH database including (black) or excluding (red) the atoms of the amino acid residue side-chains.

## 3.1. Protein Structural Classification

We built different RCC representations of protein structures to identify the best distance criterion to classify 3D protein structures. These included distances at 5, 6, 7, 8, 9, 10 and 15 Å; we originally used only a distance of 5 Å [11]; hence, this exercise allowed us to compare the efficiency of RCCs previously reported. In addition, we also included a variant in the construction of RCCs: the inclusion or exclusion of the amino acid side-chain atoms. This variation reproduces the preponderant role of the backbone in visual protein structure classification and consequently would allow for testing whether this representation is sufficient to learn 3D protein structure classification. Finally, we searched for the best model using an automatic approach based on the optimization algorithm implemented in AutoWeka, and to test the reliability of our model's classification, we conducted cross-validations; the labels of structural classes of proteins were derived from the CATH database (see Methods and Methods).

We observed that the best models were obtained using RCC representations with a contact distance threshold of 7 or 8 Å without side chains (see Figure 5). Table 2 summarizes the best model performance compared with previous results by Corral and collaborators.

**Table 2.** Parameters for best 3D protein classification.

| CATH Level | Mean Cross-Validation Accuracy * (Corral et al) | Mean Cross-Validation Accuracy ** (Current) |
|:---:|:---:|:---:|
| C | 0.96 | 0.98 |
| A | 0.88 | 0.89 |

* This accuracy was reported using a random forest implementation in Sklearn in Python language. **This accuracy was obtained using a random forest implementation in Weka in Java language.

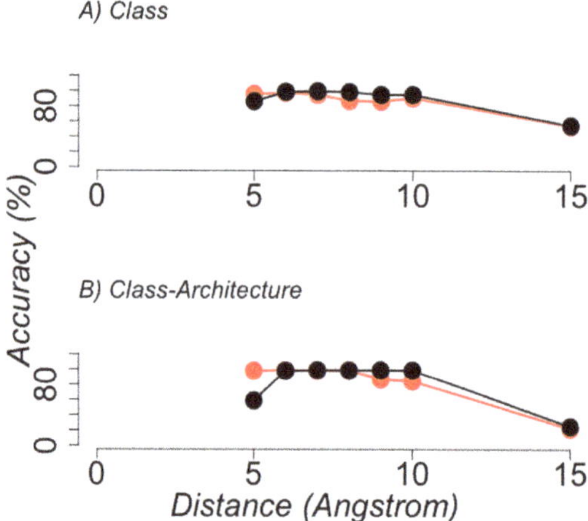


**Figure 5. Best classifier performance on the structural classification of proteins**. The average accuracy (y axis) achieved by the best classifiers after a 10-fold cross-validation test (see Methods and Methods) is shown for the different distance cutoff values used to build RCC (x axis), in the task of annotating the CATH structural classification of proteins. RCCs built using side-chain atoms are shown in red circles; RCCs built without side-chain atoms are shown in black circles. (**A**) Shows the performance when learning the class annotation from CATH classification, and (**B**) the class-architecture annotation from CATH classification.

### 3.2. Protein Functional Classification

We used the same protein structure representations described for structural classification for protein function classification; to include or not the side-chain atoms here represent an exploration of the relevance of side-chain contacts for protein function. For the functional annotation, we used the Gene Ontology (GO) annotations, consisting of three main classes: molecular function (F), biological process (P) and cellular localization (C). As described before, AutoWeka and cross-validation were used to identify the best models to learn GO functional annotation from RCC representations.

The best models were again observed with 7 or 8 Å of distance between the atoms of residues with or without side-chain atoms (see Figure 6). Table 3 summarizes the best model performance compared with the statistics reported for the second version of the Critical Assessment of protein Function Annotation contest (CAFA2) for the best models [20]. Fmax is the maximum harmonic representation of the precision and recall achieved by a set of models; Fmax = 1 is for a perfect predictor. It is important to note that the CAFA experiment attempts to unify the prediction of protein function, but does not use 3D protein structure; hence, 15% of all sequences included in CAFA2 were included in our datasets (data not shown). The best models in CAFA2 used sequence alignments and ML models that incorporated diverse proteins features, while our models only used RCCs—features derived exclusively from the protein structure. The purpose of this comparison is not to show better performance than CAFA2 models, but to note the level of reliability of our predictions in comparison with the function predictors known to be the best; that is, the best predictors in CAFA2 were close to Fmax = 0.5, so to our models. Hence, these results show that RCC achieved reliable classifications in both protein structure and protein function.

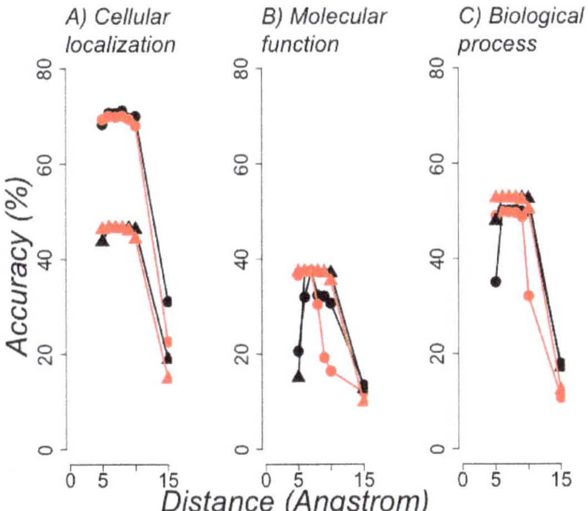



**Figure 6. Best functional classification of proteins.** The average accuracy (y axis) achieved by the best classifiers after a 10-fold cross-validation test (see Methods) is shown for the different distance cutoff values used to build RCCs (x axis), in the task of annotating the GO functional classifications of proteins. Results obtained with RCC built using side chains are in red and those without side chains in black; circle symbols indicate that functional annotation was done for all proteins; triangle symbols are for proteins with single domains. (**A**) Presents the results for predictions of cellular localization; (**B**) molecular function; (**C**) biological process, as annotated in GO.

**Table 3.** Parameters for best protein functional classification.

| GO Function | CAFA2* Fmax | Fmax** | Fmax*** |
|:---:|:---:|:---:|:---:|
| C | 0.46 | 0.44 | 0.58 |
| F | 0.59 | 0.24 | 0.48 |
| P | 0.37 | 0.41 | 0.54 |

* This accuracy was obtained from the reported Fmax values of the second version of CAFA [20]. ** This Fmax value corresponds to all proteins with an RCC computed in this study. *** This Fmax value considered only proteins with a single chain in PDB or single domain proteins.

An important aspect in functional annotation is the biased compositions of different classes; such a bias may cause predictors to classify proteins according to the most abundant functional classification. To rule out the possibility that such a bias may have affected the reliability of our predictions, we conduced a set of calculations using the ZeroR classifier in Weka; this classifier only predicts the most frequent class; hence, any machine-learning algorithm must achieve a performance better than this value to be reliable. In Figure 7 (A: cellular localization; B: molecular function and C: biological process) we show the percentages of correctly classified instances by the ZeroR classifier (square symbols) in comparison with those achieved by the best models (circle symbols) using RCCs built with different distance criteria. All the best models were above the baseline ZeroR predictions.

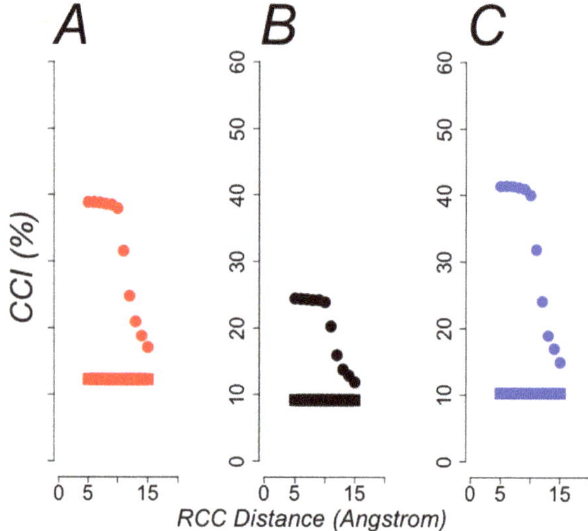



**Figure 7. Baseline performance for the best models**. The percentage of correctly classified instances (CCI (%)) is plotted for every model built from RCCs without side chains at different distances (x axis). The baseline performance achieved by ZeroR classifier (square symbols) is compared with that observed for the best models (circle symbols). (**A**) The comparison for cellular localization (red symbols), (**B**) for molecular function (black symbols) and (**C**) for biological process (blue symbols).

## 4. Discussion

Including the efficient implementation of Tomita's algorithm into the RCC calculation rendered a constant time performance independent of the protein lengths analyzed; we also noted that the time required to compute the contacting distances and get the RCCs was less than reading the input file. In terms of computing time, assuming these calculations would be done on a single central processing unit or CPU, the global time required to compute a given RCC is divided in the time required to load the protein's 3D structure or PDB file (elapsed time) and the time to obtain the RCC (CPU time), which includes constructing the contact map of residues and identifying the maximal cliques. Our results indicate that most of the time is taken by reading the PDB file; this may be improved by implementing a non-serial reading function [21]. This in turn would benefit from the improvement of current CPU technologies [22]. Additionally, considering that computing RCCs from a large database is an embarrassingly parallel problem and our current code has been implemented to deal only with concurrent computing (local computing, as opposed to distributed computing), we anticipate that there is still room for improvement in terms of using a distributed computing scheme, such as the ACTOR formalism [23].

The latest Critical Assessment for Structure Prediction (CASP) competition for the first time showed that ML (AlphaFold) improved on previous methods using protein sequence features [24], yet the features used by AlphaFold are not the same used by the best models of protein function reported in CAFA contest [8]. Thus, RCC is the first representation of proteins that allows for the efficient modeling of both fundamental aspects of proteins.

Our screening to build RCCs reveals that the backbone contains enough information to represent both 3D structure and function. This does not imply that side chains are not relevant for 3D protein structure or function; after all the backbone conformation depends on the side chains. The relevance of building RCCs without side chains is that on the one hand, such a representation does not need high-resolution structures to build a useful model; this would be relevant to further exploration: what is the range of protein structure resolution that renders a reliable model for protein structural and functional classification? On the other hand, having a model that concerns only on the protein backbone may facilitate the development of methods to predict 3D protein structure based on RCCs.

Having the same representation of proteins to model 3D structure and function would allow one to analyze the possible co-localization of structural and functional classes in the 26-dimensional space of RCCs, for instance; this would eventually lead to a better understanding of the structure–function relationship of proteins. RCCs would allow exploring for regions in this 26-dimensional space where no examples of 3D protein structure or function are known, and potentially, designing new proteins.

In summary, in this work we distribute a computationally efficient implementation with which to compute RCCs from a 3D protein structure—a representation of proteins that allows for effective modeling of both 3D protein structure and functional classification.

**Supplementary Materials:** The code and datasets described in this work are available online at https://github.com/C3-Consensus/RCC.

**Author Contributions:** Conceptualization, G.D.R. and F.F.; methodology, G.D.R. and F.F.; software, G.D.R. and F.F.; validation, F.F.; resources, G.D.R.; data curation, F.F.; writing—original draft preparation, G.D.R. and F.F.; writing—review and editing, G.D.R. and F.F.; visualization, G.D.R.; supervision, G.D.R.; funding acquisition, G.D.R. All authors have read and agreed to the published version of the manuscript.

**Funding:** This research was funded by CONACyT (CB252316) and PAPIIT (IN208817).

**Acknowledgments:** To Maria Teresa Lara Ortiz for her technical assistance required for the development of this work.

**Conflicts of Interest:** The authors declare no conflict of interest. The funders had no role in the design of the study; in the collection, analyses, or interpretation of data; in the writing of the manuscript, or in the decision to publish the results.

## References

1. Baker, D. Protein Structure Prediction and Structural Genomics. *Science* **2001**, *294*, 93–96. [CrossRef] [PubMed]

2. Nagarajan, R.; Archana, A.; Thangakani, A.M.; Jemimah, S.; Velmurugan, D.; Gromiha, M.M. PDBparam: Online Resource for Computing Structural Parameters of Proteins. *Bioinform. Boil. Insights* **2016**, *10*, 73–80. [CrossRef] [PubMed]

3. Walker, J.M. *The Proteomics Protocols Handbook*; Humana Press, Inc.: Totowa, NJ, USA, 2005. [CrossRef]

4. Zhang, Y.; Wen, J.; Yau, S.S.-T. Phylogenetic analysis of protein sequences based on a novel k-mer natural vector method. *Genomics* **2019**, *111*, 1298–1305. [CrossRef] [PubMed]

5. Juan, D.; Pazos, F.; Valencia, A. Emerging methods in protein co-evolution. *Nat. Rev. Genet.* **2013**, *14*, 249–261. [CrossRef] [PubMed]

6. Sahraeian, S.M.; Luo, K.R.; Brenner, S.E. SIFTER search: A web server for accurate phylogeny-based protein function prediction. *Nucleic Acids Res.* **2015**, *43*, W141–W147. [CrossRef] [PubMed]

7. AlQuraishi, M. AlphaFold at CASP13. *Bioinformatics* **2019**, *35*, 4862–4865. [CrossRef]

8. Zhou, N.; Jiang, Y.; Bergquist, T.R.; Lee, A.J.; Kacsoh, B.Z.; Crocker, A.W.; Lewis, K.A.; Georghiou, G.; Nguyen, H.N.; Hamid, N.; et al. The CAFA challenge reports improved protein function prediction and new functional annotations for hundreds of genes through experimental screens. *Genome Boil.* **2019**, *20*, 1–23. [CrossRef] [PubMed]

9. Kulmanov, M.; Hoehndorf, R. DeepGOPlus: Improved protein function prediction from sequence. *Bioinformatics* **2019**, *36*, 422–429. [CrossRef] [PubMed]

10. Yang, J.; Yan, R.; Roy, A.; Xu, N.; Poisson, J.; Zhang, Y. The I-TASSER Suite: Protein structure and function prediction. *Nat. Methods* **2014**, *12*, 7–8. [CrossRef] [PubMed]

11. Corral, R.C.; Chávez, E.; Del Rio, G. Machine Learnable Fold Space Representation based on Residue Cluster Classes. *Comput. Boil. Chem.* **2015**, *59*, 1–7. [CrossRef] [PubMed]

12. Vehlow, C.; Stehr, H.; Winkelmann, M.; Duarte, J.M.; Petzold, L.; Dinse, J.; Lappe, M. CMView: Interactive contact map visualization and analysis. *Bioinformatics* **2011**, *27*, 1573–1574. [CrossRef] [PubMed]

13. Geng, C. DrawGridBox-PyMOLWiki. 2016. Available online: https://pymolwiki.org/index.php/DrawGridBox (accessed on 26 February 2020).

14. Tomita, E.; Tanaka, A.; Takahashi, H. The worst-case time complexity for generating all maximal cliques and computational experiments. *Theor. Comput. Sci.* **2006**, *363*, 28–42. [CrossRef]

15. Eppstein, D.; Löffler, M.; Strash, D. Listing All Maximal Cliques in Sparse Graphs in Near-Optimal Time. In *Computer Vision*; Springer Science and Business Media LLC: Berlin, Germany, 2010; Volume 6506, pp. 403–414.

16. Berman, H.M. The Protein Data Bank. *Nucleic Acids Res.* **2000**, *28*, 235–242. [CrossRef] [PubMed]

17. Kotthoff, L.; Thornton, C.; Hoos, H.H.; Hutter, F.; Leyton-Brown, K. Auto-WEKA: Automatic Model Selection and Hyperparameter Optimization in WEKA. *The NIPS '17 Competition Build. Intell. Syst.* **2019**, *18*, 81–95.

18. Hall, M.; Frank, E.; Holmes, G.; Pfahringer, B.; Reutemann, P.; Witten, I.H. The WEKA data mining software. *ACM SIGKDD Explor. Newsl.* **2009**, *11*, 10–18. [CrossRef]

19. Burley, S.K.; Berman, H.M.; Bhikadiya, C.; Bi, C.; Chen, L.; Di Costanzo, L.; Christie, C.; Dalenberg, K.; Duarte, J.M.; Dutta, S.; et al. RCSB Protein Data Bank: Biological macromolecular structures enabling research and education in fundamental biology, biomedicine, biotechnology and energy. *Nucleic Acids Res.* **2018**, *47*, D464–D474. [CrossRef] [PubMed]

20. Jiang, Y.; Oron, T.R.; Clark, W.T.; Bankapur, A.R.; D'Andrea, D.; Lepore, R.; Funk, C.S.; Kahanda, I.; Verspoor, K.M.; Ben-Hur, A.; et al. An expanded evaluation of protein function prediction methods shows an improvement in accuracy. *Genome Boil.* **2016**, *17*, 184. [CrossRef] [PubMed]

21. Ching; Choudhary; Liao, W.-K.; Ross; Gropp, W. Efficient structured data access in parallel file systems. In Proceedings of the IEEE International Conference on Cluster Computing CLUSTR-03, Hong Kong, China, 1–4 December 2003; pp. 326–335. [CrossRef]

22. Markov, I.L. Limits on fundamental limits to computation. *Nature* **2014**, *512*, 147–154. [CrossRef] [PubMed]

23. Hewitt, C. Actor Model of Computation. 2010. Available online: http://arxiv.org/abs/1008.1459http://carlhewitt.info (accessed on 3 February 2020).

24. Senior, A.W.; Evans, R.; Jumper, J.; Kirkpatrick, J.; Sifre, L.; Green, T.; Qin, C.; Žídek, A.; Nelson, A.W.R.; Bridgland, A.; et al. Improved protein structure prediction using potentials from deep learning. *Nature* **2020**, *577*, 706–710. [CrossRef] [PubMed]

*Article*

# A Two-Stage Mutual Information Based Bayesian Lasso Algorithm for Multi-Locus Genome-Wide Association Studies


**Hongping Guo [1,2], Zuguo Yu [1,3,*], Jiyuan An [4], Guosheng Han [1], Yuanlin Ma [1] and Runbin Tang [1]**

[1]  Key Laboratory of Intelligent Computing and Information Processing of Ministry of Education and Hunan Key Laboratory for Computation and Simulation in Science and Engineering, Xiangtan University, Xiangtan 411105, China; guohongping0501@163.com (H.G.); hangs@xtu.edu.cn (G.H.); 201590110068@smail.xtu.edu.cn (Y.M.); 201831510085@smail.xtu.edu.cn (R.T.)

[2]  School of Mathematics and Computer Science, Hanjiang Normal University, Shiyan 442000, China

[3]  School of Electrical Engineering and Computer Science, Queensland University of Technology, Brisbane, QLD 4001, Australia

[4]  Centre for Tropical Crops and Biocommodities, Queensland University of Technology, Brisbane, QLD 4001, Australia; j.an@qut.edu.au

*  Correspondence: yuzg@xtu.edu.cn


Received: 01 February 2020; Accepted: 10 March 2020; Published: 13 March 2020


**Abstract:** Genome-wide association study (GWAS) has turned out to be an essential technology for exploring the genetic mechanism of complex traits. To reduce the complexity of computation, it is well accepted to remove unrelated single nucleotide polymorphisms (SNPs) before GWAS, e.g., by using iterative sure independence screening expectation-maximization Bayesian Lasso (ISIS EM-BLASSO) method. In this work, a modified version of ISIS EM-BLASSO is proposed, which reduces the number of SNPs by a screening methodology based on Pearson correlation and mutual information, then estimates the effects via EM-Bayesian Lasso (EM-BLASSO), and finally detects the true quantitative trait nucleotides (QTNs) through likelihood ratio test. We call our method a two-stage mutual information based Bayesian Lasso (MBLASSO). Under three simulation scenarios, MBLASSO improves the statistical power and retains the higher effect estimation accuracy when comparing with three other algorithms. Moreover, MBLASSO performs best on model fitting, the accuracy of detected associations is the highest, and 21 genes can only be detected by MBLASSO in *Arabidopsis thaliana* datasets.

**Keywords:** GWAS; Pearson correlation; mutual information; feature screening; Bayesian Lasso


---

## 1. Introduction

Genome-wide association study (GWAS) has evolved to be an essential technology for exploring the genetic mechanism of complex traits [1]. It concentrates on identifying the significant single nucleotide polymorphisms (SNPs) associated with the given traits. In past years, several single-locus GWAS methods have been developed [1–5], and have detected a few variants among various traits successfully. However, they still have some drawbacks, such as the combined effects of multiple loci are ignored and the threshold in multiple test correction is hard to be determined [6].

To address these drawbacks, some classical high-dimensional statistical methods were well used in GWAS when the number of SNPs is not far more than that of individuals, such as the least absolute shrinkage and selector operator (Lasso) [7], the elastic net [8], and Bayesian Lasso [9,10]. However, the current situation is the opposite, because the number of SNPs is much larger than that of individuals. In the case of ultrahigh-dimensional data, the aforementioned methods will fail due to the internal computational complexity. Fortunately, Fan and Lv [11] proposed a two-stage feature screening (or variable selection) method. The main idea of this method is: The dimension of features are firstly cut down by sure independence screening (SIS), and then a certain popular high-dimensional feature screening method (such as Lasso, the smoothly clipped absolute deviation (SCAD) [12], or the adaptive Lasso [13]) is used to select significant features and estimate regression coefficients simultaneously. The extension of SIS is iterative sure independence screening (ISIS), which can revive those non-negligible features that are single uncorrelated while indirectly correlated to the respond variables [11]. Instead of the Pearson correlation based SIS, statisticians have exploited some other SIS methods from different measurements, such as rank correlation [14], the distance correlation [15], the partial correlation [16] and so on. Among these methods, Pearson correlation and distance correlation based screening have been applied in GWAS successfully [6,17], and some genes associated with crop quantitative traits such as rice salt-tolerance and poplar growth have been identified [18,19]. ISIS expectation-maximization Bayesian LASSO (ISIS EM-BLASSO) [6] selects potentially associated SNPs in single-objective screening methodology based on the Pearson correlation between the SNPs and phenotype. In reality, the intrinsic heterogeneity is likely to be present in big data [20], thus the characterization of correlations via multi-objective method can bring higher power [21]. Although two-side high-dimensional genome-wide association studies (2HiGWAS) [17] efficiently selects the associated SNPs by combining Pearson correlation and distance correlation, the computational burden of constructing distance correlation is very high.

Since mutual information can detect broader classes of relationships [22], and the computational complexity is relatively low [23]. We propose to modify the screening method in the first stage of ISIS EM-BLASSO to a multi-objective one, which is based on the combination of Pearson correlation and mutual information. Then EM-Bayesian Lasso (EM-BLASSO) [10] is applied to further select SNPs and estimate the effects by shrinking the weak effects to zero, and likelihood ratio test is used to identify the true quantitative trait nucleotides (QTNs), these procedures are the same as those in the second stage of ISIS EM-BLASSO (also denoteds EM-BLASSO). We call our method a two-stage mutual information based Bayesian Lasso (MBLASSO). In order to validate the effectiveness of our method, we compare it with three GWAS methods, EM-BLASSO [10], ISIS EM-BLASSO [6] and genome-wide efficient mixed model association (GEMMA) [5]. EM-BLASSO represents the single-stage GWAS method without pre-screening, ISIS EM-BLASSO is a typical two-stage GWAS method using only Pearson correlation screening, and GEMMA is a golden standard GWAS method widely used for comparison.

## 2. Materials and Methods

### 2.1. Statistical Framework

In this study, we consider the linear mixed genetic model [6] as follows:

$$\mathbf{y} = \mathbf{1}\mu + Q\alpha + X\beta + \varepsilon \tag{1}$$

where $\mathbf{y}$ is a $n \times 1$ phenotypic vector of quantitative trait, and $n$ is the number of individuals; $\mathbf{1}$ is a $n \times 1$ vector in which every element is equal to 1, and $\mu$ is the overall mean; $Q = (Q_1, Q_2, \ldots, Q_q)$ is a $n \times q$ matrix of fixed effects, such as the population structure, and $q$ is the number of fix effects; $\alpha$ is a $q \times 1$ vector of fixed effects; and $X$ is a $n \times p$ matrix of SNP genotype values. For each SNP, homozygous genotype are coded as 1, and $-1$, respectively, and the heterozygous ones are indicated by 0. $p$ is the number of presumed QTNs, $\beta$ is the QTN effects, and $\varepsilon \sim MVN_n(0, \sigma_e^2 I)$ is a $n \times 1$ vector of residual error.

### 2.2. Simulation Experiments

To assess the performance of methods, we considered simulation scenarios based on the *Arabidopsis thaliana* datasets consisting of 216,130 SNPs, 199 accessions, and 107 phenotype traits [24]. For genotype simulation, we randomly selected 10,000 SNPs, 2000 for each of the five chromosomes, i.e., 11,226,256–12,038,776 bp on Chr.1, 5,045,828–6,6412,875 bp on Chr.2, 1,916,588–3,196,442 bp on Chr.3, 2,232,796–3,3143,893 bp on Chr.4, and 19,999,868–21,039,406 bp on Chr.5. Additionally, we generated the phenotype simulation data with sample size 199 from three different scenarios, and undertook 1000 times for each simulation. Six QTNs were assumed to be genuine; their heritabilities were set as 0.10, 0.05, 0.05, 0.15, 0.05, and 0.05, respectively; and their allelic frequencies we are all nearly 0.30. Both the overall mean and residual variance we are set as 10.0, and the positions and effects of the six QTNs are shown in Tables S1–S3. The genotype and phenotype simulations were the same as those used by Wang et al. [25].

The first model (only six QTNs' additive effects) is: $\mathbf{y} = \mu + \sum_{i=1}^{6} x_i b_i + \varepsilon$, $\varepsilon \sim MVN_n(0, \sigma_e^2 \mathbf{I})$. The second model (six QTNs' additive effects plus polygenic effect) is: $\mathbf{y} = \mu + \sum_{i=1}^{6} x_i b_i + \mathbf{u} + \varepsilon$, $\mathbf{u} \sim MVN_n(0, \sigma_{pg}^2 \mathbf{K})$, $\varepsilon \sim MVN_n(0, \sigma_e^2 \mathbf{I})$, and $\mathbf{K}$ is the kinship matrix. Set $\sigma_{pg}^2 = 2$, thus $h_{pg}^2 = 0.092$. The third model (six QTNs' additive effects plus three other pairs of QTNs' epistatic effects) is: $\mathbf{y} = \mu + \sum_{i=1}^{6} x_i b_i + \sum_{j=1}^{3} (A_j \# B_j) b_{jj} + \varepsilon$, $\varepsilon \sim MVN_n(0, \sigma_e^2 \mathbf{I})$, # denotes Hadamard product (element-wise multiplication), three other pairs of epistatic QTNs (unrelated to the six true QTNs) are placed on 3063784bp (Chr.4) and 5227063bp (Chr.2), 5986135bp (Chr.2) and 2031781bp (Chr.3), and 2668059bp (Chr.3) and 11824678bp (Chr.1), respectively. Each pair of QTNs was set with $\sigma_{epi}^2 = 1.25$, thus $h_{epi}^2 = 0.05$.

*2.3. Real Data and Preprocessing*

We used four flowering-time related traits of *Arabidopsis thalina* datasets [24,26] for analysis. The four traits are days to flowering time under long days with vernalization (LDV), days to flowering time under short days with vernalization (SDV), days to flowering time under long days with two weeks vernalization (2W), and days to flowering time under long days with four weeks vernalization (4W), respectively. We removed the SNPs with minor allele frequency (MAF) less than 0.01, and 178376 SNPs remained ultimately. For phenotypes, we deleted the individuals with missing phenotype value, thus 168, 159, 152 and 119 individuals were reserved for each of the four traits LDV, SDV, 2W and 4W, respectively, and then a logarithmic transformation was performed to each phenotype value. Due to the strong population structure in *Arabidopsis thaliana*, we were obliged to eliminate the impact of population structure. We reorganized the SNP genotype data via the software PLINK (Version 1.09) [27] at first, then chose a suitable value for population number $q$ from 1 to 5 with the minimum cross-validation error, and calculated the population structure matrix $Q$ synchronously by using the software ADMIXTURE (Version 1.3) [28], and finally corrected the primary phenotype vector $\mathbf{y}$ by $Q_j, j = 1, 2, \ldots, q$, whose effects $\hat{\alpha}_j$ were estimated by least-square method. Therefore, the corrected phenotype vector is:

$$\mathbf{y}' = \mathbf{y} - \sum_{j=1}^{q} Q_j \hat{\alpha}_j = \mathbf{1}\mu + \sum_{i=1}^{p} X_i \beta_i + \varepsilon \tag{2}$$

*2.4. Mutual Information*

Mutual information proposed by Shannon [29] is based on the concept of entropy and has been widely used in feature selection [23]. Given two discrete random variables $X$ and $Y$, the mutual information of $X$ and $Y$ is defined as:

$$I(X; Y) = H(X) + H(Y) - H(X, Y) \tag{3}$$

where $H(X)$ is the entropy of $X$ and $H(X, Y)$ is the joint entropy of $X$ and $Y$. They can be specified as:

$$H(X) = -\sum_{x} p(x) \cdot \log p(x) \tag{4}$$

$$H(X, Y) = -\sum_{x,y} p(x, y) \cdot \log p(x, y) \tag{5}$$

where $p(x) = P(X = x)$ is the marginal probability density function, and $p(x, y) = P(X = x, Y = y)$ is the joint probability density function. Mutual information can also be defined as:

$$I(X, Y) = H(X) - H(X|Y) = H(Y) - H(Y|X) = \sum_{x,y} p(x, y) \cdot \log \frac{p(x, y)}{p(x)p(y)} \tag{6}$$

where $H(X|Y)$ is the conditional entropy of $X$ given $Y$. We calculated the mutual information by using the matlab package "MutualInfo" (Version 0.9) written by Peng et al. [23].

In fact, mutual information can be illustrated as the amount of information one random variable contained in another random variable. The larger the mutual information is, the stronger correlation between the two random variables is. In GWAS, we consider the phenotypic vector as one random variable, and the genotype vector of a SNP as another random variable. In this way, we can calculate the mutual information between each of the SNPs and phenotype.

*2.5. SCAD*

SCAD is a penalized likelihood approach that enables to selecting variables and estimating coefficients simultaneously due to its Oracle properties [12]. The objective function $\xi$ is:

$$\xi_{\lambda,\gamma}(\beta) = argmin_{\beta} \left( \sum_{i=1}^{n} (y_i - \sum_{j=1}^{p} (X_{ij}\beta_j))^2 + \sum_{j=1}^{p} \rho_{\lambda,\gamma}(|\beta_j|) \right) \tag{7}$$

where $\beta = (\beta_1, \beta_2, \ldots, \beta_p)^T$ is the regression coefficient vector to be estimated and, $\lambda$ and $\gamma$ are penalty and shrinkage parameter, respectively, both of them are greater than 0. The former term of Equation (7) is the loss function, and the latter term is the penalty function defined by:

$$\rho_{\lambda,\gamma}(\beta_j) = \begin{cases} \lambda|\beta_j|, & \text{if} \quad |\beta_j| < \lambda, \\ \frac{-(|\beta_j|^2 - 2\gamma\lambda|\beta_j| + \lambda^2)}{2(\gamma-1)}, & \text{if} \quad \lambda \leq |\beta_j| < \gamma\lambda \quad \text{and} \quad \gamma > 2, \\ \frac{(\gamma+1)\lambda^2}{2}, & \text{if} \quad |\beta_j| \geq \gamma\lambda. \end{cases} \tag{8}$$

$\gamma = 3.7$ as suggested in the original study [12]. We performed SCAD by using the R package "ncvreg" from https://CRAN.R-project.org/package=ncvreg.

*2.6. Likelihood Ratio Test*

Likelihood ratio test is to compare the maximum of likelihood function in null hypothesis $H_0$ and alternative hypothesis $H_1$, and further determine whether the hypothesis is effective. LOD (log of odds) score is a statistic criterion used in likelihood ratio test. The definition is:

$$LOD = log_{10}(\frac{l_0}{l_1}) = \frac{-2(L_0 - L_1)}{4.6052} \tag{9}$$

$l_0 = e^{L_0}$, $l_1 = e^{L_1}$, $L_0 = L(\theta_{-k})$ and $L_1 = L(\theta)$ are the natural logarithms of the likelihood functions for null hypothesis $H_0 : \beta_k = 0$ and alternative hypothesis $H_1$, respectively, $\theta_{-k} = \{\beta_1, \ldots, \beta_{k-1}, \beta_{k+1}, \ldots, \beta_o\}$ and $\theta = \{\beta_1, \ldots, \beta_o\}$, and $o$ is the number of markers potentially associated with the trait. $LOD \geq 3$ was proposed to be the significant criterion in multi-locus GWAS [25], which is slightly stringent and equivalent to $P = Pr(\chi_1^2 > 3 \times 4.6052) \approx 0.0002$. Under $H_0$, $LOD \times 4.6052$ follows a $\chi^2$ distribution with one degree of freedom. We set the significant criterion of MBLASSO, ISIS EM-BLASSO, and EM-BLASSO as $LOD \geq 3$, which is Bonferroni correction for GEMMA by referring the published study [30].

### 2.7. A Two-Stage Mutual Information Based Bayesian Lasso (MBLASSO) Method

On the whole, this procedure is a two-stage strategy for multi-locus GWAS. In the first phase, we used a modified ISIS approach based on Pearson correlation and mutual information to obtain a subset of SNPs, the elements of which can be divided into two types, separately. As to Pearson correlation screening, Type I includes those SNPs with strong correlated to phenotype, and Type II consists of those SNPs weak correlated while indirectly correlated to phenotype with some SNPs from Type I. For mutual information screening, Types I and II are similar as those in Pearson correlation screening. The first phase of our method can be considered to select SNPs from two different measurements. In the second phase, we adopted EM-BLASSO [10] to estimate the effects and select the SNPs with nonzero effect ($\geq 10^{-5}$) to further likelihood ratio test procedure. We call this method MBLASSO. The flow chart is shown in Figure 1.

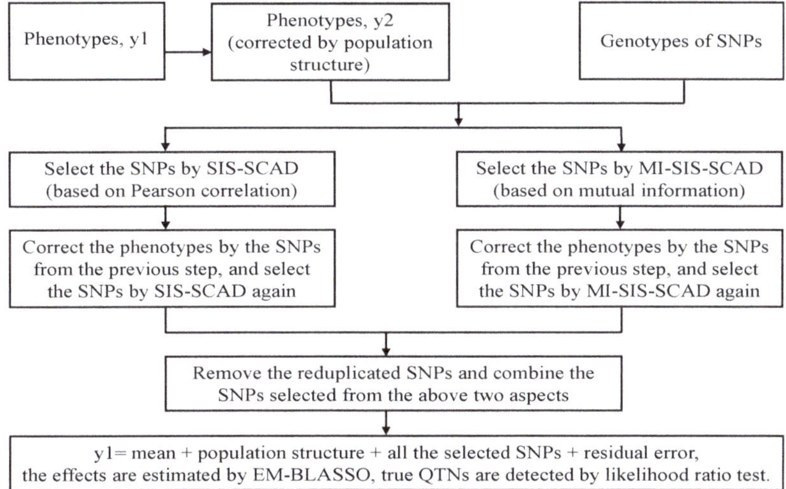

**Figure 1.** A flow chart of MBLASSO method.

More specifically, MBLASSO works as follows:

- Step 1: Correct the initial phenotype vector ($\mathbf{y}$) by the fixed effects, which indicate the population structure in our model.
- Step 2: Calculate the Pearson correlation of the $i$th SNP with the corrected phenotype ($\mathbf{y}'$), that is,

$$\omega_i = \rho_{X_i,y'} = \frac{\sum\limits_{j=1}^{n} (x_{ji} - \bar{x}_i)(y'_j - \bar{\mathbf{y}}')}{\sqrt{\sum\limits_{j=1}^{n} (x_{ji} - \bar{x}_i)^2} \cdot \sqrt{\sum\limits_{j=1}^{n} (y'_j - \bar{\mathbf{y}}')^2}} \tag{10}$$

where $x_{ji}$ is the $i$th SNP genotype value of the $j$th individual, $y'_j$ is the corrected phenotype value of the $j$th individual, $\bar{x}_i$ is the average of the genotype value of the $i$th SNP, $\bar{\mathbf{y}}'$ is the mean of the corrected phenotype value of all individuals, and $\omega = (\omega_1, \omega_2, \ldots, \omega_p)^T$ is a vector of Pearson correlation coefficients.

- Step 3: Sort the components of vector $\omega$ in descending order and define a subset:

$$\Omega = \{1 \leq i \leq p : |\omega_i| \text{ is among the } (n-1) \text{ largest of all}\} \tag{11}$$

where $n - 1$ is one of the two sizes recommended by Fan and Lv [11], and it is more appropriate in our work. Suppose that there are $k_1$ SNPs corresponding to $\Omega$, $k_1 \geq n - 1$, for the reason that more than one SNP may share a common Pearson correlation coefficient; the subset consisting of these SNPs is denoted as $\mathcal{A}_1 = \{X_{jm_1}, X_{jm_2}, \ldots, X_{jm_{k_1}}\}$, $m_1, m_2, \ldots, m_{k_1}$ are the orders of the $k_1$ selected SNPs in all the $p$ SNPs. Then implement SCAD to estimate the effects. Select the SNPs with nonzero effect to form another subset $\mathcal{A}_2 = \{X_{jl_1}, X_{jl_2}, \ldots, X_{jl_{k_2}}\} \subseteq \mathcal{A}_1$, $k_2 \leq k_1$, and $\{l_1, l_2, \ldots, l_{k_2}\} \subseteq \{m_1, m_2, \ldots, m_{k_1}\}$. The SNPs in $\mathcal{A}_2$ correspond to Type I in Pearson correlation screening. This Pearson correlation based SIS followed by SCAD is called SIS-SCAD [11].

- Step 4: Undertake ISIS-SCAD [11] to revive those non-negligible SNPs that are single uncorrelated but jointly correlated with phenotype, only one iteration is implemented here. Firstly correct the phenotype in Step 1 ($\mathbf{y}'$) by the $k_2$ SNPs selected by SIS-SCAD in Step 3, that is,

$$\mathbf{y}'' = \mathbf{y}' - \sum_{t=1}^{k_2} X_{l_t} \beta_{l_t} \tag{12}$$

where $\beta_{l_t}$ is estimated by SCAD, and then repeat SIS-SCAD to the rest of the $p - k_2$ SNPs, which results in another subset of $k_3$ SNPs, $\mathcal{A}_3 = \{X_{js_1}, X_{js_2}, \ldots, X_{js_{k_3}}\}$. The SNPs in $\mathcal{A}_3$ correspond to Type II in Pearson correlation screening. The union of the two disjoint subsets $\mathcal{A}_2$ and $\mathcal{A}_3$ is denoted as $\mathcal{A}$, $\mathcal{A} = \mathcal{A}_2 \bigcup \mathcal{A}_3$, whose size is $k$, $k = k_2 + k_3$.

- Step 5: Under the same conditions as in Step 2, calculate the mutual information of the $i$th SNP and the corrected phenotype ($\mathbf{y}'$) by

$$\psi_i = I(X_i, \mathbf{y}') = \sum_{j=1}^{n} p(x_{ji}, y_j') \cdot \log \frac{p(x_{ji}, y_j')}{p(x_{ji}) p(y_j')} \tag{13}$$

and $\psi = (\psi_1, \psi_2, \ldots, \psi_p)^T$ is a vector of mutual information for all of the $p$ SNPs with the corrected phenotype. $p(x_{ji}, y_j')$ is the joint probability, $p(x_{ji})$ and $p(y_j')$ are the marginal probabilities of $x_{ji}$ and $y_j'$, separately.

- Step 6: Similar to Step 3, sort the components of vector $\psi$ in descending order and define another subset:

$$\Psi = \{1 \leq i \leq p : \psi_i \text{ is among the } (n-1) \text{ largest of all}\} \tag{14}$$

Assume that $\tau_1$ SNPs corresponding to $\Psi$, $\tau_1 \geq n - 1$, because more than one SNP may share a public mutual information with phenotype. The subset is $\mathcal{B}_1 = \{X_{jh_1}, X_{jh_2}, \ldots, X_{jh_{\tau_1}}\}$. Then use SCAD to estimate the effects of SNPs in $\mathcal{B}_1$ and select the SNPs with nonzero effect to constitute a new subset $\mathcal{B}_2 = \{X_{jr_1}, X_{jr_2}, \ldots, X_{jr_{\tau_2}}\} \subseteq \mathcal{B}_1$, $\tau_2 \leq \tau_1$, and $\{r_1, r_2, \ldots, r_{\tau_2}\} \subseteq \{h_1, h_2, \ldots, h_{\tau_1}\}$. The SNPs in $\mathcal{B}_2$ correspond to Type I in mutual information screening. We call this mutual information based SIS followed by SCAD as MI-SIS-SCAD.

- Step 7: Refering to Step 4, correct the phenotype in Step 1 ($\mathbf{y}'$) by $\tau_2$ SNPs selected by MI-SIS-SCAD, and repeat MI-SIS-SCAD once for to the remaining of the $p - \tau_2$ SNPs, which generates a subset of $\tau_3$ SNPs, $\mathcal{B}_3 = \{X_{jt_1}, X_{jt_2}, \ldots, X_{jt_{\tau_3}}\}$. The SNPs in $\mathcal{B}_3$ correspond to Type II in mutual information screening. The union of the disjoint subsets $\mathcal{B}_2$ and $\mathcal{B}_3$ is denoted as $\mathcal{B}$, $\mathcal{B} = \mathcal{B}_2 \bigcup \mathcal{B}_3$, the size of which is $\tau$, $\tau = \tau_2 + \tau_3$. We call this process as MI-ISIS-SCAD.

- Step 8: Gather the SNPs selected from Steps 4 and 7 and remove the reduplicated ones. Then obtain a new subset of SNPs, that is, $\mathcal{C} = \mathcal{A} \bigcup \mathcal{B}$, the size of which is $\nu = k + \tau$.

- Step 9: Use EM-BLASSO to estimate the effect of the $\nu$ SNPs from $\mathcal{C}$ and further eliminate the SNPs with zero effect, the source code for EM-BLASSO can be found at https://CRAN.R-project. org/package=mrMLM, where we can also download the program of ISIS EM-BLASSO. Note that the phenotype vector in this step refers to the original one ($\mathbf{y}$).

- Step 10: Apply the likelihood ratio test to identify the true QTNs, and set the significant criterion as $LOD \geq 3$.

## 3. Results

### *3.1. The Overlap Ratio between Pearson Correlation and Mutual Information Based Screening in MBLASSO*

To illustrate the necessity of considering the correlation measured in mutual information between the SNPs and phenotype, we calculated the overlap ratio and average number of SNPs selected by Pearson correlation and mutual information in the first variable selection stage. The SNPs selected by Pearson correlation and mutual information can be divided into two types (Types I and Type II), respectively. We found that each type of screening obtains phenotype-associated SNPs without large overlapping (Table 1), which suggests that the SNPs from our MBLASSO method may have more associations with phenotype than ISIS EM-BLASSO.

**Table 1.** Screening results based on Pearson correlation and mutual information in MBLASSO under three simulation scenarios (each cell includes the overlap ratio and average number of SNPs after screening in the parentheses).

| Simulations | Pearson Correlation Screening | | | Mutual Information Screening | | |
|:---:|:---:|:---:|:---:|:---:|:---:|:---:|
| | Type I | Type II | Total | Type I | Type II | Total |
| 1 | 0.470 (15.8) | 0.086 (50.4) | 0.184 (66.2) | 0.417 (18.2) | 0.298 (15.5) | 0.356 (33.7) |
| 2 | 0.452 (16.6) | 0.091 (50.3) | 0.181 (66.9) | 0.398 (19.0) | 0.285 (17.5) | 0.334 (36.5) |
| 3 | 0.457 (14.6) | 0.090 (50.8) | 0.173 (65.4) | 0.383 (18.4) | 0.278 (17.4) | 0.323 (35.8) |

### *3.2. Statistical Power for QTN Detection*

The power for the $i$th QTN is: $power_i = \ell_i/1000, i = 1, 2, 3, 4, 5, 6$, where $\ell_i$ is the frequency that $i$th hypothetical QTN is successfully detected among all 1000 repetitions. A detected SNP within 1kb of the candidate QTN is regarded as true QTN [6,25,30]. In three simulations, powers of the six QTNs in MBLASSO are highest, except the second QTN powers are lower than those of EM-BLASSO (Figure 2a–c and Tables S1–S3). The average powers of MBLASSO are 72.4, 71.4, and 65.2 (%) in three simulations, respectively. They are improved by 26.4, 28.9, and 26.1 (%) compared to GEMMA; 5.6, 4.3, and 3.8 (%) compared to EM-BLASSO; and 2.2, 2.0, and 3.0 (%) compared to ISIS EM-BLASSO. We supposed four QTNs (QTN2, QTN3, QTN5, and QTN6) with the same 5% heritability, but the detection powers of QTN5 are much lower than three other values for MBLASSO, ISIS EM-BLASSO, and EM-BLASSO (Figure 2a–c and Tables S1–S3). To measure the robustness of methods, we used the standard deviation of powers across the four QTNs, which was proposed by Ren et al. [30]. In Simulation 1, the standard deviations for MBLASSO, ISIS EM-BLASSO and EM-BLASSO are 8.14, 8.16 and 13.99, respectively, indicating the best stability of MBLASSO. The stability comparisons in Simulations 2 and 3 are the same as that in Simulation 1. Therefore, MBLASSO improves the power and has best stability in different scenarios. A Violin plot of average statistical powers for MBLASSO in three simulation scenarios is shown in Figure S1a.

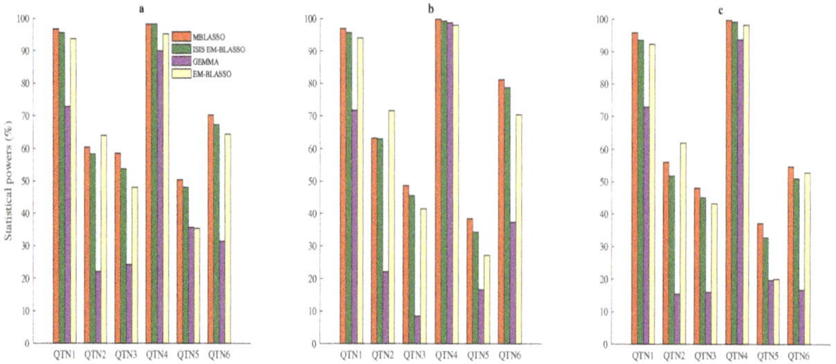

**Figure 2.** Statistical powers for the six simulated QTNs in three simulation scenarios. (**a**) only six QTNs' additive effects; (**b**) six QTNs' additive effects and polygenic background effect; and (**c**) six QTNs' additive effects and three other pairs of QTNs' epistatic effects.

### 3.3. Average Accuracy for QTN Effects

Mean squared error (MSE) was used to quantify the bias of effect estimation. The MSE of the *i*th QTN is: $MSE_i = \frac{1}{1000} \sum_{j=1}^{1000} (\hat{\beta}_{ij} - \beta_i)^2$, $i = 1, 2, 3, 4, 5, 6$, where $\hat{\beta}_{ij}$ is the effect of the *i*th QTN in the *j*th repetition, and $\beta_i$ is the theoretical effect of the *i*th QTN. The smaller the MSE is, the better the accuracy of the algorithm is. We applied the average MSE of the six QTNs to totally measure the accuracy of different algorithms. They are 0.0610, 0.0812, 0.5467 and 0.0561 for MBLASSO, ISIS EM-BLASSO, GEMMA and EM-BLASSO, respectively, in Simulation 1, the similar case is shown in Simulation 2, and the average MSE for MBLASSO is the lowest in Simulation 3 (Figure 3a–c and Tables S1–S3), indicating the better estimation accuracy of MBLASSO on the whole. A violin plot of average MSEs for MBLASSO in three simulation scenarios is shown in Figure S1b.

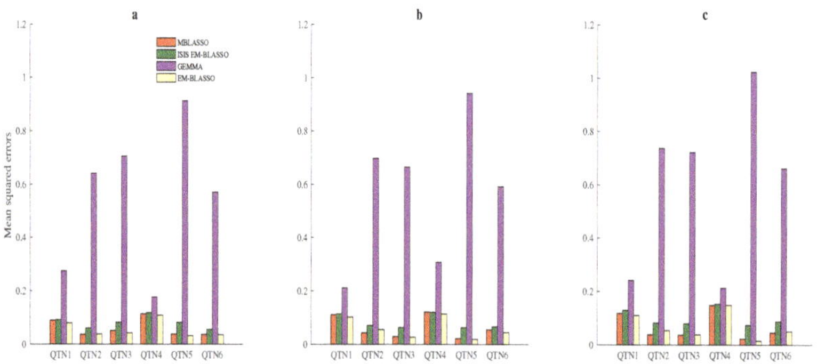

**Figure 3.** Average mean squared errors (MSEs) for the six simulated QTNs in three simulation scenarios. The description of (**a–c**) is the same as that in Figure 2.

### 3.4. Type 1 Error Ratio

Type 1 error ratio, also known as false positive ratio, is an important problem that needs to be overcome in GWAS. In Simulation 1, they are 0.0302%, 0.0325%, 0.0325% and 0.0259% for MBLASSO, ISIS EM-BLASSO, GEMMA, and EM-BLASSO, respectively, and GEMMA has the lowest Type 1 error ratio in Simulations 2 and 3 (Figure 4). Note that all Type 1 error ratios are less than 0.05% (Figure 4 and Tables S1–S3), which indicates that all the four algorithms ensure the Type 1 error is at a very low level. A violin plot of Type 1 error ratios for MBLASSO in three simulation scenarios is shown in Figure S1c.

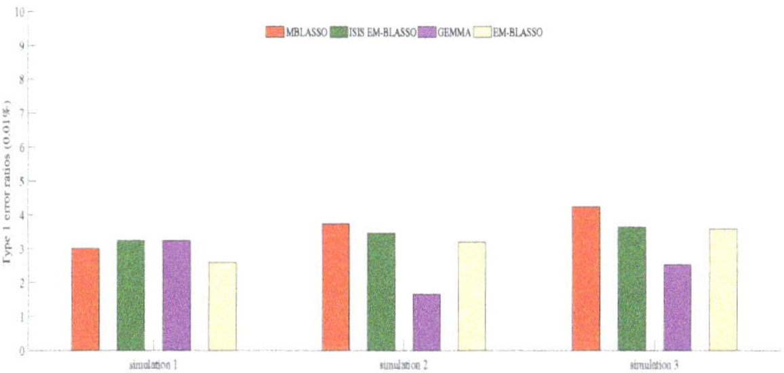


**Figure 4.** Type 1 error ratios (0.01%) in three simulation scenarios. The descriptions of Simulations 1–3 corresponding to (**a–c**) in Figure 2.

### 3.5. Computational Efficiency

The computing time of MBLASSO is longer than that of ISIS EM-BLASSO, because it needs additional computation of mutual information between all the SNPs and phenotype, but it takes less time than EM-BLASSO. For example, in Simulation 1, MBLASSO finishes the analysis of 199 individuals with 10,000 SNPs for 1000 repetitions in 4.12 h, ISIS EM-BLASSO takes 2.90 h, GEMMA spends 2.20 h, and while EM-BLASSO needs 28.86 h for the same dataset (Table S1). The specific hours spent on the other two simulations are largely identical with only minor differences to those in Simulation 1 (Tables S2 and S3), and the operations of computation are on a computer of Intel Xeon E5-2640 CPU 2.40 GHz.

### 3.6. Arabidopsis Thaliana Dataset Analysis

We analyzed four flowering-time related traits (LDV, SDV, 2W, and 4W) using by MBLASSO, ISIS EM-BLASSO, GEMMA and EM-BLASSO. Suppose that the candidate genes for the traits are in the proximity of 20 kb with the associated SNPs [6,25]; MBLASSO identifies 17, 18, 17 and 18 SNPs significant associated with each of the four traits LDV, SDV, 2W, and 4W, respectively. ISIS EM-BLASSO detects 14, 18, 19, and 16 remarkable associated SNPs; GEMMA identifies 3, 5, 1, and 2 significant SNPs; and EM-BLASSO tests 3, 0, 4, and 6 SNPs, respectively. A Venn diagram showings the overlap numbers of SNPs detected by the four algorithms in the four traits is presented in Figure S2.

To measure the model fitting degree of the detected SNPs, Akaike Information Criterion (AIC) and Bayesian Information Criterion (BIC) were computed for each trait in four various methods, where a lower value indicates a better model fitting. We can explicitly see that MBLASSO shows the lowest AIC and BIC for the four traits (Table 2), thus it is the best algorithm in model fitting, followed by ISIS EM-BLASSO, EM-BLASSO, and GEMMA.

**Table 2.** Degree of model fitting (AIC, BIC) for SNPs identified in four flowering-time related traits for *Arabidopsis thaliana.*

| Traits | MBLASSO | | ISIS EM-BLASSO | | GEMMA | | EM-BLASSO | |
|---|---|---|---|---|---|---|---|---|
| | AIC | BIC | AIC | BIC | AIC | BIC | AIC | BIC |
| LDV | −360.543 | −307.436 | −318.966 | −275.230 | 1312.693 | 1322.065 | −113.638 | −104.266 |
| SDV | −169.269 | −114.028 | −140.485 | −85.245 | 1356.907 | 1372.251 | 149.095 | 149.095 |
| 2W | −103.363 | −51.957 | −65.172 | −7.718 | 584.000 | 587.024 | 148.247 | 160.342 |
| 4W | −124.109 | −74.084 | −98.993 | −54.527 | 1253.281 | 1258.839 | 22.893 | 39.568 |

Meanwhile, by referring to the latest GO annotation [31] for *Arabidopsis thalina* genes at www.arabidopsis.org, we extracted the known genes related to flowering-time traits and found 5, 4, 2, and 3 known genes closed to the detected SNPs with MBLASSO; 3, 2, 1, and 2 known genes with ISIS

EM-BLASSO; 0, 1, 0, and 1 known genes with GEMMA; and none of known genes could be identified by using EM-BLASSO for LDV, SDV, 2W and 4W, respectively (Table 3). These results suggest that the accuracy of associations retrieved by MBLASSO are the highest.

**Table 3.** The accuracy of detected associations in four flowering-time related traits for *Arabidopsis thaliana* (the number behind slash in each cell is the count of detected SNPs, and the number in front of slash is the count of known genes in GO annotation),

| Traits | MBLASSO | ISIS EM-BLASSO | GEMMA | EM-BLASSO |
|--------|---------|----------------|-------|-----------|
| LDV | 5/17 | 3/14 | 0/3 | 0/3 |
| SDV | 4/18 | 2/18 | 1/5 | 0/0 |
| 2W | 2/17 | 1/19 | 0/1 | 0/4 |
| 4W | 3/18 | 2/16 | 1/2 | 0/6 |

In addition, totally 21 genes were only detected by MBLASSO, among which five genes (AT5G45830, AT5G45840, AT3G57230, AT5G15850, and AT5G04240) are in the 98 candidate genes [24], and AT5G45830 (alias: DOG1) is the gene with the highest frequency significant associated with flowering-time related phenotypes. Nearly all of the 23 flowering-time related phenotypes are associated with this gene [24]. Meanwhile, AT5G45840 (alias: MDIS1) is one of the Top 5 flowering-time related genes studied by researchers in *Arabidopsis thaliana* (www.arabidopsis.org). The detailed GWAS results are listed in Table S4.

About the computation speed, despite MBLASSO being is slower than ISIS EM-BLASSO and GEMMA, it is much faster than EM-BLASSO, for example, for the trait LDV, the time for MBLASSO is 2.31 min, ISIS EM-BLASSO requires 1.92 min, GEMMA takes 0.85 min and EM-BLASSO consumes to 183.6 min. We notice that the time costs of all the four flowering-time traits in MBLASSO, ISIS EM-BLASSO, and GEMMA are all less than 3 min (Table S5).

## 4. Discussion

MBLASSO is a GWAS method modified from ISIS EM-BLASSO, that is, iterative sure independence screening (ISIS) in the first stage of ISIS EM-BLASSO is replaced by a combination ISIS based on Pearson correlation and mutual information. We assume a subset of loci jointly affects the phenotype. In the first stage, we focus on selecting those SNPs that are likely to be highly associated. Considering some SNPs may have different correlations under various phenotypes, which are hard to measure only by Pearson correlation, so we adopt the mutual information to obtain the SNPs with potential correlation to phenotype. Meanwhile, since those SNPs individually irrelevant but jointly relevant to phenotype can be revived, this multi-objective screening process is a crucial component of our methodology to improve the statistical power. In the second stage, we apply the existing EM-BLASSO method [10], which is actually a single stage multi-locus GWAS strategy, to estimate the effects of selected SNPs and further filter out the SNPs with very small effect ($<10^{-5}$). Finally, we use likelihood ratio test to identify the true QTNs.

In fact, the method and criterion of hypothesis testing in different approaches may be different, e.g., the Wald test is applied in RMLM [25] and original EM-BLASSO [10], the significant level is $P = 0.01$ or 0.05, and a looser likelihood ratio test criterion $LOD \geq 2$ is employed in pLARmEB [32]. Since different significant criteria will lead to changes in results, for above three simulations, we listed the performances (average power, average MSE and Type 1 error ratio) of MBLASSO in three different significant criteria ($LOD = 3$, $LOD = 2$ and $P = 0.01$) in Table S6. We can see the average power increased with the decrease of $LOD$ value, but the Type 1 error ratio and average MSE also increased. This means that with the relaxation of significant criteria, high statistical power will be achieved, while false positives will be increased and estimation accuracy will be reduced. In addition, the performances at the significant criterion $P = 0.01$ in Wald test are between $LOD = 2$ and 3 in likelihood ratio test. GEMMA is a single-locus GWAS approach, and the significant threshold for each test is determined by Bonferroni correction ($0.05/p$, $p$ is the number of SNPs). MBLASSO, ISIS EM-BLASSO, and EM-BLASSO are multi-locus approaches and do not require multiple test correction.

We conducted paired t-test (also used in [6,25,30]) for statistical power and MSE between MBLASSO and three other methods in three simulation scenarios (Table S7). We can see it has significant improvements compared with ISIS EM-BLASSO and GEMMA. For the traits SDV and 2 W in real *Arabidopsis thaliana* datasets, the numbers of significant SNPs identified by MBLASSO are not more than ISIS EM-BLASSO, but the degrees of model fitting are better (Table 2); and the number of known candidate genes adjacent to the detected SNPs is still larger (Table 3), this phenomenon indicates MBLASSO may be more effective to capture the inherent relationship between SNPs and phenotype. The traditional EM-BLASSO [10] and GEMMA perform well in terms of Type 1 error ratio in the three simulations, but their performances in *Arabidopsis thaliana* dataset are worse than expected, not only achieving the worse model fitting performance but also fewer of genes are detected. On the whole, our algorithm MBLASSO is slightly slower than ISIS EM-BLASSO and GEMMA, but it is more effective and accurate for both simulation and real datasets.

## 5. Conclusions

Our algorithm MBLASSO is a modified version of ISIS EM-BLASSO; it integrates Pearson correlation and mutual information to the feature screening stage, and it considers different types of correlation between the SNPs and phenotype. In three different simulation scenarios, MBLASSO improves the statistical power and retains the higher effect estimation accuracy when comparing with three other methods. Meanwhile, the GWAS results in four flowering-time related traits are superior in model fitting; the accuracy of detected associations are the highest; and 21 genes can only be detected by MBLASSO.

**Supplementary Materials:** The following are available online at http://www.mdpi.com/1099-4300/22/3/329/s1.

**Author Contributions:** Conceptualization, H.G. and Z.Y.; methodology, H.G.; Writing—Original draft preparation, H.G.; Writing—Review and editing, H.G., Z.Y., J.A., G.H., Y.M. and R.T.; Supervision, Z.Y. All authors have read and agreed to the published version of the manuscript.

**Funding:** This research was funded by National Natural Science Foundation of China (Grant No. 11871061); Collaborative Research project for Overseas Scholars (including Hong Kong and Macau) of National Natural Science Foundation of China (Grant No. 61828203); Chinese Program for Changjiang Scholars and Innovative Research Team in University (PCSIRT)(Grant No. IRT_15R58); Hunan Provincial Innovation Foundation for Postgraduate (Grant No. CX2018B375); and The project for Excellent Young and Middle-aged Science and Technology Innovation Team of Hubei Province (Grant No. T201731).

**Conflicts of Interest:** The authors declare no conflict of interest.

## References

1. Yu, J.; Pressoir, G.; Briggs, W.H.; Bi, I.V.; Yamasaki, M.; Doebley, J.F.; McMullen, M.D.; Gaut, B.S.; Nielsen, D.M.; Holland, J.B. A unified mixed-model method for association mapping that accounts for multiple levels of relatedness. *Nat. Genet.* **2006**, *38*, 203–208. [CrossRef]

2. Kang, H.M.; Zaitlen, N.A.; Wade, C.M.; Kirby, A.; Heckerman, D.; Daly, M.J.; Eskin, E. Efficient control of population structure in model organism association mapping. *Genetics* **2008**, *178*, 1709–1723. [CrossRef] [PubMed]

3. Zhang, Z.; Ersoz, E.; Lai, C.Q.; Todhunter, R.J.; Tiwari, H.K.; Gore, M.A.; Bradbury, P.J.; Yu, J.; Arnett, D.K.; Ordovas, J.M.; et al. Mixed linear model approach adapted for genome-wide association studies. *Nat. Genet.* **2010**, *42*, 355–360. [CrossRef] [PubMed]

4. Lippert, C.; Listgarten, J.; Liu, Y.; Kadie, C.M.; Davidson, R.I.; Heckerman, D. FaST linear mixed models for genome-wide association studies. *Nat. Methods* **2011**, *8*, 833–835. [CrossRef]

5. Zhou, X.; Stephens, M. Genome-wide efficient mixed model analysis for association studies. *Nat. Genet.* **2012**, *44*, 821–824. [CrossRef]

6.  Tamba, C.L.; Ni, Y.L.; Zhang, Y.M. Iterative sure independence screening EM-Bayesian LASSO algorithm for multi-locus genome-wide association studies. *PLoS Comput. Biol.* **2017**, *13*, e1005357. [CrossRef]
7.  Wu, T.T.; Chen, Y.F.; Hastie, T.; Sobel, E.; Lange, K. Genome-wide association analysis by lasso penalized logistic regression. *Bioinformatics* **2009**, *25*, 714–721. [CrossRef]
8.  Cho, S.; Kim, H.; Oh, S.; Kim, K.; Taesung, P. Elastic-net regularization approaches for genome-wide association studies of rheumatoid arthritis. *BMC Proc.* **2009**, *3*, S25. [CrossRef]
9.  Li, J.; Das, K.; Fu, G.; Li, R.; Wu, R. The Bayesian lasso for genome-wide association studies. *Bioinformatics* **2011**, *27*, 516–523. [CrossRef]
10. Xu, S. An expectation-maximization algorithm for the Lasso estimation of quantitative trait locus effects. *Heredity* **2010**, *105*, 483–494. [CrossRef]
11. Fan, J.; Lv, J. Sure independence screening for ultrahigh dimensional feature space. *J. R. Stat. Soc. Ser. B* **2008**, *70*, 849–911. [CrossRef] [PubMed]
12. Fan, J.; Li, R. Variable selection via nonconcave penalized likelihood and its oracle properties. *J. Am. Stat. Assoc.* **2001**, *96*, 1348–1360. [CrossRef]
13. Zou, H. The adaptive Lasso and its oracle properties. *J. Am. Stat. Assoc.* **2006**, *101*, 1418–1429. [CrossRef]
14. Li, G.; Peng, H.; Zhang, J.; Zhu, L. Robust rank correlation based screening. *Ann. Stat.* **2012**, *40*, 1846–1877. [CrossRef]
15. Li, R.; Zhong, W.; Zhu, L. Feature screening via distance correlation learning. *J. Am. Stat. Assoc.* **2012**, *107*, 1129–1139. [CrossRef]
16. Li, R.; Liu, J.; Lou, L. Variable selection via partial correlation. *Statistica Sinica* **2017**, *27*, 983–996. [CrossRef]
17. Jiang, L.; Liu, J.; Zhu, X.; Ye, M.; Sun, L.; Lacaze, X.; Wu, R. 2HiGWAS: A unifying high-dimensional platform to infer the global genetic architecture of trait development. *Brief. Bioinform.* **2015**, *16*, 905–911. [CrossRef]
18. Cui, Y.; Zhang, F.; Zhou, Y. The application of multi-locus GWAS for the detection of salt-tolerance loci in rice. *Front. Plant Sci.* **2018**, *9*, 1464. [CrossRef]
19. Liu, J.; Ye, M.; Zhu, S.; Jiang, L.; Sang, M.; Gan, J.; Wang, Q.; Huang, M.; Wu, R. Two-stage identification of SNP effects on dynamic poplar growth. *Plant J.* **2018**, *93*, 286–296. [CrossRef]
20. Fan, J.; Han, F.; Liu, H. Challenges of big data analysis. *Nat. Sci. Rev.* **2014**, *1*, 293–314. [CrossRef]
21. Jing, P.J.; Shen, H.B. MACOED: A multi-objective ant colony optimization algorithm for SNP epistasis detection in genome-wide association studies. *Bioinformatics* **2015**, *31*, 634–641. [CrossRef] [PubMed]
22. Reshef, D.N.; Reshef, Y.A.; Finucane, H.K.; Grossman, S.R.; Mcvean, G.; Turnbaugh, P.J.; Lander, E.S.; Mitzenmacher, M.; Sabeti, P.C. Detecting novel associations in large data sets. *Science* **2011**, *334*, 1518–1524. [CrossRef] [PubMed]
23. Peng, H.; Long, F.; Ding, C. Feature selection based on mutual information: Criteria of max-dependency, max-relevance, and min-redundancy. *IEEE Trans. Pattern Anal. Mach. Intell.* **2005**, *27*, 1226–1238. [CrossRef] [PubMed]
24. Atwell, S.; Huang, Y.S.; Vilhjalmsson, B.J.; Willems, G.; Horton, M.; Li, Y.; Meng, D. Genome-wide association study of 107 phenotypes in *Arabidopsis* thaliana inbred lines. *Nature* **2010**, *465*, 627–631. [CrossRef] [PubMed]
25. Wang, S.B.; Feng, J.Y.; Ren, W.L.; Huang, B.; Zhou, L.; Wen, Y.J.; Zhang, J.; Dunwell, J.M.; Xu, S.; Zhang, Y.M. Improving power and accuracy of genome-wide association studies via a multi-locus mixed linear model methodology. *Sci. Rep.* **2016**, *6*, 19444. [CrossRef] [PubMed]
26. Togninalli, M.; Seren, Ü.; Freudenthal, J.A.; Monroe, J.G.; Meng, D.; Nordborg, M.; Weigel, D.; Borgwardt, K.; Korte, A.; Grimm, D.G. AraPheno and the AraGWAS Catalog 2020: A major database update including RNA-Seq and knockout mutation data for *Arabidopsis thaliana*. *Nucleic Acids Res.* **2019**, *48*, D1063–D1068. [CrossRef]
27. Purcell, S.; Neale, B.; Todd-Brown, K.; Thomas, L.; Ferreira, M.A.R.; Bender, D.; Maller, J.; Sklar, P.; Bakker, P.I.W.D.; Daly, M.J. PLINK: A tool set for whole-genome association and population-based linkage analyses. *Am. J. Hum. Genet.* **2007**, *81*, 559–575. [CrossRef]
28. Alexander, D.H.; Novembre, J.; Lange, K. Fast model-based estimation of ancestry in unrelated individuals. *Genome Res.* **2009**, *19*, 1655–1664. [CrossRef]
29. Shannon, C.E. A mathematical theory of communication. *Bell Syst. Tech. J.* **1948**, *27*, 379–423. [CrossRef]

30. Ren, W.L.; Wen, Y.J.; Dunwell, J.M.; Zhang, Y.M. pKWmEB: Integration of Kruskal-Wallis test with empirical Bayes under polygenic background control for multi-locus genome-wide association study. *Heredity* **2018**, *120*, 208–218. [CrossRef]

31. Berardini, T.Z.; Mundodi, S.; Reiser, L.; Huala, E.; Garcia-Hernandez, M.; Zhang, P.; Mueller, L.A.; Yoon, J.; Doyle, A.; Lander, G.; et al. Functional annotation of the Arabidopsis genome using controlled vocabularies. *Plant Physiol.* **2004**, *135*, 745–755. [CrossRef] [PubMed]

32. Zhang, J.; Feng, J.Y.; Ni, Y.L.; Wen, Y.J.; Niu, Y.; Tamba, C.L.; Yue, C.; Song, Q.; Zhang, Y.M. pLARmEB: Integration of least angle regression with empirical Bayes for multilocus genome-wide association studies. *Heredity* **2017**, *118*, 517–524. [CrossRef] [PubMed]

*Article*

# Phylogenetic Analysis of HIV-1 Genomes Based on the Position-Weighted K-mers Method

Yuanlin Ma [1,2], Zuguo Yu [1,3,*], Runbin Tang [1], Xianhua Xie [4], Guosheng Han [1] and Vo V. Anh [1,5]

[1]   Hunan Key Laboratory for Computation and Simulation in Science and Engineering and Key Laboratory of Intelligent Computing and Information Processing of Ministry of Education, Xiangtan University, Xiangtan 411105, China; 201590110068@smail.xtu.edu.cn (Y.M.); 201831510085@smail.xtu.edu.cn (R.T.); hangs@xtu.edu.cn (G.H.); vanh@swin.edu.au (V.V.A.)
[2]   School of Economics, Zhengzhou University of Aeronautics, Zhengzhou 450046, China
[3]   School of Electrical Engineering and Computer Science, Queensland University of Technology, GPO Box 2434, Brisbane, QLD 4001, Australia
[4]   School of Mathematical and Computer Science, Gannan Normal University, Ganzhou 341000, China; xiexianhua@gnnu.edu.cn
[5]   Faculty of Science, Engineering and Technology, Swinburne University of Technology, P.O. Box 218, Hawthorn, VIC 3122, Australia
*   Correspondence: yuzg@xtu.edu.cn

Received: 17 January 2020; Accepted: 20 February 2020; Published: 23 February 2020

**Abstract:** HIV-1 viruses, which are predominant in the family of HIV viruses, have strong pathogenicity and infectivity. They can evolve into many different variants in a very short time. In this study, we propose a new and effective alignment-free method for the phylogenetic analysis of HIV-1 viruses using complete genome sequences. Our method combines the position distribution information and the counts of the $k$-mers together. We also propose a metric to determine the optimal $k$ value. We name our method the *Position-Weighted k-mers (PWkmer)* method. Validation and comparison with the Robinson–Foulds distance method and the modified bootstrap method on a benchmark dataset show that our method is reliable for the phylogenetic analysis of HIV-1 viruses. *PWkmer* can resolve within-group variations for different known subtypes of Group M of HIV-1 viruses. This method is simple and computationally fast for whole genome phylogenetic analysis.

**Keywords:** Alignment-free; HIV-1 virus; phylogenetic analysis; position-weighted $k$-mers; Robinson–Foulds distance

---

## 1. Introduction

Human Immunodeficiency Viruses (HIVs) are retroviruses which are the causative agents of the global pandemic of Acquired Immunodeficiency Syndrome (AIDS) [1]. There are two types of HIVs: Type 1 (HIV-1 viruses) and Type 2 (HIV-2 viruses). HIV-1 viruses are known to originate from the Simian Immunodeficiency Viruses (SIVs) found in central and eastern African chimpanzees, which form the most common pathogenic strain of HIV viruses and have a high mortality rate [2]. Usually, HIV-1 viruses are divided into a major group (Group M) and two or more minor groups, namely Groups N, O, and possibly Group P. Group M is further divided into subtypes A, B, C, D, E, F, J, K. The subtypes A and F are further divided into sub-subtypes (A1, A2) and (F1, F2) based on differential phylogenetic clustering, respectively. Two or more HIV-1 subtypes can recombine and form Circulating Recombinant Forms (CRFs) [3]. Obviously, classification of HIV-1 strains into subtypes, sub-subtypes, and CRFs is a complex issue, which leads to major problems in the development of vaccines against HIV-1. These problems include high genetic variation, the fast evolution of different variants, and sequence diversity. The first task to solve these problems is how to obtain the phylogenetic relationships of HIV-1 genomes quickly and accurately. Traditional HIV-1 phylogenetic analysis methods are based on multiple sequence alignment. Although alignment-based methods generally yield excellent results when the sequences are closely related and can be reliably aligned, there are two limitations. Firstly, they lead to conflicting results by using different genes or genome fragments. Secondly, alignment-based methods are generally time-consuming and have high computational complexity when they are directly applied to whole-genome comparisons and phylogenetic studies [4]. Therefore, several alignment-free methods have been developed to overcome the critical limitations of alignment [5–14]. In particularly, several alignment-free methods for HIV genome comparison have been developed in the past few decades. For example, Wu et al. [5] used the complete composition vector representation proposed by Hao and Qi [15] for the phylogenetic analysis of HIV-1 genomes, and obtained some acceptable results. Pandit et al. [16] used multifractal measures to capture the genomic variation in the different retroviral species. However, this multifractal method cannot resolve the subtle variations in the subtypes of Group M of HIV-1 viruses. The first usage of $k$-mers (substring of length $k$) counts for biological sequence comparison was implemented by Blaisdell [17]. Subsequently, a lot of alignment-free methods using $k$-mers emerged. Yang and Wang [7] proposed a novel statistical measure for sequence comparison on the basis of $k$-mers counts, which removes the influence of the length of sequences, and obtained some acceptable results for the phylogenetic analysis of HIV-1 genomes. Chang et al. [8] proposed a cumulative Markov mutual information (CMMI) method which was derived from several $k$-mers distributions in different genome sequences, and reported some computational results on the HIV-1 subtyping. These results are slightly different from those reported in the NCBI (National Center for Biotechnology Information Search database). In addition, there are other alignment-free methods that may also be used for HIV-1 genome comparisons, such as the gene content-based method [18], the data compression method [19], the fractal method [20], the CVTree method [21], the inter-amino-acid distance method [10], the higher-order Markov model [11], the dynamical language model [6,12], a method using spaced-word frequencies [9], and a method based on the distribution of $k$-mer intervals [22]. All these alignment-free methods for comparing biological sequences are intended to extract hidden information from the whole genomes, but from different angles.

In this study, we present a new alignment-free method based on position-weighted $k$-mers to capture the subtle variations from the complete genome sequences of HIV-1 viruses. In our method, the effects of $k$-mers counts and $k$-mers position distributions are combined to capture more evolutionary information. On the basis of the proposed method, we report and discuss the results on the HIV-1 subtyping. More importantly, the resulting phylogenetic tree of 44 HIV genome sequences is quite consistent with the accepted taxonomy from NCBI. Our results show that the new method works as well as the conventional alignment-based phylogenetic methods and other alignment-free methods, but is simpler and requires much less computational time and resources. Moreover, our approach can be applied to study the subtype clustering and phylogenetic relationships of a large volume of genome sequences. The source codes of our method can be downloaded from https://github.com/myl446/HivStudy. The detailed information please see the Supplementary Materials.

## 2. Materials and Methods

### 2.1. Complete Genome Datasets

Twenty of the 21 genomes used in Chang et al. [8] are included in the 43 genomes used in Wu et al. [5]. For the phylogenetic analysis of HIV-1 complete genomes, we used a dataset which is composed of 44 HIV complete genomes (43 HIV complete genomic sequences used in the literature [5] and a misplaced sequence of the article categorization [8]). This dataset includes the subtypes A, B, C, D, F, G, J, K, H of the HIV-1 Groups M, O, and N, and a CPZ sequence. All of these sequences can be downloaded from the Los Alamos National Laboratory HIV Sequence Database (http://www.hiv.lanl.gov/). Specific accession, subtype, length (bp), and area are listed in Table 1.

Many studies suggested that all of the translated protein amino acid sequences from the genome is a better choice than whole genome DNA sequences and coding parts of complete genomes for genome-based phylogeny reconstruction [6,12,21,23]. However, after computational comparisons and theoretical analysis, we found that our present method is only suitable for whole genome DNA sequences.

**Table 1.** Labels of complete genome builds used for 44 HIV-1 genomes of the dataset.

| No. | Accession | Subtype | Length (bp) | Area |
|---|---|---|---|---|
| 1 | U51190 | A1 | 8999 | Uganda |
| 2 | AF004885 | A1 | 9160 | Kenya |
| 3 | AF069670 | A1 | 8813 | Somalia |
| 4 | AF484509 | A1 | 8807 | Uganda |
| 5 | AF286237 | A2 | 9060 | Cyprus |
| 6 | AF286238 | A2 | 8972 | DRC |
| 7 | AY173951 | B | 8996 | Thailand |
| 8 | AY331295 | B | 8834 | USA |
| 9 | AY423387 | B | 9359 | Netherlands |
| 10 | K03455 | B | 9719 | France |
| 11 | AF146728 | B | 8887 | Australia |
| 12 | AF067155 | C | 9002 | India |
| 13 | AY772699 | C | 9011 | South Africa |
| 14 | U46016 | C | 9031 | Ethopia |
| 15 | U52953 | C | 8959 | Brazil |
| 16 | AY371157 | D | 8379 | Cameroon |
| 17 | K03454 | D | 9176 | DRC |
| 18 | U88824 | D | 8952 | Uganda |
| 19 | AF005494 | F1 | 8968 | Brazil |
| 20 | AF075703 | F1 | 8925 | Finland |
| 21 | AF077336 | F1 | 8903 | Belgium (DRC) |
| 22 | AJ249238 | F1 | 8614 | France |
| 23 | AF377956 | F2 | 8782 | Cameroon |
| 24 | AJ249236 | F2 | 8555 | Cameroon |
| 25 | AJ249237 | F2 | 8589 | Cameroon |
| 26 | AY371158 | F2 | 8349 | Cameroon |
| 27 | AF061641 | G | 9047 | Finland(Kenya) |
| 28 | AF061642 | G | 9074 | Sweden (DRC) |
| 29 | AF084936 | G | 9707 | Belgium (DRC) |
| 30 | AF005496 | H | 8953 | Cent.Afr. Rep |
| 31 | AF190127 | H | 9056 | Belgium |
| 32 | AF190128 | H | 9707 | Belgium |
| 33 | AF082394 | J | 8943 | Sweden |
| 34 | AF082395 | J | 8953 | Sweden |
| 35 | AJ249235 | K | 8600 | DRC |
| 36 | AJ249239 | K | 8604 | Cameroon |
| 37 | AJ006022 | N | 9182 | Cameroon |
| 38 | AJ271370 | N | 9045 | Cameroon |
| 39 | AY532635 | N | 8938 | Cameroon |
| 40 | AJ302647 | O | 9829 | Senegal |
| 41 | AY169812 | O | 9110 | Cameroon |
| 42 | L20571 | O | 9793 | Cameroon |
| 43 | L20587 | O | 9754 | Cameroon |
| 44 | AF447763 | CPZ | 9326 | Tanzania |

DRC: Democratic republic of Congo

## 2.2. The Measure of Position-Weighted K-mers

Assume that $s_1 s_2 \ldots s_k$ is a $k$-mer, where $s_i \in \{A, T, C, G\}$. If the $k$-mer $s_1 s_2 \ldots s_k$ occurs in a given nucleic acid sequence $X$, then we denote by $P_{s_1 s_2 \ldots s_k}$ the vector composed of the positions of $s_1 s_2 \ldots s_k$ in

$X$ and by $P_{s_1 s_2 \ldots s_k}(i)$ its *ith* element. If $s_1 s_2 \ldots s_k$ does not exist in $X$, $P_{s_1 s_2 \ldots s_k}$ is a zero vector. For example, we consider the 2-mers position vectors for the following short nucleic acid sequence of length 20: $X = TAAGCCGCATTAGCTGGTTT$. We get $P_{AA} = (2), P_{GA} = (0), P_{GC} = (4, 7, 13) \ldots$. These $k$-mers position vectors can effectively capture the distribution information of each $k$-mer in the given sequence. For a fixed $k$, we can reverse this sequence by some $k$-mers position vector. Furthermore, if a $k$-mer exists in the given sequence, the counts of this $k$-mer in the nucleic acid sequence are equal to the length of its corresponding position vector. Therefore, we can use the following 2-mers position vector to reconstruct the nucleic acid sequence used in this example:

$$P_{AA} = (2), P_{AC} = (0), P_{AG} = (3, 12), P_{AT} = (9), P_{CA} = (0), P_{CC} = (5),$$
$$P_{CG} = (6), P_{CT} = (14), P_{GA} = (0), P_{GC} = (4, 7, 13), P_{GG} = (16),$$
$$P_{GT} = (17), P_{TA} = (1, 11), P_{TC} = (0), P_{TG} = (15), P_{TT} = (10, 18, 19).$$

The 2-mers $AC, CA, GA,$ and $TC$ do not appear in this example. Now, we reverse the given nucleic acid sequence as follows:

$$P_{TA} = (1, 11) \qquad TA \ldots TA \ldots$$
$$P_{AA} = (2) \qquad TAA \ldots TA \ldots$$
$$P_{AG} = (3, 12) \qquad TAAG \ldots TA.AG \ldots$$
$$P_{GC} = (4, 7, 13) \qquad TAAGC.GCTA.AGC \ldots$$
$$P_{CG} = (6) \qquad TAAGCGGCTA.AGC \ldots$$
$$P_{TA} = (1, 11) \qquad TAAGCGGCTATAGC \ldots$$
$$P_{TG} = (15) \qquad TAAGCGGCTATAGCTG \ldots$$
$$P_{GT} = (17) \qquad TAAGCGGCTATAGCTGGT \ldots$$
$$P_{TT} = (10, 18, 19) \qquad TAAGCGGCTATAGCTGGTTT.$$

Suppose

$$P_{s_1 s_2 \ldots s_k} = (p_1, p_2, \ldots, p_m),$$

where $m$ is the count of $s_1 s_2 \ldots s_k$ in the given nucleic acid sequence. The measure of $s_1 s_2 \ldots s_k$ based on its position in the sequence, denoted $f(s_1 s_2 \ldots s_k)$, is defined as

$$f(s_1 s_2 \ldots s_k) = \begin{cases} \dfrac{\left( \frac{p_1}{L} + \frac{p_2}{L} + \cdots + \frac{p_m}{L} \right)}{L - k + 1}, & m \neq 0, \\ 0, & m = 0, \end{cases}$$

where $L$ is the length of the given sequence.

After simplifying, the following form is obtained:

$$f(s_1 s_2 \ldots s_k) = \begin{cases} \dfrac{\sum\limits_{i=1}^{m} p_i}{(L - k + 1)L}, & m \neq 0, \\ 0, & m = 0. \end{cases} \tag{1}$$

To calculate the similarity distances between different sequences, we should assign a measure to each $k$-mer based on the $k$-mers position information. In this study, we use Formula (1) to extract evolutionary information from the nucleic acid sequence. As compared with the other $k$-mers-based methods, our method involves not only the counts of $s_1 s_2 \ldots s_k$, but also all the occurring positions of $s_1 s_2 \ldots s_k$. The method proposed here combines the position distribution information and the counts of the $k$-mers together, which can capture more phylogenetic information from sequences. For example, for two sequences $X_1 = CCAGTTGCCC, X_2 = CCCAGTTGCC$, the counts of CC in $X_1$ and $X_2$ are both 3. If we only consider the frequency of $CC$, $N_{X_1}(CC) = N_{X_2}(CC)$, the phylogenetic information of

CC captured by $N(CC)$ is not sufficient. However, when we use our measure $f(s_1 s_2 \ldots s_k)$, $f_{X_1}(CC) = 0.2$, $f_{X_2}(CC) = 0.133$. Hence, more phylogenetic information of CC can be captured by $f(CC)$.

## 2.3. Distance Calculations

There are a total of $4^k$ distinct $k$-mers for a fixed $k$. Sorting these $k$-mers in a fixed order, we can obtain a $4^k$-dimensional feature representation vector denoted by $(S_1, S_2, \ldots, S_{4^k})$. Then, according to the feature vector and our measure for $k$-mers, we obtain the corresponding vector $(f_1, f_2, \ldots, f_{4^k})$. For given $n$ nucleic acid sequences, we can get a $n \times 4^k$ feature matrix $F$ ($f_{i,j}$ represents the $jth$ feature of the sequence $i$, $i = 1, 2, \ldots, n$, $j = 1, 2, \ldots, 4^k$, $k$ is the length of $k$-mers):

$$\begin{bmatrix} f_{1,1} & f_{1,2} & \cdots & f_{1,4^k} \\ f_{2,1} & f_{2,2} & \cdots & f_{2,4^k} \\ \vdots & \vdots & \ddots & \vdots \\ f_{n,1} & f_{n,2} & \cdots & f_{n,4^k} \end{bmatrix} \tag{2}$$

There are many methods to calculate the distance between two vectors. In this paper, we use the Manhattan distance [24,25], which was commonly used to analyze similarity of biological sequences. Assuming that $Y = (f_{Y_1}, f_{Y_2}, \ldots, f_{Y_{4^k}})$ and $Z = (f_{Z_1}, f_{Z_2}, \ldots, f_{Z_{4^k}})$ represent the feature vectors of the two sequences calculated by our method, we use the following formula to calculate the Manhattan distance:

$$d(Y, Z) = \sum_{l=1}^{4^k} |f_{Y_l} - f_{Z_l}|. \tag{3}$$

For the experimental dataset, we can obtain the pairwise distance matrix based on the Manhattan distance. The distance matrix can depict the similarity information of the nucleic acid sequences. After generating the distance matrix, we use it as an input to the MEGA7 [26] and use the Neighbor-Joining (NJ) program [27] to generate the phylogenetic tree. We name this method the *Position-Weighted k-mers (PWkmer)* method.

## 2.4. Selection of the k Value

The $k$ value in our *PWkmer* method is very important to capture the subtle variation information of a genome sequence. Certainly, a larger value of $k$ will give a vector containing finer evolutionary information. However, many $k$-mers with large value of $k$ will not occur in the genome sequence. At the same time, some important information may be discarded and noise will dominate when a large value of $k$ is considered. In order to determine the optimal $k$ value, similar to the definition of the matrix in Shannon entropy by Zhao et al. [28], we consider a scoring scheme $score(k)$ to estimate the distribution of $k$-mers defined as

$$score(k) = -\frac{1}{n} \sum_{i=1}^{n} \sum_{j=1}^{4^k} f_{ij} \log f_{ij}. \tag{4}$$

Note that the larger $score(k)$ is, the more information can be extracted by the $k$-mers distribution.

The relation between some $score(k)$ and $k$ in our experiment using the dataset of HIV is given in Figure 1. It can be seen that $score(k)$ reaches the largest value at $k = 8$ and decreases after $k > 8$. This indicates that the difference between these genome sequences in 9-mers distributions is decreasing. At the same time, it will require a lot of memory to be computationally efficient when $k$ increases. Therefore, we determine $k = 8$ as the optimal value in our *PWkmer* method to distinguish these genome sequences.

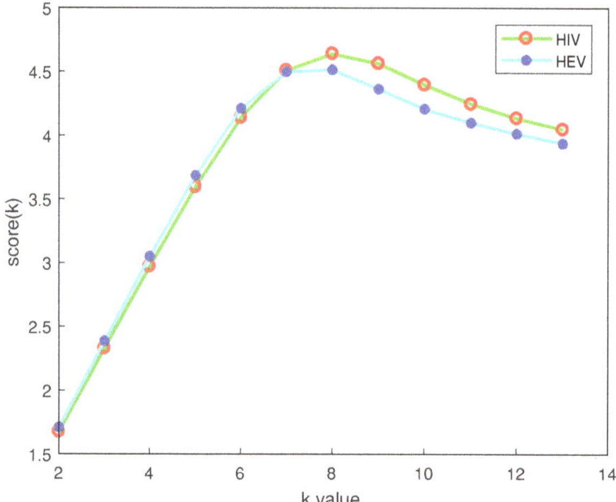

**Figure 1.** The trend chart of *k* value vs. scoring scheme *score(k)*. The red circles represent the score of the HIV dataset for different *k* values, and the blue dots represent the score of the HEV dataset for different *k* value.

*2.5. Accuracy Test of the Phylogenetic Tree Based on the Robinson–Foulds Distance and Robustness Test Using the Modified Bootstrap Method*

There are many methods to evaluate the accuracy of tree reconstruction methods. The Robinson–Foulds [29] metric is a way to measure the distance between unrooted phylogenetic trees. In this work, we use it to evaluate the accuracy of the trees we constructed. In general, subtyping of virus species is usually based on multiple sequence alignment in the field of virology. Therefore, we firstly find the reference tree of the species studied. Then, the Robinson–Foulds distance between our tree and the reference tree is implemented in the treedist program of the Phylip package [30]. The smaller the Robinson–Foulds distance is, the more accurate our tree is.

We also use the modified version of the bootstrap method proposed by Yu et al. [6] to evaluate the robustness of the trees we constructed. The workflow is as follows: first, we construct Matrix (2) with each row being the feature vector of each genome sequence. Second, we resample with repeats the $4^k$ columns to construct a new matrix. Third, we compute the Manhattan distances between any two row vectors based on the new matrix. Then, a distance matrix can be obtained based on the resampled matrix. Fourth, the same tree-building method is used to rebuild the tree. Finally, we repeat the above process a large number of times (usually 100 times). The frequency with which a particular phylogenetic branch emerges can be used as a measure of its reliability.

## 3. Results

*3.1. Subtyping of HIV-1 Based on PWkmer Feature for Complete Genome Sequences*

Using our *PWkmer* method, the phylogenetic analysis was performed on 44 HIV complete genome sequences listed in Table 1. We reconstructed the phylogenetic trees for $k = 2, 3, \ldots, 10$. The phylogenetic tree for $k = 8$ is the best among these trees, which agree with our theoretical optimal value for $k$. The obtained phylogenetic tree for $k = 8$ is shown in Figure 2. It is seen in Figure 2 that the strains from the same subtype are closely clustered together. Forty-four HIV genomes are distinctly divided into four groups: Group M is the main group of viruses in the HIV-1 global pandemic, and it contains multiple subtypes (A, B, C, D, F, G, H, J, K). Groups N and O are very distinctive forms of the viruses, which originate from other primates and then infect human beings. Group CPZ contains the closest non-human primate viruses related to HIV-1, which are the primate viruses isolated from chimpanzees. In this tree, all subtypes are clearly grouped together as distinct branches, and the closeness relationships among the subtypes are also well demonstrated. Namely, Subtypes B and D are closer to each other than to the others, and Subtype F(A) indeed contains two distinguishable Sub-subtypes F1 and F2 (A1 and A2). All these results are in very good agreement with those of previous studies [5,31].

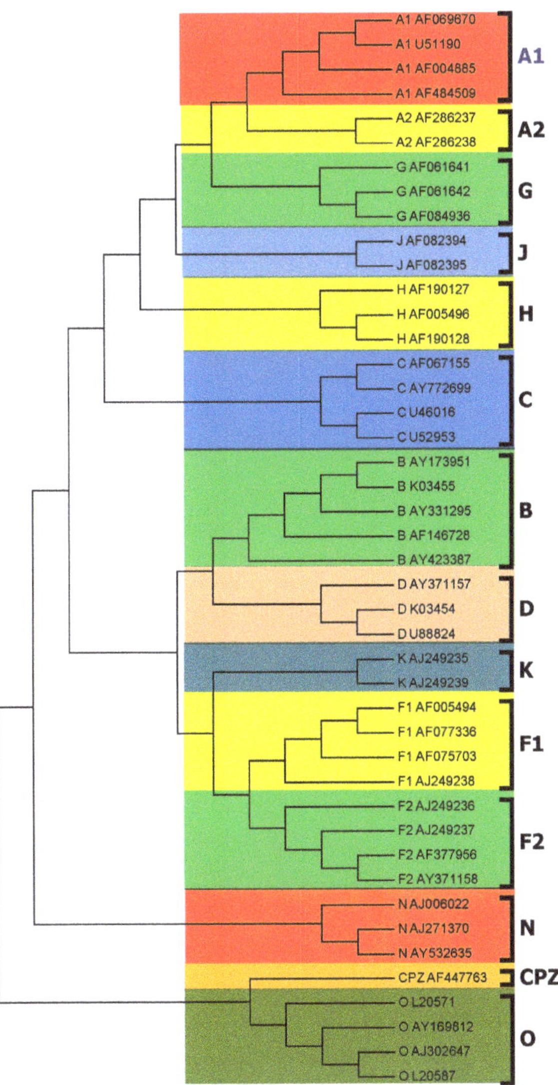



**Figure 2.** Subtyping of HIV based on position weighted $k$-mers feature for whole genome sequences. The Neighbor-Joining (NJ) tree of 44 HIV whole genomes is constructed by position weighted $k$-mers feature distance matrix ($k = 8$).

To verify the accuracy and reliability of the tree constructed by the *PWkmers* method, we used ClustalX [32], which is a multiple sequence alignment program, to construct a reference tree of 44

HIV complete genome sequences. As shown in Figure 3, this tree is quite consistent with the accepted taxonomy from NCBI. Moreover, we calculated the Robinson–Foulds distance between the tree constructed by the *PWkmer* method and the tree constructed by ClustalX. DLTree [12] and CVTree [21] are the more classical alignment-free methods in the publicized existing software of phylogenetic analysis. We also used them to construct the phylogenetic trees for 44 HIV complete genome sequences. At the same time, we computed the Robinson–Foulds distance between these trees constructed by the *PWkmers* method, CVTree [21], DLTree [12], and the tree constructed by ClustalX for 44 HIV complete genome sequences. The distances of the tree constructed by each method to the tree constructed by ClustalX are shown in Figure 4. The Robinson–Foulds distance of the *PWkmers* method is minimal, which illustrates that our results are the most closely consistent with the results of ClustalX.

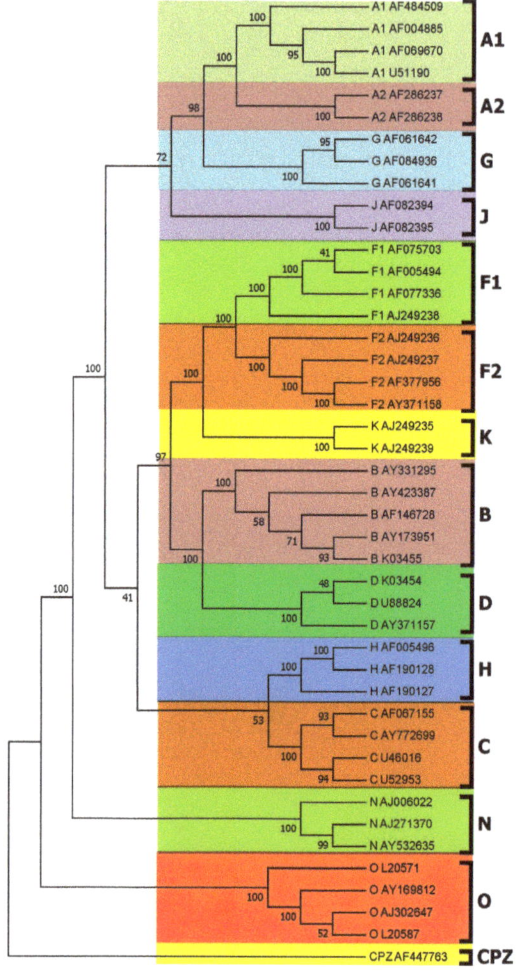

**Figure 3.** Subtyping of HIV based on alignment for whole genome sequences. The NJ tree of 44 HIV whole genomes is constructed by ClustalX.

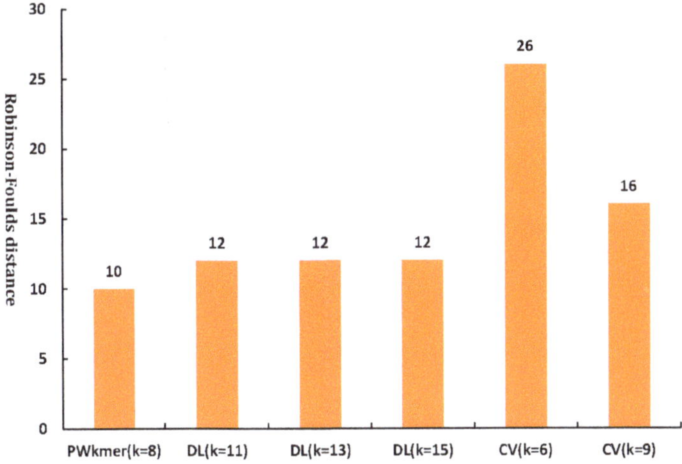

**Figure 4.** Robinson–Foulds distance between phylogenetic trees reconstructed by the *PWkmer* method, the CVTree method [20], the DLTree [12] method, and the tree reconstructed by ClustalX method for 44 HIV genome sequence in Table 1 (we selected their optimal result tree by CVTree and DLTree).

The modified bootstrap consensus tree for 44 HIV complete genome sequences is shown in Figure 5. As compared with Figure 2, the division of all HIV-1 genomes into Groups M, N, O, and CPZ is 100% supported. In Group M, each subtype branch is also 100% supported. In particular, in Subtype A and Subtype F of Group M, Sub-subtypes F1 and F2 (A1 and A2) are all 100% supported by the PWkmers. The branch of Subtype B and Subtype D is also supported by 100%. In Figure 2, Subtype C is divided into Group M, but in the consistent tree, as shown in Figure 5, Subtype C is divided out of Group M with a low supporting rate (44%).

We also compared the computational time required for our method in comparison to ClustalX [32] and DLTree [12]. On a modest PC (3.6 GHz quad core Intel Xeon processor, 4 GB RAM), for the whole genome sequences used in Table 1, it took 85 mins 54 secs for the alignment in ClustalX [32]. The DLTree model approach, which is a free-alignment method, used 20.3 secs of CPU time to get the distance matrices while the present *PWkmers* method only needs 5.8 secs of CPU time to get the distance matrices. This clearly shows the applicability of the *PWkmers* method for large datasets.

We also tested our method on three larger datasets: 867 HIV genomic sequences [5], 1625 HIV circulating recombinant form (CRF) genomic sequences, and 5596 pure subtype HIV genomic sequences from http://www.hiv.lanl.gov/ for $k = 8$, respectively. We put these three datasets on https://github.com/myl446/HivStudy. Our method on our PC only takes 70secs, 244secs, and 46mins 52 secs for each dataset, respectively. For the two datasets including 867 HIV genomic sequences and 5596 pure subtype HIV genomic sequences, all HIV-1 sequences from the same subtype are clustered together with 100% accuracy, while for the dataset including 1625 HIV CRF genomic sequences, the accuracy is 88.35%.

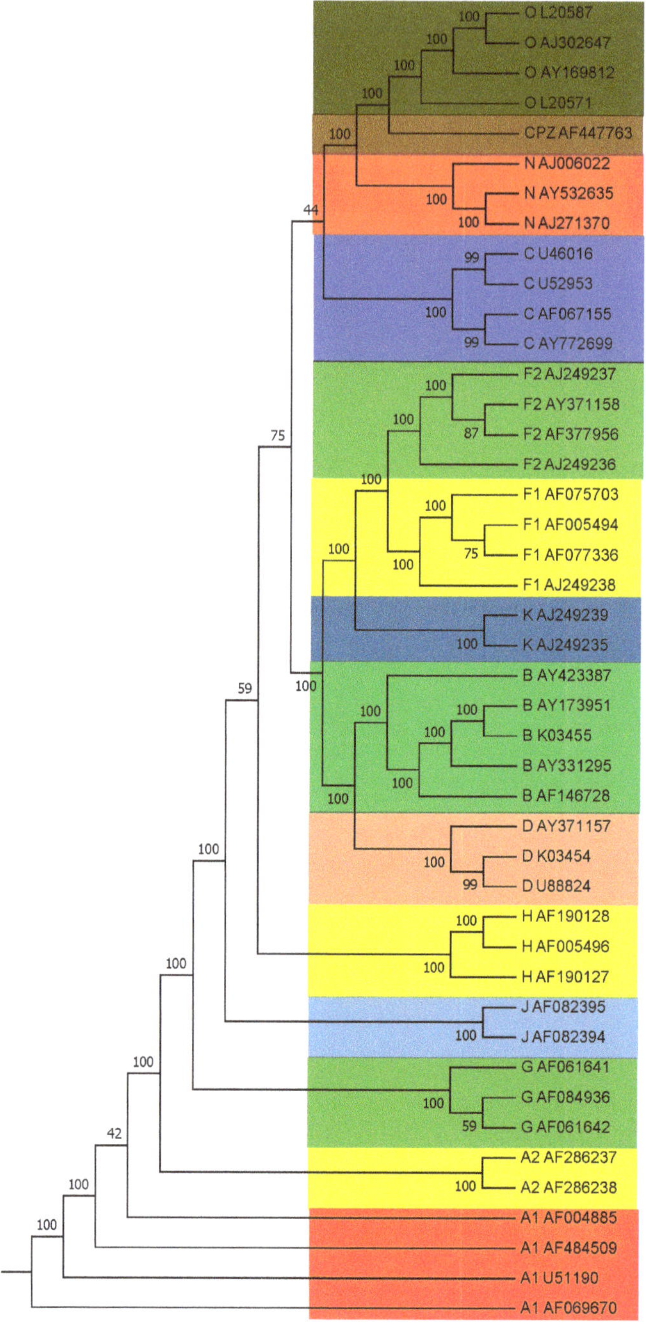

**Figure 5.** The modified bootstrap consensus tree for Figure 2 based on 100 replicates.

### 3.2. Application of Our Method on Other Datasets

We also used another benchmark dataset including 48 complete genome sequences used in previously published papers [7,8] to evaluate our *PWkmers* method. All these sequences can be downloaded from NCBI (https://www.ncbi.nlm.nih.gov/). Details of these sequences can be found in [7,8]. Hepatitis E is an inflammation of the liver caused by infection by the HEV (hepatitis E viruses). Hepatitis E is divided into four genotypes, and classification is based on the nucleotide sequences of the complete genome. Genotype 1 has been classified into five subtypes, Genotype 2 into two subtypes, and Genotypes 3 and 4 into ten and seven subtypes [33], respectively.

The tree constructed by our *PWkmers* method (not shown here) indicates that 48 HEV genomes are grouped into four branches. Genotype 1 includes Subtypes Ia, Ib, Ic, Id, and Ie. Genotype 2 contains only a complete HEV genome M1. Genotype 3 includes Subtypes IIIa, IIIb, and IIIc. Genotype 4 includes Subtypes IVa, IVb, and IVc. This shows that our results are consistent with the accepted trees [34,35] and the reference tree constructed by ClustalX.

On the HEV dataset, we also compared the computational time of our method with ClustalX [32] and DLTree [12]. For the whole genome used in 48 HEV sequences, it took 87 mins 34 secs on our computer for alignment in ClustalX [32]. The DLTree model approach used 25.7 secs of CPU time to get the distance matrices while the present *PWkmers* method only needs 6 seconds of CPU time to get the distance matrices.

## 4. Discussion

Subtype classification has always been a focus in the field of virology, especially in the classification of HIV-1 viruses. Because of the wide range of viruses, sequence diversity, and rapid evolution, the development of HIV-1 vaccines is facing enormous challenges. In this work, we propose a new method to solve the problem of HIV-1 classification.

In our *PWkmer* method, we combined the number and position distribution of $k$-mers, and sequence length to capture more sequence information than traditional methods. In fact, our method records the average position of $k$-mers on the sequence. Ding et al. [36] presented an alignment-free method based on the normalized $k$-mers average interval distance to capture evolutionary information for sequence comparison. They only extracted the number and position distribution of $k$-mers. Tang et al. [37] presented the normalized $k$-mers average relative distance to improve the method of Ding et al. [36]. Nevertheless, in their methods, the determination of the $k$ value requires empirical calculation, while we directly determine $k = 8$ by $score(k)$.

We computed the Robinson–Foulds distances between the phylogenetic trees reconstructed for different $k$ by our method and the reference tree reconstructed by ClustalX on our HIV-1 dataset, which are shown in Table 2. It can be seen from Table 2 that when $k = 8$, the Robinson–Foulds distances decrease to a lower value, which means that, with the further increase of $k$, the trees of HIV become unstable and its topological structures change little. From Figure 1 and Table 2, we can see that the relative change in the score value and the Robinson–Foulds distance is the same, which further implies the rationality of the score value defined by us. Furthermore, when $k = 8$, the distance between the tree constructed by the *PWkmers* method and that constructed by ClustalX is the minimum. Therefore, in the subtyping of HIV-1 viruses, we recommend the $k$ value of the string length to be 8.

**Table 2.** Robinson–Foulds distances between phylogenetic trees reconstructed by our method at $k = 2, 3, \ldots, 9, 10$ in Manhattan distance and the tree reconstructed by ClustalX on the HIV dataset.

| Species | $k = 2$ | $k = 3$ | $k = 4$ | $k = 5$ | $k = 6$ | $k = 7$ | $k = 8$ | $k = 9$ | $k = 10$ |
|---------|---------|---------|---------|---------|---------|---------|---------|---------|----------|
| HIV | 74 | 54 | 38 | 26 | 20 | 14 | **10** | 12 | 14 |

The HIV subtype classification method based on sequence comparison mainly relies on three gene coding proteins: gag, pol, and env. There are controversies about the spread and origin of $SIV_{CPZ}$. In this study, as can be seen from Figure 2, $SIV_{CPZ}$ is more closely related to group O, and after the bootstrap test, it has a 100% support rate, which is consistent with the classification results based on the proteins env and pol in the HIV database (http://www.hiv.lanl.gov/). However, in the HIV database, the classification results based on the protein gag are consistent with the classification results of ClustalX, and $SIV_{CPZ}$ is classified outside Groups N and O. As compared with the benchmark dataset used in many studies [5,7], we added a sequence (AF146728, Subtype B, HIV-1 isolated from Australia) which was obviously misclassified by Chang et al. [8]. In our method, we correctly grouped it in Subtype B and the cluster was 100% supported in the bootstrap test. In Pandit et al. [16], the authors concatenated the first and last of 10 sequences in the same subtype, and then classified them according to the fractal dimension. However, given a new sequence, this method cannot be used to determine which subtype it is attached to, or to which subtype it belongs. On the other hand, our method can directly calculate to determine which subtype or sub-subtype the new sequence belongs to. Our results show that the *PWkmer* method is useful and efficient.

## 5. Conclusions

The subtype classification of species in virology has always been a challenging problem. With the development of sequencing technology, more and more complete genome sequences become available. However, traditional sequence alignment tools and evolutionary models are not efficient in dealing with large-scale genome sequences. In this study, we proposed a new method to solve the problem of the subtype classification of HIV-1. Validation of the Robinson–Foulds distance method and the modified bootstrap method shows that the presented method is reliable for the phylogenetic analysis of HIV-1. At present, the common method for virus subtype classification is based on multi-sequence alignment. Compared with multi-sequence alignment, our method is fast and accurate, and can process large-scale data.

The selection of the $k$ value is very important. Specifically, if the $k$ value is too small, $k$-mers cannot capture the tiny differences in the genome of different strains; if the $k$ value is too large, it takes too much time and computer memory space for function $f$ of all $k$-mers. To determine the optimal $k$ value, we proposed a new method, which provides a quantitative index for its determination. We then found that the $k$ value is independent of the number of genome sequences in the dataset. In summary, our method can capture the $k$-mer distribution information and provide a fast tool for whole genome sequence comparison analysis. We hope that our method will be useful in the phylogenetic analysis of within-species variants using their complete genome sequences.

**Supplementary Materials:** The following are available online at http://www.mdpi.com/1099-4300/22/2/255/s1.

**Author Contributions:** Y.M. contributed to the conception and design of the study, developed the method and wrote the manuscript. Z.Y. gave the ideas and supervised the project. R.T. analyzed the data and results. All authors discussed the results and reviewed the manuscript. All authors have read and agreed to the published version of the manuscript.

**Funding:** This project was supported by the Natural Science Foundation of China (Grant No. 11871061); Collaborative Research project for Overseas Scholars (including Hong Kong and Macau) of National Natural Science Foundation of China (Grant No. 61828203); the Research Foundation of Education Commission of Hunan Province of China (Grant No. 17K090), the innovation project of Hunan Province of China (Grant No. Cx2016B252), the Science and Technology Project of Jiangxi Provincial Education Department (GJJ170820), and partially by the Australian Research Council Grant DP160101366.

**Conflicts of Interest:** The authors declare no conflict of interest.

## References

1.  Zachary, T.; Aboulafia, D.M. Review of screening guidelines for non-AIDS-defining malignancies: Evolving issues in the era of highly active antiretroviral therapy. *Aids Rev.* **2012**, *14*, 3–16.

2.  Lemey, P.; Pybus, O.G.; Rambaut, A.; Drummond, A.J.; Robertson, D.L.; Roques, P.; Worobey, M.; Vandamme, A.M. The molecular population genetics of HIV-1 group O. *Genetics* **2004**, *167*, 1059–1068. [CrossRef] [PubMed]

3.  Tebit, D.M.; Nankya, I.; Arts, E.J.; Gao, Y. HIV diversity, recombination and disease progression: How does fitness "fit" into the puzzle? *Aids Rev.* **2015**, *9*, 75–87.

4.  Herniou, E.A.; Luque, T.; Chen, X.; Vlak, J.M.; Winstanley, D.; Cory, J.S.; O'reilly, D.R. Use of whole genome sequence data to infer baculovirus phylogeny. *J. Virol.* **2001**, *75*, 8117–8126. [CrossRef]

5.  Wu, X.; Cai, Z.; Wan, X.F.; Hoang, T.; Goebel, R.; Lin, G. Nucleotide composition string selection in HIV-1 subtyping using whole genomes. *Bioinformatics* **2007**, *23*, 1744–1752. [CrossRef]

6.  Yu, Z.G.; Chu, K.H.; Li, C.P.; Vo, A.; Zhou, L.Q.; Wang, R.W. Whole-proteome phylogeny of large dsDNA viruses and parvoviruses through a composition vector method related to dynamical language model. *BMC Evol. Biol.* **2010**, *10*, 192. [CrossRef]

7.  Yang, X.; Wang, T. A novel statistical measure for sequence comparison on the basis of k-word counts. *J. Theor. Biol.* **2013**, *318*, 91–100. [CrossRef]

8.  Chang, G.; Wang, H.; Zhang, T. A novel alignment-free method for whole genome analysis: Application to HIV-1 subtyping and HEV genotyping. *Inf. Sci.* **2014**, *279*, 776–784. [CrossRef]

9.  Leimeister, C.A.; Boden, M.; Horwege, S.; Lindner, S.; Morgenstern, B. Fast alignment-free sequence comparison using spaced-word frequencies. *Bioinformatics* **2014**, *30*, 1991–1999. [CrossRef]

10. Xie, X.H.; Yu, Z.G.; Han, G.S.; Yang, W.F.; Anh, V. Whole-proteome based phylogenetic tree construction with inter-amino-acid distances and the conditional geometric distribution profiles. *Mol. Phylogenet. Evol.* **2015**, *89*, 37–45. [CrossRef] [PubMed]

11. Yang, W.F.; Yu, Z.G.; Anh, V. Whole genome/proteome based phylogeny reconstruction for prokaryotes using higher order Markov model and chaos game representation. *Mol. Phylogenet. Evol.* **2016**, *96*, 102–111. [CrossRef] [PubMed]

12. Wu, Q.; Yu, Z.G.; Yang, J. Dltree: Efficient and accurate phylogeny reconstruction using the dynamical language method. *Bioinformatics* **2017**, *33*, 2214–2215. [CrossRef] [PubMed]

13. Li, W.; Freudenberg, J. Alignment-free approaches for predicting novel Nuclear Mitochondrial Segments (NUMTs) in the human genome. *Gene.* **2019**, *691*, 141–152. [CrossRef] [PubMed]

14. Zielezinski, A.; Vinga, S.; Almeida, J.; Karlowski, W.M. Alignment-free sequence comparison: Benefits, applications, and tools. *Genome Biol.* **2017**, *18*, 186. [CrossRef] [PubMed]

15. Hao, B.; Qi, J. Prokaryote phylogeny without sequence alignment: From avoidance signature to composition distance. *J. Bioinf. Comput. Biol.* **2004**, *2*, pp. 1–19. [CrossRef] [PubMed]

16. Pandit, A.; Dasanna, A.K.; Sinha, S. Multifractal analysis of HIV-1 genomes. *Mol. Phylogenet. Evol.* **2012**, *62*, 756–763. [CrossRef]

17. Blaisdell, B.E. A measure of the similarity of sets of sequences not requiring sequence alignment. *Proc. Nat. Acad. Sci. USA* **1986**, *83*, 5155–5159. [CrossRef]

18. Snel, B.; Bork, P.; Huynen, M.A. Genomes in flux: The evolution of archaeal and proteobacterial gene content. *Genome Res.* **2002**, *12*, 17–25. [CrossRef]

19. Song, K.; Ren, J.; Reinert, G.; Deng, M.; Waterman, M.S.; Sun, F. New developments of alignment-free sequence comparison: Measures, statistics and next-generation sequencing. *Brief. Bioinf.* **2014**, *15*, 343–353. [CrossRef]

20. Yu, Z.G.; Anh, V.; Lau, K.S. Multifractal and correlation analyses of protein sequences from complete genomes. *Phys. Rev. E* **2003**, *68*, 021913. [CrossRef]

21. Zuo, G.; Hao, B. CVTree3 web server for whole-genome-based and alignment-free prokaryotic phylogeny and taxonomy. *Genom. Proteom. Bioinf.* **2015**, *13*, 321–331. [CrossRef] [PubMed]

22. Han, G.-B.; Cho, D.-H. Genome classification improvements based on k-mer intervals in sequences. *Genomics* **2019**, *111*, 1574–1582. [CrossRef] [PubMed]

23. Yu, Z.G.; Zhou, L.Q.; Anh, V.V.; Chu, K.H.; Long, S.C.; Deng, J.Q. Phylogeny of prokaryotes and chloroplasts revealed by a simple composition approach on all protein sequences from complete genomes without sequence alignment. *J. Mol. Evol.* **2005**, *60*, 538–545. [CrossRef] [PubMed]

24. Krause, E.F. Taxicab geometry: Adventure in non-euclidean geometry. *Mathematical Gazette* **1988**, *72*, 255.

25. Solis-Reyes, S.; Avino, M.; Poon, A.; Kari, L. An open-source k-mer based machine learning tool for fast and accurate subtyping of HIV-I genomes. *PLoS ONE* **2018**, *13*, e0206409. [CrossRef]

26. Kumar S.; Stecher G.; Tamura K. MEGA7: Molecular evolutionary genetics analysis version 7.0 for bigger datasets. *Mol. Biol. Evol.* **2016**, *33*, 1870–1874. [CrossRef]

27. Saitou, N.; Nei, M. The neighbor-joining method: A new method for reconstructing phylogenetic trees. *Mol. Biol. Evol.* **1987**, *4*, 406–425.

28. Zhao, Z.Q.; Han, G.S.; Yu, Z.G.; Li, J. Laplacian normalization and random walk on heterogeneous networks for disease-gene prioritization. *Comput. Biol. Chem.* **2015**, *57*, 21–28. [CrossRef]

29. Robinson, D.F.; Foulds, L.R. Comparison of phylogenetic trees. *Math. Biosci.* **1981**, *53*, 131–147. [CrossRef]

30. Felsenstein, J. Mathematics vs. evolution: Mathematical evolutionary theory. *Science* **1989**, *246*, 941–942. [CrossRef]

31. Foley, B.T.; Korber, B.T.M.; Leitner, T.K.; Apetrei, C.; Hahn, B.; Mizrachi, I.; Mullins, J.; Rambaut, A.; Wolinsky, S. HIV Sequence Compendium 2018. Available online: https://www.osti.gov/biblio/1458915 (accessed on 22 February 2020)

32. Larkin, M.A.; Blackshields, G.; Brown, N.P.; Chenna, R.; Mcgettigan, P.A.; Mcwilliam, H.;Valentin, F.; Wallace, I.M.; Wilm, A.; Lopez, R.; et al. Clustal W and clustal X version 2.0. *Bioinformatics* **2007**, *23*, 2947–2948. [CrossRef] [PubMed]

33. Manns, M.P.; Lohse, A.W.; Vergani, D. Autoimmune hepatitis-Update 2015. *J. Hepatol.* **2015**, *62*, S100–S111. [CrossRef] [PubMed]

34. Liu, Z.; Meng, J.; Sun, X. A novel feature-based method for whole genome phylogenetic analysis without alignment: Application to HEV genotyping and subtyping. *Biochem. Biophys. Res. Commun.* **2015**, *368*, 223–230. [CrossRef] [PubMed]

35. Ling, L.; Li, C.; Hagedorn, C.H. Phylogenetic analysis of global hepatitis E virus sequences: Genetic diversity, subtypes and zoonosis. *Rev. Med. Virol.* **2006**, *16*, 5–36.

36. Ding, S.; Li, Y.; Yang, X.; Wang, T. A simple *k*-word interval method for phylogenetic analysis of DNA sequences. *J. Theor. Biol.* **2013**, *317*, 192–199. [CrossRef]

37. Tang, J.; Hua, K.; Chen, M.; Zhang, R.; Xie, X. A novel *k*-word relative measure for sequence comparison. *Comput. Biol. Chem.* **2014**, *53*, 331–338. [CrossRef]

*Review*

# Application of Biological Domain Knowledge Based Feature Selection on Gene Expression Data

**Malik Yousef** [1,2,*], **Abhishek Kumar** [3,4] and **Burcu Bakir-Gungor** [5]

1    Department of Information Systems, Zefat Academic College, Zefat 13206, Israel
2    Galilee Digital Health Research Center (GDH), Zefat Academic College, Zefat 13206, Israel
3    Institute of Bioinformatics, International Technology Park, Bangalore 560066, India; abhishek@ibioinformatics.org
4    Manipal Academy of Higher Education (MAHE), Manipal 576104, India
5    Department of Computer Engineering, Faculty of Engineering, Abdullah Gul University, Kayseri 38080, Turkey; burcu.gungor@agu.edu.tr
*    Correspondence: malik.yousef@gmail.com

**Abstract:** In the last two decades, there have been massive advancements in high throughput technologies, which resulted in the exponential growth of public repositories of gene expression datasets for various phenotypes. It is possible to unravel biomarkers by comparing the gene expression levels under different conditions, such as disease vs. control, treated vs. not treated, drug A vs. drug B, etc. This problem refers to a well-studied problem in the machine learning domain, i.e., the feature selection problem. In biological data analysis, most of the computational feature selection methodologies were taken from other fields, without considering the nature of the biological data. Thus, integrative approaches that utilize the biological knowledge while performing feature selection are necessary for this kind of data. The main idea behind the integrative gene selection process is to generate a ranked list of genes considering both the statistical metrics that are applied to the gene expression data, and the biological background information which is provided as external datasets. One of the main goals of this review is to explore the existing methods that integrate different types of information in order to improve the identification of the biomolecular signatures of diseases and the discovery of new potential targets for treatment. These integrative approaches are expected to aid the prediction, diagnosis, and treatment of diseases, as well as to enlighten us on disease state dynamics, mechanisms of their onset and progression. The integration of various types of biological information will necessitate the development of novel techniques for integration and data analysis. Another aim of this review is to boost the bioinformatics community to develop new approaches for searching and determining significant groups/clusters of features based on one or more biological grouping functions.

**Keywords:** feature selection; feature ranking; grouping; clustering; biological knowledge



**Citation:** Yousef, M.; Kumar, A.; Bakir-Gungor, B. Application of Biological Domain Knowledge Based Feature Selection on Gene Expression Data. *Entropy* **2021**, *23*, 2. https://dx.doi.org/doi:10.3390/e23010002

Received: 27 November 2020
Accepted: 16 December 2020
Published: 22 December 2020

## 1. Introduction

Biological systems are massively complex and heterologous in nature. To resolve the mysteries behind complex biological systems, large-scale studies have been conducted which yielded massive volumes of biological data, including the genetic variations associated with specific phenotypes. Currently, we are encountering an -omics revolution in which genome, epigenome, transcriptome, and other -omics can be readily characterized. With advancements in various -omics approaches, it is now possible to generate multiomics data to answer various biological problems. Nowadays, several types of -omics data are considered as depicted in Figure 1, and the numbers of different -omics data types are increasing day-by-day [1]. Additionally, there are complex cascades and interactions among different -omics data types. For example, genomic and epigenomic variations have the capacity to control or modulate the transcriptome and in turn affect the proteome.

99

Here, epigenomics refers to the measurement of DNA methylation, histone modifications (methylation, acetylation, phosphorylation, DP-ribosylation, and ubiquitination), and non-coding RNAs (microRNAs, long noncoding RNAs, small interfering RNAs). Similarly, the epigenome of an organism refers to the entire collection of the molecules that modify the genome and control the genes to turn on and off. Since the epigenome shows how environmental factors influence the activity of genes, the study of the epigenome integrated with the study of the genome is crucial to fully account for phenomics. Accounting for such molecular deviations is crucial for making tangible improvements in biomarker analysis.

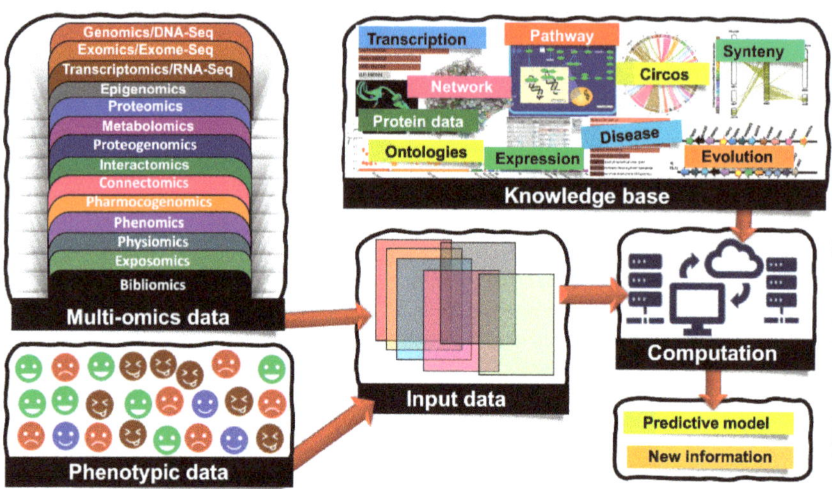


**Figure 1.** Machine learning (ML) applications that combine multi-omics and phenotypic data. Multi-omics data are classified into the following groups: genomics/DNA-Seq—the study of the genetic material for an organism, it assesses DNA sequence and structural variations including single-nucleotide polymorphisms (SNPs), insertions and deletions, copy number variations (CNVs), and inversions; epigenomics—the measurement of DNA methylation, histone modifications (methylation, acetylation, phosphorylation, DP-ribosylation, and ubiquitination), and noncoding RNAs (microRNAs, long noncoding RNAs, small interfering RNAs); transcriptomics/RNA-Seq—the study of the transcriptome of an organism; exomics/exome-seq—the study of the exome of an organism (coding regions); proteomics—the study of the total proteins within an organism; metabolomics—the study of the total metabolites; proteogenomics—combined study of genomics and proteomics; interactomics—interactions between nucleotides, proteins and metabolites; connectomics—study of the connections, neural pathways in the brain; pharmocogenomics—the application of genomics to pharmacology; phenomics—observable phenotypes; physiomics—functional behavior of an organism; exposomics—study of an organism's environment and bibliomics (the literature concerning a topic).

Traditional analyses attempted to untangle the molecular mechanisms of complex diseases using a single -omics dataset which contributes towards the identification of disease-specific mutations and epigenetic alterations. However, in the postgenomic era, it has been noticed that a single -omics dataset is not sufficient to explain disease hallmarks. It requires the combined analysis of various -omics datasets. As such, recent studies are shifting towards multi-omics data analysis, where each of these different -omics data types are critical for deciphering the molecular signatures of human diseases. Therefore, the integrated analysis of different data types has become a recent trend. For a holistic understanding of complex biological problems, it is becoming clear that integrations of different -omics data types are essential steps. However, it is a notorious task, as handling heterogeneous and noisy biological data is a challenging issue [2].

In addition to the 'omics' realm, another major reason for phenotypic differentiation is post-translational modifications (PTM). They can be both in physiologically reasonable and pathologically anomalous forms. Methods for bioinformatically incorporating PTM effects are emerging from the gradual improvement of sequence motifs, or less directly from compensatory expression patterns that emerge when an organism seeks to correct for aberrant biochemistry arising from anomalous structural modifications.

Due to the recent advancements in next-generation sequencing and microarray technologies, the cost of obtaining the gene expression profile of a sample is rapidly decreasing, and hence expression profiling has become a routine protocol in biological laboratories. The high turnaround of expression data is also coupled by the massive increase in the use of the revolutionary RNA-Seq method [3]. It is best exemplified by the large oncogenomic expression profiles hosted at The Cancer Genome Atlas (TCGA) [4]. Mutations are the core causative agents of diseases such as different cancers [5] when coupled with gene expression profiles. These datasets provide sufficient information to scientists and physicians for deciphering the disease mechanisms. It is becoming clear that the proper design of the RNA-seq can be used for mutational profiling as well as expression profiling [6]. This information also enables the design of platforms to assist diagnosis, to assess patients' prognosis, and to create patient treatment plans. For instance, van't Veer et al. had collected gene expression profiling datasets of primary breast tumors derived from a cohort of 117 young patients [7]. Machine Learning (ML) with feature selection was used to unravel a gene expression signature, which served as a signal for distant metastases, even divergent conditions such as lymph node negative [7].

Data analysis approaches to gene expression profiling have evolved rapidly as there are massive shifts from DNA microarray to RNA-seq-based profiling. The earlier methods involved clustering approaches and traditional ML approaches. Since a large volume of biological knowledge has become available, in the literature there are obvious shifts from the pure data-oriented approaches to biological domain knowledge-based integrative approaches. This fact has triggered bioinformatics researchers to suggest and develop advanced tools that consider the emerging biological knowledge, and hence they exploit this knowledge for deep analysis of the data. There are many resources of biological knowledge, such as textual knowledge, as more and more literature emerges, different databases and repositories such as miRTarBase [8] for microRNA, DNA Sequence Databases, Immunological Databases, Gene Expression Omnibus (GEO), Proteomics Resources, Protein Sequence Databases, TCGA, Gene Ontology (GO) and others.

Most feature selection algorithms that are applied on gene expression data are based on statistics and ML. However, most of them neglect the biological knowledge of the data that could contribute to perform better feature selection. R. Bellazzi and B. Zupan [9] discussed recent developments in gene expression-based analysis methods, focusing on studies (such as associations and classification) and implications (such as reverse-engineering of gene–gene networks and resulting phenotypes). Authors surveyed the clustering approaches that group the genes using different distance measures, such as Euclidean distance and/or Pearson's correlation. Moreover, incorporating biological knowledge in the clustering algorithm is a very challenging task. The GOstats package [10] allows one to define semantic similarity between the genes via incorporating the GO [11]. An additional study by Kustra and Zagdanski [12] used the incorporation of GO annotation to expression data by inducing a correlation-based dissimilarity matrix to derive a GO-based dissimilarity matrix.

The flood of -omics data and the need for more informative results urge the need for integrative approaches. The book of Ref. [13] is the first book on integrative data analysis and visualization in this area. It outlines essential techniques for the integration of data derived from multiple sources. It is one of the first systematic books that overviews the issue of biological data integration using analytical approaches. The book provides a framework for the creation and implementation of integrative analytical methods for the study of biological data on a systematic scale. Additionally, a recent review [2] describes the principles of biological data integration along with different approaches and methods indicating the importance of utilizing ML for biomedical datasets. However, to the best of our knowledge, in the literature, there is no comprehensive survey on biological domain knowledge-based feature selection methods, except from the study of Perscheid et al. [14]

that compares the performances of traditional gene selection methods against integrative ones. Moreover, authors also proposed a straightforward method to integrate external biological knowledge with traditional gene selection approaches. They introduced a framework for the automatic integration of external knowledge for selected genes and their evaluation. Herein, we aimed to provide a selective review on such gene selection approaches. In this regard, for the analysis of gene expression datasets, we present the traditional and integrative gene selection approaches in Sections 2.1 and 2.2, respectively.

On the other hand, multi-dimensional biological data are another challenge as these data are often derived from limited numbers of samples because of the associated costs of biological data generation. This is an example of the 'curse of dimensionality' problem, as initially reported by Bellman [15] in 1961. It means that the dimensions of gene features and/or functional parameters are critical input variables, and there are requirements of a minimal number of samples for the estimation of an arbitrary function, where an increase in sample size improves the chance of function prediction. In Section 3 of this manuscript, we present a prospective solution to this problem by defining a novel grouping of features and estimating their contribution to the machine learning model for two-class classification problems. We evaluate the methods that select the features using a classifier in a traditional way in Section 3.1, and we present integrative approaches that incorporate biological domain knowledge into ML to group and rank the genes in Section 3.2. In Section 4, we conclude our review with discussions and future prospects.

## 2. Gene Selection Approaches for Gene Expression Datasets

Gene selection approaches for gene expression datasets can be mainly categorized into two classes, such as traditional gene selection and integrative gene selection. While traditional gene selection approaches are solely based on statistical and computational analyses of the expression levels, integrative gene selection approaches incorporate domain knowledge from external biological resources during gene selection.

### 2.1. Traditional Gene Selection

Traditional gene selection approaches are heavily based on statistical and computational analyses of the actual expression levels. Recent reviews have summarized various methods for describing the selection process of disease-specific features from large gene expression datasets [16,17]. Primarily, these approaches are classified into three major classes, as (i) filtering-based, (ii) wrapping-based, and (iii) embedding-based approaches. Briefly, the filtering approaches are based on F-statistic (ANOVA, *t*-test, etc.), not based on ML. Wrapping-based approaches are primarily learning techniques and these are used for the exploration of usefulness of features, whereas embedding-based approaches are combining the feature selection and the classifier construction. Wei Pan carried out a comparative study on different filtering methods in Ref. [17] and he summarized similar and dissimilar points among three main methods (namely *t*-test method, regression modeling approach and mixture model approach).

Additional comparisons of filtering techniques are available in Ref. [16]. I. Inza [17] also carried out a comparison between filter metrics and the wrapper sequential search procedure, which are both applied on gene expression datasets. Additionally, hybrid, and ensemble approaches, which combine multiple approaches, are two additional categories of gene selection. Cindy et al. [14] presents an overview of the recent gene selection methods, where each method is classified according to these five categories.

The traditional gene selection approach has several drawbacks. For example, the filtering approach evaluates the significance of each gene individually without considering the relationships and the interactions between the genes. Although the wrapping-based approaches can find the optimal set, it might be specific to the model used, such as SVM, decision trees or other models. In other words, it might be overfitting the data [18]. The main disadvantages of such methods are their difficulties for biological interpretation, and they are unlikely to generate new biological knowledge.

### 2.2. Integrative Gene Selection

Although the traditional gene selection approaches became popular for a long time, they have several drawbacks when one needs to precisely identify the underlying biological processes. Alternatively, integrative gene selection approaches incorporate domain knowledge from external biological resources during gene selection [9,18], which improves interpretability and predictive performance. One of the widely used external ontology resources is the Gene Ontology (GO) [19], which provides (i) cellular component (CC), (ii) molecular function (MF), and (iii) biological process (BP) terms for the products of each gene. GO captures biological knowledge in a computable form that consists of a set of concepts and their relationships to each other. The first attempt to integrate biological background into a statistical analysis/ML analyses was to incorporate Gene Ontology (GO) [19] in clustering gene expression data [10]. Another widely used external ontology resource is the Kyoto Encyclopedia of Genes and Genomes (KEGG), which is a pathway knowledge-base providing manually curated pathways [20]. Yet another widely used external biological resource is DisGeNET, which is a meta knowledge-base on gene–disease–variant associations [20].

One example of the integrative gene selection approach is proposed by Qi and Tang, where they utilize the power of biological information contained in GO annotations to rank the genes [21]. The algorithm is designed in an iterative manner that starts by applying Information Gain (IG) to compute discriminative scores for each gene. The genes that have a score of zero are removed from the analysis. The second step is to integrate the biological knowledge, which is achieved by annotating those surviving genes with a GO term. The third step is to score the GO terms as the mean of their associated genes' discriminative scores, which were computed before using IG. The final gene set is created as follows: Starting from the highest ranked GO terms, the genes with the highest discriminative scores are chosen. These genes are removed from the annotated genes and this procedure is repeated until the final gene set is complete. Using multiple cancer datasets, Qi and Tang showed that their proposed method can achieve better results, as compared to using IG only.

Another example of the integrative gene selection approach is SoFoCles [22], which uses GO terms to find semantically similar genes. In order to assign a discriminative score to each gene, SoFoCles utilizes a classic filter approach, such as $\chi 2$, ReliefF, or IG. The initial set of candidate genes is composed of the top n ranked genes. Genes receive a similarity score based on their associated GO terms. Then, the genes which have a high similarity score, i.e., the genes that are semantically very similar to the candidate genes, are added to the set of candidates. The experiments conducted on SoFoCles showed that the incorporation of biological knowledge into the gene selection process improves the results.

Yet another study by Fang et al. [18] combines KEGG and GO terms with IG. The authors initially apply IG on the dataset as the filtering step and then check the GO and KEGG annotations of the remaining genes. Then, the authors use association mining and calculate the interestingness of the frequent itemsets by averaging the original discriminative scores (from IG) of the included genes. The final gene set is generated via selecting the highest ranked genes from the top n frequent itemsets. They evaluated this method using GO, using KEGG, and using both terms against IG only and against Qi and Tang's approach. Although their proposed approach slightly increased the overall accuracy, the main advantage of this approach was that it used a much lower number of genes.

The integrative gene selection approach that is proposed by Raghu et al. [23] makes use of KEGG, DisGeNET, and further genetic meta information [20]. In their approach, for each gene, (i) the importance score and (ii) the gene distance metrics are computed. The importance score is calculated via combining a gene–disease association score from DisGeNET with the gene expression levels in the data. The gene distance is defined as the physical distance between two genes (in terms of their chromosomal locations) and their associations to the same diseases. Both of the scores (importance score and gene distance) are later used to find maximally relevant and diverse gene sets. As compared to variance-based gene selection techniques, the use of the top n genes according to the importance score resulted in a slightly better performance in predictive modeling task.

The integrative approach of Quanz et al. aims to map genes into KEGG pathways and then uses these pathways as features for further pattern mining [24]. In their approach, they make use of a global test to extract KEGG pathways which are related to the phenotypes of a dataset. In their feature extraction step, the genes in each pathway are then transformed into one single feature by applying mean normalization or logistic regression. In this way, the data are represented as the number of pathways, which can be considered as a feature reduction step and it provides dramatic reduction. For instance, for the diabetes data, 17 pathways, out of approximately 300 pathways, are selected and thus for the classification task the dimensionality is reduced from 22,283 to 17. Even though this approach was not tested on multiclass problems such as cancer (sub-) type classification, the experiments on binary classification problems showed an improved performance over different traditional approaches.

Mitra et al. adopted the clustering large applications based upon randomized search (CLARANS) method to the feature (gene) selection problem via utilizing biological knowledge [25]. Their reduced feature set is composed of gene clusters, which are the medoids of biologically enriched sets. Later on, the authors attempted to use a fuzzy clustering technique instead of CLARANS, and developed a technique called FCLARANS for feature selection [26].

In Ref. [27], the authors proposed an integrative gene (feature) selection approach based on the sample clustering technique, which utilizes gene annotation information from GO. On the generated gene–GO term matrix, they applied Partitioning Around Medoids clustering. In their method, the optimal number of clusters (k) is chosen by comparing their silhouette index values. For the selected k number of clusters, the medoids are used as the selected gene subset. They reported that the integration of biological knowledge during the gene selection process not only reduces the dimensionality of the feature space, but also increases the accuracy of sample classification.

The related studies that are presented until this point are highly specific to a single knowledge-base, e.g., KEGG pathway or GO terms. On the other hand, Perscheid et al. [14] proposed an approach that can flexibly combine traditional gene selection approaches with several knowledge-bases. They comparatively evaluated the performance of traditional gene selection approaches with integrative gene selection approaches. Their study concluded that the integration of external data especially improves on simple traditional filter approaches, e.g., information gain. Once external biological data are integrated, such traditional filter approaches become compatible with more complex machine learning approaches at very similar classification accuracies, but far lower computational running times and a more transparent and thus interpretable computation processes.

The above-mentioned studies proposed predictive models, but most of the time, instead of obtaining high predictive accuracies in these models, the scientists are curious about the biological meaning of the predictive model. The 'black box' nature of the predictive model can hamper its interpretation. The information excerpted from the model may require further processing, and careful interpretation with corresponding biological knowledge may be needed. The interpretation of the complicated cases may be quite challenging, and such an interpretation may currently be out of reach. Although the joint analysis of multiple biological data types has the potential to enlighten our understanding of complex biological phenomena, the data integration is challenging due to the heterogeneity of different data types. For example, an expression profile, as obtained from a transcriptomic study, is a vector of real values and the length of a vector is equal to the number of genes in the genome. However, the genetic variants as obtained from a genomic study are categorical, and they have different vector lengths. While different studies [1,4] proposed several strategies for data integration, the best practices by which -omics data types can be integrated and information on how to integrate these biological data are still needed.

Feature selection and discovering the molecular explanation of diseases describe the same process, where the first one is a computer science term and the second one is used in the biomedical sciences. In 2007, Yousef et al. proposed a new feature selection method, support vector machines–recursive cluster elimination (SVM-RCE), to group/cluster genes for gene expression data analysis. This study invented the "recursive cluster elimination" phrase for the first time in the machine-learning domain and introduced it to the computational community. As such, this study became a pioneer study in this field. Interests in this approach have increased over time and several studies have successfully applied the SVM-RCE approach to identify the features/genes that are directly associated with a disease/condition [28]. This growing interest is based on the reconsideration of how feature selection in biological datasets can benefit from incorporating the biomedical relationships of the features in the selection process. The usefulness of SVM-RCE then led to the development of maTE [29], which uses the same approach based on the interactions of microRNAs (miRNA) and their gene targets. Additionally, in the literature, the biological information buried in genetic interaction networks is utilized for classification studies. For example, SVM-RNE (SVM with recursive network elimination) integrates network information with recursive feature elimination based on SVM [30]. It is shown that SVM-RNE has a good performance and also improves the biological interpretability of the results. Studies similar to SVM-RCE and SVM-RNE were later carried out by different groups [31,32], which indicates the importance and the merit of the SVM-RCE approach. The study of Ref. [33] has a slightly modified SVM-RCE algorithm in the disease state prediction step. Additionally, they used the already invented term of "recursive cluster elimination".

The study of Zhao, X. et al. [34] has used the SVM-RCE tool for comparison and used expression profiles for identifying microRNAs related to venous metastasis in hepatocellular carcinoma. Another similar study to SVM-RNE is carried out by Johannes M. et al. [35] for integration of pathway knowledge into a reweighted recursive feature elimination approach for the risk stratification of cancer patients. A recent tool, SVM-RCE-R [36], is an updated version of SVM-RCE, which is implemented in Knime [37], and uses a random forest classifier with additional important features such as suggesting a new approach of ranking the clusters.

The term "knowledge-driven variable selection (KDVS)" is a similar term to "integration of biological knowledge", and both of them are used in the process of feature selection. An additional similar study that applied KDVS to SVM-RNE is presented by Ref. [38], in which the authors proposed a framework that uses a priori biological knowledge in high-throughput data analysis.

The RCE algorithm [28] considers similar features/genes and applies a rank function to the feature group. Since it uses k-means as the clustering algorithm, we refer to these groups as clusters, but it could include other biological or more general functions combined with the features, as was suggested in several studies [29,30]. In the original paper of SVM-RCE, the contribution to the accuracy is achieved in distinguishing specific classes for ranking the clusters. The data for that ranking are divided into training and testing, with the data represented by each gene/feature being assigned to a specific cluster of features. The rank function is then applied as the mean of m times repeats of the training–testing performance while recording different measurements of accuracy (sensitivity, specificity, etc.).

In Table 1, we summarize the specifications, advantages and disadvantages of the presented integrative gene selection approaches.

**Table 1.** Summary table of the presented methodologies that integrate biological knowledge. While "A" refers to the advantages, "D" refers to the disadvantages of the methods.

| Tool Name | Incorporated Biological Knowledge | Methodology | Advantage/Disadvantage | Ref |
|---|---|---|---|---|
| N/A | GO | Rank the genes uses information gain (IF) incorporated with Gene Ontology GO terms | **A:** The novelty of this work is to evaluate genes based on not only their individual discriminative powers but also the powers of GO terms that annotate them. | [21] |
| N/A | GO | χ2, ReliefF, or IG | **A:** Including biological knowledge in the gene selection process improves results. | [22] |
| N/A | Combines KEGG and GO terms | Utilizes graphical causal modeling IG as an initial filter search for GO and KEGG annotations' frequent items | **A:** Method is capable of intelligently selecting genes for learning effective causal networks. **D:** No significant improvement in accuracy. | [18] |
| N/A | KEGG, DisGeNET, and further genetic meta information | Gene–disease association score from DisGeNET Gene distance metrics | | [23] |
| N/A | KEGG pathways | Uses these pathways as features for further pattern mining | **A:** Reduce the dimension of the data by transforming to KEGG feature space. **A:** Improved performance over different traditional approaches. | [24] |
| N/A | Gene ontology (GO) | Randomized search (CLARANS) | **A:** Reducing the dimension dramatically. | [25] |
| SVM-RCE | Genes related are correlated | SVM and K-means | **A:** Discover significant of clusters. **D:** Might lose important genes because they were in lower-ranked clusters. | [28,36] |
| SVM-RNE | GXNA for creating subnetworks from gene expression | SVM, GXNA | **A:** Reducing the dimension of the data by considering subnetworks. **D:** The subnetworks are created as a prediction of the gene expressions data. | [30] |
| maTE | microRNA genes targets | Random forest groups the genes that associated with microRNA | **A:** A novel approach of integrating microRNA into gene expression. **D:** The size of the groups might be large and might rank these groups highly as a result of that. | [29] |
| CogNet | | Random forest, based on pathFindR tool | **A:** Improve the results of the pathFindR tool by ranking its groups. | [39] |
| miRcorrNet | | Random forest based on the correlation with miRN expressions | **A:** Novel approach for integrating miRNE and mRNA expressions using machine learning. | [40] |

## 3. Grouping and Ranking of the Genes for Classification Problem

The genes that are involved in the same biological process are likely to be co-expressed [41]. Therefore, one potential way of discovering gene function is to group genes with a similar expression profile. Thus, different clustering algorithms [42] were considered to perform the grouping step. This was the first approach, and more advanced approaches that use biological information in order to group the genes are later proposed. In this section, we will introduce a generic approach to grouping that is accompanied by ranking and classification. The presented model is used by different studies and other similar studies are still ongoing.

The main aim of the generic approach is to search for and determine significant groups/clusters of features based on one or more biological grouping function (will be referred as *bioF()* throughout the rest of this paper) that are integrated with the ML algorithms. The generic approach is presented in Figure 2. The advantage of those systems is that the grouping of the genes/features is in the hand of the researcher, that is, it is actually based on available biological knowledge. The researcher will provide how genes or features should be grouped and then the algorithm will proceed to score and rank those groups in terms of the classification problem. The final model will be built from the top n groups according to the researcher's settings. The outcome of the algorithm is different from the traditional current approaches (such as SVM-RFE [43]), where the algorithm takes as input the data of gene expression with class labels. Then the outcome is just a list of significant genes that are able to distinguish the two classes. With the integration framework, the researcher will get a more informative list of significant groups/clusters with its genes list that is able to distinguish the two classes. Additionally, the researcher can use the computational approach of grouping that is based on clustering approaches such as k-means or others, and specify different measurements for ranking the groups/clusters based on their interest and their research aims. The outcome of the algorithm will be more specific to the researcher's interest.

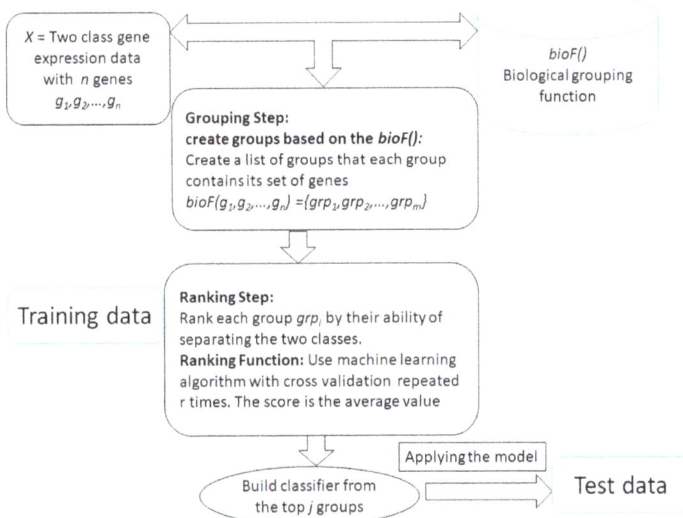


**Figure 2.** The generic framework of the algorithm that is based on biological integration for grouping, ranking and classification.

The generic approach mainly consists of two main components. The first component is the grouping step relying on the *bioF()* function that is based on biological knowledge to group the genes into groups. For example, *bioF()* might be disease-related genes; then the function will group the genes into groups where each group is associated with one disease. Another possibility is grouping the genes that are targeted by specific miRNAs, such as in the maTE [29] tool. One interesting use of *bioF()* is that it allows one to create different biological groupings, such as creating groups related to miRNA, groups related to disease, groups related to KEGG pathways, and others. However, the grouping can also be based on clustering algorithms such as k-means, as suggested in SVM-RCE [28] for grouping correlated genes. Similarly, SVM-RNE [30] incorporates another tool, GXNA [44], to create the groups. GXNA utilizes gene expression profiles and prior biological information to suggest differentially expressed pathways or gene networks.

*3.1. Traditional Approach of Feature/Gene Selection Using a Classifier*

There are many classifiers that were used to fit the data in order to rank the features and perform the process of features selection. The most simple one is the linear model, where the coefficients of the variable/features can serve as a measurement of the feature's rank. We will be describing the first approach that suggests the RFE procedure. RFE refers to recursive feature elimination. The approach uses SVM as the linear model.

SVM-RFE (Support Vector Machines with Recursive Feature Elimination)

Let us assume that we are given a set of points called S. S consists of m points xi ∈ Rd in dimension d. Let us assume that we have two class labels, denoted by yi ∈ { −1, +1}. We call S a linear separable set if there is a hyperplane of equation w·x + w0 = 0 (we refer to it as hyperplane (w, w0)) that separates the points with label + 1 from the points with label −1. The signed distance di of a point xi to the separating hyperplane (w, w0) is given by di = (w·xi + w0)/ | | w | | .

For simplicity, let us define the optimal separating hyperplane to be f(x) = w0 + w1 × 1 + w2 × 2 + ... + wd × d, where the (x1, x2, ... , xd) is the features and w = (w1, w2, ... , wd) is the corresponding weights. SVM is actually the solution of finding the optimal linear function, as developed by Vapnik [45]. It is obvious that the contribution of features with lower weights is non-significant to the sign of the f(x). So one can consider removing those features in order to perform dimension reduction of the feature space.

SVM-RFE, which stands for support vector machines with recursive feature elimination, was firstly introduced by Isabelle Guyon et al. [43] and applied to gene expression data. The primary goal of SVM-RFE is to use SVM (linear SVM) to compute the weights of the features. The weights are actually the ranks of the features. SVM-RFE performs an iterative step to remove features with low ranks. The RFE procedure can be described as:

1.  Train the classifier on the given data;
2.  Assign rank for each feature as its weight;
3.  Remove one feature or percentage (10%) with the smallest weight;
4.  Repeat steps 1–3 until reaching a predefined number of genes.

Different studies [46] have emerged as extended variations of the original SVM-RFE algorithm. However, SVM-RFE has some limitations that other studies have reported to suggest an improvement approach. One such limitation is that SVM-RFE is designed as a greedy method and tries to find out superlative possible combinations leading to binary classification, where these combinations may not be biologically significant. To overcome this limitation, a novel feature selection algorithm, sigFeature [47], based on SVM and t statistic, was developed.

*3.2. Biological Domain Knowledge Based ML Approaches*

ML is becoming a very powerful computational approach in the field of bioinformatics. In this respect, the first step in utilizing ML is to apply unsupervised and supervised algorithms on biological data. However, in the era of big biological data, we come up with emerging approaches that integrate biological knowledge with ML. A recent review [48] on ML and complex biological data discusses the challenges and hurdles of the analysis and discovery of complex biological data. They predict that in the very near future, more researchers will be interested in applying ML to complex biological data. In this section, we will review different approaches that consider the biological structure or knowledge for the process of feature selection. The following approaches follow the generic approach presented in Figure 2.

3.2.1. SVM-RCE (Support Vector Machines with Recursive Cluster Elimination)

SVM-RCE [28,36] is based on the concept of grouping and ranking, where the k-means clustering algorithm is used for performing the related grouping of *bioF()*. Correlated genes are hypothesized to have similar biological functions. Then, the rank component is applied to assign a score for each cluster, indicating its significance in terms of the classification of the two given classes. In order to perform the rank component, each cluster is considered by representing the data based on the genes that belong to it while keeping the class labels of the original data. Now the data are transferred to cluster genes representation. The rank component performs internal cross-validation and aggregates the performance outcome as the score of the cluster. The RCE procedure is applied to remove the lowest ranked groups. The SVM-RCE results show that the classification accuracy is superior to other approaches, suggesting that the classification results are more interpretable, and this creates new hypotheses for future investigation.

### 3.2.2. SVM-RNE (Support Vector Machines with Recursive Network Elimination)

SVM-RNE [30] is an extended version of the SVM-RCE approach that uses the tool GXNA [42] as the *bioF()* for grouping the genes into subnetworks of genes. Then, a similar procedure of ranking is applied as described in SVM-RCE. SVM-RNE also performs the recursive elimination procedure by ranking firstly all the groups, and then removing the least significant groups. The algorithm proceeds by applying again the GXNA to suggest groups. This process is repeated until satisfying some predefined constraints on the number of groups.

### 3.2.3. MaTE

Disease development mechanisms mainly involve changes in the transcript levels and protein abundance. MicroRNAs (miRNAs) are instrumental in regulating the gene expression, and hence they affect transcript levels and protein abundance. The fact that microRNAs target more than one mRNA helps us to group the genes into groups where each group consists of the list of genes targeted by a specific microRNA. In other words, the *bioF()* biological grouping function here is the biological association between microRNA and its set of targets. The *bioF()* grouping function is based on the database mirTarBase [8]. mirTarBase has accumulated more than three hundred and sixty thousand miRNA-target interactions. Thus, a novel approach called maTE [29] has been developed.

Table 2 presents partially the result of applying *bioF()* on mirTarBase. Additionally, it performs a computation procedure in order to score/rank the importance for each group for the classification tasks.

**Table 2.** Example of microRNA and their targets list.

| MicroRNA Group Name | Target Genes List |
| --- | --- |
| HSA-MIR-147A | VEGFA, ACVR1C, MCM3, NDUFA4, PSMA3, HIF3A, SLC22A3, MCM3, NDUFA4, PSMA3, HIF3A, VEGFA, ACVR1C, MCM3, NDUFA4, PSMA3, HIF3A, SLC22A3 |
| HSA-MIR-18B-5P | ESR1, MDM2, CTGF, TNRC6B, HIF1A, SMAD2, FOXN1, IGF1, IGF1, CTGF, HIF1A, SMAD2, FOXN1, ESR1, MDM2, CTGF, TNRC6B, HIF1A, SMAD2, FOXN1, IGF1, IGF1 |
| HSA-MIR-19B-3P | BACE1, PTEN, PTEN, PTEN, ATXN1, HIPK3, ARID4B, MYLIP, ESR1, KAT2B, SOCS1, BCL2L11, BCL2L11, TGFBR2, TGFBR2, BMPR2, BMPR2, TLR2, PPP2R5E, PPP2R5E, CYP19A1, GCM1, HIPK1, SMAD4, MYCN, MXD1, BCL3, DNMT1, TNFAIP3, PKNOX1, MTUS1, PITX1, PTEN, PTEN, PTEN, ATXN1, ESR1, NCOA3, KAT2B, SOCS1, TGFBR2, BMPR2, CUL5, TLR2, HIPK1, MXD1, BCL3, TNFAIP3, MTUS1, PITX1, BACE1, PTEN, PTEN, PTEN, PTEN, ATXN1, HIPK3, ARID4B, MYLIP, ESR1, NCOA3, KAT2B, SOCS1, BCL2L11, BCL2L11, TGFBR2, TGFBR2, BMPR2, BMPR2, CUL5, TLR2, PPP2R5E, PPP2R5E, CYP19A1, GCM1, HIPK1, SMAD4, MYCN, MXD1, BCL3, DNMT1, TNFAIP3, PKNOX1, MTUS1, PITX1 |
| HSA-MIR-210-5P | CFB |

The inputs to the maTE tool are the gene expression data, and the list of microRNAs and its target genes. The main function of the tool is to produce a group of genes based on the miRNA target information and then rank each group by applying random forest with cross-validation, which is repeated r times. The average of the accuracy for each iteration is actually the rank for a specific group. Then the groups are ranked according to the rank values. The model will be built considering the genes on the top $j$ groups. The default value of $j$ is 2. We apply to the training part of the data a $t$-test statistics in order to remove noisy genes. The test part of the data is used in order to estimate the performance of the tool.

### 3.2.4. CogNet

The CogNet tool is a classification of gene expression data based on ranked active-subnetwork-oriented KEGG pathway enrichment analysis [39]. CogNet is based on biological knowledge as a function for grouping the genes for the task of ranks and classification. The pathfindR tool serves to be the biological grouping function allowing the main algorithm to rank active-subnetwork-oriented KEGG pathway enrichment analysis [49]. CogNet was tested on 13 gene expression datasets of different diseases. In these experiments, CogNet was shown to outperform maTE and obtain similar performance results with SVM-RCE.

CogNet provides a list of significant KEGG pathways, including its genes that are able to separate the classes of the data. The list would serve the biology researcher for deep analysis and better interpretability of the role of KEGG pathways in the data, or the case that is being studied. As a future work, we would develop CogNet to explore the effectiveness of different combinations of the KEGG pathways in the data. In the current version, we treat each KEGG pathway individually.

### 3.2.5. MiRcorrNet

Due to the advances in technology, both mRNA and microRNA expression profiles can be generated allowing integrative analysis aiming to uncover the functional effects of RNA expression in complex diseases, such as cancer. Most of the approaches that integrate miRNA and mRNA are based on statistical methods, such as Pearson correlation, combined with enrichment analysis approaches. In this study [40], a novel tool is used called miRcorrNet, which performs machine learning-based integration to analyze miRNA and mRNA gene expression profiles. miRcorrNet groups mRNA genes based on their correlation to miRNA expression. Then, these groups are subjected to a rank function for classification. We have tested our tool on TCGA data miRNA-seq and mRNA-seq expression compared to other tools. The performance results show that the tool works as well as other tools in terms of accuracy measurements, reaching an AUC above 95%. Moreover, we conducted a deep biological analysis to explore the list of significant miRNAs. Accumulated results suggest that miRcorrNet is able to accurately prioritize pan-cancer-regulating high-confidence miRNAs.

## 4. Conclusions

As we have more advanced high-throughput technologies, big transcriptomic datasets become available, and extracting insights from long lists of differentially expressed genes becomes a challenge. Since the gene expression data typically have small samples size but high dimensions and noise, the major challenge is the detection of disease-related information from vast amounts of redundant data and noise. As such, the gene (feature) selection and the removal of redundant/irrelevant genes has been a key step to address this problem. For gene expression data analysis, most of the existing feature selection methods rely on expression values alone to select the genes, and biological knowledge is integrated at the end of the analysis in order to gain biological insights or to support the initial findings. However, lately, the gene selection process has shifted from being purely data-centric to more incorporative analysis with additional biological knowledge. Integrative gene selection approaches incorporate domain knowledge from external biological resources during gene selection [9,18], which improves interpretability and predictive performance. One of the more widely used external ontology resources is GO [19], which captures biological knowledge in a computable form that consists of a set of concepts and their relationships to each other. As another alternative, pathway-based analysis approaches aim to investigate the aggregation of the genes that are part of a functional unit, where these functional units are predefined by prior biological knowledge. These pathway-based methods rely on statistical tests that aim to detect damaged functionalities, which may result in disease phenotype. Several studies reported that the genetic variations occurring at multiple loci often disturb signal transduction, and regulatory and metabolic pathways, which causes severe changes in phenotype [18]. In this regard, a widely used external ontology resource is KEGG, which is a knowledge-base of manually curated pathways [20]. Yet another widely used external biological resource is DisGeNET, which is a meta knowledge-base for gene–disease–variant associations [20].

High-throughput profiling technologies currently enable us to concurrently measure gene expression levels for tens of thousands of genes in a single experiment, but they have some drawbacks. The high dimensionality of the gene expression data and relatively small sample sizes make the interpretation of the data a complicated, and often overwhelming, task. Although sample sizes have continued to grow in recent years, new and efficient feature selection algorithms are still needed to overcome the challenges in the existing methods [4]. As such, this is an active research topic in the field of bioinformatics.

At present, ML is applied to specific data in order to explain and answer a specific biological query in the biological knowledge domain. One of the challenges of future integrative model-based ML is the ability to combine different biological resources to enhance our understanding of multiple biological questions. Once the full potential of the available data is achieved, they can be used in the development of gene-based diagnostic tests, drug discovery studies and in the development of therapeutic strategies for improving public health.

To sum up, since biological systems are quite complex and they have an interconnected nature, a single model that is trained on a single dataset can only benefit from a small portion of the entire biomedical knowledge. For this reason, in order to get the complete picture of molecular biology and medicine, the integration of diverse biological resources and multi-omics data is crucial. In the field of gene expression data analysis, there are still many challenges that the community needs to solve, such as the integration of gene expression datasets that are generated by different research groups for the same phenotype (which will help to overcome the batch effect), and an additional obstacle is the integration of non-similar data, wherein each dataset tackles a specific disease.

**Author Contributions:** All authors contributed equally to the manuscript. All authors have read and agreed to the published version of the manuscript.

**Funding:** This research received no external funding.

**Acknowledgments:** A.K. is recipient of Ramalingaswami Re-Retry Faculty Fellowship (Grant; BT/RLF/Re-entry/38/2017). The work of B.B.G. has been supported by the Abdullah Gul University Support Foundation (AGUV). The work of M.Y. has been supported by the Zefat Academic College.

**Conflicts of Interest:** The authors declare no conflict of interest.

## References

1. Hasin, Y.; Seldin, M.; Lusis, A. Multi-omics approaches to disease. *Genome Biol.* **2017**, *18*, 83. [CrossRef] [PubMed]
2. Zitnik, M.; Nguyen, F.; Wang, B.; Leskovec, J.; Goldenberg, A.; Hoffman, M.M. Machine learning for integrating data in biology and medicine: Principles, practice, and opportunities. *Inf. Fusion* **2019**, *50*, 71–91. [CrossRef] [PubMed]
3. Wang, Z.; Gerstein, M.; Snyder, M. RNA-Seq: A revolutionary tool for transcriptomics. *Nat. Rev. Genet.* **2009**, *10*, 57–63. [CrossRef] [PubMed]
4. Tomczak, K.; Czerwińska, P.; Wiznerowicz, M. The Cancer Genome Atlas (TCGA): An immeasurable source of knowledge. *Contemp. Oncol. Poznan Pol.* **2015**, *19*, A68–A77. [CrossRef]
5. Fiala, C.; Diamandis, E.P. Mutations in normal tissues—some diagnostic and clinical implications. *BMC Med.* **2020**, *18*, 283. [CrossRef]
6. Sheng, Q.; Zhao, S.; Li, C.-I.; Shyr, Y.; Guo, Y. Practicability of detecting somatic point mutation from RNA high throughput sequencing data. *Genomics* **2016**, *107*, 163–169. [CrossRef]
7. Veer, L.J.V.; Laura, J.; Dai, H.; van de Vijver, M.J.; He, Y.D.; Hart, A.A.M.; Mao, M.; Peterse, H.L.; van der Kooy, K.; Marton, M.J.; et al. Gene expression profiling predicts clinical outcome of breast cancer. *Nature* **2002**. [CrossRef]
8. Chou, C.; Chang, N.; Shrestha, S.; Hsu, S.; Lin, Y.; Lee, W.; Yang, C.; Hong, H.; Wei, T.; Tu, S.; et al. miRTarBase 2016: Updates to the experimentally validated miRNA-target interactions database. *Nucleic Acids Res.* **2016**, *44*. [CrossRef]
9. Bellazzi, R.; Zupan, B. Towards knowledge-based gene expression data mining. *J. Biomed. Inform.* **2007**, *40*, 787–802. [CrossRef]
10. Falcon, S.; Gentleman, R. Using GOstats to test gene lists for GO term association. *Bioinformatics* **2007**, *23*, 257–258. [CrossRef]
11. Consortium, T.G.O. Gene ontology: Tool for the unification of biology. *Gene Ontol. Consort.* **2000**, *25*, 25–29.
12. Kustra, R.; Zagdanski, A. Incorporating Gene Ontology in Clustering Gene Expression Data. In Proceedings of the 19th IEEE Symposium on Computer-Based Medical Systems (CBMS'06), Salt Lake City, UT, USA, 22–23 June 2006; pp. 555–563. [CrossRef]

13. Azuaje, F.; Dopazo, J. (Eds.) *Data Analysis and Visualization in Genomics and Proteomics*; John Wiley: Hoboken, NJ, USA, 2005.
14. Perscheid, C.; Grasnick, B.; Uflacker, M. Integrative Gene Selection on Gene Expression Data: Providing Biological Context to Traditional Approaches. *J. Integr. Bioinform.* **2019**, *16*. [CrossRef] [PubMed]
15. Bellman, R. *Adaptive Control Processes: A Guided Tour. (A RAND Corporation Research Study)*; Princeton University Press: London, UK, 1961.
16. Lazar, C.; Taminau, J.; Meganck, S.; Steenhoff, D.; Coletta, A.; Molter, C.; de Schaetzen, V.; Duque, R.; Bersini, H.; Nowe, A. A survey on filter techniques for feature selection in gene expression microarray analysis. *IEEEACM Trans. Comput. Biol. Bioinform. IEEE ACM* **2012**, *9*, 1106–1119. [CrossRef]
17. Inza, I.; Larrañaga, P.; Blanco, R.; Cerrolaza, A.J. Filter versus wrapper gene selection approaches in DNA microarray domains. *Artif. Intell. Med.* **2004**, *31*, 91–103. [CrossRef] [PubMed]
18. Fang, O.H.; Mustapha, N.; Sulaiman, M.N. An integrative gene selection with association analysis for microarray data classification. *Intell. Data Anal.* **2014**, *18*, 739–758. [CrossRef]
19. Ashburner, M.; Ball, C.A.; Blake, J.A.; Botstein, D.; Butler, H.; Cherry, J.M.; Davis, A.P.; Dolinski, K.; Dwight, S.S.; Eppig, J.T.; et al. Gene Ontology: Tool for the unification of biology. *Nat. Genet.* **2000**, *25*, 25–29. [CrossRef]
20. Piñero, J.; Bravo, À.; Queralt-Rosinach, N.; Gutiérrez-Sacristán, A.; Deu-Pons, J.; Centeno, E.; García-García, J.; Sanz, F.; Furlong, L.I. DisGeNET: A comprehensive platform integrating information on human disease-associated genes and variants. *Nucleic Acids Res.* **2017**, *45*, D833–D839. [CrossRef]
21. Qi, J.; Tang, J. Integrating gene ontology into discriminative powers of genes for feature selection in microarray data. In Proceedings of the 2007 ACM symposium on Applied computing—SAC'07, Seoul, Korea, 11–15 March 2007; p. 430. [CrossRef]
22. Papachristoudis, G.; Diplaris, S.; Mitkas, P.A. SoFoCles: Feature filtering for microarray classification based on Gene Ontology. *J. Biomed. Inform.* **2010**, *43*, 1–14. [CrossRef]
23. Raghu, V.K.; Ge, X.; Chrysanthis, P.K.; Benos, P.V. Integrated Theory-and Data-Driven Feature Selection in Gene Expression Data Analysis. In Proceedings of the 2017 IEEE 33rd International Conference on Data Engineering (ICDE), San Diego, CA, USA, 19–22 April 2017; pp. 1525–1532. [CrossRef]
24. Quanz, B.; Park, M.; Huan, J. Biological pathways as features for microarray data classification. In *2nd International Workshop on Data and Text Mining in Bioinformatics—DTMBIO'08*; ACM Press: Napa Valley, CA, USA, 2008; p. 5. [CrossRef]
25. Mitra, S.; Ghosh, S. Feature Selection and Clustering of Gene Expression Profiles Using Biological Knowledge. *IEEE Trans. Syst. Man Cybern. Part C Appl. Rev.* **2012**, *42*, 1590–1599. [CrossRef]
26. Ghosh, S.; Mitra, S. Gene selection using biological knowledge and fuzzy clustering. In Proceedings of the 2012 IEEE International Conference on Fuzzy Systems, Brisbane, Australia, 10–15 June 2012; pp. 1–9. [CrossRef]
27. Acharya, S.; Saha, S.; Nikhil, N. Unsupervised gene selection using biological knowledge: Application in sample clustering. *BMC Bioinform.* **2017**, *18*, 513. [CrossRef]
28. Yousef, M.; Jung, S.; Showe, L.C.; Showe, M.K. Recursive Cluster Elimination (RCE) for classification and feature selection from gene expression data. *BMC Bioinform.* **2007**, *8*. [CrossRef] [PubMed]
29. Yousef, M.; Abdallah, L.; Allmer, J. maTE: Discovering expressed interactions between microRNAs and their targets. *Bioinformatics* **2019**, *35*, 4020–4028. [CrossRef] [PubMed]
30. Yousef, M.; Ketany, M.; Manevitz, L.; Showe, L.C.; Showe, M.K. Classification and biomarker identification using gene network modules and support vector machines. *BMC Bioinform.* **2009**, *10*, 337. [CrossRef] [PubMed]
31. Harris, D.; Niekerk, A.V. Feature clustering and ranking for selecting stable features from high dimensional remotely sensed data. *Int. J. Remote Sens.* **2018**, *39*, 8934–8949. [CrossRef]
32. Lazzarini, N.; Bacardit, J. RGIFE: A ranked guided iterative feature elimination heuristic for the identification of biomarkers. *BMC Bioinform.* **2017**. [CrossRef]
33. Deshpande, G.; Li, Z.; Santhanam, P.; Coles, C.D.; Lynch, M.E.; Hamann, S.; Hu, X. Recursive cluster elimination based support vector machine for disease state prediction using resting state functional and effective brain connectivity. *PLoS ONE* **2010**, *5*, e14277. [CrossRef]
34. Zhao, X.; Wang, L.; Chen, G. Joint Covariate Detection on Expression Profiles for Identifying MicroRNAs Related to Venous Metastasis in Hepatocellular Carcinoma. *Sci. Rep.* **2017**, *7*, 5349. [CrossRef]
35. Johannes, M.; Brase, J.; Fröhlich, H.; Gade, S.; Gehrmann, M.; Fälth, M.; Sültmann, H.; Beißbarth, T. Integration of pathway knowledge into a reweighted recursive feature elimination approach for risk stratification of cancer patients. *Bioinformatics* **2010**. [CrossRef]
36. Yousef, M.; Bakir-Gungor, B.; Jabeer, A.; Goy, G.; Qureshi, R.; Showe, L.C. Recursive Cluster Elimination based Rank Function (SVM-RCE-R) implemented in KNIME. *F1000Research* **2020**, *9*, 1255. [CrossRef]
37. Berthold, M.R.; Cebron, N.; Dill, F.; Gabriel, T.R.; Kötter, T.; Meinl, T.; Ohl, P.; Thiel, K.; Wiswedel, B. KNIME—The Konstanz Information Miner. *SIGKDD Explor.* **2009**, *11*, 26–31. [CrossRef]
38. Zycinski, G.; Barla, A.; Squillario, M.; Sanavia, T.; di Camillo, B.; Verri, A. Knowledge Driven Variable Selection (KDVS)—A new approach to enrichment analysis of gene signatures obtained from high-throughput data. *Source Code Biol. Med.* **2013**. [CrossRef] [PubMed]
39. Yousef, M.; Ulgen, E.; Ozisik, O.; Sezerman, O.U. CogNet: Classification of gene expression data based on ranked active-subnetwork-oriented KEGG pathway enrichment analysis. *PeerJ* **2020**. [CrossRef]

40. Yousef, M.; Goy, G.; Mitra, R.; Eischen, C.M.; Amhar, J.; Burcu, B. miRcorrNet: Integrated microRNA Gene Expression and mRNA Expression Based Machine Learning combined with Features Grouping and Ranking. Unpublished Work, 2020; in submit.
41. Eisen, M.B.; Spellman, P.T.; Brown, P.O.; Botstein, D. *Cluster Analysis and Display of Genome-Wide Expression Patterns*; National Academy of Sciences: Washington, DC, USA, 1998; Volume 95.
42. Wang, J.; Li, H.; Zhu, Y.; Yousef, M.; Nebozhyn, M.; Showe, M.; Showe, L.; Xuan, J.; Clarke, R.; Wang, Y. VISDA: An open-source caBIG$^{TM}$ analytical tool for data clustering and beyond. *Bioinformatics* **2007**, *23*. [CrossRef]
43. Guyon, J.W.I.; Stephen, B.; Vladimir, V. Gene Selection for Cancer Classification using Support Vector Machines, Machine Learning. *Mach. Learn.* **2002**, *46*, 389–422. [CrossRef]
44. Nacu, S.; Critchley-Thorne, R.; Lee, P.; Holmes, S. Gene expression network analysis and applications to immunology. *Bioinformatics* **2007**, *23*, 850–858. [CrossRef]
45. Sain, S.R.; Vapnik, V.N. The Nature of Statistical Learning Theory. *Technometrics* **1996**. [CrossRef]
46. Duan, K.-B.; Rajapakse, J.C.; Wang, H.; Azuaje, F. Multiple SVM-RFE for Gene Selection in Cancer Classification With Expression Data. *IEEE Trans. Nanobiosci.* **2005**, *4*, 228–234. [CrossRef]
47. Das, P.; Roychowdhury, A.; Das, S.; Roychoudhury, S.; Tripathy, S. sigFeature: Novel Significant Feature Selection Method for Classification of Gene Expression Data Using Support Vector Machine and t Statistic. *Front. Genet.* **2020**, *11*, 247. [CrossRef]
48. Xu, C.; Jackson, S.A. Machine learning and complex biological data. *Genome Biol.* **2019**, *20*, 76. [CrossRef]
49. Ulgen, E.; Ozisik, O.; Sezerman, O.U. PathfindR: An R package for comprehensive identification of enriched pathways in omics data through active subnetworks. *Front. Genet.* **2019**, *10*, 858. [CrossRef]

*Article*

# Radiomics Analysis on Contrast-Enhanced Spectral Mammography Images for Breast Cancer Diagnosis: A Pilot Study

Liliana Losurdo [1,†], Annarita Fanizzi [1,*,†], Teresa Maria A. Basile [2], Roberto Bellotti [2], Ubaldo Bottigli [3], Rosalba Dentamaro [1], Vittorio Didonna [1], Vito Lorusso [1], Raffaella Massafra [1], Pasquale Tamborra [1], Alberto Tagliafico [4], Sabina Tangaro [5] and Daniele La Forgia [1]

[1] Istituto Tumori "Giovanni Paolo II" I.R.C.C.S., 70124 Bari, Italy; lilianalosurdo@gmail.com (L.L.); r.dentamaro@oncologico.bari.it (R.D.); v.didonna@oncologico.bari.it (V.D.); v.lorusso@oncologico.bari.it (V.L.); massafraraffaella@gmail.com (R.M.); pasqualetamborra@gmail.com (P.T.); d.laforgia@oncologico.bari.it (D.L.F.)

[2] Dip. Interateneo di Fisica "M. Merlin", Università degli Studi di Bari "A. Moro", 70125 Bari, Italy; teresamaria.basile@uniba.it (T.M.A.B.); roberto.bellotti@uniba.it (R.B.)

[3] Dip. di Scienze Fisiche, della Terra e dell'Ambiente, Università degli Studi di Siena, 53100 Siena, Italy; ubaldo.bottigli@unisi.it

[4] Dip. di Scienza della Salute, Università degli Studi di Genova, 16132 Genova, Italy; alberto.tagliafico@unige.it

[5] INFN—Istituto Nazionale di Fisica Nucleare, 70126 Bari, Italy; Sonia.Tangaro@ba.infn.it

[*] Correspondence: annarita.fanizzi.af@gmail.com; Tel.: +39-080-555-5111

[†] These authors contributed equally to this work.

Received: 8 October 2019; Accepted: 11 November 2019; Published: 13 November 2019

**Abstract:** Contrast-enhanced spectral mammography is one of the latest diagnostic tool for breast care; therefore, the literature is poor in radiomics image analysis useful to drive the development of automatic diagnostic support systems. In this work, we propose a preliminary exploratory analysis to evaluate the impact of different sets of textural features in the discrimination of benign and malignant breast lesions. The analysis is performed on 55 ROIs extracted from 51 patients referred to Istituto Tumori "Giovanni Paolo II" of Bari (Italy) from the breast cancer screening phase between March 2017 and June 2018. We extracted feature sets by calculating statistical measures on original ROIs, gradiented images, Haar decompositions of the same original ROIs, and on gray-level co-occurrence matrices of the each sub-ROI obtained by Haar transform. First, we evaluated the overall impact of each feature set on the diagnosis through a principal component analysis by training a support vector machine classifier. Then, in order to identify a sub-set for each set of features with higher diagnostic power, we developed a feature importance analysis by means of wrapper and embedded methods. Finally, we trained an SVM classifier on each sub-set of previously selected features to compare their classification performances with respect to those of the overall set. We found a sub-set of significant features extracted from the original ROIs with a diagnostic accuracy greater than 80%. The features extracted from each sub-ROI decomposed by two levels of Haar transform were predictive only when they were all used without any selection, reaching the best mean accuracy of about 80%. Moreover, most of the significant features calculated by HAAR decompositions and their GLCMs were extracted from recombined CESM images. Our pilot study suggested that textural features could provide complementary information about the characterization of breast lesions. In particular, we found a sub-set of significant features extracted from the original ROIs, gradiented ROI images, and GLCMs calculated from each sub-ROI previously decomposed by the Haar transform.

**Keywords:** breast cancer; radiomics analysis; feature extraction; feature selection; Haar wavelet decomposition; gray-level co-occurrence matrix; contrast-enhanced spectral mammography

---

## 1. Introduction

In image processing, feature extraction plays a very important role: it allows obtaining quantitative information (features) from medical images that cannot be detected by means of simple visual observation by the operator, using appropriate mathematical methods and computers [1]. This discipline, known as radiomics, is an emerging translational research topic in cancer studies. Radiomics analysis on different medical images is often inserted in a framework of pattern recognition tasks for characterizing lesions of various natures and on different imaging modalities [2,3]. Indeed, although clinicians are trained for visual pattern recognition, it is still a subjective evaluation. Several studies have investigated the usefulness and reliability of radiomics to discriminate benign breast lesions from cancers, evaluate prognosis, or response to therapies, demonstrating that it could potentially improve diagnosis and characterization of lesions: automatic recognition tools can provide objective information to support clinical decision-making or improve the radiologist's confidence in the challenging diagnostic task [4,5].

For this purpose, many techniques may be applied. Depending on the clinical utility of the research and the type of medical images, different typologies of features can be extracted from them, such as statistical, textural, morphological, and shape features [6–9], each of which provides particular information useful to describe a specific aspect of a lesion. Some radiomics works are based on texture analysis, since the texture may be defined as the pattern of information or arrangement of structure found in an image [10–12].

Texture analysis aims to describe the fundamental characteristics of textures and to represent them in a simpler, but distinctive form, in order to use them for a robust and accurate classification and segmentation of objects [13]. There are two types of textural feature measurements, first and second order: in the first order, texture measurements are statistics directly calculated from an individual pixel and do not consider pixel neighbor relationships (i.e., the intensity histogram and intensity features); in the second order, measurements consider the relationship between neighboring pixels. Moreover, when the image is analyzed and decomposed into different frequency sub-bands by means of wavelet transforms or co-occurrence matrices, this technique may be more effective [13,14].

In general, this approach can catch information about the characteristics of a tumor missed by a human reader, therefore providing details with a significant diagnostic value. Textural features capture spatial and spectral frequency patterns, as well as characterize the relationships between different intensity levels within the lesion; they might not be immediately visible to radiologists and thus have the potential to complete their diagnostic skills. Moreover, this analysis can be performed in an automated way without any human intervention, not as well as for morphological/shape features or BIRADS [15] descriptors.

With particular reference to breast lesions, in order to characterize masses and microcalcifications, textural features are often extracted from mammographic images after having applied the so-called Gray-Level Co-occurrence Matrix (GLCM), which correlates the intensity of the gray levels of neighboring pixel pairs in different directions [16–19]. Some works also consider the possibility to decompose each image into sub-images by means of 2D and discrete wavelet transforms [18,20,21], Gabor filters [19,22,23], and the image gradient [24] before calculating co-occurrence matrices to detect defects in the image texture. Then, they make a comparison between these feature extraction methods in order to estimate the most appropriate method of feature extraction from mammograms. To analyze breast Magnetic Resonance (MR) images, textural features extracted by GLCMs [25,26] or wavelet transforms [27] are used, while in some other works, a combination of textural and statistical [28,29] or morphological [30] features is preferred.

Nevertheless, in recent years, new radiological imaging equipment has been developed in order to increase the diagnostic performances, especially when breast is dense. Among these new techniques, Contrast-Enhanced Spectral Mammography (CESM) [31–33] combines the principles of mammography with the injection of an intravenous iodinated Contrast Medium (CM), which allows, as in MR images, a contrastographic evaluation of the breast: this highlights the areas that capture the contrast medium, the typical expression of neo-angiogenesis neoplasm. As in MR, CESM images may present enhancement of normal breast parenchyma after CM injection, known as Background Parenchymal Enhancement (BPE) [15,34]. On the contrary, CESM is less influenced by hormonal status than MR [35], and this could provide important additional information on the detection of lesions in patients with a high BPE in degree where distinguishing a lesion from the non-enhanced background is objectively difficult. Moreover, CESM is less expensive and more tolerated by patients than MR [36].

In the literature, several analyses are aimed at comparing CESM performances to mammography [37–39] and MR [40–42] ones by the reading medical images by expert radiologists. Differently, there are few works in which a radiomics analysis has been performed. The first approaches to develop computerized algorithms addressed to increase the diagnostic performances on CESM images were reported only in [43–45]. In these works, several features used, such as morphological and BIRADS descriptors, presented limits of subjectivity in the feature extraction process due to the fact that they depended on the judgment of the radiologist who manually segmented the lesions and determined their benign/malignant nature based on his/her experience. Moreover, some standard textural features were extracted on original and pre-processed images by using GLCMs of original images, Gabor filter banks, and Laplacian of Gaussian histograms. However, no comparative analysis was performed between the different features used.

In our work, we propose a preliminary radiomics analysis aimed to explore the usefulness of quantitative information extracted from CESM images, both in original and pre-processed format, to support the radiologist in the diagnosis of breast cancer. Specifically, in order to make the lesion classification more objective and operator independent, once the radiologist has identified the Regions Of Interest (ROIs), the characteristic features are extracted in an automated manner. The aim of our work is to understand the behavior of each different set of well known textural features automatically extracted from CESM images and to compare them with each other. An important role is played by the feature selection processes used to describe and characterize ROIs: starting from the initial feature set, a sub-set of these features, characterized by a higher discriminating power, is selected for a more manageable data processing [46]. Then, we select the most important features by developing two different approaches of feature importance, such as embedded and wrapper methods.

## 2. Materials and Methods

### 2.1. Materials

#### 2.1.1. CESM Examination

CESM is a technique allowing the acquisition of multiple views of both breasts by producing two types of images. A typical example of CESM images is shown in Figure 1, where it is clear that a Low Energy (LE) image may be overlapped completely by a 2D digital mammography image (a), while High Energy (HE) images are not displayable in the reporting monitor (b); instead, a ReCombined (RC) image highlights contrast medium uptake, as a breast MR image (c).

The CESM technique consists of the acquisition of low and high energy digital mammograms, both in CranioCaudal (CC) and MedioLateral Oblique (MLO) views, with the dual energy technique after the administration of an intravenous iodinated CM by an automated injector to ensure a constant flow. In order to reduce the so-called anatomical noise due to the typical tissue overlap especially of mammographically "dense" fibro-glandular breasts, a combined mammographic image, where only the CM is highlighted, is produced by means of spectral subtraction.

On these RC images, some motion blur could be sometimes observed because of movements between the acquisition of different images; however, this dual energy subtraction technique is less sensitive to movement artifacts than traditional temporal subtraction.

In this study, a modified digital mammography system derived from a standard Senographe Essential (GE Healthcare) was used for all CESM exams. CESM images were all in DICOMformat and were evaluated by two dedicated radiologists with more than 10 years of experience in reading mammography and breast MR images and trained in reading contrast enhanced images.

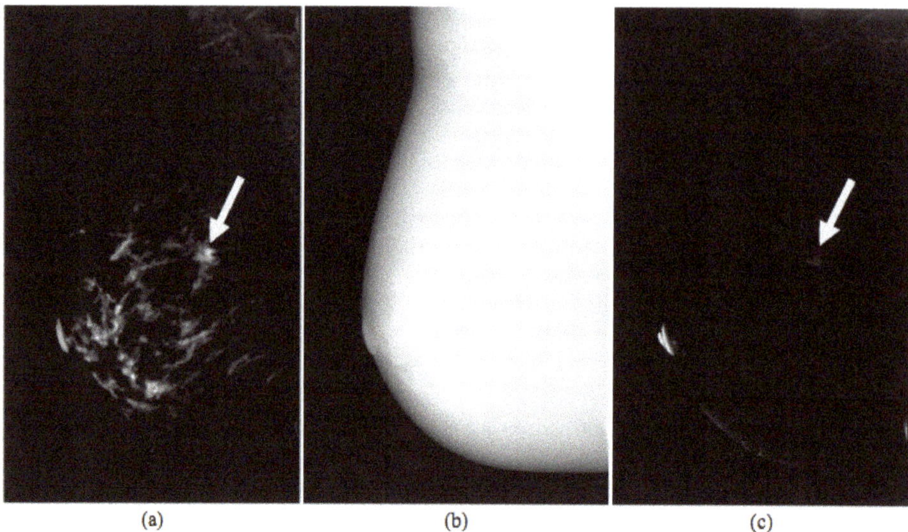


**Figure 1.** Images produced by CESM instrumentation. Typical example of low energy (**a**), high energy (**b**), and recombined (**c**) images [37]. The white arrow points to a suspicious lesion.

### 2.1.2. Experimental Dataset

From March 2017 to June 2018, we collected CESM images of patients referred to Istituto Tumori "Giovanni Paolo II" I.R.C.C.S. of Bari (Italy) from the breast cancer screening phase. Patients undergoing CESM had indications for breast MR, but they could not perform it due to several contraindications or impossibility. Therefore, in our Institute, the use of this method is applied only as a second alternative to MR, even for patients who have to perform urgent MR for therapies or programmed surgery, but that have not found access to MR, as indicated by the European guidelines on CESM [47]. Our observational study was approved by the Medical Ethics Committee of the Institute, and written informed consent prior to undergoing CESM examination was signed by all eligible patients.

We selected images in MLO or CC view of 51 patients aged between 38 and 80 years (with a mean of $52.3 \pm 9.9$ years), resulting in being positive in the methods for the presence of at least one finding after histological examination.

In order to avoid the BPE degree being a confusing factor for the purposes of evaluating the diagnostic capacity of the features, the sample was selected in such a way as to have a fair distribution of benign and malignant lesions for each BPE class.

Two of our radiologists dedicated to senologic diagnostics identified and classified a total of 55 primary and, if present, also secondary lesions (29 benign and 26 malignant) from 0.5 to 13.5 cm according to the BIRADS classification [48]: lesions belonging to BIRADS 2 and 3 classes were labeled as benign, while lesions belonging to BIRADS 4 and 5 classes were considered as malignant. Then, the histological diagnosis based on bioptic sampling established that 29 ROIs contained benign lesions and 26 ROIs included malignant ones. All ROIs were extracted both from LE and RC images.

## 2.2. Methods

### 2.2.1. Feature Extraction

In this paper, we present a comparison of several feature sets in order to establish an order of importance among them in the classification of lesions on CESM images. For this purpose, some properties of the image texture [49], such as gray-tone distribution and spatial dependencies, were considered. MATLAB R2017a (MathWorks, Inc., Natick, MA, USA) software was used for all analysis steps.

The considered sets of features extracted from the original ROIs (both LE and RC), their gradient, and wavelet decompositions using some methods are listed below. Figure 2 summarizes the feature extraction process, from the identification of ROIs to each extracted feature set.

STAT set: Several standard statistical features (mean, standard deviation and their ratio, variance, skewness, entropy, relative smoothness, and kurtosis) were directly extracted from each original ROI. Moreover, the minimum and maximum values of gray-level and their difference were computed. These 11 features were extracted both from LE and RC original ROIs, forming a vector of 22 features and giving some relevant and objective information of each ROI.

GRAD set: The gradient of an image is represented as a two-component vector ($x$- and $y$-derivative) defined at each pixel [50]: they can be computed by the convolution with a kernel, such as the Sobel or Prewitt operator, since the image is a discrete function for which the derivatives are not defined. For each vector, the magnitude $Gmag$ shows how quickly the intensity of each pixel is changing in the neighborhood of pixel $(x, y)$ in the direction of the gradient, while the direction $Gdir$ represents the orientation of greatest intensity change in the neighborhood of pixel $(x, y)$. They are given by $\sqrt{f_x^2 + f_y^2}$ and $\arctan(f_y/f_x)$, respectively, where $f_x$ and $f_y$ are the components of the vector. If the original images are obtained under different conditions (i.e., exposure energy), it is possible that the pixel values are drastically different, even though they represent the same characteristics (e.g., a benign or malignant lesion). The gradiented images are less susceptible to these factors and therefore are usually used for robust feature and texture matching. For this feature set, mean, variance, skewness, entropy, relative smoothness, and kurtosis were extracted from the gradient's magnitude and direction of each LE and RC original ROI by using a Sobel kernel, thus obtaining a total of 24 features.

HAAR set: As a fundamental property of the image texture, the scale at which the image is observed and analyzed was exploited by using a wavelet transform based on texture analysis approach, such as the Haar wavelet transform [50,51]. This allows decomposing the image by using an orthonormal basis composed by scaled and translated functions. Conceptually, the scaled function represents the low frequency component of the scaling function in 2 dimensions, obtaining one 2D scaling function. On the contrary, the translated function includes three different components (horizontal, vertical, and diagonal). Since this wavelet transform is separable, four combinations of these functions may be obtained by means of low and high filters. The Haar transform is considered the first known wavelet basis and widely used as a teaching example. In particular, the 2D Haar wavelet decomposes an image first with a low-pass filter obtaining a downscaled Low-Low ($LL$) sub-image and then with a high-pass filter for each component of the translated function obtaining the corresponding High-Low ($HL$), Low-High ($LH$), and High-High ($HH$) sub-images. In general, low and medium frequencies match image content while high ones usually emphasize noise or texture areas. Therefore, in the wavelet domain, noise and image content or image regions of different complexity (quasi-homogeneous, textural, containing borders or objects, etc.) can be distinguished and used in noise parameters' estimation, filtering, compression, etc. Then, the image decomposition can be iterated at successive levels applying the Haar wavelet transform on the first downscaled sub-image. In this work, we performed 2D Haar transform at two levels of decomposition on each ROI; hence, we extracted mean, variance, skewness, entropy, relative smoothness, and kurtosis from each sub-ROI, both LE and RC, thus obtaining a set of 96 features.

GLCM set: The approach considers how many times the gray-level intensity value of a reference pixel is associated with another gray-level intensity value on each neighbor pixel in a specific spatial relationship obtaining the Gray-Level Co-occurrence Matrix (GLCM) [10,13] for each ROI. This spatial relationship, known as offset, is given by the distance between a pixel and its neighbors according to a specified direction ($dir1 = 0°$, $dir2 = 45°$, $dir3 = 90°$, $dir4 = 135°$).

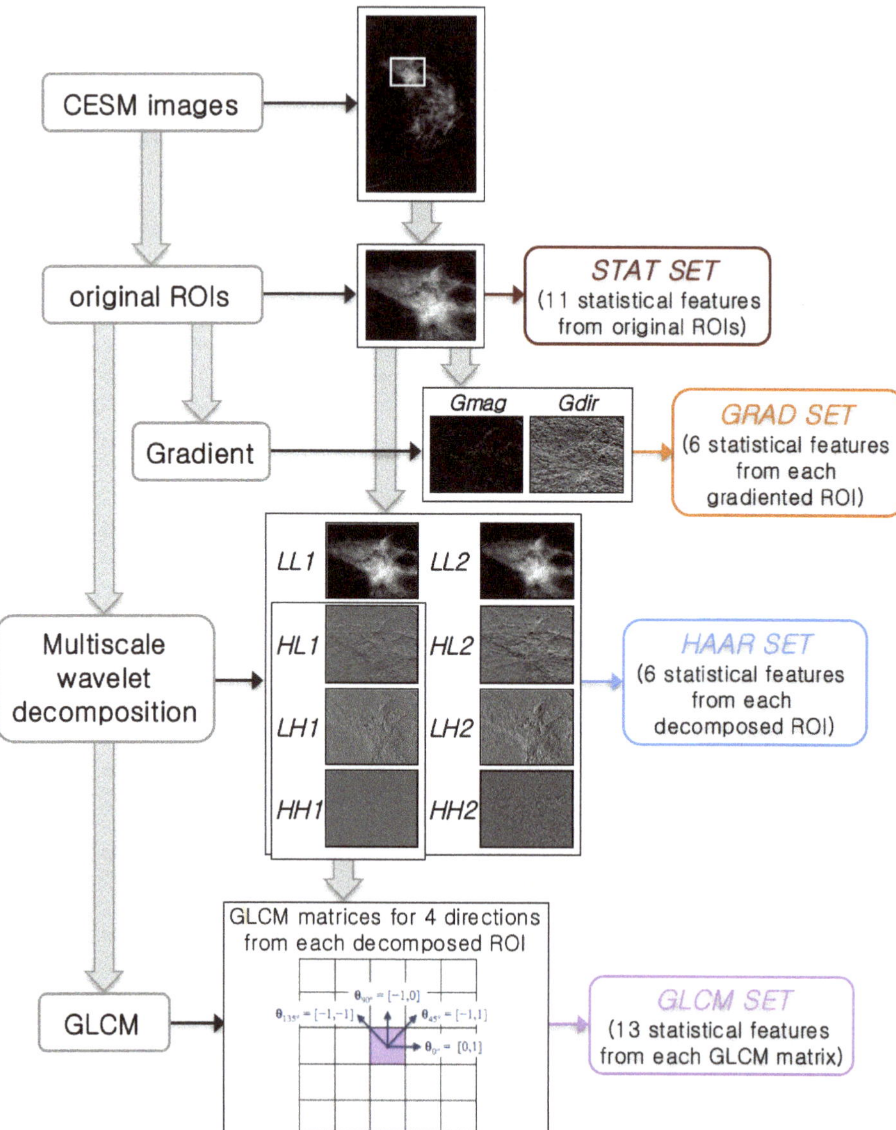


**Figure 2.** Scheme of feature extraction. Feature extraction process going from the identified original ROI to each extracted feature set. This scheme is shown starting from a low energy image, but it is also performed for recombined images.

Thus, this last set of statistical features (contrast, correlation, cluster prominence, cluster shade, dissimilarity, energy, entropy, homogeneity, sum average, sum variance, sum entropy, difference entropy, and normalized inverse difference moment) was extracted from the co-occurrence matrices of each sub-ROI previously decomposed by the Haar transform only at Level 1 (*HL, LH,* and *HH*) in the four directions, obtaining 156 features. These particular measurements have invariant properties under some image transformations because they are calculated by GLCM and allow better detecting any defects in the image texture. Then, they could determine the location and the range of the pixels having a structure with considerable deviation in their values of intensity or spatial arrangement with respect to the background texture. Since we have taken into account both LE and RC images, this set was totally formed by 312 features.

## 2.2.2. Feature Reduction and Importance Analysis

The aim of this work is to explore the discriminating power of feature sets extracted by different techniques, as described above, to characterize the benign and malignant breast lesions. For this purpose, we present a multi-parametric analysis approach to evaluate how these individual typologies of features behave on images of this still unexplored imaging technique (i.e., CESM). Specifically, we analyzed the features of each feature set jointly in order to solve the benign vs. malignant classification problem by means of two different approaches, reduction of the feature number and selection of the most discriminating features. In Figure 3, a schematic overview of our feature analysis approach is shown.

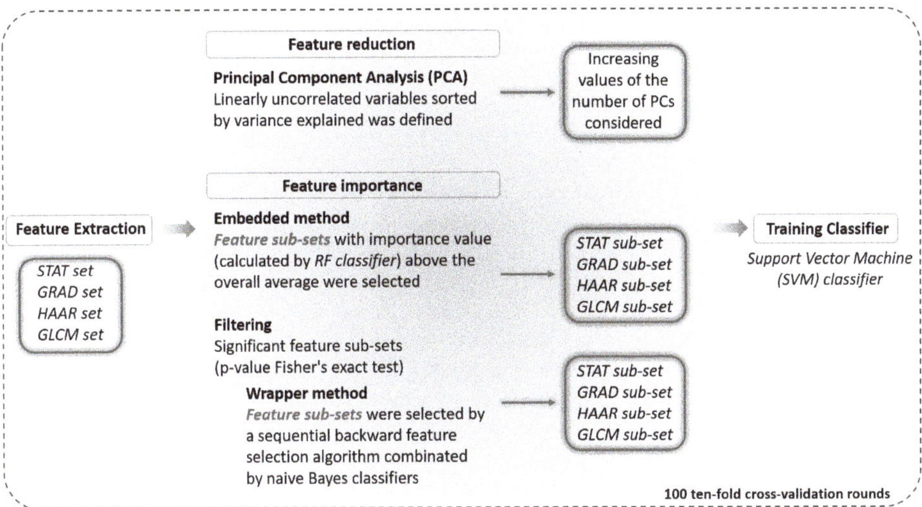


**Figure 3.** Schematic overview of the radiomic analysis approach. Textural features automatically extracted from each ROI in the first step (Figure 2) are analyzed by using a principal components analysis and a feature importance process by means of two different approaches (wrapper and embedded techniques). The discrimination performances of an SVM classifier trained on the feature subsets are evaluated on 100 ten-fold cross-validation rounds.

First, we evaluated the overall impact of each feature set on the diagnosis through a Principal Component Analysis (PCA) [52]. This method of feature reduction performs a linear mapping of the data to a lower dimensional space in such a way that the variance of the data in the low-dimensional representation is maximized, allowing removing redundant information. In this way, the number of features is reported in the same number of linearly uncorrelated latent variables: this technique performs a linear transformation of the features that projects the original ones into a new Cartesian system where the variables are sorted in descending order with respect to the overall variance percentage explained. Therefore, we evaluated the classification performance of each feature set considering an increasing number of Principal Components (PCs) to train a Support Vector Machine (SVM) classifier.

Moreover, in order to identify for each feature set extracted a sub-set of them with higher diagnostic power, we developed two approaches for the task of feature importance evaluation, such as wrapper and embedded methods [46].

Wrapper methods measure the usefulness of features based on the classifier performance. They solve the "real" problem using the predictor as a black box and its performance as an objective function to evaluate the variable sub-set [53]. The sequential backward feature selection algorithm identifies the features that have best predicted the expected result by sequentially removing features from the initial candidate set until a removal increases the error or the accuracy decreases significantly; therefore, it stops when a local maximum is found. In particular, in this work, we developed two steps of feature selection wrapper algorithm: first, we identified a significant feature sub-set correlated with the variable to be predicted ($p$-value Wilcoxon–Mann-Whitney test $< 0.05$); then, on this selected sub-set of features, we implemented a sequential backward feature selection algorithm combined with a naive Bayes classifier.

Embedded methods allow an optimization between the interaction of the selected features and the machine learning algorithm used for the classification, because the selection criterion is grafted onto it. Indeed, they combine the qualities both of filter and wrapper methods. Random Forests (RF) are among the most popular machine learning algorithms because they generally provide good predictive performance and low over-fitting [54]. The RF classifier processes an analysis of feature importance with respect to its expected result; therefore, these methods essentially fulfill the goal, i.e., the optimization of the classification performances. In particular, the tree-based strategy used by RF naturally ranks by how well they improve the purity of the node: nodes with the greatest decrease in impurity are at the start of the trees, while nodes with the least decrease in impurity occur at the end of trees measured by Gini's diversity index.

Thus, in this work, the RF algorithm allowed estimating predictor importance values by permuting out-of-bag observations among the trees, and features with an importance value above the overall average were selected.

For each feature set, we identified the significant feature sub-set with a selection frequency different from chance ($p$-value Fisher's exact test $< 0.05$). Specifically, we tested that the occurrence frequency of the features selected by each of two methods (i.e., embedded and wrapper) was significantly different from that obtained after permuting the diagnostic target in the dataset [55].

Finally, in order to compare the classification performances of the features selected by the two methods, we trained a binary SVM classifier. The performances of the prediction model were evaluated on 100 ten-fold cross-validation rounds [56] in terms of:

$$Accuracy = (TP + TN)/(TP + TN + FP + FN),$$

$$Sensitivity = TP/(TP + FN),$$

$$Specificity = TN/(TN + FP),$$

where $TP$ and $TN$ stand for True Positive (number of true malignant ROIs identified) and True Negative (number of true benign ROIs identified) cases, while $FP$ (number of benign ROIs identified as malignant) and $FN$ (number of malignant ROIs identified as benign) are the False Positive and False Negative ones, respectively.

## 3. Results

The goal of our analysis was to investigate the impact of textural features extracted in an automated manner through different techniques in discriminating benign and malignant ROIs. The main interest of this study was to understand the total behavior of these textural features in order to have an overall view of the goodness of the approach used for the feature reduction. First, we evaluated the overall discriminant power of each feature set performing a principal component analysis as the feature reduction technique. Then, for each feature set, we searched for those that contributed the most to discriminating benign and malignant ROIs.

### 3.1. Principal Component Analysis

A preliminary analysis, reported in Figure 4, shows that in each feature set, there were significant, often very strong correlations between them. This suggested that a feature reduction approach can be useful to analyze the overall diagnostic power of each feature set, specifically for more numerous sets, like HAAR and GLCM sets. Therefore, we performed a PCA for each suitably standardized set of features.

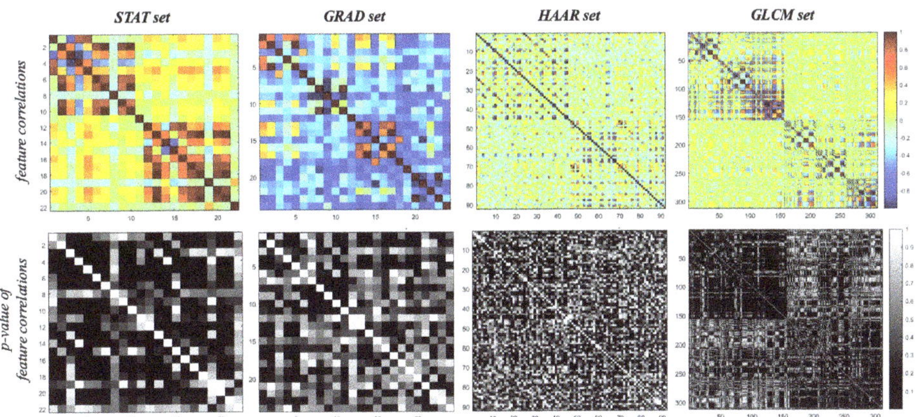

**Figure 4.** Correlation graphs. Graphs of feature correlations (**top**) and their relative *p*-values (**bottom**) for each set.

In Figure 5, we show the performance measurements obtained by training an SVM classifier for increasing values of the number of PCs used that were previously ordered with respect to the overall variance explained by the same principal components. There are several criteria in order to select this number of PCs that guarantees the lowest possible loss of information, but they could lead to different results. Since the purpose of the work was to provide an overall assessment of the diagnostic power of the extracted features and not to optimize the classification problem, we show the global performance trend.

The STAT and HAAR sets showed the best mean accuracy (about 80%) on 100 ten-fold cross-validation rounds; on the contrary, the GLCM set was the one with the lowest diagnostic power (the best mean accuracy reached about 64%), while the GRAD set achieved a maximum mean accuracy of about 72%. Specifically, in accordance with what has been observed about the presence of significant correlations between features in each set, the best predictive performance was achieved with no more than 10 principal components.

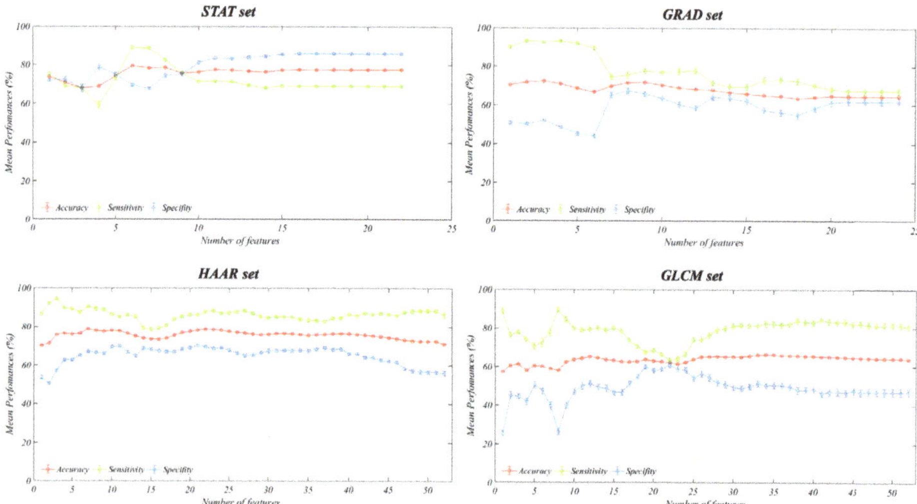

**Figure 5.** Classification performances. Mean classification performances evaluated on 100 ten-fold cross-validation rounds for each feature set with respect to the number of Principal Components (PCs) used to train an SVM classifier.

*3.2. Feature Importance Analysis*

In Figure 6, we summarize the significant features whose occurrence frequency was significantly different from chance by testing as described above.

Regarding the STAT set, 19 of the 22 features were statistically significant, and only six of these were selected by both methods implemented.

Among the 24 features of the GRAD set, only 13 were significant, and eight of them were selected by both implemented methods.

The HAAR and GLCM sets were the groups with the largest number of features. The HAAR set consisted of 96 features, and only 43 were statistically significant, while no more than 19 features were selected by both methods implemented. The GLCM set was that with the greater reduction in the number of selected features. Indeed, it consisted of 312 features, and only 51 resulted in being statistically significant. The features selected by the wrapper method were 39, and only seven of them were also selected by the embedded method.

The diagnostic performances of the sub-set formed by only significant features selected for each set were evaluated by training an SVM classifier and obtaining on 100 ten-fold cross-validation rounds in terms of accuracy, sensitivity, and specificity, as summarized in Table 1. We report the mean value and relative confidence interval of each performance measure obtained in the cross-validation process to provide an idea of our results' reliability. During the study and development of the present work, we evaluated several classifiers, such as SVM, Bayes, and k-nearest neighbors, different from those used in the feature selection step, in order to have results that were not biased by the same selection technique. However, SVM was the best performing classifier; therefore, we preferred to report only this result without overloading the reading of this paper.

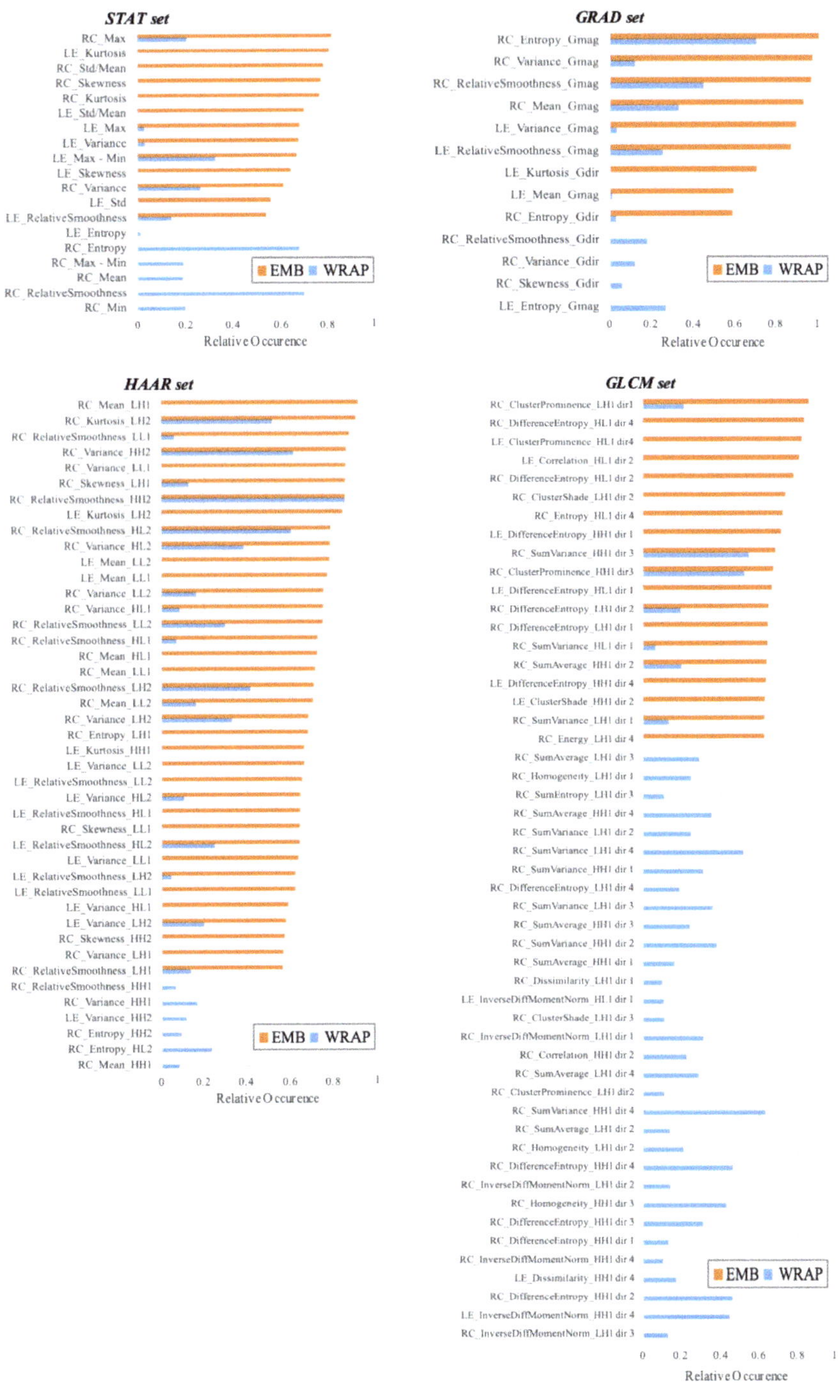

**Figure 6.** Occurrence frequency of the selected features. Occurrence frequency of the features selected by the Embedded (EMB) and Wrapper (WRAP) methods that is significantly different from chance ($p$-value of Fisher's exact test $\leq 0.05$) for each feature set.

**Table 1.** Classification performances of the SVM classifier trained on sub-sets of significant features identified by two methods of feature selection. Furthermore, the total performances obtained taking into account the significant sub-sets selected from both methods are shown.

| Method of Feature Selection | Feature Set (# of Selected Features) | Accuracy (%) Mean [CI 95%] | Sensitivity (%) Mean [CI 95%] | Specificity (%) Mean [CI 95%] |
|---|---|---|---|---|
| Embedded | STAT (13) | 80.69 [80.42, 80.96] | 86.38 [85.59, 87.17] | 75.00 [74.10, 75.90] |
| | GRAD (9) | 76.02 [75.65, 76.39] | 81.28 [80.60, 81.96] | 70.76 [70.19, 71.33] |
| | HAAR (37) | 59.22 [58.25, 60.19] | 70.59 [66.65, 74.53] | 47.86 [44.18, 51.70] |
| | GLCM (19) | 73.86 [73.37, 74.35] | 85.62 [83.68, 87.56] | 62.10 [59.96, 64.24] |
| Wrapper | STAT (10) | 80.91 [80.65, 81.16] | 90.28 [89.75, 90.81] | 71.55 [71.09, 72.01] |
| | GRAD (12) | 74.66 [73.97, 75.35] | 82.86 [80.83, 84.89] | 66.45 [64.44, 68.46] |
| | HAAR (25) | 60.76 [59.91, 61.61] | 54.79 [50.16, 59.42] | 66.72 [62.24, 71.20] |
| | GLCM (39) | 76.50 [75.92, 77.08] | 81.93 [80.75, 87.94] | 71.07 [69.67, 66.22] |
| Embedded + Wrapper | STAT (19) | 79.43 [79.25, 79.61] | 82.83 [82.10, 83.56] | 76.03 [75.25, 76.81] |
| | GRAD (13) | 74.47 [73.87, 75.06] | 83.17 [81.49, 84.85] | 65.76 [63.82, 67.70] |
| | HAAR (43) | 58.83 [57.90, 59.76] | 61.76 [56.94, 66.58] | 55.90 [51.52, 60.28] |
| | GLCM (51) | 75.95 [75.36, 76.54] | 80.59 [79.20, 81.98] | 71.31 [69.41, 73.21] |

Generally, the significant feature sub-set selected for the STAT and GRAD sets reproduced the average performance obtained by PCA, and therefore, they were the ones with the highest discrimination power. Specifically, the feature sub-set selected with the embedded approach was preferable to that identified with the wrapper method because it reduced the trade-off between sensitivity and specificity. For the model trained on the STAT sub-set selected by the embedded method, these were 86.38% and 75.00%, respectively, with respect to the 90.28% and 71.55% obtained with features selected by the wrapper method; similarly, the model trained on the GRAD sub-set selected by the embedded method reached a sensitivity and a specificity of 81.28% and 70.76%, respectively, compared to the 82.86% and 66.45% obtained with features selected by the wrapper method. For both sets, the classification performances decreased when the model was trained on all significant features. As described above, the average classification accuracy obtained by the PCs of the GLCM set did not exceed 64%. Nevertheless, using the sub-set of significant features identified by the wrapper method, the model reached a mean accuracy of 76.50%, a mean sensitivity of 81.93%, and a mean specificity of 71.07%.

On the contrary, the HAAR set of features used globally were the ones with the highest discriminating power, but the performances reached with this set trained the model only on the significant features, collapsing by about 20 percentage points, regardless of the sub-set of features used. The best performance obtained by the HAAR set of features was reached by using the significant sub-set selected by the wrapper method with a mean accuracy, sensitivity, and specificity of 60.76%, 54.79%, and 66.72%, respectively.

## 4. Discussion

In this paper, several methods were used to extract features from CESM images and to analyze them. This automatic extraction process was performed by taking into account computational simplicity, invariance properties, and noise sensitivity. Moreover, we focused our work on textural features, since they highlighted the relationships between different levels of intensity within the lesion and captured spatial and spectral frequency patterns. For these purposes, we extracted feature sets calculating statistical measures on the original ROIs (STAT set) and on their manipulations by filters (GRAD set), wavelet functions (HAAR set), or considering the relationships between neighboring pixels and their gray levels (GLCM set).

Currently, the literature is poor regarding radiomic analysis of breast cancer on CESM images useful to develop systems as diagnostic support tools. In [43], a first approach was proposed to analyze morphological descriptors on mass (shape, margins, pattern, and degree of internal enhancement) and non-mass (distribution, pattern, and degree of internal enhancement) lesions on CESM images aimed to assess their impact on the discrimination between benign and malignant ones.

In [45], the authors developed a convolutional neural network based decision support system combining CESM pixel information with BIRADS descriptors provided by radiologists. Nevertheless, these works presented limits of subjectivity of the feature extraction process, in particular of morphological features and BIRADS descriptors, because of their dependence on the radiologist's experience.

Instead, in [44], a multi-parametric feature analysis approach aimed to construct a computer aided diagnosis tool to increase the diagnostic performances of the CESM technique was presented. In this work, a set of morphologic and textural features was extracted by GLCMs from original images, Gabor filter banks, Laplacian of Gaussian histograms, local binary pattern, and discrete orthonormal Stockwell transform, and after that, an expert breast radiologist manually outlined lesion boundaries on each image. However, an analysis of the individual types of features used was not performed.

Our goal was to evaluate several well known textural feature sets in biomedical image analysis whose extraction does not require the intervention of radiologists, in order to find quantitative additional information that may be integrated with the experience of human readers to enhance diagnostic accuracy.

For this purpose, firstly, we evaluated the overall diagnostic power of each feature set by means of a principal component analysis. Then, we implemented two different feature selection techniques, such as wrapper and embedded; these methods are quite similar since they are used to optimize the performance of a learning algorithm or model. However, they differ in the fact that only for the embedded methods, an intrinsic model building metric is used during training, while the wrapper ones operate by iteratively selecting the insertion or removal of a feature and evaluating the results obtained. Finally, the accuracy classification of PCs calculated and also each feature subsets identified were evaluated in the cross-validation process with a well known state-of-the-art classifier, such as SVM, independent of those used in the selection phase.

The experimental results showed that for the STAT set, the wrapper method has more frequently selected features related to entropy and relative smoothness, as well as absolute variability measures; on the contrary, the embedded method has more frequently selected features linked to the shape of the gray levels' distribution, such as kurtosis and skewness, as well as absolute and relative variability measures. With regard to the significant features of the GRAD set, it emerged that the most important features were measures calculated on the gradient magnitude of ROIs both on LE and RC images. The significant feature sub-set selected by the two approaches used in this work for STAT and GRAD sets was actually that with the highest discriminating power, because it could reproduce the average performances obtained by using it globally through the PCs.

Most of the significant features of the HAAR and GLCM sets were calculated on RC images, unlike what happened for the other two feature sets. In particular, among the 43 and 51 significant features of HAAR and GLCM, only 15 and 9 features, respectively, were calculated on LE images. For these two sets, the classification performances underwent a trend reversal when they were globally used by means of a PCA or when only the significant features identified were used. Indeed, the average classification accuracy obtained using the PCs of the GLCM set did not exceed 64%. Nevertheless, when we trained the classifier on the sub-set of significant features, the classification accuracy grew at least about 10 percentage points; therefore, there were evidently some features that introduced some distortions. Instead, the features of the HAAR set with the highest discriminating power were those used globally. However, the performances achieved in this way of training the model on the significant feature sub-set fell by about 20 percentage points. This indicated that the other features of the set contributed to a lesser extent to the resolution of the diagnostic problem.

## 5. Conclusions

Recent feasibility studies suggested that CESM is an useful investigation tool and that it can provide pre-operative staging and accurate treatment planning in breast cancer patients with an accuracy no less than MR [40]. CESM showed interesting results in terms of diagnostic sensitivity, compatible with those obtained by MR: in [39], the sensitivity was 100% for both techniques, while in [41], it was 100% by CESM and 93% by breast MR. Moreover, on the basis of state-of-the-art comparative results, CESM also had better tolerance and less discomfort than MR, as shown in [36,57,58]. Thus, this new imaging technique can represent a valid alternative to MR, also due to its better tolerance and lower discomfort with respect to the latter [36,57]. Nevertheless, CESM, as MR, has presented false positive cases [31], and it can still be considered a method that is subjective and dependent on the operator's experience due to the current lack of objective diagnostic support systems.

In this work, we proposed some preliminary results of a radiomics analysis useful to drive future works about automatic radiological support systems for the diagnosis of breast lesions by means of CESM images. We performed an extraction process of textural features, and then, we evaluated the diagnostic power of four feature sets extracted by using different techniques.

Textural features have the potential to support the diagnostic skills of radiologists because they capture spatial and spectral frequency patterns, often not easily visible to the human reader.

We found a sub-set of significant features extracted from the original ROIs, gradiented ones, and GLCMs calculated from each sub-ROI previously decomposed by the Haar transform. Nevertheless, the feature sets extracted from each sub-ROI decomposed by two levels of Haar transform were reliable in differentiating benign from malignant breast lesions when all of these were used without any selection. Moreover, most of the significant features calculated on HAAR decompositions and their GLCMs were extracted from RC images.

Future works include a validation study in which we will test the robustness of the significant features identified in a larger population, also with respect to the histological results of each lesion. Moreover, we will develop a computer aided diagnosis system combining them in order to optimize the classification performances.

**Author Contributions:** Conceptualization, L.L., A.F., T.M.A.B., S.T., and D.L.F.; data curation, L.L., A.F., and D.L.F.; formal analysis, L.L., A.F., T.M.A.B., and S.T.; Funding acquisition, V.L., and D.L.F.; methodology, L.L., A.F., and T.M.A.B.; software, L.L.and A.F.; supervision, S.T. and D.L.F.; validation, A.F.; writing, original draft, L.L., A.F., T.M.A.B., S.T., and D.L.F.; writing, review and editing, R.B., U.B., R.D., V.D., V.L., R.M., P.T., and A.T.

**Funding:** This research received no external funding.

**Acknowledgments:** This work was supported by funding from the Italian Ministry of Health "Ricerca Corrente 2018–2020".

**Conflicts of Interest:** The authors declare no conflict of interest. The funders had no role in the design of the study; in the collection, analyses, or interpretation of data; in the writing of the manuscript; nor in the decision to publish the results.

## Abbreviations

The following abbreviations are used in this manuscript:

| | |
|---|---|
| BPE | Background Parenchymal Enhancement |
| CC | CranioCaudal |
| CESM | Contrast-Enhanced Spectral Mammography |
| CI | Confidence Interval |
| CM | Contrast Medium |
| dir1 | Direction 1 (0°) |
| dir2 | Direction 2 (45°) |

dir3  Direction 3 (90°)
dir4  Direction 4 (135°)
EMB  Embedded
FN  False Negative
FP  False Positive
Gdir  Gradient direction
Gmag  Gradient magnitude
GLCM  Gray-Level Co-occurrence Matrix
HE  High Energy
HH  High-High
HL  High-Low
LDA  Linear Discriminant Analysis
LE  Low Energy
LH  Low-High
LL  Low-Low
MLO  MedioLateral Oblique
MR  Magnetic Resonance
PC(A)  Principal Component (Analysis)
RC  ReCombined
RF  Random Forest
ROI  Region Of Interest
SD  Standard Deviation
SVM  Support Vector Machine
TN  True Negative
TP  True Positive
WRAP  Wrapper

## References


1. Kumar, G.; Bhatia, P.K. A detailed review of feature extraction in image processing systems. In Proceedings of the 2014 Fourth International Conference on Advanced Computing & Communication Technologies, Rohtak, India, 8–9 February 2014; pp. 5–12.
2. Larue, R.T.; Defraene, G.; De Ruysscher, D.; Lambin, P.; Van Elmpt, W. Quantitative radiomics studies for tissue characterization: A review of technology and methodological procedures. *Br. J. Radiol.* **2017**, *90*, 20160665. [CrossRef] [PubMed]
3. Turani, Z.; Fatemizadeh, E.; Blumetti, T.; Daveluy, S.; Moraes, A.F.; Chen, W.; Mehregan, D.; Andersen, P.E.; Nasiriavanaki, M. Optical Radiomic Signatures Derived from Optical Coherence Tomography Images Improve Identification of Melanoma. *Cancer Res.* **2019**, *79*, 2021–2030. [CrossRef] [PubMed]
4. Valdora, F.; Houssami, N.; Rossi, F.; Calabrese, M.; Tagliafico, A.S. Rapid review: radiomics and breast cancer. *Breast Cancer Res. Treat.* **2018**, *169*, 217–229. [CrossRef] [PubMed]
5. Crivelli, P.; Ledda, R.E.; Parascandolo, N.; Fara, A.; Soro, D.; Conti, M. A new challenge for radiologists: Radiomics in breast cancer. *BioMed Res. Int.* **2018**, *2018*, 6120703. [CrossRef] [PubMed]
6. Li, H.; Mendel, K.R.; Lan, L.; Sheth, D.; Giger, M.L. Digital mammography in breast cancer: Additive value of radiomics of breast parenchyma. *Radiology* **2019**, *291*, 15–20. [CrossRef]
7. Guo, Y.; Hu, Y.; Qiao, M.; Wang, Y.; Yu, J.; Li, J.; Chang, C. Radiomics analysis on ultrasound for prediction of biologic behavior in breast invasive ductal carcinoma. *Clin. Breast Cancer* **2018**, *18*, e335–e344. [CrossRef]
8. Fan, M.; Li, H.; Wang, S.; Zheng, B.; Zhang, J.; Li, L. Radiomic analysis reveals DCE-MRI features for prediction of molecular subtypes of breast cancer. *PLoS ONE* **2017**, *12*, e0171683. [CrossRef]
9. Ha, S.; Park, S.; Bang, J.I.; Kim, E.K.; Lee, H.Y. Metabolic radiomics for pretreatment 18 F-FDG PET/CT to characterize locally advanced breast cancer: histopathologic characteristics, response to neoadjuvant chemotherapy, and prognosis. *Sci. Rep.* **2017**, *7*, 1556. [CrossRef]
10. Pathak, B.; Barooah, D. Texture analysis based on the gray-level co-occurrence matrix considering possible orientations. *Int. J. Adv. Res. Electr. Electron. Instrum. Eng.* **2013**, *2*, 4206–4212.

11. Lee, S.E.; Han, K.; Kwak, J.Y.; Lee, E.; Kim, E.K. Radiomics of US texture features in differential diagnosis between triple-negative breast cancer and fibroadenoma. *Sci. Rep.* **2018**, *8*, 13546. [CrossRef]

12. Adabi, S.; Hosseinzadeh, M.; Noei, S.; Conforto, S.; Daveluy, S.; Clayton, A.; Mehregan, D.; Nasiriavanaki, M. Universal in vivo textural model for human skin based on optical coherence tomograms. *Sci. Rep.* **2017**, *7*, 17912. [CrossRef]

13. Mohanaiah, P.; Sathyanarayana, P.; GuruKumar, L. Image texture feature extraction using GLCM approach. *Int. J. Sci. Res. Publ.* **2013**, *3*, 1.

14. Selvarajah, S.; Kodituwakku, S. Analysis and comparison of texture features for content based image retrieval. *Energy* **2011**, *1*, 1.

15. Morris, E.; Comstock, C.; Lee, C. ACR BI-RADS Magnetic Resonance Imaging. In *ACR BI-RADS, Breast Imaging Reporting and Data System*; American College of Radiology: Reston, VA, USA, 2013.

16. Vaidehi, K.; Subashini, T. Automatic characterization of benign and malignant masses in mammography. *Procedia Comput. Sci.* **2015**, *46*, 1762–1769. [CrossRef]

17. Zyout, I.; Abdel-Qader, I. Classification of microcalcification clusters via pso-knn heuristic parameter selection and glcm features. *Int. J. Comput. Appl.* **2011**, *31*, 34–39.

18. Berbar, M.A. Hybrid methods for feature extraction for breast masses classification. *Egypt. Inform. J.* **2018**, *19*, 63–73. [CrossRef]

19. Kitanovski, I.; Jankulovski, B.; Dimitrovski, I.; Loskovska, S. Comparison of feature extraction algorithms for mammography images. In Proceedings of the 2011 4th International Congress on Image and Signal Processing, Shanghai, China, 15–17 October 2011; Volume 2, pp. 888–892.

20. Ramos, R.P.; do Nascimento, M.Z.; Pereira, D.C. Texture extraction: An evaluation of ridgelet, wavelet and co-occurrence based methods applied to mammograms. *Expert Syst. Appl.* **2012**, *39*, 11036–11047. [CrossRef]

21. Losurdo, L.; Fanizzi, A.; Basile, T.M.; Bellotti, R.; Bottigli, U.; Dentamaro, R.; Didonna, V.; Fausto, A.; Massafra, R.; Monaco, A.; et al. A Combined Approach of Multiscale Texture Analysis and Interest Point/Corner Detectors for Microcalcifications Diagnosis. In Proceedings of the International Conference on Bioinformatics and Biomedical Engineering, Barcelona, Spain, 29–30 October 2018; pp. 302–313.

22. Khan, S.; Hussain, M.; Aboalsamh, H.; Bebis, G. A comparison of different Gabor feature extraction approaches for mass classification in mammography. *Multimed. Tools Appl.* **2017**, *76*, 33–57. [CrossRef]

23. Buciu, I.; Gacsadi, A. Directional features for automatic tumor classification of mammogram images. *Biomed. Signal Process. Control* **2011**, *6*, 370–378. [CrossRef]

24. Al-Shamlan, H.; El-Zaart, A. Feature extraction values for breast cancer mammography images. In Proceedings of the 2010 International Conference on Bioinformatics and Biomedical Technology, Chengdu, China, 16–18 April 2010; pp. 335–340.

25. Karahaliou, A.; Vassiou, K.; Arikidis, N.; Skiadopoulos, S.; Kanavou, T.; Costaridou, L. Assessing heterogeneity of lesion enhancement kinetics in dynamic contrast-enhanced MRI for breast cancer diagnosis. *Br. J. Radiol.* **2010**, *83*, 296–309. [CrossRef]

26. Nagarajan, M.B.; Huber, M.B.; Schlossbauer, T.; Leinsinger, G.; Krol, A.; Wismüller, A. Classification of small lesions in breast MRI: Evaluating the role of dynamically extracted texture features through feature selection. *J. Med. Biol. Eng.* **2013**, *33*. [CrossRef] [PubMed]

27. Hassanien, A.E.; Kim, T.H. Breast cancer MRI diagnosis approach using support vector machine and pulse coupled neural networks. *J. Appl. Logic* **2012**, *10*, 277–284. [CrossRef]

28. Hassanien, A.E.; Moftah, H.M.; Azar, A.T.; Shoman, M. MRI breast cancer diagnosis hybrid approach using adaptive ant-based segmentation and multilayer perceptron neural networks classifier. *Appl. Soft Comput.* **2014**, *14*, 62–71. [CrossRef]

29. Wang, J.; Kato, F.; Oyama-Manabe, N.; Li, R.; Cui, Y.; Tha, K.K.; Yamashita, H.; Kudo, K.; Shirato, H. Identifying triple-negative breast cancer using background parenchymal enhancement heterogeneity on dynamic contrast-enhanced MRI: A pilot radiomics study. *PLoS ONE* **2015**, *10*, e0143308. [CrossRef]

30. Sutton, E.J.; Dashevsky, B.Z.; Oh, J.H.; Veeraraghavan, H.; Apte, A.P.; Thakur, S.B.; Morris, E.A.; Deasy, J.O. Breast cancer molecular subtype classifier that incorporates MRI features. *J. Magn. Reson. Imaging* **2016**, *44*, 122–129. [CrossRef]

31. Patel, B.K.; Lobbes, M.; Lewin, J. Contrast enhanced spectral mammography: A review. In *Seminars in Ultrasound, CT and MRI*; Elsevier: Amsterdam, The Netherlands, 2018; Volume 39, pp. 70–79.

32. James, J.; Tennant, S. Contrast-enhanced spectral mammography (CESM). *Clin. Radiol.* **2018**, *73*, 715–723. [CrossRef]

33. Tagliafico, A.S.; Bignotti, B.; Rossi, F.; Signori, A.; Sormani, M.P.; Valdora, F.; Calabrese, M.; Houssami, N. Diagnostic performance of contrast-enhanced spectral mammography: systematic review and meta-analysis. *Breast* **2016**, *28*, 13–19. [CrossRef]

34. Losurdo, L.; Basile, T.M.A.; Fanizzi, A.; Bellotti, R.; Bottigli, U.; Carbonara, R.; Dentamaro, R.; Diacono, D.; Didonna, V.; Lombardi, A.; et al. A Gradient-Based Approach for Breast DCE-MRI Analysis. *BioMed Res. Int.* **2018**, *2018*, 9032408. [CrossRef]

35. Sogani, J.; Morris, E.A.; Kaplan, J.B.; D'Alessio, D.; Goldman, D.; Moskowitz, C.S.; Jochelson, M.S. Comparison of background parenchymal enhancement at contrast-enhanced spectral mammography and breast MR imaging. *Radiology* **2016**, *282*, 63–73. [CrossRef]

36. Phillips, J.; Miller, M.M.; Mehta, T.S.; Fein-Zachary, V.; Nathanson, A.; Hori, W.; Monahan-Earley, R.; Slanetz, P.J. Contrast-enhanced spectral mammography (CESM) versus MRI in the high-risk screening setting: patient preferences and attitudes. *Clin. Imaging* **2017**, *42*, 193–197. [CrossRef]

37. Lalji, U.; Jeukens, C.; Houben, I.; Nelemans, P.; van Engen, R.; van Wylick, E.; Beets-Tan, R.; Wildberger, J.; Paulis, L.; Lobbes, M. Evaluation of low-energy contrast-enhanced spectral mammography images by comparing them to full-field digital mammography using EUREF image quality criteria. *Eur. Radiol.* **2015**, *25*, 2813–2820. [CrossRef] [PubMed]

38. Fallenberg, E.M.; Dromain, C.; Diekmann, F.; Renz, D.M.; Amer, H.; Ingold-Heppner, B.; Neumann, A.U.; Winzer, K.J.; Bick, U.; Hamm, B.; et al. Contrast-enhanced spectral mammography: does mammography provide additional clinical benefits or can some radiation exposure be avoided? *Breast Cancer Res. Treat.* **2014**, *146*, 371–381. [CrossRef] [PubMed]

39. Li, L.; Roth, R.; Germaine, P.; Ren, S.; Lee, M.; Hunter, K.; Tinney, E.; Liao, L. Contrast-enhanced spectral mammography (CESM) versus breast magnetic resonance imaging (MRI): A retrospective comparison in 66 breast lesions. *Diagn. Interv. Imaging* **2017**, *98*, 113–123. [CrossRef] [PubMed]

40. Fallenberg, E.; Dromain, C.; Diekmann, F.; Engelken, F.; Krohn, M.; Singh, J.; Ingold-Heppner, B.; Winzer, K.; Bick, U.; Renz, D.M. Contrast-enhanced spectral mammography versus MRI: initial results in the detection of breast cancer and assessment of tumour size. *Eur. Radiol.* **2014**, *24*, 256–264. [CrossRef]

41. Łuczyńska, E.; Heinze-Paluchowska, S.; Hendrick, E.; Dyczek, S.; Ryś, J.; Herman, K.; Blecharz, P.; Jakubowicz, J. Comparison between breast MRI and contrast-enhanced spectral mammography. *Med. Sci. Monit. Int. Med. J. Exp. Clin. Res.* **2015**, *21*, 1358.

42. Fallenberg, E.M.; Schmitzberger, F.F.; Amer, H.; Ingold-Heppner, B.; Balleyguier, C.; Diekmann, F.; Engelken, F.; Mann, R.M.; Renz, D.M.; Bick, U.; et al. Contrast-enhanced spectral mammography vs. mammography and MRI–clinical performance in a multi-reader evaluation. *Eur. Radiol.* **2017**, *27*, 2752–2764. [CrossRef]

43. Kamal, R.M.; Helal, M.H.; Wessam, R.; Mansour, S.M.; Godda, I.; Alieldin, N. Contrast-enhanced spectral mammography: Impact of the qualitative morphology descriptors on the diagnosis of breast lesions. *Eur. J. Radiol.* **2015**, *84*, 1049–1055. [CrossRef]

44. Patel, B.K.; Ranjbar, S.; Wu, T.; Pockaj, B.A.; Li, J.; Zhang, N.; Lobbes, M.; Zhang, B.; Mitchell, J.R. Computer-aided diagnosis of contrast-enhanced spectral mammography: A feasibility study. *Eur. J. Radiol.* **2018**, *98*, 207–213. [CrossRef]

45. Perek, S.; Kiryati, N.; Zimmerman-Moreno, G.; Sklair-Levy, M.; Konen, E.; Mayer, A. Classification of contrast-enhanced spectral mammography (CESM) images. *Int. J. Comput. Assist. Radiol. Surg.* **2019**, *14*, 249–257. [CrossRef]

46. Saeys, Y.; Inza, I.; Larrañaga, P. A review of feature selection techniques in bioinformatics. *Bioinformatics* **2007**, *23*, 2507–2517. [CrossRef]

47. Sardanelli, F.; Fallenberg, E.M.; Clauser, P.; Trimboli, R.M.; Camps-Herrero, J.; Helbich, T.H.; Forrai, G.; European Society of Breast Imaging (EUSOBI). Mammography: an update of the EUSOBI recommendations on information for women. *Insights Imaging* **2017**, *8*, 11–18. [CrossRef] [PubMed]

48. D'Orsi, C.; Sickles, E.; Mendelson, E.; Morris, E. *2013 ACR BI-RADS Atlas: Breast Imaging Reporting and Data System*; American College of Radiology: Reston, VA, USA, 2014.

49. Haralick, R.M.; Shanmugam, K.; Dinstein, I. Textural features for image classification. *IEEE Trans. Syst. Man Cybern.* **1973**, *SMC-3*, 610–621. [CrossRef]

50. Gonzalez, R.C.; Woods, R.E. Image processing. *Digit. Image Process.* **2007**, *2*, 1.
51. Mallat, S.G. A theory for multiresolution signal decomposition: The wavelet representation. *IEEE Trans. Pattern Anal. Mach. Intell.* **1989**, *11*, 674–693. [CrossRef]
52. Jolliffe, I. *Principal Component Analysis*; Springer: Cham, Switzerland, 2011.
53. Kohavi, R.; John, G.H. Wrappers for feature subset selection. *Artif. Intell.* **1997**, *97*, 273–324. [CrossRef]
54. Breiman, L. Random forests. *Mach. Learn.* **2001**, *45*, 5–32. [CrossRef]
55. Annis, D.H. Permutation, Parametric, and Bootstrap Tests of Hypotheses. *J. Am. Stat. Assoc.* **2005**, *100*, 1457–1458. [CrossRef]
56. Hastie, T.; Tibshirani, R.; Friedman, J.; Franklin, J. The elements of statistical learning: Data mining, inference and prediction. *Math. Intell.* **2005**, *27*, 83–85.
57. Hobbs, M.M.; Taylor, D.B.; Buzynski, S.; Peake, R.E. Contrast-enhanced spectral mammography (CESM) and contrast enhanced MRI (CEMRI): Patient preferences and tolerance. *J. Med. Imaging Radiat. Oncol.* **2015**, *59*, 300–305. [CrossRef]
58. Lobbes, M.B.; Lalji, U.; Houwers, J.; Nijssen, E.C.; Nelemans, P.J.; van Roozendaal, L.; Smidt, M.L.; Heuts, E.; Wildberger, J.E. Contrast-enhanced spectral mammography in patients referred from the breast cancer screening programme. *Eur. Radiol.* **2014**, *24*, 1668–1676. [CrossRef]

*Review*

# Models of the Gene Must Inform Data-Mining Strategies in Genomics

Łukasz Huminiecki

Department of Molecular Biology, Institute of Genetics and Animal Biotechnology, Polish Academy of Sciences, 00-901 Warsaw, Poland; l.huminiecki@igbzpan.pl

Received: 13 July 2020; Accepted: 22 August 2020; Published: 27 August 2020

**Abstract:** The gene is a fundamental concept of genetics, which emerged with the Mendelian paradigm of heredity at the beginning of the 20th century. However, the concept has since diversified. Somewhat different narratives and models of the gene developed in several sub-disciplines of genetics, that is in classical genetics, population genetics, molecular genetics, genomics, and, recently, also, in systems genetics. Here, I ask how the diversity of the concept impacts data-integration and data-mining strategies for bioinformatics, genomics, statistical genetics, and data science. I also consider theoretical background of the concept of the gene in the ideas of empiricism and experimentalism, as well as reductionist and anti-reductionist narratives on the concept. Finally, a few strategies of analysis from published examples of data-mining projects are discussed. Moreover, the examples are re-interpreted in the light of the theoretical material. I argue that the choice of an optimal level of abstraction for the gene is vital for a successful genome analysis.

**Keywords:** gene concept; scientific method; experimentalism; reductionism; anti-reductionism; data-mining

## 1. Introduction

The gene is one of the most fundamental concepts in genetics (It is as important to biology, as the atom is to physics or the molecule to chemistry.). The concept was born with the Mendelian paradigm of heredity, and fundamentally influenced genetics over 150 years [1]. However, the concept also diversified in the course of its long evolution, giving rise to rather separate traditions in several sub-disciplines of genetics [2]. In effect, somewhat different narratives about the gene (and models of the gene) developed in classical genetics, population genetics, molecular genetics, genomics, statistical genetics, and, recently also, in systems genetics.

The fundamental goal of this paper is to summarize the intellectual history of the gene, asking how the diversity of the concept impacts on data integration and data-mining strategies in genomics. I hope to show that many practical decisions of a statistician, a bioinformatician, or a computational biologist reflect key theoretical controversies that permeated the field of genetics for over a century (Many practical tasks must be informed by the theory.). When integrating data or designing databases or writing software, an analyst must make smart decisions about the architecture of their system. For example, a genomics database may focus on low-level concepts such as exons, or individual transcripts. Alternatively, a database may focus on higher-level concepts such as gene families, pathways, or networks.

There are two parts to this text. In the first part, which is theoretical, I discuss the background of the gene concept in history and ideas. The theoretical part will set the ground for more practical considerations. In the second part, several practical examples of genome analyses from my own work will be given (The examples will be re-interpreted in the light of the theoretical concepts introduced in the first part.).

*1.1. Ideas and Concepts—The Development of the Mendelian Paradigm of Heredity*

The story of the gene started with Gregor Mendel, a Moravian monk, scientist, and later abbot, who hypothesized an innovative model of particulate heredity. In Mendel's model, independent intracellular elements determine differentiation of cells into visible traits. Importantly, Mendel's particles of heredity do not mix with each other, instead, they persist over generations (However, new varieties can evolve through mutations).

In many ways, Mendel was a prophet of genetics in the 19th century. In the first place, he hypothesized a particulate paradigm of heredity (The paradigm dominated genetics in the 20th century, complementing Darwin's theory of evolution, and developing further with molecular biology and genomics.). Moreover Mendel was an early experimentalist, relying on prospective experimental procedures to test his hypothesis [1]. This was a rather modern approach in the century still dominated by an observational tradition. Interestingly, Mendel's laboratory was exceedingly simple—it was just the vegetable garden of his monastery. At the same time, Mendel excelled in the scientific method (including logico-mathematical analysis). This enabled him to design just the right simple experiment and to interpret its results insightfully.

At first, however, the Mendelian model of the gene was just a black box that could not be looked into to discern the molecular mechanism. The black box was perfectly non-transparent—nothing was formally postulated about parts of the model, or how these parts interacted. The model specified that the gene did certain things for the cell, but not how. In effect, the gene model was a mere instrument of analysis. It worked best as an intermediate variable in the context of breeding or medical genetics.

Mechanistic details about the gene were only provided when molecular biology revolutionized classical genetics. As a result, genetical phenomena became more akin physical phenomena. This precipitated an avalanche of discoveries. After a physical form was given to Mendel's abstract gene concept, biochemical details of molecular processes such as transcription and translation, as well as of their regulation, were being revealed [3].

In addition to molecular biology, ingenious experimental models were also crucial. This commonly meant a careful choice and laboratory upkeep of a particular animal species, a plant species, or a microorganism that facilitated a certain class of genetical experiments. For example, Max Delbrück, who was a physicist with interest in genetics [4], developed simple experimental models in viruses and bacteria (Note that the term *model* in this case has no mathematical or conceptual meaning, but instead signifies a simple biological system established for the purposes of experimental study. That the above was achieved by a physicist is not surprising, as there had been a long tradition of working with simple but enlightening models in physics.).

*1.2. Genomics Technology and the Gene*

It might be generally argued that progress in genetics was facilitated by enabling experimental technologies. This was the case for protocols of molecular biology as mentioned above, however, the same may also be argued for genomics. Like molecular biology before, genomics also challenged many orthodox views of the Mendelian gene [5,6]. For example, genomics revealed that genes are not always independent of each other, as they may overlap in complex loci [7].

Presently, assays based on next generation sequencing, followed by bioinformatics, and high-dimensional statistical analysis are drivers for progress in the emergent discipline of systems genetics. At the same time, the high-throughput assays are not the panacea. Next generation sequencing may promote a technology-focused culture, with most attention invested in data generation, data post-processing and storage, as well as descriptive analysis of high-throughput datasets. With the focus on the high-throughput technology, formulation and critique of novel biological concepts receives less attention. In effect, high-throughput datasets are increasingly easy to generate but difficult to interpret.

### 1.3. Empiricism

Let me now introduce several methodological strategies grouped under the umbrella of the method of science. This is crucial to deeper intellectual understanding of the story of the gene. The strategies are as follows: empiricism, experimentalism, reductionism, anti-reductionism, statistical data analysis, and scientific model building.

Empiricism is a philosophical theory that was instrumental to the emergence of the scientific method and the success of science. According to empiricism, all knowledge, including scientific knowledge, can only have solid grounding in the generalizations of practical experiences derived from the senses. Empiricism is skeptical about the value of non-scientific prior knowledge such as systems of beliefs associated with religions, or even views and theories of philosophical schools.

In biology, the empirical tradition of observing and collecting samples from the natural world goes all the way back to the first biological studies of Aristotle [8], which focused on animals (their parts, movements, generation, and development). Aristotle was an ancient but pragmatic philosopher in the Academy of Classical Athens (4th century BC). He was one of the most able students of idealistic Plato. However, Aristotle was much more empirically minded than Plato. Plato himself was skeptical about the value of sensory observations favoring the sources of knowledge that were alternative to empiricism but still based on rational thinking. Plato favored either logical deductions from abstract theories, or instinctive reasoning known presently as intuition (Note that intuition is built from past experiences using unconscious functionalities of the brain that are probably analogous to machine learning.). On balance, Aristotelianism proved more important for the development of empiricism in genetics than Platonism. However, Aristotelian logical doctrines eventually became synonymous with medieval intellectual stagnation, scholasticism, and a lack of scientific creativity.

### 1.4. The Baconian Method

Francis Bacon, a 17th century English philosopher and statesman, formulated an ideological and political manifesto calling for the abandonment of Aristotelianism. His proposal was contained in the treaty *Novum Organum* [9]. In particular, Bacon called for the rejection of futile deductions, unproductive syllogisms, and scholasticism of late Aristotelianism. Instead, Bacon affirmed systematic empiricism on a grand scale, supported by the state politically and financially (Bacon made many insights into the political organization of the scientific system. Indeed, Bacon's motivation for promoting empiricism politically might have been to lay the grounds for the economic development of the English state.). Interestingly, Bacon is not known for making grand original scientific discoveries like Copernicus, Galileo, or Newton. Instead, his main legacy lies in the vision of science as a systematic body of empirical knowledge and a political program.

Methodologically, Bacon underlined that only unbiased empirical observations can yield certain knowledge. Facts might be gathered selectively to support prior ideas if empiricism is not performed according to a strict plan. Moreover, generalizations, proceeding by the method of induction (named so in contrast with Aristotelian deduction), must not be made beyond what the facts truly demonstrate. As the reader may easily guess, the repeating cycles of: (1) systematic empirical observations, and (2) inductive generalizations—taken together—form the basis of the Baconian method. However, the transition to the method of Bacon was a slow and gradual process. This was probably because, the Baconian enterprise required the professionalization of science and the development of state funding, which became possible only in the 20th century (The observational tradition, rather than experimentalism, continued to dominate the study of nature in the 19th century. Empirical evidence was given, but it tended to be unsystematic and anecdotal, chosen arbitrarily, frequently in the form of samples casually collected by gentlemanly hobbyists. There was little or statistical analysis of data. Moreover, 19th century scientific theories tended to be sweeping and somewhat over-ambitious in relation to limited empirical support. Examples include Charles Lyell's and Charles Darwin's overarching treaties, as both the authors were upper-class men whose inspirations were mostly travels, conversations, reading, or correspondence.).

*1.5. The Baconian Method and Integrative Genomics: On the Importance of Identifying and Avoiding Bias*

In our age, genomics is a new example of the need for the Baconian principle of unbiased empiricism. Note that bias could be introduced on many levels. One type of bias may result from temporarily-, spatially-, or taxonomically-limited sampling of the genome space. For example, scientists sequencing extant genomes are only probing the biosphere as it exists at present. Their generalizations may not be applicable to conditions on Earth in the near past, e.g., during the Last Glacial Period, when different selection forces acted on the population. Even more so, their generalizations are unlikely to be applicable to the distant past when the composition of the atmosphere (and, therefore, of the biosphere) was substantially different. For example, recall that oxygen levels in the atmosphere where at the maximum of around 30% about 280 million years ago, and practically zero before the Great Oxygenation Event approximately 2.4 billion years ago. Even if one focuses exclusively on the present conditions, one should appreciate that certain spatial regions of the biosphere are under-sampled. For example, the deepest parts of the ocean and its floor, or deeper parts of Earth's crust are still physically difficult to access. Finally, care should be taken not over-generalize beyond the taxa from which genomic data were obtained, or for which it is parsimonious to extend the findings based on phylogenetic relationships.

Other sources of bias may result from the limitations of the experimental technologies of genomics. For example, the technological platform of microarrays, used for expression profiling, can only measure expression of genes for which probes were pre-selected by the manufacturer of the platform. Until relatively recently, this meant a set of protein-coding genes of a given species. (At present, however, that there are also commercial microarrays targeting *micro*RNAs). Moreover, as the features of the microarray chip are arbitrarily pre-selected, there could be a bias towards well-studied and highly transcriptionally active genes (Such *active* genes are also called highly expressed.). It is logical that genes that are weakly, or temporarily expressed, or expressed in a very tissue-specific manner were less likely to be discovered using conventional techniques of molecular biology (and pre-selected for the inclusion on the chip). Additionally, most microarray chips are not designed to discern between alternative transcripts (In practice, this usually means that the probe or probes on the chip are designed to target the longest transcript from a reference collection, such as RefSeq [10].).

*1.6. Experimentalism and the Laboratory*

In the 20th century, there was a social and methodological change in empiricism. Most scientists became professionals, working in academic or governmental laboratories. Research became routinely funded through dedicated grant agencies, which were becoming independent of the government. Women were increasingly involved in the profession. Moreover, casual observations were increasingly replaced by systematic scientific experiments. Unlike observations of naturalists, scientific experiments were planned in advance, being designed to test an explicitly stated hypothesis. The setting was controlled, either in field conditions, or in carefully managed laboratories. Moreover, confounding variables were controlled for by assigning test subjects randomly to either experimental groups or the control. There also emerged technical protocols for biochemistry and molecular biology, highly specialized laboratory chemicals and biological reagents, as well as advanced laboratory instrumentation (for example, an electron microscope). Thus, natural philosophy was being replaced by experimental natural science.

With the rise of experimental biology, a new type of heroic scientific figure emerged: an experimentalist. One of the most outstanding experimentalists in genetics was Thomas Hunt Morgan. He was originally an embryologist rather than a geneticist. However, Morgan moved later on in his career to test experimentally several components of the chromosomal theory of heredity. Morgan's greatest strength was in the ability to set up sophisticated genetical experiments in well-chosen and expertly run animal models. For example, Morgan worked on genetical problems in several experimental animal models that he mastered in his laboratory, especially in the fruit fly [11]. A further strength of these experiments lied in the fact that they were quantitative; for example, Morgan not only demonstrated inheritance linked to sex chromosomes, but also constructed first quantitative chromosomal maps (Morgan's student used frequencies of crossing-over between six sex-linked loci on the X chromosome of the fly as the proxy measure of chromosomal distances [12].).

*1.7. The Theory of Experimentalism*

The work of practical experimentalists such as Morgan was complemented by the development of the theory of empirical knowledge. A set of increasingly well-understood practices, known presently as the scientific method, was being developed, studied, and codified. The scientific method integrated a range of disparate tasks, including: (1) constructing scientific instruments, (2) formulating hypotheses and designing experiments, (3) making observations and recording results, (4) statistical data analysis and interpretation, (5) modeling and formulating generalizations or new scientific laws, and (6) developing theories about how science works.

Note that early theoretical insights into experimentalism tended to be made by applied statisticians. For example, an English statistician, Sir Ronald Aylmer Fisher pioneered many practical methods for the analysis of experimental datasets, and made advances in the theory of the design of experiments, randomization, and the optimal sample size [13].

In parallel, a group of German-speaking philosophers known presently as logical empiricists [14] formed an academic and social movement devoted to promoting science as a social cause and a set of methodological doctrines (The members of these social groups met in 1920s and 1930s in the so-called Vienna and Berlin circles. Later, many of them emigrated to Anglo-Saxon countries, where they were active in 1940s.). Logical empiricists promoted the idea that experiments must be intertwined with logical analyses [14]. In other words, there was a broad understanding among logical empiricists that experimentalism was no mindless collecting of facts. According to logical empiricists, a scientist advances new hypotheses through logical analyses of old theories and data. The best hypotheses are then prioritized for experimental testing. Logic is then employed again to develop insightful interpretations of results.

Of course, the theory of experimentalism has continued to develop since the peak of the activity of logical empiricists. Philosophers of the following generations argued critically that logical empiricism presented a sort of sterile, overly idealized vision of the scientific method. In particular, Thomas Kuhn argued that social and historical factors must also be taken into account. He argued that science effectively developed in a series of socially-conditioned paradigm shifts [15], in which dominant personalities, fashions, or the socio-economical context could be just as important as scientific methods and facts.

At the end of this section, I would like to argue that the need for interplay between empiricism and logical analysis is well illustrated in the story of the gene. For example, Mendel's experimental results would have probably been completely overlooked if there were not accompanied step-by-step by insightful logical analyses. First, Mendel proposed, inspired by intuition, that discrete particles were the bearers of hereditary information. In the second step, the experiments on hybrids of varieties of *Pisum sativum* were designed. These two initial steps were logico-analytical. In the next step, the model was verified empirically. In the final step, the implications of confirmatory results were logically considered by Mendel, leading to the crystallization of his paradigm of heredity.

Moreover, Mendel's theory is also an example of a socially-conditioned scientific paradigm shift of the type postulated by Thomas Kuhn. The Mendelian theory was a radical intellectual revolution replacing pre-scientific views on heredity dominant until the 19th century, in particular the blending theory of inheritance [16]. Gregor Mendel was uniquely positioned to lead the revolution having had education in physics and philosophy, and social background in agriculture. The monastery supported his education and experiments. Note that Mendel's mathematical skills were probably developed when studying physics. In contrast, the strengths of Charles Darwin and his fellow English naturalists were not strong in mathematical skills and invested their energies in travels and sample collecting.

Presently, some theoreticians like to argue that a new paradigm shift in genetics is on the horizon. According to this view, the Mendelian paradigm may need to be updated to accommodate new genome-wide evidence for adaptive mutations, as well as data generated in the field of trans-generational epigenetic inheritance [16]. There can be little doubt that data-mining of genomic datasets will play a key role in this process.

### 1.8. Reductionism

In the history of 20th century genetics, molecular biology played the role of a reducing theory with respect to classical genetics [17,18]. In the case of molecular biology, reductionism [2] means physicalism (Note that there is a related idea in reductionism that breaking a system down into smaller and smaller modules will enable a biologist to understand the large system in fullness. For this to work, the system under consideration needs to be modular, i.e., composed of independent and self-contained units.). Physicalism is a claim and a research program suggesting that biological phenomena can be described and explained first as chemical facts, and in a further reductive step, using the laws of physics, also as physical facts.

Moreover, physicalism was interpreted as the basis for scientific and social progress by the movement for the Unity of Science (this was a group within logical empiricism led by socially-active Otto Neurath). In this radical perspective, empiricism marches towards a comprehensive scientific conception of the world, where all experimental knowledge is unified by low-level chemical and physical principles. If this were true, the molecular interpretation of the gene would need to have the power to explain everything that the classical gene could, but do this with more detail and correctness. However, is this really the case? The difficulty lies in the fact that radical reductionism is a research program with a fundamentalist agenda. The agenda extends beyond a pragmatic need for providing useful molecular facts illuminating old biological problems. Radical proponents of reductionism would like for all biological explanations and research programs to progressively follow the reductionist pattern set by biochemistry and molecular biology. Ultimately, radical reductionists desire all scientific theories to be based exclusively on physicalism. In the opinion of most theoreticians and philosophers, this is neither possible nor desirable in genetics. The following section explains why this is the case.

### 1.9. Anti-Reductionism

Anti-reductionists, who include most philosophers of biology, oppose the ideology of radical reductionism. In particular, anti-reductionists view physicalist explanations as mostly unnecessary, frequently unrealistic, and sometimes even dangerously misleading. Anti-reductionists claim that a scientist dogmatically following the physicalist principle would ultimately develop a kind of conceptual myopia. That is to say, a radical follower of the ideology of reductionism would obsess about details, but fail to see the big picture. To use a popular idiomatic expression, they could not see the forest for the trees. In fact, a dogmatic reductionist might be tempted to claim that there is no such thing as a forest, that every tree must be considered in isolation and on its own.

It is, therefore, clear that by logical extension the ideology of radical reductionism would lead to a kind of conceptual eliminativism toxic to scientific discourse. For example, an eliminativist might claim that the concept of the gene should be abandoned altogether, in favor of the exclusive focus on the function of individual DNA base pairs. Metaphorically speaking, the elimination of all higher-level concepts would be akin to a massive cerebral infarction. Genetics based on such principles would be merely a mindless collecting of facts. In genetics, there are many examples of useful high-level concepts that are too far removed from the physics of a single atom or the chemistry of a single molecule for reductionism to be a useful approach. Examples include quantitative traits, multi-genic diseases, enzymatic pathways, signaling networks, pleiotropy, and complex loci.

Anti-reductionists can forcefully claim that the high-level concepts must not only be retained, but become the focus of analysis in order to interpret irreducible complexity. Presently, anti-reductionism is becoming equal in importance in genetics to the reductionism of molecular biology. This is leading to the development of a new sub-discipline—systems genetics. Systems geneticists use high-level concepts such as signaling pathways and networks to interpret genome-wide association studies.

### 1.10. Statistical Data Analysis

Statistical tools are essential for the organization, display, analysis, and interpretation of empirical data. It is, of course, beyond the scope of this paper to discuss the development of statistics. Historical material is widely available [19]. I will limit myself to noting that early tools of the biometric school, developed by Francis Galton and Karl Pearson, were conceptually better suited for the analysis of observational studies rather than experiments. Indeed, these tools were developed for surveys of demographic data, or mapping of human characteristics (such as height or weight) in very large populations.

In contrast, biological experiments presented with themselves a different set of challenges. The essence of a statistical analysis of an experiment is to compare sets of observations (i.e., experimental groups versus control). In biological experiments, there quickly emerged a problem of the small numbers of individual observations in the sets compared. The numbers tended to be small for practical reasons, such as limited budgets, limited manpower, lack of space in experimental facilities, or difficulty in sourcing biological specimens. As such, small samples are associated with sampling errors (in addition to the measurement error and biological randomness). As a result, biometric methods, developed for large populations, were both too complicated and too inaccurate to be useful for biological experiments. The problem became known as the small sample problem. The problem was only properly addressed by the second generation of statistical tools, pioneered by William Sealy Gosset, and developed further by Ronald Fisher [13] (Fisher also initiated work on statistical tools for the analysis of experimental groups employing the analysis of variance—ANOVA—rather than more established analysis of means. Moreover, he begun work on a theoretical framework—Fisher information—for the prediction of minimal sample sizes necessary to detect experimental effects of a given magnitude.).

Note that many contemporary genomic applications work well within the Fisherian framework of small-sample-problem-aware parametric statistics. This is the case when one encounters typical measurement and sampling errors (for example, in microarray studies with several replicates in each experimental group).

In other situations, data analyzed are not a random sample from any population. Indeed, individual genome projects are not at all experiments. They are more similar to maps in cartography. In other words, sequences of individual genomes present themselves as they are—without any sampling error. That is to say, there is only one consensus genome sequence for the human species, and the goal of a genome project is to map the genome sequence completely and accurately. As there is no population of sequences that is sampled, the most fundamental assumptions of Fisherian parametric statistics are grossly violated.

Fortunately, genomics datasets emerged at the end of the 20th century, when substantial computational resources were easily available—even on office computers. Presently, even laptop computers are frequently sufficient to analyze genomics datasets. This makes it rather practical to apply permutation and randomization methods, which are free from the assumption of random sampling. Such methods were known since Fisher's times, but were initially rarely used due to their computational intensity [20]. In the case of observational studies, one can apply a permutation procedure to compute the distribution of the test statistic under the null hypothesis. Typically, a subset of genomic features is compared against the background of the features in the whole genome (In the case of experimental studies, one can employ a randomization test [21], in which observations are randomly assigned to experimental groups, or the control.).

The problem of multiple testing was another statistical challenge frequently met in functional genomics. For example, there are as many tests as microarray features when comparing sets of microarrays to identify differentially expressed genes. The problem was quickly recognized in the early days of functional genomics at the beginning of the 21$^{st}$ century. As a response, robust and well-characterized solutions for the problem of multiple testing emerged, such as procedures that control the rate of false discovery [20].

The current trend in genomics is to transform it to a data science. Increasingly, the most recent computer science algorithms, such as deep learning, are applied to genomics datasets. We must however be aware that data science emphasizes practical algorithms, but neglects statistical theory and statistical inference. On a positive side, we can enjoy pragmatic benefits delivered by powerful industrial-grade algorithms. On a negative side, there is relatively little understanding of how exactly these algorithms work, what is their sensitivity and specificity, or against what kinds of inputs they are likely to fail.

*1.11. Scientific Model Building*

A scientist is frequently attempting to construct a model for an unknown mechanism [2], for example to interpret observations or experiments. Scientific modeling is a creative process—an art for which there are no strict guidelines or firmly established rules. Note that models are never perfect representations of the material world. They contain many assumptions, idealizations, and simplifications. Nonetheless, models can be very useful in science if used skillfully for the right purpose.

A model is a physical, conceptual, mathematical, or probabilistic approximation of a real natural phenomenon that is of interest to a scientist. In general, the advantage of constructing a model in science is that it is easier to work with than the mechanism being modeled. One can understand, visualize, study, or manipulate a model easier than the raw mechanism. One can also use a model to make abstract mechanisms more concrete. Moreover, a model can be used to discover, that is to get to know further mechanistic details of the phenomenon under investigation. Finally, a model can be used to predict a future behavior of the system.

I will now give an example of a very simple but, nonetheless, extremely influential model in physics. This is the Copernican model of the solar system. It is frequently said that Copernicus completely changed the paradigm of astronomy by putting the sun in the center of the solar system [15,22]. This changed the long established orthodoxy of the geocentric model of the Universe due to an ancient astronomer and mathematician—Claudius Ptolemy (The Ptolemaic model had Earth at the center.). However, it is less known that the model of Copernicus was so simple that it did not actually produce better predictions for astronomical data. At the same time, a major advantage of the Copernican system lied precisely in its geometrical frugality (That is to say, the Copernican model explained observed data in more parsimonious terms.).

Intellectually, the achievement of Copernicus lied in having the courage to question a long established orthodoxy that was favored by dominant non-scientific ideologies of the time. Only later, astronomers proficient at empirical observations and mathematics, such as Galileo, Kepler, and Newton, provided solid results in support of the heliocentric model. Moreover, additional details of the model were provided. Circular orbits were replaced with elliptical ones, and the force of gravity was proposed to explain the movements of planets.

Unfortunately, scientific modeling is less established in biology than in astronomy and physics. Moreover, the term *model* is loosely defined in biology. There is also an overlap between how the terms *model* and *concept* are used. A key to being a successful biologist may lie in choosing an appropriate modeling technique for the mechanism of interest, and for the purposes to which the model is to be applied. For example, when applied to the gene concept, modeling techniques differ in the level of mechanistic detail included (see Table 1). At one end of the spectrum, the models could be like the Mendelian black box—where nothing is known about the mechanism hidden inside. Alternatively, the models could be like rough sketches characteristic of classical genetics. Finally, there are also detailed see-through models characteristic of molecular biology (such models resemble transparent glass boxes).

**Table 1.** Several modeling approaches have been applied to the gene concept.

| Gene Concept | Modeling Approach | Empirical Evidence |
| --- | --- | --- |
| Mendel's cellular elements. | A black box model. The gene has certain functions, but nothing is known about the components of the genic mechanism, or how they interact. | Influenced by Darwinian natural history and associated 19th century evolutionary debates. Mendel performed experiments, but might selectively reported them [23]. |
| The gene of classical genetics, and the chromosomal theory of heredity. | A sketch (i.e., a semi-transparent glass box). Mechanistic details are fuzzy, but the gene has several well-defined and empirically proven properties. | Initially, the evidence came from field experiments in plant genetics. Increasingly, there were also experiments in animal genetics for which specialized model systems were set up in laboratories (such as Morgan's fruit fly model). |
| The gene of population genetics. | Mathematical equations. Statistical analyses. Probability theory. | The observations of genetic variability in natural or artificial populations. Some experiments, especially in the context of artificial evolution. |
| The molecular gene. | A transparent glass box. | Experiments and genetic engineering focusing on simplest organisms, to establish the basic principles. Next, the basic principles were extended to other species. |
| The gene of genomics. | A hierarchy of domains and functional units. | High-throughput screens from the surveys of populations, or from experimental groups. |
| The evolutionary gene. | The model includes information on gene's evolutionary history, in particular on the pattern of duplications and speciation events. Data-mining strategies can be anti-reductionist, for example genes can be grouped and analyzed as gene families. | Morphological or sequence characters; gene and protein sequences can be aligned and phylogenetics trees can be constructed. |
| The virtual gene. | Computational data integration; data storage in relational databases. Data-mining strategies can be anti-reductionist, for example genes can be grouped and analyzed as pathways or networks. | Many kinds of empirical data can integrated using bioinformatics pipelines and databases. Data can be browsed using genome browsers, or data-mined using statistical tools and visualization. |

Nonetheless, models can be very useful in biology as well. For example, Watson and Crick famously used modeling to discover the structure of DNA. At first, Watson and Crick constructed two-dimensional models of individual DNA bases. They manipulated the bases manually—seeking to understand if and how they could pair. Later, they also constructed a three-dimensional (3D) model of the double helix to visualize the pairing of the strands of DNA (Admittedly, it is true that the model of Watson and Crick was critically informed by Rosalind Franklin's original data on X-ray diffraction patterns generated by purified DNA [24]. Nevertheless, Watson and Crick did show more intellectual courage and good judgment when constructing their comprehensive 3D model of DNA.). This, in turn, suggested a likely mechanism of DNA replication. This is just one example of how important modeling can be as a part of logical analysis, providing generalizations and concepts as added value to raw empirical observations.

## 2. Practical Examples of Genomic Analyses

A few published examples will be given and discussed. The examples will be re-interpreted in the light of the theoretical material discussed in Part I. In particular, I will ask whether a genomic analysis under consideration followed reductionist or anti-reductionist principles. I will also inquire whether the empirical method employed was an observational survey—conceptually analogous to a map. Alternatively, an analysis might have been more akin an experiment—conceptually analogous to a test of hypothesis (A summary of the examples is given in Table 2.).

**Table 2.** Examples of data-mining of genomics datasets. Examples of reductionist and anti-reductionist data-mining strategies in genomics are given. Analytical strategy and the focus of the analysis are specified, as well as the type of evidence and the main result.

| Strategy | Focus. (Evidence) | Main Result | Reference |
|---|---|---|---|
| Reductionist | Promoter (An integrative survey of functional genomic datasets.) | A correlation between the size of promoter architecture and the breadth of expression was detected. Transcription factors with the strongest contribution to housekeeping expression were identified. | [25,26] |
| | Gene (A meta-analysis with experimental follow-up.) | First, an integrative data-mining procedure was used to clone the most endothelial-specific genes. The procedure was combined with experimental verification. Subsequently, one of the most endothelial-specific genes — ROBO4 — was found to be expressed in sites of active angiogenesis. | [27,28] |
| Anti-reductionist | Gene family (An integrative survey of multiple genomics databases.) | The analysis of the gene family of roundabouts suggested that magic roundabout (ROBO4) is an endothelial-specific ohnolog of ROBO1. ROBO4 neo-functionalized to an essential new role in angiogenesis | [29] |
| | Signaling pathway (A survey of 33 animal genomes.) | The evolution of the TGF-beta signaling pathway in the animal kingdom was analyzed. The components of the pathway were found to have emerged with the first animals, and diversified in vertebrates. | [30] |
| | Signaling network (An integrative analysis of an evolutionary database and a signaling network.) | 2R-WGD was found to have remodeled the signaling network of vertebrates. This macro-mutation facilitated the evolution of key vertebrate evolutionary novelties and environmental adaptations. | [31] |

*2.1. A Survey of Individual Endothelial-Specific Genes (Which Followed Reductionist Principles)*

Let me start with some background information necessary to understand this example. Vertebrates, a clade that includes the human species, are animals with complex body plans and hundreds of different cell-types. Although, evolutionary biologists do not like to designate any taxon as more advanced than others, it is a fact that vertebrate animals have more tissue- and cell-types than any other group of organisms on Earth. How do vertebrate cell types differ between each other? The answer is that all cells in the human body have the same genome, however, different cell types follow different differentiation trajectories. During differentiation, the epigenome is modified and different sets of genes are sequentially activated to be transcriptionally expressed. Thus, differential expression of genes defines individual cell types. This is true both during development and in terminally differentiated somatic cells.

The example under consideration in this section focuses on one of somatic cell types. This is the endothelial cell (EC) type, which is spread throughout the body, being present in all tissue types. The endothelium is a single layer of cells lining the lumen of the cavities of blood vessels. Note that evolutionarily ECs are unique to vertebrates, as there is no true endothelium in invertebrates [32] (Remarkably, the endothelium is present in every vertebrate species without exception. This is because the endothelium emerged along with the pressurized circulatory system characteristic of the vertebrate clade.).

The endothelium plays a primary structural role in maintaining the integrity of blood vessels. At the same time, ECs are not just a passive structural barrier. These cells have functional roles in addition to their structural role within the vasculature. For example, ECs are a primary instrument of angiogenesis—that is the process of the sprouting and growth of new blood vessels from pre-existing blood vessels. However, the EC cell type is remarkably active, taking part in many physiological processes besides angiogenesis. These processes include the regulation of vascular permeability, the control of hemostasis, the regulation of blood pressure, as well as the recruitment of immune cells. The versatility of ECs is reflected in a rich set of endothelial transcripts, many of which are preferentially expressed in ECs or even entirely specific to this cell type.

Inspired by the fact that the rich repertoire of genes expressed in ECs characterizes this cell type, Huminiecki and Bicknell set out to identify through transcriptomics the most endothelial-specific genes. Broadly speaking, their strategy was an integrative meta-analysis of functional genomic databases followed by experimental verification [27]. In technical terms, the analysis consisted of two parts. The first part was a computational meta-analysis of pooled datasets generated from a number of libraries based on several different genomic technologies. The datasets were available in the public domain (Note that the datasets analysed in the computational part included genomic surveys, as well as genomic experiments. For example, one of the experiments in the meta-analysis was a comparison of transcriptomes from the cell culture of human microvascular ECs with or without angiogenic stimulation. The angiogenic stimulation consisted of cell culture in the medium with added vascular endothelial growth factor. In the surveys, only libraries from ECs cultured in standard conditions were analysed: there were no additional experimental factors.). The goal of the meta-analysis was to computationally identify consensus endothelial-specific genes. The consensus predictions were then prioritized for empirical verification. In the empirical verification, RNAs from a small panel of endothelial and non-EC cell types were used to test whether the consensus predictions were indeed endothelial-specific in their spatial expression pattern.

Conceptually, I argue that Huminiecki and Bicknell took a reductionist approach to the analysis of the endothelial transcriptome. This is because the authors strived to identify individual genes that were expressed most specifically in this cell type. Crucially for the reductionist argument, the authors assumed that the essence of the endothelial transcriptome could be discerned by looking at individual genes independently of each other—rather than by looking at signaling or metabolic pathways or networks. Moreover, Huminiecki and Bicknell assumed that sufficient insights could be derived just by looking at finally differentiated ECs, without an analysis of progenitor cells or differentiation trajectories (Note that an analysis of differentiation would be technically much more difficult. Such an analysis would have to include time courses and multiple cell types).

### 2.2. A New Genomic Technology for the Analysis of Individual Promoters

A reductionist wants to divide a problem into smaller and smaller parts. This strategy is frequently facilitated by the emergence of a new laboratory technology, which typically allows one to look at a given biological mechanism in more detail.

For example, surveying the mechanism of gene expression at a greater resolution recently changed scientists' view on the diversity of the transcriptome. Namely, a new technology for the detection of transcription, called Cap Analysis of Gene Expression (CAGE), allowed scientists to map gene expression at the level of individual base pairs [33]. Previously, microarrays typically only measured the expression signal of whole genes, ignoring the fact that most genes have multiple alternative promoters and transcriptional start sites—TSSes. Thus, the essence of the technological advancement brought about by CAGE is that one can characterize individual TSSes across the entire genome. Therefore, there is more detail and no bias is introduced by pre-selecting genomic features.

In recent practice, CAGE technology was employed by an international research consortium for the functional annotation of the mammalian genome (FANTOM5), led from Japan, to map gene expression in human and mouse genomes [34]. The survey performed by FANTOM5 was comprehensive; consortium's expression data included profiles of 952 human and 396 mouse tissues, primary cells, and cancer cell lines. The map of transcription generated by the consortium demonstrated that most mammalian genes have multiple TSSes (and accompanying promoters located in the brackets of 3000 base pairs upstream/downstream of the TSS [26]). Moreover, alternative promoters can differ in expression patterns they drive [34], and in the structure of resulting transcripts (Interestingly, some of the variability in transcripts produced was previously known in the literature. But the variability was attributed to alternative splicing rather than to the existence of alternative TSSes.).

Furthermore, the map of gene expression generated by FANTOM5 became a reductionist research tool for dozens of other projects. For example, researchers used this dataset to map a correlation between the size of the architecture of a promoter, and the breadth of expression pattern the promoter drives [25]. In a further reductionist step, it was even technically feasible to survey separate contributions made to the housekeeping gene expression pattern by individual kinds of transcription factor binding sites located within the proximal promoter [26,35]. This, again, underlines the power of reductionist strategies in genomics. Indeed, many insights can be achieved by dividing the problem at hand into smaller and smaller parts, and analyzing it at a greater resolution.

### 2.3. An Anti-Reductionist Analysis of the Evolution of an Entire Signalling Pathway

In the next example, Huminiecki et al. [30] followed anti-reductionist principles in an evolutionary context. The approach could be dubbed: evolutionary systems genetics. The empirical method employed was a survey of genomes, conducted from an evolutionary perspective (rather than a test of an experimental hypothesis). Specifically, the paper focused on the emergence, development, and diversification of the transforming growth factor-$\beta$ (TGF-$\beta$) signaling pathway.

Some background information is necessary to illuminate the narrative of the paper. It must first be mentioned that most genes belong to gene families, which are derived through consecutive cycles of gene duplication. In animals, gene duplication is the most important source of new genes and new cellular functions for the evolutionary process. However, genes usually duplicate individually: that is one at a time. Whole genome duplication (WGD) is a rare and dramatic evolutionary event, a so-called macro-mutation. In a WGD, all genes duplicate simultaneously (this is known as polyploidization). In 1970, Susumu Ohno suggested that a WGD event, termed 2R-WGD, occurred at the base of the evolutionary tree of vertebrates [36]. More recently, the 2R-WGD hypothesis received a lot of empirical support from genome sequences analyzed using sophisticated algorithms aiming to detect WGDs [37].

Huminiecki et al. also investigated the 2R-WGD hypothesis using genome sequences, but the authors focused exclusively on the components of the TGF-$\beta$ pathway. Altogether, there are eight intra-cellular transducers (Smads) for the pathway in the human genome, accompanied by twelve different TGF-$\beta$ receptors. After a survey of homologs of these human genes in 33 animal genomes, Huminiecki et al. deduced using the principles of parsimony that the evolutionary emergence of the TGF-$\beta$ pathway paralleled the emergence of first animals. The pathway can be inferred to have initially existed in a simplified ancestral form, including just four trans-membrane receptors, and four Smads. This simple repertoire of the components of the ancestral pathway is similar to that observed in the extant genome of a primitive tablet animal in the species *Trichoplax adhaerens*. However, the pathway expanded in the evolutionary lineage leading to humans following 2R-WGD.

The interpretation of the above genomic screen focused on the following evolutionary hypothesis. The increase in the number of components of the pathway probably paralleled an increase in complexity of biochemical functions that the pathway could carry out, as well as an increase in cellular/organismal processes in which the pathway played a role. Note that progenitors of first animals were probably colony-forming organisms, similar to colonial choanoflagellates, with little specialization of cell-types. Accordingly, the primitive TGF-$\beta$ pathway was probably only involved in sensing nutrients, or in mediating adhesion and attachment to solids (This ancestral function is probably still present in taxa Placozoa and Porifera.). However, the pathway gained an important role in cellular transfer of signals as numerous specialized cell types that emerged in true multicellular animals. (This function still corresponds to the role fulfilled by the pathway in invertebrates.) The TGF-$\beta$ pathway further gained in complexity with the emergence of vertebrate animals—becoming a sort of super-signaling engine capable of communicating diverse stimuli and fulfilling a bewildering variety of biological roles. Indeed, the versatile vertebrate version of the pathway functions in many diverse physiological processes. These processes range from development, through organogenesis, to the control of stem cells, and even in immunity.

Note that the conclusions reached by Huminiecki et al. were only possible after the anti-reductionist analysis. The entire TGF-$\beta$ pathway had to be analyzed in parallel—consisting of all its receptors and intra-cellular signal transducers in as many animal genomes as possible. A reductionist analysis of individual components, gene-by-gene or exon-by-exon, performed no matter with how much attention to detail, could not deliver the synthesis and relevant insights.

Another methodological point is that an evolutionary analysis of the type undertaken by Huminiecki et al. is not an experiment but instead a type of systematic and deliberate observational study, carried out on sequences from extant genomes. Needless to say, an experimental test of a hypothesis on such extensive evolutionary timescales would be impossible.

*2.4. An Anti-Reductionist Analysis of the Evolution of the Entire Vertebrate Signalling Network*

Following the aforementioned anti-reductionist analysis of the TGF-$\beta$ pathway, Huminiecki and Heldin broadened their investigations. That is to say, the impact of 2R-WGD on the evolution of the entire vertebrate signaling network was analyzed in a follow-up paper [31]. Specifically, Huminiecki characterized in the follow-up paper the impact of 2R-WGD on the functionality of the whole cell (and the organism). What matters most for the purpose of the current review is that the above analysis had to be anti-reductionist. The global trends could have never been discovered by looking at individual genes in isolation.

The authors identified functional classes of genes that 2R-WGD had greatest impact on. In terms of their biochemical roles, these were signaling genes (i.e., ligands, receptors, intracellular transducers), as well as transcription factors effecting the responses of signaling pathways. In terms of the corresponding biological processes, genes preferentially retained following 2R-WGD provided emergent vertebrates with their specific evolutionary novelties. These included (1) the finely-tunable machinery of cell-cycle, (2) multi-compartment brains wired by neurons endowed with versatile synapses, (3) a pressurized circulatory system and a heart that powers it, (4) dynamic musculature and bones which facilitate active locomotion, and (5) adipose tissue that facilitates thermoregulation. Clearly, the above set of evolutionary novelties powered the radiation of vertebrates, and kick started their subsequent evolution on land.

Can the above trends be generalized to a universal scientific law? It turns out that the answer is 'yes', but only partially. Specifically, preferential retention of signaling genes and transcription factors after WGDs is a general law. This is because similar conclusions could be reached following analyses of WGDs in animals, plants, yeasts, and protozoans [37]. WGDs rather than duplications of individual genes facilitate the evolution of cellular network hubs and rewiring of the cellular network. At the same time, genomic evidence for WGDs can be only rarely observed in animals, but rather frequently in plants (This is probably due to reproductive differences.).

## 3. Conclusions

The take-home message is that the choice of an appropriate model of the gene strongly depends on the goals of a multidimensional analysis. This is because alternative models of the gene are on a broad spectrum strongly differing in the level of physical and chemical detail being modeled. The right approach will vary depending on many factors, for example depending on the sub-discipline of genetics, the hypothesis being tested, the design of the experiment, statistical methods used, and whether the study is of purely academic, translational, or industrial interest. Indeed, it is critical to choose the optimal level of abstraction for various genetical concepts in a computational project. The scientist and the statistician should discuss their strategy in advance. Such decisions are of equal importance to the choice of an optimal experimental design and a sensitive statistical test.

Indeed the alternatives available are broadly varied. The optimal choice of the model of the gene may fall somewhere on the spectrum close to radical reductionism (which is similar in spirit to the methods of molecular biology, biochemistry, or even organic chemistry). Such approaches will strive to directly model chemical and physical phenomena for individual molecules, or even atoms. At the extreme end of the spectrum, a computational geneticist might even strive to model the quantum phenomena within individual atoms (For example, if the intention was to understand how mutagens interact with the stacked arrangement of bases in nucleic acids).

Alternatively, it might be preferable to employ an anti-reductionist strategy. I illustrated this with examples from my own work, focusing on the TGF-$\beta$ signaling pathway (example 3) or on the vertebrate signaling network (example 4). The examples served to illustrate the principle but were by no means comprehensive in their representation of the field. Indeed, various anti-reductionist strategies are currently widely employed in both biology and medicine. Analyses of different kinds of biological networks are being applied to either theoretical or practical problems. The networks include not only signaling pathways, but also protein interactions, metabolic pathways, or transcriptional networks. Indeed, there emerged whole new fields of research such as systems genetics, network biology [38], or network medicine [39,40]. For example, in systems genetics, databases of cellular protein interactions and signaling pathways are being used to model the inheritance of complex traits and to interpret the results of genome-wide association studies [41,42]. In cancer research, mutations are being put in network context to predict tumor subtypes, or to identify key signaling hubs that might be attractive targets for pharmacological anti-cancer interventions [43] (Note also that a cancer signaling map [44] was re-used by me in an evolutionary context in example 4.). Indeed, the anti-reductionist approaches vary widely over different inputs, across varying biological networks, and depending on accepted performance metrics. Therefore, benchmarking studies comparing the performance of different algorithms proved useful in highlighting their respective strengths and weaknesses [45,46].

Unfortunately, there is still not enough recognition in the literature of the theoretical importance of the choice of either a reductionist or an anti-reductionist agenda for data-mining. Moreover, there are few formal guidelines for the choice of the suitable model of the gene to fit the purposes of a given genomic analysis. This review hopes to be of some help in defining the challenge and setting up the stage for fuller theoretical and statistical considerations in the future.

**Funding:** L.H. was funded to perform this work by the National Science Centre, Poland, grant POLONEZ 2 (grant number 2016/21/P/NZ2/03926). This project has received funding from the European Union's Horizon 2020 research and innovation programme under the Marie-Sklodowska-Curie grant agreement No. 665778. The funders had no role in study design, data collection and analysis, decision to publish, or preparation of the manuscript.

**Conflicts of Interest:** The author declares no conflict of interest.

## References

1. Mendel, J.G. Versuche über Pflanzenhybriden. *Verhandlungen Naturforschenden Vereines Brünn* **1866**, *IV*, 3–47.
2. Griffiths, P.S.K. Gene. In *The Cambridge Companion to the Philosophy of Biology*; Hull, D.L., Ruse, M., Eds.; Cambridge University Press: Cambridge, UK, 2007.
3. Ptashne, M.G.A. *Genes and Signals*; Cold Spring Harbor Laboratory Press: New York, NY, USA, 2002; p. 192.
4. Hayes, W. Max Ludwig Henning Delbruck—September 4, 1906-March 10, 1981. *Biogr. Mem. Natl. Acad. Sci.* **1992**, *62*, 67–117. [PubMed]
5. Griffiths, P.E.; Stotz, K. Genes in the postgenomic era. *Theor. Med. Bioeth.* **2006**, *27*, 499–521. [CrossRef]
6. Fogle, T. Are Genes Units of Inheritance. *Biol. Philos.* **1990**, *5*, 349–371. [CrossRef]

7. Engstrom, P.G.; Suzuki, H.; Ninomiya, N.; Akalin, A.; Sessa, L.; Lavorgna, G.; Brozzi, A.; Luzi, L.; Tan, S.L.; Yang, L.; et al. Complex Loci in human and mouse genomes. *PLoS Genet.* **2006**, *2*, e47. [CrossRef] [PubMed]

8. Aristotle; Barnes, J. *The Complete Works of Aristotle: The Revised Oxford Translation*; Princeton University Press: Princeton, NJ, USA, 1995.

9. Bacon, F. George Fabyan Collection (Library of Congress). In *Francisci de Verulamio, Summi Angliae Cancellarii, Instauratio Magna*; Apud Joannem Billium, Typographum Regium: London, UK, 1620.

10. Wheeler, D.L.; Barrett, T.; Benson, D.A.; Bryant, S.H.; Canese, K.; Chetvernin, V.; Church, D.M.; DiCuccio, M.; Edgar, R.; Federhen, S.; et al. Database resources of the National Center for Biotechnology Information. *Nucleic Acids Res.* **2007**, *35*, D5–D12. [CrossRef] [PubMed]

11. Kenney, D.E.; Borisy, G.G. Thomas Hunt Morgan at the marine biological laboratory: Naturalist and experimentalist. *Genetics* **2009**, *181*, 841–846. [CrossRef]

12. Sturtevant, A.H. The linear arrangement of six sex-linked factors in Drosophila, as shown by their mode of association. *J. Exp. Zool.* **1913**, *14*, 43–59. [CrossRef]

13. Fisher, R.A.; Bennett, J.H. *Statistical Methods, Experimental Design, and Scientific Inference*; Oxford University Press: Oxford, UK; New York, NY, USA, 1990.

14. Uebel, T.E.; Richardson, A.W. *The Cambridge Companion to Logical Empiricism*; Cambridge University Press: Cambridge, UK, 2007.

15. Kuhn, T.S. *The Structure of Scientific Revolutions*; University of Chicago Press: Chicago, IL, USA, 1962.

16. Portin, P. The Development of Genetics in the Light of Thomas Kuhn's Theory of Scientific Revolutions. *Recent Adv. DNA Gene Seq.* **2015**, *9*, 14–25. [CrossRef]

17. Nagel, E. *The Structure of Science: Problems in the Logic of Scientific Explanation*; Routledge & Kegan Paul: London, UK, 1961.

18. Schaffner, K.F. *Discovery and Explanation in Biology and Medicine*; University of Chicago Press: Chicago, IL, USA, 1993.

19. Lehmann, E.L. *Fisher, Neyman, and the Creation of Classical Statistics*; Springer: Berlin, Germany, 2011.

20. Efron, B.; Hastie, T. *Computer Age Statistical Inference: Algorithms, Evidence, and Data Science*; Cambridge University Press: Cambridge, UK, 2016.

21. Edgington, E.S. *Randomization Tests*; CRC Press: Boca Raton, FL, USA, 2007.

22. Kuhn, T.S. *The Copernican Revolution—Planetary Astronomy in the Development of Western Thought*; Harvard University Press: Cambridge, MI, USA, 1985.

23. Weeden, N.F. Are Mendel's Data Reliable? The Perspective of a Pea Geneticist. *J. Hered.* **2016**, *107*, 635–646. [CrossRef]

24. Klug, A. Rosalind Franklin and the double helix. *Nature* **1974**, *248*, 787–788. [CrossRef] [PubMed]

25. Hurst, L.D.; Sachenkova, O.; Daub, C.; Forrest, A.; Huminiecki, L. A simple metric of promoter architecture robustly predicts expression breadth of human genes suggesting that most transcription factors are positive regulators. *Genome Biol.* **2014**, *15*, 413. [CrossRef] [PubMed]

26. Huminiecki, L. Modelling of the breadth of expression from promoter architectures identifies pro-housekeeping transcription factors. *PLoS ONE* **2018**, *13*, e0198961. [CrossRef] [PubMed]

27. Huminiecki, L.; Bicknell, R. In silico cloning of novel endothelial-specific genes. *Genome Res.* **2000**, *10*, 1796–1806. [CrossRef] [PubMed]

28. Huminiecki, L.; Gorn, M.; Suchting, S.; Poulsom, R.; Bicknell, R. Magic roundabout is a new member of the roundabout receptor family that is endothelial specific and expressed at sites of active angiogenesis. *Genomics* **2002**, *79*, 547–552. [CrossRef]

29. Huminiecki, L. Magic roundabout is an endothelial-specific ohnolog of ROBO1 which neo-functionalized to an essential new role in angiogenesis. *PLoS ONE* **2019**, *14*, e0208952. [CrossRef]

30. Huminiecki, L.; Goldovsky, L.; Freilich, S.; Moustakas, A.; Ouzounis, C.; Heldin, C.H. Emergence, development and diversification of the TGF-beta signalling pathway within the animal kingdom. *BMC Evol. Biol.* **2009**, *9*, 28. [CrossRef]

31. Huminiecki, L.; Heldin, C.H. 2R and remodeling of vertebrate signal transduction engine. *BMC Biol.* **2010**, *8*, 146. [CrossRef]

32. Monahan-Earley, R.; Dvorak, A.M.; Aird, W.C. Evolutionary origins of the blood vascular system and endothelium. *J. Thromb. Haemost.* **2013**, *11*, 46–66. [CrossRef]

33. FANTOM5 Presentation of CAGE Technology. Available online: http://fantom.gsc.riken.jp/protocols/ (accessed on 25 August 2020).
34. Forrest, A.R.; Kawaji, H.; Rehli, M.; Baillie, J.K.; de Hoon, M.J.; Lassmann, T.; Itoh, M.; Summers, K.M.; Suzuki, H.; Daub, C.O.; et al. A promoter-level mammalian expression atlas. *Nature* **2014**, *507*, 462–470.
35. Huminiecki, L.; Horbanczuk, J. Can We Predict Gene Expression by Understanding Proximal Promoter Architecture? *Trends Biotechnol.* **2017**, *35*, 530–546. [CrossRef]
36. Ohno, S. *Evolution by Gene Duplication*; Springer: Berlin, Germany, 2013.
37. Conant, G.C.; Wolfe, K.H. Turning a hobby into a job: How duplicated genes find new functions. *Nat. Rev. Genet.* **2008**, *9*, 938–950. [CrossRef] [PubMed]
38. Barabasi, A.L.; Oltvai, Z.N. Network biology: Understanding the cell's functional organization. *Nat. Rev. Genet.* **2004**, *5*, 101–113. [CrossRef] [PubMed]
39. Barabasi, A.L.; Gulbahce, N.; Loscalzo, J. Network medicine: A network-based approach to human disease. *Nat. Rev. Genet.* **2011**, *12*, 56–68. [CrossRef] [PubMed]
40. Barabasi, A.L. Network medicine—From obesity to the "diseasome". *N. Engl. J. Med.* **2007**, *357*, 404–407. [CrossRef] [PubMed]
41. Nam, D.; Kim, J.; Kim, S.Y.; Kim, S. GSA-SNP: A general approach for gene set analysis of polymorphisms. *Nucleic Acids Res.* **2010**, *38*, W749–W754. [CrossRef]
42. Yoon, S.; Nguyen, H.C.T.; Yoo, Y.J.; Kim, J.; Baik, B.; Kim, S.; Nam, D. Efficient pathway enrichment and network analysis of GWAS summary data using GSA-SNP2. *Nucleic Acids Res.* **2018**, *46*, e60. [CrossRef]
43. Li, L.; Tibiche, C.; Fu, C.; Kaneko, T.; Moran, M.F.; Schiller, M.R.; Li, S.S.; Wang, E. The human phosphotyrosine signaling network: Evolution and hotspots of hijacking in cancer. *Genome Res.* **2012**, *22*, 1222–1230. [CrossRef]
44. Cui, Q.; Ma, Y.; Jaramillo, M.; Bari, H.; Awan, A.; Yang, S.; Zhang, S.; Liu, L.; Lu, M.; O'Connor-McCourt, M.; et al. A map of human cancer signaling. *Mol. Syst. Biol.* **2007**, *3*, 152. [CrossRef]
45. Picart-Armada, S.; Barrett, S.J.; Wille, D.R.; Perera-Lluna, A.; Gutteridge, A.; Dessailly, B.H. Benchmarking network propagation methods for disease gene identification. *PLoS Comput. Biol.* **2019**, *15*, e1007276. [CrossRef]
46. Hill, A.; Gleim, S.; Kiefer, F.; Sigoillot, F.; Loureiro, J.; Jenkins, J.; Morris, M.K. Benchmarking network algorithms for contextualizing genes of interest. *PLoS Comput. Biol.* **2019**, *15*, e1007403. [CrossRef] [PubMed]

*Review*

# Fifteen Years of Gene Set Analysis for High-Throughput Genomic Data: A Review of Statistical Approaches and Future Challenges

**Samarendra Das** [1,2,3], **Craig J. McClain** [4,5,6,7,8] **and Shesh N. Rai** [2,3,5,6,9,*]

1. Division of Statistical Genetics, ICAR-Indian Agricultural Statistics Research Institute, New Delhi 110012, India; samarendra.das@louisville.edu
2. School of Interdisciplinary and Graduate Studies, University of Louisville, Louisville, KY 40292, USA
3. Biostatistics and Bioinformatics Facility, JG Brown Cancer Center, University of Louisville, Louisville, KY 40202, USA
4. Department of Medicine, University of Louisville, Louisville, KY 40202, USA; craig.mcclain@louisville.edu
5. Hepatobiology and Toxicology Center, University of Louisville, Louisville, KY 40202, USA
6. Alcohol Research Center, University of Louisville, Louisville, KY 40202, USA
7. Department of Pharmacology and Toxicology, University of Louisville, Louisville, KY 40202, USA
8. Robley Rex Louisville VAMC, Louisville, KY 40206, USA
9. Department of Bioinformatics and Biostatistics, University of Louisville, Louisville, KY 40202, USA
* Correspondence: shesh.rai@louisville.edu; Tel.: +1-502-426-0016

Received: 24 February 2020; Accepted: 3 April 2020; Published: 10 April 2020

**Abstract:** Over the last decade, gene set analysis has become the first choice for gaining insights into underlying complex biology of diseases through gene expression and gene association studies. It also reduces the complexity of statistical analysis and enhances the explanatory power of the obtained results. Although gene set analysis approaches are extensively used in gene expression and genome wide association data analysis, the statistical structure and steps common to these approaches have not yet been comprehensively discussed, which limits their utility. In this article, we provide a comprehensive overview, statistical structure and steps of gene set analysis approaches used for microarrays, RNA-sequencing and genome wide association data analysis. Further, we also classify the gene set analysis approaches and tools by the type of genomic study, null hypothesis, sampling model and nature of the test statistic, etc. Rather than reviewing the gene set analysis approaches individually, we provide the generation-wise evolution of such approaches for microarrays, RNA-sequencing and genome wide association studies and discuss their relative merits and limitations. Here, we identify the key biological and statistical challenges in current gene set analysis, which will be addressed by statisticians and biologists collectively in order to develop the next generation of gene set analysis approaches. Further, this study will serve as a catalog and provide guidelines to genome researchers and experimental biologists for choosing the proper gene set analysis approach based on several factors.

**Keywords:** gene set analysis; microarrays; RNA-sequencing; genome wide association study; competitive; self-contained; sampling model; null hypothesis

---

## 1. Background

The advancement in genome sequencing technologies has led to the generation of tremendous volume of high-throughput and high-dimensional biological data [1]. Further, exploiting these data and drawing valid biological insights has posed a great challenge to researchers across the globe. For instance, in a gene expression (GE) study, the expression levels of several thousand(s) of genes are measured in a single experiment and further used for identifying the groups of genes which are relevant to the condition/trait under study [2–4]. Earlier, biologists considered this differential expression (DE) study as the end of their analysis [5]. However, such analysis is the starting point of a complex process of drawing valid biological insights into high-throughput genomic data [6]. Further, the DE analysis produces a list of associated genes ranked by the ascending or descending order of the magnitude of computed test statistic(s)/*p-values* (e.g., Z-score, fold change, t-test, etc.) [3–5]. This is a crucial step undertaken by the experimental biologists and genome researchers to select the informative genes as well as to obtain a global view of expression changes. Further, to put the long list of gene-level results into a broader biological context and to further reduce the complexity of analysis, secondary analytical approaches have been developed by grouping the long list of genes into smaller sets of related genes. One such approach is gene set analysis (GSA), and one of its popular forms is called as pathway analysis [7].

In the last decade, GSA has completely shifted the focus in GE and association data analysis from individual gene to gene set level [7–11]. Further, GSA has been extensively used in complex disease biology due to the polygenic nature of these disorders. GSA involves testing for association of sets of functionally related variants/genes, and can provide biological context for multiple genetic risk factors [12]. Recently, GSA was able to provide biological insights into mechanisms and possible treatment targets for complex diseases, including schizophrenia [13], bipolar disorder [14], Crohn's disease [15], rheumatoid arthritis [16], breast cancer [17], and obesity [18]. Moreover, GSA has also been applied in plant biology to understand the abiotic stress response mechanisms in *Arabidopsis thaliana*, *Oryza sativa*, *Zea mays*, and *Gossypium raimondii* [9,10]. The GSA applications have led to novel biological hypotheses about the diseases/stress responses, and have suggested new avenues for molecular drug designing/crop breeding intervention [6,7,10,19–22].

Numerous statistical approaches and tools for GSA are now available for analysis of high-throughput genomic datasets. This includes GE data from microarrays and RNA-sequencing (RNA-seq) studies and single nucleotide polymorphism (SNP) data from genome wide association studies (GWAS). However, many researchers have tried to review the available GSA approaches in different times, but these are limited to only specific genomic studies. There is no comprehensive review of GSA approaches and tools meant for these broad spectra of datasets. Further, without sufficient understanding of the underlying statistical principles of GSA approaches, we may risk drawing erroneous biological interpretations and statistical conclusions. Moreover, there are minimal studies on grouping the available GSA approaches. Therefore, in this article, we aim to provide a comprehensive overview, statistical structure and steps concerning GSA approaches used for high-throughput genomic data analysis. Further, we classify the GSA approaches and tools based on the type of genomic study, null hypothesis, sampling model, nature of test statistic(s), etc. We also provide an overview of the evolution of GSA approaches in terms of different generations rather than reviewing them individually, along with their relative merits and demerits. Here, we address the key biological and statistical challenges in current GSA, which need to be addressed to develop the next generation of GSA approaches and tools.

## 2. Structure of Gene Set Analysis

The term GSA refers to an analysis of set of genes and does not specifically mean modelling of the relations among genes in the gene set. Formally, GSA is defined as a secondary statistical approach used to test the involvement/enrichment of the gene sets with any biological process or pre-existing bio-knowledge base or quantitative trait. In other words, genes are aggregated to gene sets based on shared biological or functional properties or any pre-existing bio-knowledge base or quantitative trait [6]. These bio-knowledge bases include databases of molecular knowledge, i.e., molecular interactions, regulation, molecular product(s), and even phenotype associations or quantitative traits. A list of available bio-knowledge bases is given in Supplementary Table S1. In other words, GE and SNP datasets are used as input for GSA (in the presence of a annotation database) to provide valid biological insights into various complex diseases (Figure 1 and Figure S1) [7,23]. In fact, GSA has the potential to be used for all genomic data analysis, where the output is a long list of genes or transcripts. For instance, that long list of genes can even come from any upstream analysis including signatures of co-expressed genes from weighted gene co-expression network analysis [4].

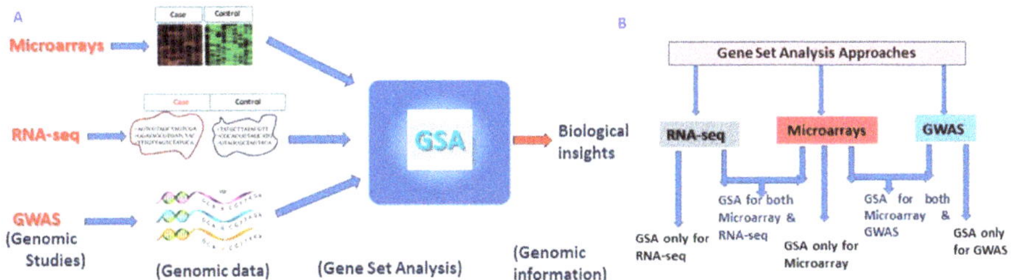


**Figure 1.** Outlines and classification of gene set analysis approaches. (**A**): Outlines of gene set analysis approaches; (**B**): Classification of gene set analysis approaches for high-throughput sequencing studies.

## 2.1. Units of Gene Set Analysis

The functional unit of GSA is the gene set, which can be defined as any group of genes that share a particular property, i.e., involvement in a common biological process or any pre-existing bio-knowledge base [7,12]. Through GSA, a gene set that shares a common property is tested for its association with the trait or phenotype under study [24]. For this purpose, a wide range of GSA approaches and tools are available for high-throughput sequencing studies. These tools have differences in underlying statistical principles and practices, but there are similarities among the available tools in terms of statistical structure. For instance, GSA for GE studies has a two-tier structure [12,25]: (a) computation of gene level statistic(s); and (b) bi-variate statistical testing to compute the test statistic or *p-value* for the gene set. However, GSA for GWAS has a three-tier structure: (a) computation of SNP level statistics; (b) associating SNPs to genes and computing gene-level statistics from SNP statistics; and (c) computation of enrichment statistic or *p-value or* False Discovery Rate (FDR) for the gene set.

## 2.2. Hypotheses in Gene Set Analysis

The available statistical approaches for GSA greatly vary with respect to underlying statistical tests and hence depend on the formulation of the null hypothesis [6,11,23]. These null hypotheses can be grouped as self-contained and competitive [26]. In the usual set up of GE studies (or GWAS), genes (or SNPs) that are significantly associated with a trait/phenotype are identified and then evaluated, whether the significantly associated genes (or SNPs) tend to cluster in predefined gene sets or not. For instance, the self-contained null hypothesis can be framed as, $H_0$: genes/SNPs in predefined gene sets are not associated with the underlying trait (phenotype) against alternate $H_1$: genes/SNPs in predefined gene sets are associated with the trait (phenotype). The statistical approaches with a self-contained null hypothesis are called as self-contained approaches of GSA and they only consider the genes (SNPs) in the predefined gene sets. Statistical tests of GSA with a competitive null hypothesis are known as competitive GSA approaches, and the underlying null hypothesis can be expressed as, $H_0$: genes/SNPs in predefined gene sets are associated with the underlying trait (phenotype) as much as are genes/SNPs outside the predefined gene set, against $H_1$: genes/SNPs in predefined gene sets are more associated with the trait (phenotype) than genes outside predefined gene set. Here, the competitive GSA approaches consider genes (SNPs) from both the predefined gene set and the outside gene set [6,10]. The self-contained null hypothesis is invariably more restrictive than the competitive null hypothesis.

### 2.3. Sampling Models in Gene Set Analysis

The enrichment significance of a gene set is assessed through *p-value or* adjusted *p-value or* FDR after multiple testing correction (i.e., lower values indicate more enrichment and *vice-versa*) computed from a statistical test. Further, these statistical tests are commonly based on experimental designs having subjects/genes as units. On such statistical designs, different sampling procedures are rigorously used to obtain the distribution of the test statistic(s). Here, two types of sampling models are used in GSA: (i) subject sampling model; and (ii) gene sampling model.

#### 2.3.1. Subject Sampling Model

Classical statistical tests are based on an experimental design having microarray/RNA-seq samples as subjects, where each subject has the same set of (GE) measurements [6,10,24]. In the usual supervised setting, the sampling model consists of $M$ independent realizations (for $M$ subjects) of $(X_1, y_1)$, $(X_2, y_2)$, ..., $(X_s, y_s)$, ..., $(X_M, y_M)$, where, $X_s$ represents the $N$-dimensional vector ($N$: total number of genes) of the GE levels for *s-th* subject and $y_s$ is the corresponding class label (e.g., case: +1 vs. control: −1), $s = 1, 2, ... , M$. Therefore, $M$ expression levels of different subjects are assumed to be independently and identically distributed (iid), but expression levels of genes within the same subject may be correlated for a given condition. Usually, resampling procedures like bootstrap and permutation procedures are used on such models for gene [4,27] as well as gene set testing [6,28]. The statistical combination of subject sampling model and a self-contained null hypothesis provides a reliable platform for valid computation of *p-values* with easy interpretation and close relation(s) with single gene (or SNP) testing [29].

#### 2.3.2. Gene Sampling Model

In GSA, $2 \times 2$ tables are extensively used to statistically fit a Hypergeometric distribution [6,30]. The underlying model of a $2 \times 2$ table is a gene sampling model. Further, each cell of such a table is filled with a sample of genes, each of which is drawn at random from the gene space (i.e., set of genes in the data). Here, in this sampling model, each sampling unit (i.e., gene) can be subjected to two fixed set of indicator measurements, i.e., $(A, B)$, where, (i) $A$ (1 or 0) indicates whether the gene is a part of the predefined gene set or not and (ii) $B$ (1 or 0) indicates whether that gene is in the list of differentially expressed genes or not [6,10]. Further, the gene space can be formalized into a population having $N$ units (for $N$ genes) and shown as: $(A_1, B_1)$, $(A_2, B_2)$, ..., $(A_i, B_i)$, ..., $(A_N, B_N)$. The competitive null hypothesis is popular and easy to formulate in a gene-sampling model setup [23]. Here, the gene sampling model may be considered as a mirror image of classical subject sampling model [27]. The gene sampling model considers the sampling units as iid, which assumes that genes are independent. Such assumptions are highly unrealistic, and the *p-values* computed using such models are statistically invalid for further interpretations. Hence, gene sampling models are quite complex and delicate as compared to a subject sampling model and need the utmost care while using.

### 3. GSA Approaches for High-Throughput Genomic Studies

The GSA approaches can be grouped based on different high-throughput genomic studies, as the underlying nature and distributions of the datasets are different (Supplementary Table S2). A classification of GSA approaches with respect to their application to genomic studies is shown in Figure 1. Initially, the GSA approaches were developed for microarrays (i.e., microarrays GSA) and subsequently extended to RNA-seq and GWAS data analysis (Figure 1). For instance, gene set enrichment analysis (GSEA) was originally developed for microarrays, and subsequent extensions of GSEA, i.e., SeqGSEA and GSEA-SNP were introduced to analyze RNA-seq and SNP datasets respectively.

### 3.1. Microarrays GSA

Huge amounts of GE data from microarrays are available in public domain databases (Supplementary Table S3), which need to be analyzed for drawing valid biological insights into such datasets. Therefore, several GSA methodologies have been developed for this purpose. The classification of GSA microarrays is shown in Figure 2, which illustrates the evolution of GSA approaches over time in terms of the requirement of annotation information, sampling model, various null hypotheses under statistical tests. Moreover, the work on GSA started with the immediate need for functional analysis of microarray data based on gene ontology (GO) that gave rise to over representation analysis (ORA), which evaluates the statistical significance of gene sets in a particular pathway/functional category [21]. It is also referred to as a $2 \times 2$ table method [6], due to the fact that ORA approaches are mostly based on $2 \times 2$ tables and gene sampling models. The most commonly used statistical tests in ORA approaches/tools are hypergeometric, chi-square or binomial tests [20,31,32] (Supplementary Document S1). However, despite the extreme popularity and ease of execution, the ORA approaches also suffer from limitations, as listed in Table 1. The ORA form of analysis of gene sets can also be labelled as first generation of microarrays GSA.

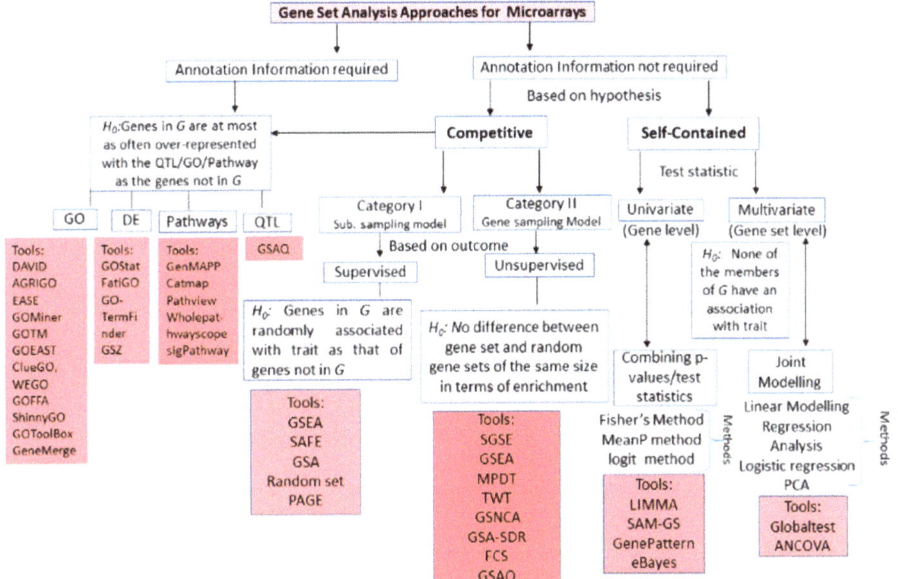


**Figure 2.** Classification of gene set analysis approaches and tools available for microarrays. Schematic representation of the breakup of GSA methods available for microarrays data analysis based on statistical tests (i.e., null hypothesis, test statistic(s)) and requirement of annotation databases. *G*: Gene set; * Tools require normalization of data prior to application.

In most of the cases, the gene annotation information is either incomplete or totally unavailable; therefore, another class of GSA approach was developed. These approaches include the Enrichment Score (ES) form of GSA [33], starting with the landmark work on enrichment analysis of gene sets (i.e., GSEA) [8,24]. Subsequently, several other statistical approaches, algorithms and tools were developed for assessing the significance of gene sets in interpreting the high-throughput microarray data. The ES based GSA approaches greatly vary among themselves with respect to underlying statistical tests and sampling models. However, there are also commonalities among these ES based approaches in terms of execution, which is given in Supplementary Figures S1 and S2. The major steps for such approaches include initial computation of the gene-level statistic(s) using GE data under two contrasting conditions (Figures S1 and S2). For instance, correlation of expression measurements with phenotypes/traits [34], ANOVA [35], Q-statistic [26], signal-to-noise ratio [24], t-statistic [3], fold change [36], Z-score [37], etc., are implemented in contemporary ES based tools. There is a wider choice for gene-level statistic(s), ranging from parametric to non-parametric, for GSA. However, the selection of a gene-level statistic has a negligible effect on identification of significantly enriched gene sets [30]. When there are few biological replicates available, a regularized statistic may be preferred [30]. The second step is aggregation of gene-level statistic(s) for all genes in a gene set into a single gene-set level statistic (Figure 3). This includes the computation of gene-set level statistic using multivariate or univariate techniques (Figure 2). The former accounts for interdependencies among genes, while the latter disregards the same among genes distributed across the gene set. The currently available ES based GSA approaches/tools include Kolmogorov-Smirnov (KS) statistic, weighted KS statistic [24,33], sum, mean, or median of gene-level statistic [38], Wilcoxon rank sum [39], Max-mean statistic [8], etc. under univariate category. Moreover, multivariate category includes global test, ANCOVA, etc. for computing gene-set level statistic [26]. Interestingly, multivariate statistic(s) are expected to have higher statistical power, but univariate statistic(s) actually show more power at a higher level of significance (e.g., 0.1%) in real biological data, and equal power as the former at lower level of significance (e.g., 5%) [40].

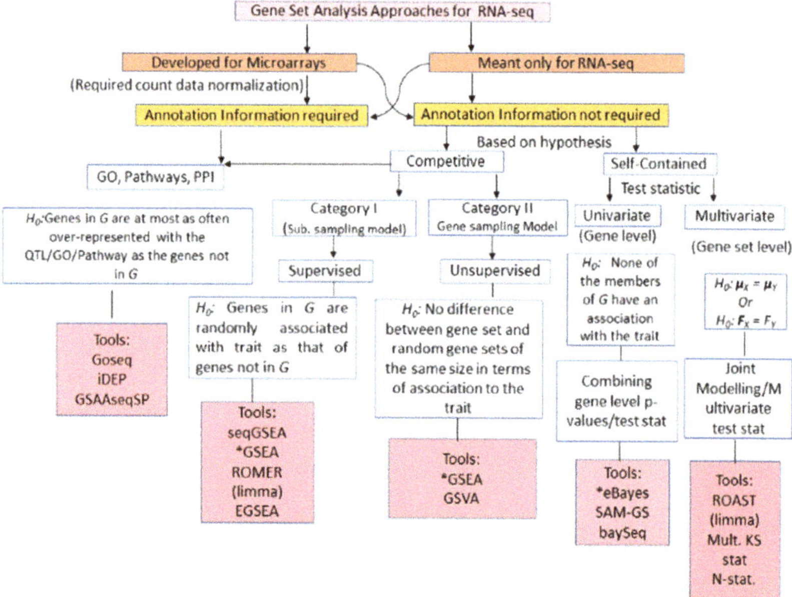



**Figure 3.** Classification of gene set analysis approaches and tools available for RNA-seq data analysis. Schematic representation of the breakup of GSA methods available for RNA-seq data analysis based on statistical tests (i.e., null hypothesis, test statistic(s)) and requirement of annotation databases. The first level of branching of the GSA methods based on their adaption from Microarrays practice to fit RNA-seq data as well as those specifically designed for RNA-seq. Subsequent branching depends on the different null hypotheses they test. G: Gene set. * Tools require normalization of data prior to application.

**Table 1.** Generation-wise evolution of GSA approaches for microarray studies.

| Approach | Methodology | Advantages | Limitations | Tools/Algorithms |
|---|---|---|---|---|
| **Over Representation Analysis (First generation microarrays GSA)** | Hypergeometric distribution/Fisher's test Binomial distribution, Chi-square distribution, etc. | • Easiness in execution. <br>• Assigns easily interpretable measure like p-values to the whole gene set. | • Highly dependent on threshold/cutoff value, which is at user's discretion and hard to determine. <br>• Test statistic independent of genes differential expression score. <br>• Uses only most significant genes based on hard threshold and discards others, lead to information loss. <br>• Assumes each gene contribute equally to phenotype/trait. <br>• Assumes each gene as independent and ignores the correlation or redundancy among genes in gene set. <br>• Assumes that each predefined gene set is independent of others, which is erroneous. | DAVID [41], AgriGO [32], Onto-Express [21], GenMAPP [42], GoMiner [43], FatiGO [44], GOstat [20], FuncAssociate [19], GOToolBox [45], GeneMerge [46], GOEAST [47], ClueGO [48], FunSpec [49], GARBAN [50], GO:TermFinder [22], WebGestalt [51], GOFFA [52], WEGO [53], GOTM [54], EASE, GSAQ [10], Pathview [55], Wholepathwayscope [56], ShinnyGO |
| **Enrichment Statistic Analysis (Second generation microarrays GSA)** | Wilcoxon signed rank test, Sum, Mean, or Median of gene-level statistic(s), Wilcoxon signed rank sum, Max-Mean Statistic | • Do not require a threshold/cutoff value for dividing gene space into selected and non-selected part. <br>• Considers dependence among genes in gene set. <br>• Test statistic is based on the differential GE score of genes in gene set. | • Analyzes each gene set independently. <br>• Considers only the number of genes in a gene set (pathway) for performing GSA but ignores the additional information available from the bio-knowledge bases. <br>• Assumes the predefined gene sets mutually exclusive, but in biology, these gene sets are overlapping. <br>• Most ESA methods use differential GE to rank genes/compute test statistic but discard this information from further analysis. | GSEA [24], SAFE [39], GSA [8], Random set [57], sigPathway, Category, GlobalTest [26], PCOT2 [58], SAM-GS [59], LIMMA [60], Catmap [61], T-profiler [62], FunCluster [63], GeneTrail [64], Gazer [65], GSAQ [10], ANCOVA test, CAMERA [66], PAGE [37], GAGE [67], SGSE [68], GSNCA [69], GSA-SDR [70], GenePattern [71], plantGSEA [9], GSAR [29] |
| **Topology Analysis (Third generation microarrays GSA)** | Graph/network theory | • Considers both genes relation /dependency with other genes as well as experimental condition changes. <br>• Considers the topology of the pathways/gene sets in modeling. | • Dependent on the type of cell due to cell-specific GE profiles and condition being studied, which is rarely available. <br>• Not so popular as require more rarely available information and computationally intensive. <br>• Unable to consider interactions between gene sets (pathways). <br>• Heavily dependent on annotations. | PathwayExpress [72], ScorePAGE [73], SPIA [74], NetGSA [75], TopoGSA [76], CliPPER [77] |

The third step is computation of statistical significance (*p-value*) or adjusted *p-value* or FDR to assess the enrichment of gene sets (for gene-set level statistic) (Figure S1). This step requires the formulation, as well as testing of the null hypothesis against alternate one. Based on the null hypothesis, the ES-based GSA approaches can be broadly divided into: (i) competitive approaches and (ii) self-contained approaches (Figure 2). Moreover, the competitive approaches can be further subdivided into two categories based on the available outcome information of class: (i) supervised approaches and (ii) unsupervised approaches (Figure 2). Mostly, the supervised competitive approaches use the subject sampling model to randomly sample the class labels of each sample and compare the genes in the gene set with those of its complement. Here, it may be noted that the supervised term is used as the class labels are known and these approaches use these class labels for sampling purposes. However, unsupervised competitive approaches used the gene sampling model to compute the *p-value* through comparing genes in gene set with the genes outside gene set. But self-contained ES-based GSA approaches use the permutation procedure to compute the *p-values* by permuting the class labels for each sample and comparing the genes in the gene set with itself, while ignoring the genes outside gene set. Here, it is evident that competitive ES-based GSA approaches have more statistical power as compared to self-contained approaches [8]. This may be due to the fact that competitive approaches require information on both genes in the gene set as well as genes not in the gene set [6]. Furthermore, the ES form of analysis of gene sets may constitute the second generation of microarrays GSA (Table 1). The background methodologies for the various generations of GSA is given in Supplementary Document S1.

### 3.2. RNA-seq GSA

Recently, transcriptome deep sequencing i.e., RNA-Seq has surpassed microarrays by providing better quantification of GE for very high and low expressed genes (in terms of read counts), and higher levels of accuracy and reproducibility [11,78,79]. Hence, it is highly pertinent to adapt the existing microarrays GSA to RNA-seq data with the help of data transformation along with new approaches being developed (Figure 1B). The first approach of GSA for RNA-seq data (i.e., RNA-seq GSA), i.e., GOseq was suggested by Young et al. a decade ago [80]. It performs over-representation of GO categories enriched with a long list of highly expressed genes in RNA-Seq data. Further, an easy-to-use web application, integrated differential expression and pathway (iDEP) analysis was developed for in-depth analysis of RNA-seq data [81]. Detailed descriptions of the available RNA-seq GSA approaches, background methodologies, execution tools, and their features are listed in Table 2 and Supplementary Document S1. Moreover, the ORA-based RNA-seq GSA may be considered as the first generation of RNA-seq GSA.

To tackle the limitations of ORA approaches (Table 2), ES-based RNA-seq GSA approaches are developed, which constitute the second generation of RNA-seq GSA. Further, the major steps for RNA-seq GSA approaches are shown in Figures S1 and S3. Here, the read counts are given as input for computation of different test statistic(s) for GSA, which depend on the nature and distribution of the data. For instance, microarrays GSA (i.e., ES-based GSA) deal with continuous data expected to follow a Gaussian distribution (Supplementary Table S2) [78]. However, RNA-seq involves measurements that are non-negative counts ranging from zero to millions and are expected to follow negative binomial distribution (Supplementary Table S2) [11,79]. Therefore, microarrays GSA approaches may not be directly applicable to RNA-Seq data. Hence, some authors suggested normalization of the count data prior to the use of microarrays GSA [11]. For instance, VOOM-normalization is used for normalizing the read counts for sequence-depths, then microarrays GSA are applied on the normalized RNA-seq data [82]. The Goeman and Buhlmann formulation can be applied to classify the ES-based RNA-seq GSA approaches into either competitive or self-contained [6], based on the underlying null hypotheses (Figure 3). Further, a competitive GSA approach, i.e., gene set variation analysis (GSVA), was developed and demonstrated highly correlated results between microarrays and RNA-Seq sets for samples of lympho-blastoids cell lines [83]. This high correlation may be due to the fact that GSVA as a non-parametric approach does not depend on the distributional nature of data obtained from the studies. Fridley et al. proposed a GSA approach, i.e., gamma method (GM), with a soft truncation threshold to determine the significant gene set, while a generalized linear model is used to assess significance [84]. Subsequently, GSEA, the first ever competitive approach of RNA-seq GSA, was used for RNA-seq data analysis after normalization of the count data [84]. Thereafter, several modifications were made in GSEA by integrating both DE and differential splicing (DS) information in the analyses to develop SeqGSEA and has better performance over GSEA [28].

**Table 2.** Generation-wise evolution of GSA approaches for RNA-sequencing studies.

| Approach | Methodology | Advantages | Limitations | Tools |
|---|---|---|---|---|
| **Over Representation Analysis (First generation RNA-seq GSA)** | Hypergeometric distribution, Fisher's exact test | • Simple to use.<br>• Assigns easily interpretable measure like *p-value* to the whole gene set.<br>• Less time consuming to interpret huge RNA-seq data. | • Use hard threshold approach to select gene sets.<br>• Assumes each transcript as independent and ignores the correlation or gene-gene interaction.<br>• Mostly dependent on annotation bases, but RNA-seq transcripts are not well annotated. | GoSeq [80], iDEP [81] |
| **GS Enrichment Analysis (Second generation of RNA-seq GSA)** | Wilcoxon signed rank test, Max-Mean Statistic (with count normalization technique) | • Do not require a threshold for dividing gene space into selected and non-selected part.<br>• Considers dependence among genes in gene set. | • Use normalization technique to get microarray like data, hence, loss of the count nature of RNA-seq data<br>• Through data transformation, dispersion and other inherent nature of RNA-seq data are lost<br>• ES based tools/algorithms use differential score to prepare ranked transcript list but ignore this information for gene set testing.<br>• GSEA based tools like seqGSEA are computationally intensive, time consuming and and only offers the single gene set-level statistic.<br>• GSVA is not designed for gene set-based differential expression analysis<br>• between two phenotypically distinct sample groups.<br>• ES based GSA approaches do not consider the inherent zero inflation in the RNA-seq data. | AbsFilterGSEA [85], GSAAseqSP [86], seqGSEA [87], ssGSEA, EGSEA [88], GSVA [83], GSEPD [89], RNA-Enrich [90] |

The self-contained GSA approaches can be divided into (a) univariate or gene-level; and (b) multivariate or gene set-level based on the distributional nature of the test statistic (Figure 3). The gene-level GSA approaches test a null hypothesis that the gene-set associated score does not differ between phenotypes/traits. Further, the univariate approaches are executed in two steps: (i) computation of gene level statistic(s) from the count data; and (ii) combining gene-level statistics to compute gene set level statistic or *p-value* or adjusted *p-value*. For the former case, the gene-level test statistic(s) of microarrays GSA were used in a recent study for RNA-seq GSA [84], which is quite straight forward and easy to implement. For the latter step, the gene-level statistic(s) can be combined into a single gene set statistic/*p-value* through Fisher's method, Stoufer's method, Meanp, logit method, etc. [10]. Moreover, the self-contained multivariate GSA approaches jointly model the genes to compute the gene set-level statistic(s) (Figure 3). These tests include multivariate generalization of the KS statistic [24,33], N-statistic [78], ROAST [82], etc. Further, the application of these tests requires the normalization of the RNA-seq data over varying sequencing depths [82]. Moreover, statistical significance is computed by comparing the observed statistics of gene sets with its null distribution, obtained by permuting the sample labels. Then, the enrichment significance of the gene set is assessed through the computed *p-value or* adjusted *p-value* or FDR after multiple testing correction.

### 3.3. GWAS GSA

GWAS has been successfully applied to identify many novel loci for complex traits, which are quantitative (polygenic) in nature [17–22,41]. Therefore, to understand the underlying genetic architecture, GSA approaches have been used that place GWAS results in a broader biological context [91]. Initially, GSA methods for GWAS (i.e., GWAS GSA) were borrowed from microarrays [24,33] and subsequent new approaches were developed exclusively for GWAS (Figure 1). The classification of GWAS GSA approaches is shown in Figure 4. The first step for classification of GWAS GSA approaches can be their source of origin, including: (i) GSA microarrays adapted to GWAS; and (ii) those developed exclusively for GWAS (Figures 1 and 4. Further, based on the requirement of annotation libraries, the GWAS GSA approaches can also be classified as: (a) GSA requiring pre-defined gene sets; or (b) GSA which does not require pre-defined gene sets. These approaches are based on the principle of over-representation of genes in those predefined gene sets obtained from different bio-knowledge bases (Table S1). Moreover, such ORA approaches constitute the first generation of GWAS GSA.

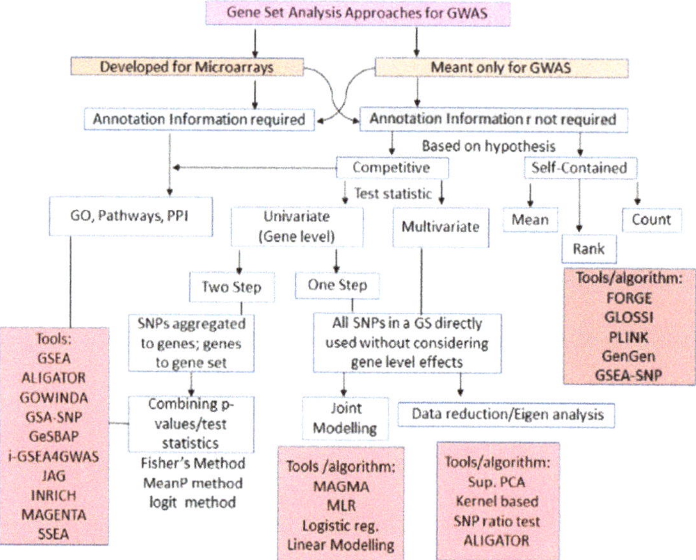


**Figure 4.** Classification of gene set analysis approaches and tools available for SNP data analysis. Schematic representation of the breakup of GSA methods available for SNP data analysis based on statistical tests and requirement of annotation databases. The first level of branching of the GSA methods based on their adaption from Microarrays to fit SNP data as well as those specifically designed for SNP data analysis. Subsequent branching depends on the different null hypotheses they test (i.e., null hypothesis, test statistic(s)). G: Gene set.

Due to the limitations of ORA-based GWAS GSA approaches, ES-based GWAS GSA approaches came into use, which we may call the second generation of GSA in GWAS. Their operational procedures and major analytical steps are given in Supplementary Figures S1 and S4. Further, the second generation of GWAS GSA starts with the enrichment analysis of gene sets for SNP data, i.e., GSEA-SNP [25,92] using weighted KS statistics [93]. Later approaches, based on other tests, *viz.* weighted-sum test [94], simple-sum test [95], collapsing test in combined multivariate and collapsing method [96] and sequence kernel association test [97], are used for computation of the gene-set enrichment score. Moreover, varieties of ES-based methods with similar ideas have been developed, such as the gene set based testing of polymorphism [98], GSA-SNP [92], SNP-ratio test [99], etc.

A class of GWAS GSA approaches have been developed by considering the topology of the gene sets/pathways, and this constitutes the third generation of GWAS GSA. This includes methods to parse the internal information of the pathway (e.g., signaling pathway impact analysis (SPIA) [74] and CliPPER [77]). Further, the second and third generation GWAS GSA methods focus on statistical results such as *p-values* or ES, as input rather than original data. Thus, the fourth generation of GWAS GSA approaches are developed by providing original data as input. Further, the underlying principle of these approaches is testing of the multivariate distribution of the multi-loci data or extracting the principal components from the original data. This includes linear combination test [100], supervised principal component analysis (SPCA) [100], Smoothed functional PCA [101], etc. Other model-based methods include LRpath [102], a logistic regression-based method, and MAGMA [103], linear model based method. Recently, the Generalized Berk-Jones (GBJ) statistic, a permutation-free parametric framework, was used for GSA [103], and this incorporates information from multiple signals in the same gene. The descriptions of the available GWAS GSA approaches, tools, their background methodologies pertaining to various generations are listed in Table 3 and Supplementary Document S1.

**Table 3.** Generation-wise evolution of GWAS GSA approaches for SNP data analysis.

| Approach | Methodology | Advantages | Limitations | Tools/Algorithm |
|---|---|---|---|---|
| **Over Representation Analysis (First generation GWAS GSA)** | Hypergeometric distribution, Fisher's exact test, Binomial test | • Simple to use and easy to interpret<br>• Assigns statistically convincing measure like p-value for SNP set, which is biologically meaningful<br>• Computationally not so expensive | • Hard threshold (arbitrary) divides the SNP list into selected and not selected SNP set. For instance, if threshold value for p-value is 0.05, means SNP with value 0.051 is not included in SNP list<br>• Uses only most significant SNP and discards others, lead to information loss<br>• Test statistic is independent of SNP data (based on only SNP count), and ignores the strength of association<br>• Considers each SNP independent and ignores the linkage disequilibrium<br>• Assumes each SNP contribute equally, which is not true as there are common and rare variants<br>• Dependent on pre-defined bio-knowledge base, which is mostly incomplete or unavailable | SNPtoGO [104], ALIGATOR [105], ATRP [106], MetaCore [107], PARIS [108], SET SCREEN test [109], SNP ratio test [99], GLOSSI, GeSBAP [98], INRICH [110], GeneSetDB [111], MAGENTA [112], KGG-HYST [113], PLINK [114], JAG [115], FORGE [116] |
| **Enrichment Statistic(s) Analysis (Second generation GWAS GSA)** | Wilcoxon signed rank test, Sum test, Weighted Sum test (Enrichment score like statistic) | • Do not require hard threshold for dividing SNP list into selected and non-selected part<br>• Jointly consider multiple contributing factors in the same gene set, might complement the most-significant SNPs/genes approach<br>• Test statistic is computed from the SNP data considering linkage disequilibrium | • Analyzes each gene set independently.<br>• Only considers data for selecting SNPs and after ignores the data from gene-set testing.<br>• Treat all genes in a gene set independently and do not account for the relationships between genes. | GSA-SNP [92], GSA-SNP2, GSEA-SNP [117], GSEA-P [118] GenGen [15], ICSNPathway [119], i-GSEA4GWAS [120], i-GSEA4GWAS2 [121] |
| **Topology Analysis (Third generation GWAS GSA)** | Graph/Network theory | • Relationships between genes are used to assign different levels of "importance" to genes in the set<br>• Helps in integrate gene set membership information with interaction data from a separate source | • Difficult to generalize<br>• True topology is dependent on the type of cell and experimental condition, which are rarely available<br>• Cannot model the dynamicity of the cellular system<br>• Heavily dependent on annotations, which is either missing or incomplete | dmGWAS [122], Ingenuity Pathway Analysis (IPA) [123], PINBPA [124], PathVisio [125], Cytoscape [126] |
| **Multivariate/Model/ Regression Analysis (Fourth generation GWAS GSA)** | Linear regression Model, Ridge regression, Logistic regression, Linear models | • Consider both SNP and gene set information simultaneously in same model<br>• Jointly consider linkage disequilibrium and gene-gene interaction in gene set for modeling<br>• Future behavior of the system can be predicted<br>• Dynamicity of the biological system can also be modeled and studied | • Computationally intensive<br>• High dimensionality of genomic data raises serious concerns<br>• Ignores the non-linear interactions among biomolecules | LRpath [102], SPCA [100], SFPCA [101], MAGMA [127], GRASS, GeneralizedBerk-Jones statistic [103], |

The formulations based on underlying statistical tests [6] can also be used for classifying GSA GWAS, i.e., self-contained and competitive approaches (Figure 4). Self-contained GWAS GSA considers only the SNPs in the gene set and tests the null hypothesis that none of those SNPs are associated with the phenotype. Competitive GSA considers all SNPs in the data and tests the null hypothesis that the genes in the gene set are no more strongly associated with the phenotype than other genes [128]. Further, the competitive GWAS GSA approaches can be divided into: (i) two-step approach(s), in which SNPs (in each gene) are first used to evaluate association with the gene, then gene-level statistic(s) are aggregated to gene-set level enrichment value to test its association with the phenotype; and (ii) a one-step approach, in which all SNPs in a gene set are simultaneously considered in the analysis without consideration of gene-level effects (e.g., MAGMA) (Figure 4). For the former categories the univariate statistical approaches are used, while multivariate techniques such as joint modelling are used for latter. Moreover, the self-contained GWAS GSA approaches can also be grouped based on the type of gene-set test statistic used for testing (Figure 4). This can be broadly subdivided into three classes: (i) mean-based, (i.e., mean or sum of the gene-association scores); (ii) count-based, (i.e., classifying genes as 'significant' or 'not significant' by applying a threshold to the gene-association scores and using the number of 'significant' genes in the gene set as a test statistic); and; (iii) rank-based, first ranking the genes according to their gene-association score and computing overrepresentation of the gene-set genes at the top of that ranking.

## 4. Limitations and Future Challenges of GSA

Here, we report the existing limitations as well as the key challenges observed in the available GSA approaches that should be kept in mind while using them. These existing limitations and challenges can be divided into two broad categories: (i) biological annotation challenges and (ii) methodological challenges.

### 4.1. Biological Annotation Challenges

The classification of GSA approaches for high-throughput genomic studies (Figures 2–4) shows that GSA approaches require annotation information for analyzing gene sets. It is expected that the next generation GSA will require improvement of the existing annotations as well as new high-throughput annotation information [30,58]. Therefore, it is important to create accurate, high resolution bio-knowledge bases with specific emphasis on cell dynamics and condition, along with tissue information to annotate genes studied in an experiment. These knowledge bases will allow us to model the inherent organism's response to any extraneous condition as a dynamic system and will help in predicting the system's behavior at different times as well as in relation to various factors (e.g., mutation, disease, environmental conditions, etc.).

*Limited annotation information*: The contemporary GSA approaches mostly use GO and pathways information for analyzing gene sets [9,20,32,41,43,44,80,104,105], but there is enough other annotation information available or will soon be available in public domain databases that can be effectively used for GSA to gain biological insights into the etiology of complex diseases in humans as well as other organisms. A list of alternate annotation information along with possible hypotheses are listed in Supplementary Table S4. For instance, Das et al. used the quantitative trait loci (QTL) data as annotation information to develop a GSA approach to analyze the gene sets obtained from microarrays [10]. This approach has immense use for performing trait/QTL enrichment analysis of gene sets and further, QTL enriched gene sets can be used for molecular breeding programs for biotic/abiotic stress engineering in plants. Moreover, this annotation information can also be used in the future for developing new generation GSA approaches for analysis of RNA-seq and GWAS data. Such advances in GSA will open new avenues to understand the molecular complexity behind complex diseases in humans and other organisms including crop plants.

*Low resolution knowledge bases*: Recent advancement in genomics and proteomics leads to a paradigm shift in data generation, with unprecedented high resolution. At the same time, there is a demand for high resolution annotation bio-knowledge bases to perform GSA. For instance, during the early period of GE genomics, microarrays were the key experiment to obtain a global view of GE in the human genome. To perform GSA, GO [129] and KEGG [130] annotation bases were developed in parallel and implemented in several web tools. Further, such databases specify which genes (in terms of probe id/Enetrez id) are active in each GO category/pathway/any predefined gene sets. However, microarray technology has been replaced with RNA-seq and single cell RNA-seq (scRNA-seq) technologies. Hence, the current annotation databases need to be updated with respect to these high-resolution techniques. It is essential that they also begin specifying other information, such as transcripts (or scRNA-seq transcript) and SNPs that are active in each predefined pathway, GO category, etc.

*Missing or incomplete annotation*: Although enormous annotation bases are available in the public domain, some annotations are either missing or incomplete for certain genes. For instance, the current release of GO contained entries for 19,649 human genes annotated with at least one GO term. Many of these genes are hypothetical, predicted or pseudogenes. For example, the number of protein-coding genes in the human genome is estimated to be 20,000–25,000 [52], which shows that annotation information of hundred(s) of genes is still missing, and this may have a crucial role in various diseases. In addition to the missing annotations, most of the current databases have lower resolution (i.e., lesser information on transcript and SNP) [30,131], which leads to biased results from GSA. Further, current knowledge bases are built by curating experiments performed in different cell types at different time points under different conditions/locations. However, these details are typically not available in these knowledge bases. Thus, these databases need to be updated for future dynamic or cell specific GSA.

*4.2. Methodological Challenges*

*Lack of benchmark/gold standard*: In simulation, it is expected that multivariate approaches outperform the univariate counterparts, as the former considers inter-variable correlations. However, in biology, it is observed that univariate statistic(s) are equal to or better than multivariate statistic(s) [40]. This observation raises several questions about the performance assessment of GSA approaches using simulated datasets as a benchmark. It is likely that biology is more complicated than simulated scenarios and is influenced by factors such as the absence of exclusive division into classes, presence of outliers, experimental or technical hidden factors, environmental influence(s), random errors, etc. Therefore, one way to handle such a situation is to use benchmark/gold standard datasets with a valid biological basis. For instance, Ballard et al. (2010) compared two GSA methods based on their applications to three Crohn's disease benchmark GWAS datasets with well-known biological basis [12,15,23]. Further, a combination of both benchmark biological datasets with statistically strong criteria can provide a suitable platform for comparative performance analysis of GSA approaches.

*Criteria for comparing GSA approaches*: When the performance of a GSA approach is assessed, it is expected to have certain proportions of false positives from the test. The ES-based GSA approaches compare the observed ES statistic with its null distribution as generated by random sampling/permuting the sample labels/disease outcomes or permuting genes/genotypes information [7,103]. Usually, through permutation, *p-values* are computed for assessing the enrichment significance of gene sets [6,26]. Then, $-\log10(p\text{-}value)$ and power of the statistical tests are used to assess the performance of GSA approaches [10]. However, alternate measures may also be used for comparative performance analysis of GSA approaches. In one such measure, the above computed *p-values* may be used to plot the histogram for the null gene sets, and that is expected to follow a uniform distribution. This phenomenon may be used to compute type-I error rates for GSA approaches, which can then be used as an efficient criterion for performance analysis of GSA approaches along with statistical power and FDR. In other words, GSA approaches with lower type-I error rates will be considered as better and *vice-versa*. These criteria can be computed on benchmark/gold standard datasets, which will provide a suitable platform to compare GSA approaches.

*Improvement in terms of statistical power*: In ORA-based GSA approaches, the test statistic(s) are computed by treating each gene equally. But in biology, some genes contribute more toward the disease/trait development. Treating all genes as equal in computing the test statistic reduces the statistical power of the GSA approach. Hence, one powerful strategy may be to consider the DE scores of genes [24,33,132,133] or ranks of the genes in a gene list while constructing the test statistic(s). This mechanism will attribute more statistical power to GSA approaches as compared to the existing ones. This approach needs to be well studied on benchmark data in future to assess its rigor and reproducibility. Further, other *a priori* biological information, *viz.* eQTL, network topology, co-expression scores, etc., can be used as auxiliary information in GSA approaches to improve their performance.

*Selection of null hypotheses:* The competitive GSA approaches use a gene sampling model to compute the *p-values* for gene sets [6,26]. In gene sampling model, it is assumed that genes are iid, which is highly unrealistic from a biological standpoint. Hence, the test statistic computed based on such assumptions from the gene sampling model leads to biased and misleading results. Therefore, methods, such as GSEA [24,33] and SAFE [39] use a hybrid concept, i.e., compute their test statistic(s) based on a gene-sampling model but calculate their *p-values* using the subject sampling model. The discrepancy between these two models makes the statistical properties of the test unclear and its interpretation very difficult. These problems are unavoidable, as the definition of the competitive null hypothesis is intimately tied to the gene-sampling model, whereas valid *p-values* are easily available for subject sampling only. This type of problem may provide impetus to future research in GSA.

*Inability to model and analyze a dynamic response*: It is well known that biological systems are dynamic. There has been a long debate about the feasibility of using static models to model the inherent dynamics of biological systems. However, in GSA, only static approaches (linear, gamma, generalized linear and regression models) [80,98,99] have been used so far. This raises a serious concern for the use of GSA approach in assessing living systems. The lack of methods that analyze gene sets as a dynamic system is partly due to the limitations of current molecular measurement technologies. These technologies can only quantify a snapshot of a biological system because they are unable to: (i) determine the protein states in a high-throughput fashion, or are severely restricted in this regard; and (ii) detect signals that propagate without affecting GE. Therefore, we encourage researchers in the future to use dynamic models such as time-series models, auto-regressive models, dynamic Bayesian models, etc. for GSA from time-dependent GE or association data.

*Redundancy among genes in gene sets*: In GE data analysis, redundancy among genes (i.e., genes may not be related to a case/disease but ranked in the top due to high correlation with other top ranked genes) is a serious issue [27]. During the process of ranked gene list preparation, redundant genes may be included and further, do not give valid *p-values* for the gene set testing, as genes in gene lists are correlated. In other words, *p-values* may easily be falsely significant when the genes in the gene set are correlated, even when none of the genes is truly significant. One strategy may be to use such a GE data analysis approach, (i.e., MRMR, Boot-MRMR [27]) which minimizes the redundancy among genes during the gene ranked list preparation. Other approaches may include avoiding the use of gene-sampling models in gene set testing for *p-value* computation. For this purpose, Goeman and Buhlman developed a subject-sampling $2 \times 2$ table method alternate to the gene sampling model to compute valid *p-values* for gene sets [6].

*Develop threshold-free approach(s):* ORA based GSA approaches are mostly threshold dependent [25]. Further, other GSA methods like mGSZ (based on Gene Set Z-scoring function) requires a threshold value for DE score to divide the ranked gene list into member genes and non-member genes (i.e., two gene groups) [132]. Gene set testing (e.g., Z-test) is then performed on these gene groups [15,24,33,132]. The determination of an optimal threshold is often a cumbersome task. Therefore, the obtained analytical results from such approach are unstable and irreproducible [24,25,93]. Hence, researchers use a set of threshold values to compute enrichment significance of gene sets and then select the threshold that gives the most significant results [6,134]. This approach seems inelegant. A more comprehensive and computationally intensive approach for choosing a threshold will be a reasonable compromise among power, type I error and reproducibility of results, using a cross validation technique. Another strategy may be development of threshold-free GSA approaches to improve the stability of results.

*Proper permutation procedure*: Current GSA approaches mostly use permutation procedures that compute *p-values* by comparing the observed test statistic with its null distribution generated from the permuted datasets [6,8,73,134]. It is expected to reflect chance-based confounding effects, including biases introduced by the gene set. However, the permutation procedures (if not designed properly) can produce misleading results and introduce bias in the resulting inference. For instance, permutation of SNPs, which is often used in *p-value* based approaches, may disrupt the linkage disequilibrium pattern and may not generate the correct null distribution. For gene-based approaches, permutation of sample labels may not generate the correct null distribution, as the samples are generated from tissues of same or related individuals [23,135]. Moreover, when the SNPs or genes or phenotypes are being permuted, the sampling units are assumed to be iid, which may not be the case; SNPs may be correlated due to linkage disequilibrium or gene-gene interactions. Therefore, proper care should be taken before choosing the permutation procedure for computing the *p-values* for gene sets.

*GSA approach(s) for alternate annotations*: The existing ORA-based GSA approaches have mostly focused on whether the selected gene sets are over-represented by known pathways or GO terms [9,20,32,41,43,44,80,104,105]. However, in plant and complex disease biology, such approaches may not able to establish any formal relation between the underlying genotypes and the trait/phenotype, as most of the traits are quantitative in nature and controlled by polygenes [10,12–14]. For this purpose, a statistical approach and R package of GSA with QTL has recently been developed [10], which is useful for obtaining QTL-enriched gene sets. Moreover, *like* QTL, there is a lot of genomic annotation information (Supplementary Table S4) available in public domain databases which can be used to develop new and innovative GSA approaches and tools.

*Stability of gene set testing results*: The statistical power and FDR are used for performance analysis of GSA approaches [7,8,11,78]. It is well known that different samples (on which the test is based) would give different results due to sampling errors. One way to deal with such a problem is to draw different sub-samples from a relative homogenous population, and the approach with small variance and uniform results over sub-samples can be termed as stable approach [16]. This principle can be applied to GSA, i.e., first, sub-samples can be taken from all samples, and then GSA can be applied on each sub-sample to compute the *p-value* for the gene sets. Finally, one can evaluate the stability of the approach by comparing a change in ranks over different sub-samples. The approach with the least change in ranks can be termed as the stable approach and can be easily implemented in simulation analysis. In biology, several factors may be responsible for causing instabilities to the results; these include, gene-gene correlations, genetic heterogeneity, and patient-to-patient variability. To address this problem, several researchers have hypothesized that testing gene sets rather than individual gene/marker will be more stable across different samples [8,136,137]. More relevant and specialized studies and methodologies are needed to validate such claims.

## 5. Discussion

In the last 15 years since its inception, GSA has become an extremely popular approach for secondary analysis of genome wide expression as well as association data. It has been successfully used to gain biological insights into the etiology of various complex diseases in humans as well as model organisms, including mammals, and other cellular organisms [9,10,13,14,138]. GSA has immense benefits in terms of biological interpretation of results, as well as numerous computational advantages over single gene studies [57]. It also enhances biologically meaningful interpretation of results and reproducibility of important gene lists yielded by independent studies, etc. [7–11]. In other words, the cumulative effects of the genetic variants (SNPs) or genes distributed in a gene set is considered in a single analysis and has more statistical power as compared to the univariate counterparts [8]. Despite of their usefulness, there are limited number of studies found in the literature, which consider the wider gamut of high throughput genomic studies from the GSA perspective. Hence, we have summarized the commonalities of GSA approaches used in three key genomic studies in terms of their execution, underlying null hypotheses, nature of test statistic, sampling models, etc. Further, the structure and key analytical steps common to most of the GSA approaches are discussed in this study.

Over the past few years, a diverse set of methods for performing GSA has been proposed for microarrays, RNA-seq and GWAS data analysis and the increased application of these methods has exposed several factors that affect the interpretations of GSA results. These factors include the null hypothesis being tested, the underlying sampling/permutation procedure, and the nature and distribution of test statistic(s). All of these factors play a significant role for choosing proper GSA for the data analysis. Researchers have also identified a variety of circumstances that can lead to faulty findings; hence, proper care is suggested to avoid misleading results. Several individual studies have been conducted over time to summarize GSA approaches for each type of genomic study [5–123]. Here, we summarize a comprehensive review of GSA approaches in terms of statistical structure, execution and classification for three different high-throughput genomic studies. Several approaches and tools have evolved over time, individually for each type of genomic study. Thus, instead of individually reviewing them, we present the classification of GSA approaches for microarrays, RNA-seq and GWAS into different generations along with underlying statistical methodologies/tests and special features. Many earlier reviews of GSA are data independent studies [6,11,23], but our study is data dependent and comprehensive.

This study will serve as a catalogue and provide guidelines to genome researchers and experimental biologists for choosing the proper GSA based on several factors. In this study, we reported several challenges which need to be addressed by statisticians and biologists collectively to develop the next generation of GSA approaches. These new approaches will be able to analyze high-throughput genomic data more efficiently in order to better understand the biological systems and to increase the specificity, sensitivity, utility, and relevance of GSA.

**Supplementary Materials:** The following are available online at http://www.mdpi.com/1099-4300/22/4/427/s1, Document S1: Background methodologies of GSA approaches and tools for different generation, Figure S1: Operational procedures for gene set analysis followed in microarrays, RNA-seq, and GWAS data analysis, Figure S2: Analytical steps of GSA for microarray data analysis, Figure S3: Analytical steps of GSA for RNA-seq data analysis. Figure S4. Analytical steps of GSA for SNP (GWAS) data analysis. Table S1: List of available bio-knowledge bases used for Gene Set Analysis, Table S2: Nature and distribution of genomic datasets, Table S3: Available microarray datasets in NCBI, Table S4: Alternate annotation information for possible gene set analysis.

**Author Contributions:** Conceived and designed the study: S.D. Contributed materials: S.D., S.N.R. Drafted the manuscript: S.D. Corrected the manuscript: S.D., C.J.M., S.N.R. Funding Acquisition: C.J.M. All authors have read and agreed to the published version of the manuscript.

**Funding:** This study was supported in part by National Institutes of Health (NIH), USA grants, P20GM113226, and 5P50AA024337 (CJM). It was also supported by Netaji Subhas-ICAR International Fellowship, OM No. 18(02)/2016-EQR/Edn. (SD) of Indian Council of Agricultural Research (ICAR), New Delhi, India and partly by Wendell Cherry Chair (SNR) in Clinical Trial Research, University of Louisville, USA. The content is solely the responsibility of the authors and does not necessarily represent the official views of NIH or ICAR.

**Acknowledgments:** Authors acknowledge the support obtained from ICAR-Indian Agricultural Statistics Research Institute, New Delhi, India.

**Conflicts of Interest:** The authors declare no conflicts of interest.

## References

1. Marx, V. The big challenges of big data. *Nature* **2013**, *498*, 255–260. [CrossRef] [PubMed]
2. Wang, J.; Chen, L.; Wang, Y.; Zhang, J.; Liang, Y.; Xu, D. A Computational Systems Biology Study for Understanding Salt Tolerance Mechanism in Rice. *PLoS ONE* **2013**, *8*, e64929. [CrossRef]
3. Cui, X.; Churchill, G.A. Statistical tests for differential expression in cDNA microarray experiments. *Genome Biol.* **2003**. [CrossRef] [PubMed]

4.  Das, S.; Meher, P.K.; Rai, A.; Bhar, L.M.; Mandal, B.N. Statistical Approaches for Gene Selection, Hub Gene Identification and Module Interaction in Gene Co-Expression Network Analysis: An Application to Aluminum Stress in Soybean (Glycine max L.). *PLoS ONE* **2017**, *12*, e0169605. [CrossRef] [PubMed]

5.  Liang, Y.; Zhang, F.; Wang, J.; Joshi, T.; Wang, Y.; Xu, D. Prediction of Drought-Resistant Genes in Arabidopsis thaliana Using SVM-RFE. *PLoS ONE* **2011**, *6*, e21750. [CrossRef] [PubMed]

6.  Goeman, J.J.; Buhlmann, P. Analyzing gene expression data in terms of gene sets: Methodological issues. *Bioinformatics* **2007**, *23*, 980–987. [CrossRef] [PubMed]

7.  de Leeuw, C.A.; Neale, B.M.; Heskes, T.; Posthuma, D. The statistical properties of gene-set analysis. *Nat. Rev. Genet.* **2016**, *17*, 353–364. [CrossRef]

8.  Efron, B.; Tibshirani, R. On testing the significance of sets of genes. *Ann. Appl. Stat.* **2007**, *1*, 107–129. [CrossRef]

9.  Yi, X.; Du, Z.; Su, Z. PlantGSEA: A gene set enrichment analysis toolkit for plant community. *Nucleic Acids Res.* **2013**. [CrossRef]

10. Das, S.; Rai, A.; Mishra, D.C.; Rai, S.N. Statistical Approach for Gene Set Analysis with Trait Specific Quantitative Trait Loci. *Sci. Rep.* **2018**, *8*, 2391. [CrossRef]

11. Rahmatallah, Y.; Emmert-Streib, F.; Glazko, G. Gene set analysis approaches for RNA-seq data: Performance evaluation and application guideline. *Brief Bioinform.* **2016**, *17*, 393–407. [CrossRef] [PubMed]

12. Mooney, M.A.; Wilmot, B. Gene set analysis: A step-by-step guide. *Am. J. Med. Genet. Part B Neuropsychiatr. Genet.* **2015**. [CrossRef] [PubMed]

13. Sullivan, P.F.; Posthuma, D. Biological pathways and networks implicated in psychiatric disorders. *Curr. Opin. Behav. Sci.* **2015**, *2*, 58–68. [CrossRef]

14. Nurnberger, J.I.; Koller, D.L.; Jung, J.; Edenberg, H.J.; Foroud, T.; Guella, I.; Vawter, M.P.; Kelsoe, J.R. Identification of Pathways for Bipolar Disorder. *JAMA Psychiatry* **2014**, *71*, 657. [CrossRef] [PubMed]

15. Wang, K.; Zhang, H.; Kugathasan, S.; Annese, V.; Bradfield, J.P.; Russell, R.K.; Sleiman, P.M.; Imielinski, M.; Glessner, J.; Hou, C.; et al. Diverse Genome-wide Association Studies Associate the IL12/IL23 Pathway with Crohn Disease. *Am. J. Hum. Genet.* **2009**. [CrossRef]

16. Eleftherohorinou, H.; Hoggart, C.J.; Wright, V.J.; Levin, M.; Coin, L.J.M. Pathway-driven gene stability selection of two rheumatoid arthritis GWAS identifies and validates new susceptibility genes in receptor mediated signalling pathways. *Hum. Mol. Genet.* **2011**. [CrossRef]

17. Menashe, I.; Maeder, D.; Garcia-Closas, M.; Figueroa, J.D.; Bhattacharjee, S.; Rotunno, M.; Kraft, P.; Hunter, D.J.; Chanock, S.J.; Rosenberg, P.S.; et al. Pathway analysis of breast cancer genome-wide association study highlights three pathways and one canonical signaling cascade. *Cancer Res.* **2010**. [CrossRef]

18. Locke, A.E.; Kahali, B.; Berndt, S.I.; Justice, A.E.; Pers, T.H.; Day, F.R.; Powell, C.; Vedantam, S.; Buchkovich, M.L.; Yang, J.; et al. Genetic studies of body mass index yield new insights for obesity biology. *Nature* **2015**, *518*, 197–206. [CrossRef]

19. Berriz, G.F.; King, O.D.; Bryant, B.; Sander, C.; Roth, F.P. Characterizing gene sets with FuncAssociate. *Bioinformatics* **2003**. [CrossRef]

20. Beißbarth, T.; Speed, T.P. GOstat: Find statistically overrepresented Gene Ontologies with a group of genes. *Bioinformatics* **2004**. [CrossRef]

21. Khatri, P.; Draghici, S.; Ostermeier, G.C.; Krawetz, S.A. Profiling Gene Expression Using Onto-Express. *Genomics* **2002**, *79*, 266–270. [CrossRef] [PubMed]

22. Boyle, E.I.; Weng, S.; Gollub, J.; Jin, H.; Botstein, D.; Cherry, J.M.; Sherlock, G. GO::TermFinder–open source software for accessing Gene Ontology information and finding significantly enriched Gene Ontology terms associated with a list of genes. *Bioinformatics* **2004**, *20*, 3710–3715. [CrossRef] [PubMed]

23. Fridley, B.L.; Patch, C. Gene set analysis of SNP data: Benefits, challenges, and future directions. *Eur. J. Hum. Genet.* **2011**, *19*, 837–843. [CrossRef] [PubMed]

24. Subramanian, A.; Tamayo, P.; Mootha, V.K.; Mukherjee, S.; Ebert, B.L.; Gillette, M.A.; Paulovich, A.; Pomeroy, S.L.; Golub, T.R.; Lander, E.S.; et al. Gene set enrichment analysis: A knowledge-based approach for interpreting genome-wide expression profiles. *Proc. Natl. Acad. Sci. USA* **2005**, *102*, 15545–15550. [CrossRef] [PubMed]

25. Wang, L.; Jia, P.; Wolfinger, R.D.; Chen, X.; Zhao, Z. Gene set analysis of genome-wide association studies: Methodological issues and perspectives. *Genomics* **2011**, *98*, 1–8. [CrossRef]

26. Goeman, J.J.; Van de Geer, S.; De Kort, F.; van Houwellingen, H.C. A global test for groups fo genes: Testing association with a clinical outcome. *Bioinformatics* **2004**. [CrossRef]

27. Das, S.; Rai, A.; Mishra, D.C.; Rai, S.N. Statistical approach for selection of biologically informative genes. *Gene* **2018**, *655*. [CrossRef]

28. Wang, X.; Cairns, M.J. Gene set enrichment analysis of RNA-Seq data: Integrating differential expression and splicing. *BMC Bioinform.* **2013**, *14*, S16. [CrossRef]

29. Rahmatallah, Y.; Zybailov, B.; Emmert-Streib, F.; Glazko, G. GSAR: Bioconductor package for Gene Set analysis in R. *BMC Bioinform.* **2017**. [CrossRef]

30. Khatri, P.; Sirota, M.; Butte, A.J. Ten Years of Pathway Analysis: Current Approaches and Outstanding Challenges. *PLoS Comput. Biol.* **2012**, *8*, e1002375. [CrossRef]

31. Dennis, G.; Sherman, B.T.; Hosack, D.A.; Yang, J.; Gao, W.; Lane, H.C.; Lempicki, R.A. DAVID: Database for Annotation, Visualization, and Integrated Discovery. *Genome Biol.* **2003**, *4*, R60. [CrossRef]

32. Tian, T.; Liu, Y.; Yan, H.; You, Q.; Yi, X.; Du, Z.; Xu, W.; Su, Z. AgriGO v2.0: A GO analysis toolkit for the agricultural community, 2017 update. *Nucleic Acids Res.* **2017**. [CrossRef] [PubMed]

33. Mootha, V.K.; Lindgren, C.M.; Eriksson, K.-F.; Subramanian, A.; Sihag, S.; Lehar, J.; Puigserver, P.; Carlsson, E.; Ridderstråle, M.; Laurila, E.; et al. PGC-1α-responsive genes involved in oxidative phosphorylation are coordinately downregulated in human diabetes. *Nat. Genet.* **2003**, *34*, 267–273. [CrossRef] [PubMed]

34. Pavlidis, P.; Qin, J.; Arango, V.; Mann, J.J.; Sibille, E. Using the Gene Ontology for Microarray Data Mining: A Comparison of Methods and Application to Age Effects in Human Prefrontal Cortex. *Neurochem. Res.* **2004**, *29*, 1213–1222. [CrossRef]

35. Al-Shahrour, F.; Diaz-Uriarte, R.; Dopazo, J. Discovering molecular functions significantly related to phenotypes by combining gene expression data and biological information. *Bioinformatics* **2005**, *21*, 2988–2993. [CrossRef] [PubMed]

36. Tian, L.; Greenberg, S.A.; Kong, S.W.; Altschuler, J.; Kohane, I.S.; Park, P.J. Discovering statistically significant pathways in expression profiling studies. *Proc. Natl. Acad. Sci. USA* **2005**, *102*, 13544–13549. [CrossRef]

37. Kim, S.Y.; Volsky, D.J. PAGE: Parametric analysis of gene set enrichment. *BMC Bioinform.* **2005**. [CrossRef]

38. Jiang, Z.; Gentleman, R. Extensions to gene set enrichment. *Bioinformatics* **2007**. [CrossRef]

39. Barry, W.T.; Nobel, A.B.; Wright, F.A. Significance analysis of functional categories in gene expression studies: A structured permutation approach. *Bioinformatics* **2005**. [CrossRef]

40. Glazko, G.V.; Emmert-Streib, F. Unite and conquer: Univariate and multivariate approaches for finding differentially expressed gene sets. *Bioinformatics* **2009**. [CrossRef]

41. Huang, D.W.; Sherman, B.T.; Tan, Q.; Kir, J.; Liu, D.; Bryant, D.; Guo, Y.; Stephens, R.; Baseler, M.W.; Lane, H.C.; et al. DAVID Bioinformatics Resources: Expanded annotation database and novel algorithms to better extract biology from large gene lists. *Nucleic Acids Res.* **2007**, *35*, W169–W175. [CrossRef] [PubMed]

42. Dahlquist, K.D.; Salomonis, N.; Vranizan, K.; Lawlor, S.C.; Conklin, B.R. GenMAPP, a new tool for viewing and analyzing microarray data on biological pathways. *Nat Genet.* **2002**, *31*, 19–20. [CrossRef] [PubMed]

43. Zeeberg, B.R.; Feng, W.; Wang, G.; Wang, M.D.; Fojo, A.T.; Sunshine, M.; Narasimhan, S.; Kane, D.W.; Reinhold, W.C.; Lababidi, S.; et al. GoMiner: A resource for biological interpretation of genomic and proteomic data. *Genome Biol.* **2003**, *4*, R28. [CrossRef] [PubMed]

44. Al-Shahrour, F.; Díaz-Uriarte, R.; Dopazo, J. FatiGO: A web tool for finding significant associations of Gene Ontology terms with groups of genes. *Bioinformatics* **2004**. [CrossRef] [PubMed]

45. Martin, D.; Brun, C.; Remy, E.; Mouren, P.; Thieffry, D.; Jacq, B. GOToolBox: Functional analysis of gene datasets based on Gene Ontology. *Genome Biol.* **2004**. [CrossRef]

46. Castillo-Davis, C.I.; Hartl, D.L. GeneMerge-Post-genomic analysis, data mining, and hypothesis testing. *Bioinformatics* **2003**. [CrossRef]

47. Zheng, Q.; Wang, X.J. GOEAST: A web-based software toolkit for Gene Ontology enrichment analysis. *Nucleic Acids Res.* **2008**. [CrossRef]

48. Bindea, G.; Mlecnik, B.; Hackl, H.; Charoentong, P.; Tosolini, M.; Kirilovsky, A.; Fridman, W.-H.; Pagès, F.; Trajanoski, Z.; Galon, J. ClueGO: A Cytoscape plug-in to decipher functionally grouped gene ontology and pathway annotation networks. *Bioinformatics* **2009**. [CrossRef]

49. Robinson, M.D.; Grigull, J.; Mohammad, N.; Hughes, T.R. FunSpec: A web-based cluster interpreter for yeast. *BMC Bioinform.* **2002**. [CrossRef]

50. Martínez-Cruz, L.A.; Rubio, A.; Martínez-Chantar, M.L.; Labarga, A.; Barrio, I.; Podhorski, A.; Segura, V.; Campo, J.L.S.; Avila, M.A.; Mato, J.M. GARBAN: Genomic analysis and rapid biological annotation of cDNA microarray and proteomic data. *Bioinformatics* **2003**. [CrossRef]

51. Wang, J.; Duncan, D.; Shi, Z.; Zhang, B. WEB-based GEne SeT AnaLysis Toolkit (WebGestalt): Update 2013. *Nucleic Acids Res.* **2013**. [CrossRef] [PubMed]

52. Sun, H.; Fang, H.; Chen, T.; Perkins, R.; Tong, W. GOFFA: Gene Ontology for Functional Analysis—A FDA Gene Ontology tool for analysis of genomic and proteomic data. *BMC Bioinform.* **2006**. [CrossRef] [PubMed]

53. Ye, J.; Fang, L.; Zheng, H.; Zhang, Y.; Chen, J.; Zhang, Z.; Wang, J.; Li, S.; Li, R.; Bolund, L.; et al. WEGO: A web tool for plotting GO annotations. *Nucleic Acids Res.* **2006**, *34*, W293–W297. [CrossRef] [PubMed]

54. Zhang, B.; Schmoyer, D.; Kirov, S.; Snoddy, J. GOTree Machine (GOTM): A web-based platform for interpreting sets of interesting genes using Gene Ontology hierarchies. *BMC Bioinform.* **2004**. [CrossRef]

55. Luo, W.; Brouwer, C. Pathview: An R/Bioconductor package for pathway-based data integration and visualization. *Bioinformatics* **2013**. [CrossRef]

56. Yi, M.; Horton, J.D.; Cohen, J.C.; Hobbs, H.H.; Stephens, R.M. WholePathwayScope: A comprehensive pathway-based analysis tool for high-throughput data. *BMC Bioinform.* **2006**. [CrossRef]

57. Newton, M.A.; Quintana, F.A.; den Boon, J.A.; Sengupta, S.; Ahlquist, P. Random-set methods identify distinct aspects of the enrichment signal in gene-set analysis. *Ann. Appl. Stat.* **2007**. [CrossRef]

58. Cao, W.; Li, Y.; Liu, D.; Chen, C.; Xu, Y. Statistical and Biological Evaluation of Different Gene Set Analysis Methods. *Procedia Environ. Sci.* **2011**, *8*, 693–699. [CrossRef]

59. Dinu, I.; Potter, J.D.; Mueller, T.; Liu, Q.; Adewale, A.J.; Jhangri, G.S.; Einecke, G.; Famulski, K.S.; Halloran, P.; Yasui, Y. Improving gene set analysis of microarray data by SAM-GS. *BMC Bioinform.* **2007**. [CrossRef]

60. Smyth, G.K.; Ritchie, M.; Thorne, N.; Wettenhall, J. limma: Linear Models for Microarray Data. Bioinformatics and Computational Biology Solutions Using R and Bioconductor. *Stat. Biol. Health* **2005**. [CrossRef]

61. Breslin, T.; Edén, P.; Krogh, M. Comparing functional annotation analyses with Catmap. *BMC Bioinform.* **2004**. [CrossRef] [PubMed]

62. Boorsma, A.; Foat, B.C.; Vis, D.; Klis, F.; Bussemaker, H.J. T-profiler: Scoring the activity of predefined groups of genes using gene expression data. *Nucleic Acids Res.* **2005**. [CrossRef] [PubMed]

63. Henegar, C.; Cancello, R.; Rome, S.; Vidal, H.; Clément, K.; Zucker, J.-D. Clustering biological annotations and gene expression data to identify putatively co-regulated biological processes. *J. Bioinform. Comput. Biol.* **2006**, *4*, 833–852. [CrossRef] [PubMed]

64. Backes, C.; Keller, A.; Kuentzer, J.; Kneissl, B.; Comtesse, N.; Elnakady, Y.A.; Müller, R.; Meese, E.; Lenhof, H.-P. GeneTrail-advanced gene set enrichment analysis. *Nucleic Acids Res.* **2007**. [CrossRef]

65. Kim, S.-B.; Yang, S.; Kim, S.-K.; Kim, S.C.; Woo, H.G.; Volsky, D.J.; Kim, S.Y.; Chu, I.-S. GAzer: Gene set analyzer. *Bioinformatics* **2007**, *23*, 1697–1699. [CrossRef]

66. Wu, D.; Smyth, G.K. Camera: A competitive gene set test accounting for inter-gene correlation. *Nucleic Acids Res.* **2012**. [CrossRef]

67. Luo, W.; Friedman, M.S.; Shedden, K.; Hankenson, K.D.; Woolf, P.J. GAGE: Generally applicable gene set enrichment for pathway analysis. *BMC Bioinform.* **2009**. [CrossRef]

68. Frost, H.R.; Li, Z.; Moore, J.H. Spectral gene set enrichment (SGSE). *BMC Bioinform.* **2015**, *16*, 70. [CrossRef]

69. Rahmatallah, Y.; Emmert-Streib, F.; Glazko, G. Gene Sets Net Correlations Analysis (GSNCA): A multivariate differential coexpression test for gene sets. *Bioinformatics* **2014**. [CrossRef]

70. Hsueh, H.M.; Tsai, C.A. Gene set analysis using sufficient dimension reduction. *BMC Bioinform.* **2016**. [CrossRef]

71. Reich, M.; Liefeld, T.; Gould, J.; Lerner, J.; Tamayo, P.; Mesirov, J.P. Gene Pattern 2.0. *Nat Genet.* **2006**, *38*, 500–501. [CrossRef] [PubMed]

72. Wu, X.; Hasan MAl Chen, J.Y. Pathway and network analysis in proteomics. *J. Theor. Biol.* **2014**. [CrossRef] [PubMed]

73. Rahnenführer, J.; Domingues, F.S.; Maydt, J.; Lengauer, T. Calculating the Statistical Significance of Changes in Pathway Activity From Gene Expression Data. *Stat. Appl. Genet. Mol. Biol.* **2005**. [CrossRef] [PubMed]

74. Tarca, A.L.; Draghici, S.; Khatri, P.; Hassan, S.S.; Mittal, P.; Kim, J.S.; Kim, C.J.; Kusanovic, J.P.; Romero, R. A novel signaling pathway impact analysis. *Bioinformatics* **2009**. [CrossRef] [PubMed]

75. Alexeyenko, A.; Lee, W.; Pernemalm, M.; Guegan, J.; Dessen, P.; Lazar, V.; Lehtiö, J.; Pawitan, Y. Network enrichment analysis: Extension of gene-set enrichment analysis to gene networks. *BMC Bioinform.* **2012**. [CrossRef] [PubMed]

76. Glaab, E.; Baudot, A.; Krasnogor, N.; Valencia, A. TopoGSA: Network topological gene set analysis. *Bioinformatics* **2010**. [CrossRef] [PubMed]

77. Martini, P.; Sales, G.; Massa, M.S.; Chiogna, M.; Romualdi, C. Along signal paths: An empirical gene set approach exploiting pathway topology. *Nucleic Acids Res.* **2013**, *41*, e19. [CrossRef]

78. Rahmatallah, Y.; Emmert-Streib, F.; Glazko, G. Comparative evaluation of gene set analysis approaches for RNA-Seq data. *BMC Bioinform.* **2014**, *15*, 397. [CrossRef]

79. Conesa, A.; Madrigal, P.; Tarazona, S.; Gomez-Cabrero, D.; Cervera, A.; McPherson, A.; Szcześniak, M.W.; Gaffney, D.J.; Elo, L.L.; Zhang, X.; et al. A survey of best practices for RNA-seq data analysis. *Genome Biol.* **2016**. [CrossRef]

80. Young, M.D.; Davidson, N.; Wakefield, M.J.; Smyth, G.K.; Oshlack, A. goseq: Gene Ontology testing for RNA-seq datasets. *R Bioconductor* **2012**, *8*, 1–25.

81. Ge, S.X.; Son, E.W.; Yao, R. iDEP: An integrated web application for differential expression and pathway analysis of RNA-Seq data. *BMC Bioinform.* **2018**. [CrossRef] [PubMed]

82. Wu, D.; Lim, E.; Vaillant, F.; Asselin-Labat, M.L.; Visvader, J.E.; Smyth, G.K. ROAST: Rotation gene set tests for complex microarray experiments. *Bioinformatics* **2010**. [CrossRef] [PubMed]

83. Hänzelmann, S.; Castelo, R.; Guinney, J. GSVA: Gene set variation analysis for microarray and RNA-Seq data. *BMC Bioinform.* **2013**. [CrossRef] [PubMed]

84. Fridley, B.L.; Jenkins, G.D.; Grill, D.E.; Kennedy, R.B.; Poland, G.A.; Oberg, A.L. Soft truncation thresholding for gene set analysis of RNA-seq data: Application to a vaccine study. *Sci. Rep.* **2013**. [CrossRef]

85. oon, S.; Kim, S.Y.; Nam, D. Improving gene-set enrichment analysis of RNA-Seq data with small replicates. *PLoS ONE* **2016**. [CrossRef]

86. Xiong, Q.; Mukherjee, S.; Furey, T.S. GSAASeqSP: A toolset for gene set association analysis of RNA-Seq data. *Sci. Rep.* **2014**. [CrossRef]

87. Wang, X.; Cairns, M.J. SeqGSEA: A Bioconductor package for gene set enrichment analysis of RNA-Seq data integrating differential expression and splicing. *Bioinformatics* **2014**. [CrossRef]

88. Alhamdoosh, M.; Ng, M.; Wilson, N.J.; Sheridan, J.M.; Huynh, H.; Wilson, M.J.; Ritchie, M.E. Combining multiple tools outperforms individual methods in gene set enrichment analyses. *Bioinformatics* **2017**. [CrossRef]

89. Stamm, K.; Tomita-Mitchell, A.; Bozdag, S. GSEPD: A Bioconductor package for RNA-seq gene set enrichment and projection display. *BMC Bioinform.* **2019**. [CrossRef]

90. Lee, C.; Patil, S. Sartor MA. RNA-Enrich: A cut-off free functional enrichment testing method for RNA-seq with improved detection power. *Bioinformatics* **2016**. [CrossRef]

91. Wu, M.C.; Kraft, P.; Epstein, M.P.; Taylor, D.M.; Chanock, S.J.; Hunter, D.J.; Lin, X. Powerful SNP-Set Analysis for Case-Control Genome-wide Association Studies. *Am. J. Hum. Genet.* **2010**. [CrossRef] [PubMed]

92. Nam, D.; Kim, J.; Kim, S.-Y.; Kim, S. GSA-SNP: A general approach for gene set analysis of polymorphisms. *Nucleic Acids Res.* **2010**, *38*, W749–W754. [CrossRef] [PubMed]

93. Wang, K.; Li, M.; Bucan, M. Pathway-Based Approaches for Analysis of Genomewide Association Studies. *Am. J. Hum. Genet.* **2007**. [CrossRef]

94. Madsen, B.E.; Browning, S.R. A Groupwise Association Test for Rare Mutations Using a Weighted Sum Statistic. *PLoS Genet.* **2009**, *5*, e1000384. [CrossRef] [PubMed]

95. Morris, A.P.; Zeggini, E. An evaluation of statistical approaches to rare variant analysis in genetic association studies. *Genet. Epidemiol.* **2010**, *34*, 188–194. [CrossRef] [PubMed]

96. Li, B.; Leal, S.M. Methods for Detecting Associations with Rare Variants for Common Diseases: Application to Analysis of Sequence Data. *Am. J. Hum. Genet.* **2008**. [CrossRef]

97. Wu, M.C.; Lee, S.; Cai, T.; Li, Y.; Boehnke, M.; Lin, X. Rare-variant association testing for sequencing data with the sequence kernel association test. *Am. J. Hum. Genet.* **2011**. [CrossRef]

98. Medina, I.; Montaner, D.; Bonifaci, N.; Pujana, M.A.; Carbonell, J.; Tarraga, J.; Al-Shahrour, F.; Dopazo, J. Gene set-based analysis of polymorphisms: Finding pathways or biological processes associated to traits in genome-wide association studies. *Nucleic Acids Res.* **2009**, *37* (Suppl. 2), W340–W344. [CrossRef]

99. O'Dushlaine, C.; Kenny, E.; Heron, E.A.; Segurado, R.; Gill, M.; Morris, D.W.; Corvin, A. The SNP ratio test: Pathway analysis of genome-wide association datasets. *Bioinformatics* **2009**. [CrossRef]

100. Chen, X.; Wang, L.; Hu, B.; Guo, M.; Barnard, J.; Zhu, X. Pathway-based analysis for genome-wide association studies using supervised principal components. *Genet. Epidemiol.* **2010**, *34*, 716–724. [CrossRef]

101. Luo, L.; Zhu, Y.; Xiong, M. Smoothed functional principal component analysis for testing association of the entire allelic spectrum of genetic variation. *Eur. J. Hum. Genet.* **2013**, *21*, 217–224. [CrossRef] [PubMed]

102. Kim, J.H.; Karnovsky, A.; Mahavisno, V.; Weymouth, T.; Pande, M.; Dolinoy, D.C.; Rozek, L.S.; Sartor, M.A. LRpath analysis reveals common pathways dysregulated via DNA methylation across cancer types. *BMC Genom.* **2012**. [CrossRef] [PubMed]

103. Sun, R.; Hui, S.; Bader, G.D.; Lin, X.; Kraft, P. Powerful gene set analysis in GWAS with the Generalized Berk-Jones statistic. *PLOS Genet.* **2019**, *15*, e1007530. [CrossRef] [PubMed]

104. Schwarz, D.F.; Hädicke, O.; Erdmann, J.; Ziegler, A.; Bayer, D.; Möller, S. SNPtoGO: Characterizing SNPs by enriched GO terms. *Bioinformatics* **2008**. [CrossRef]

105. Holmans, P.; Green, E.K.; Pahwa, J.S.; Ferreira, M.A.R.; Purcell, S.M.; Sklar, P.; The Wellcome Trust Case-Control Consortium; Owen, M.J.; O'Donovan, M.C.; Craddock, N. Gene Ontology Analysis of GWA Study Data Sets Provides Insights into the Biology of Bipolar Disorder. *Am. J. Hum. Genet.* **2009**. [CrossRef]

106. Yu, K.; Li, Q.; Bergen, A.W.; Pfeiffer, R.M.; Rosenberg, P.S.; Caporaso, N.; Kraft, P.; Chatterjee, N. Pathway analysis by adaptive combination of P-values. *Genet Epidemiol.* **2009**. [CrossRef]

107. Bessarabova, M.; Ishkin, A.; JeBailey, L.; Nikolskaya, T.; Nikolsky, Y. Knowledge-based analysis of proteomics data. *BMC Bioinform.* **2012**, *13*, S13. [CrossRef]

108. Yaspan, B.L.; Bush, W.S.; Torstenson, E.S.; Ma, D.; Pericak-Vance, M.A.; Ritchie, M.D.; Sutcliffe, J.S.; Haines, J.L. Genetic analysis of biological pathway data through genomic randomization. *Hum Genet.* **2011**. [CrossRef]

109. Moskvina, V.; O'Dushlaine, C.; Purcell, S.; Craddock, N.; Holmans, P.; O'Donovan, M.C. Evaluation of an approximation method for assessment of overall significance of multiple-dependent tests in a genomewide association study. *Genet Epidemiol.* **2011**. [CrossRef]

110. Lee, P.H.; O'dushlaine, C.; Thomas, B.; Purcell, S.M. INRICH: Interval-based enrichment analysis for genome-wide association studies. *Bioinformatics* **2012**. [CrossRef]

111. Araki, H.; Knapp, C.; Tsai, P.; Print, C. GeneSetDB: A comprehensive meta-database, statistical and visualisation framework for gene set analysis. *FEBS Open Bio* **2012**, *2*, 76–82. [CrossRef] [PubMed]

112. Ayellet, V.S.; Groop, L.; Mootha, V.K.; Daly, M.J.; Altshuler, D. Common inherited variation in mitochondrial genes is not enriched for associations with type 2 diabetes or related glycemic traits. *PLoS Genet.* **2010**. [CrossRef]

113. Li, M.X.; Kwan, J.S.H.; Sham, P.C. HYST: A hybrid set-based test for genome-wide association studies, with application to protein-protein interaction-based association analysis. *Am. J. Hum. Genet.* **2012**. [CrossRef] [PubMed]

114. Purcell, S.; Neale, B.; Todd-Brown, K.; Thomas, L.; Ferreira, M.A.R.; Bender, D.; Maller, J.; Sklar, P.; Bakker, P.I.W.; Daly, M.J.; et al. PLINK: A tool set for whole-genome association and population-based linkage analyses. *Am. J. Hum. Genet.* **2007**, *81*, 559–575. [CrossRef]

115. Lips, E.S.; Cornelisse, L.N.; Toonen, R.F.; Min, J.L.; Hultman, C.M.; Holmans, P.A.; O'Donovan, M.C.; Purcell, S.M.; Smit, A.B.; Verhage, M.; et al. Functional gene group analysis identifies synaptic gene groups as risk factor for schizophrenia. *Mol. Psychiatry* **2012**, *17*, 996–1006. [CrossRef]

116. Pedroso, I.; Lourdusamy, A.; Rietschel, M.; Nöthen, M.M.; Cichon, S.; McGuffin, P.; AI-Chalabi, A.; Barnes, M.R.; Breen, G. Common genetic variants and gene-expression changes associated with bipolar disorder are over-represented in brain signaling pathway genes. *Biol. Psychiatry* **2012**. [CrossRef]

117. Holden, M.; Deng, S.; Wojnowski, L.; Kulle, B. GSEA-SNP: Applying gene set enrichment analysis to SNP data from genome-wide association studies. *Bioinformatics* **2008**, *24*, 2784–2785. [CrossRef]

118. Subramanian, A.; Kuehn, H.; Gould, J.; Tamayo, P.; Mesirov, J.P. GSEA-P: A desktop application for Gene Set Enrichment Analysis. *Bioinformatics* **2007**, *23*, 3251–3253. [CrossRef]

119. Zhang, K.; Chang, S.; Cui, S.; Guo, L.; Zhang, L.; Wang, J. ICSNPathway: Identify candidate causal SNPs and pathways from genome-wide association study by one analytical framework. *Nucleic Acids Res.* **2011**, *39*, W437–W443. [CrossRef]

120. Zhang, K.; Cui, S.; Chang, S.; Zhang, L.; Wang, J. i-GSEA4GWAS: A web server for identification of pathways/gene sets associated with traits by applying an improved gene set enrichment analysis to genome-wide association study. *Nucleic Acids Res.* **2010**, *38*, W90–W95. [CrossRef]

121. Zhang, K.; Chang, S.; Guo, L.; Wang, J. I-GSEA4GWAS v2: A web server for functional analysis of SNPs in trait-associated pathways identified from genome-wide association study. *Protein Cell* **2015**, *6*, 221–224. [CrossRef] [PubMed]

122. Jia, P.; Zheng, S.; Long, J.; Zheng, W.; Zhao, Z. dmGWAS: Dense module searching for genome-wide association studies in protein–protein interaction networks. *Bioinformatics* **2011**, *27*, 95–102. [CrossRef] [PubMed]

123. Krämer, A.; Green, J.; Pollard, J.; Tugendreich, S. Causal analysis approaches in Ingenuity Pathway Analysis. *Bioinformatics* **2014**, *30*, 523–530. [CrossRef] [PubMed]

124. Wang, L.; Matsushita, T.; Madireddy, L.; Mousavi, P.; Baranzini, S.E. PINBPA: Cytoscape app for network analysis of GWAS data. *Bioinformatics* **2015**, *31*, 262–264. [CrossRef]

125. Kutmon, M.; van Iersel, M.P.; Bohler, A.; Kelder, T.; Nunes, N.; Pico, A.R.; Evelo, C.T. PathVisio 3: An Extendable Pathway Analysis Toolbox. *PLOS Comput Biol.* **2015**, *11*, e1004085. [CrossRef]

126. Smoot, M.E.; Ono, K.; Ruscheinski, J.; Wang, P.-L.; Ideker, T. Cytoscape 2.8: New features for data integration and network visualization. *Bioinformatics* **2011**, *27*, 431–432. [CrossRef]

127. de Leeuw, C.A.; Mooij, J.M.; Heskes, T.; Posthuma, D. MAGMA: Generalized Gene-Set Analysis of GWAS Data. *PLoS Comput. Biol.* **2015**. [CrossRef]

128. Maciejewski, H. Gene set analysis methods: Statistical models and methodological differences. *Brief Bioinform.* **2014**, *15*, 504–518. [CrossRef]

129. Ashburner, M.; Ball, C.A.; Blake, J.A.; Botstein, D.; Butler, H.; Cherry, J.M.; Davis, V.P.; Tarver, L.I.; Kasarakis, A.; Lewis, S.; et al. Gene Ontology: Tool for the unification of biology. *Nat Genet.* **2000**, *25*, 25–29. [CrossRef]

130. Kanehisa, M. The KEGG resource for deciphering the genome. *Nucleic Acids Res.* **2004**, *32*, D277–D280. [CrossRef]

131. Carbon, S.; Dietze, H.; Lewis, S.E.; Mungall, C.J.; Munoz-Torres, M.C.; Basu, S.; Chisholm, R.L.; Dodson, R.J.; Fey, P.; Thomas, P.D.; et al. Expansion of the Gene Ontology knowledgebase and resources. *Nucleic Acids Res.* **2017**, *45*, D331–D338. [CrossRef]

132. Mishra, P.; Törönen, P.; Leino, Y.; Holm, L. Gene set analysis: Limitations in popular existing methods and proposed improvements. *Bioinformatics* **2014**. [CrossRef] [PubMed]

133. Abatangelo, L.; Maglietta, R.; Distaso, A.; D'Addabbo, A.; Creanza, T.M.; Mukherjee, S.; Ancona, N. Comparative study of gene set enrichment methods. *BMC Bioinform.* **2009**. [CrossRef] [PubMed]

134. Tarca, A.L.; Bhatti, G.; Romero, R. A comparison of gene set analysis methods in terms of sensitivity, prioritization and specificity. *PLoS ONE* **2013**. [CrossRef]

135. Pers, T.H. Gene set analysis for interpreting genetic studies. *Hum. Mol. Genet.* **2016**. [CrossRef] [PubMed]

136. Tamayo, P.; Steinhardt, G.; Liberzon, A.; Mesirov, J.P. The limitations of simple gene set enrichment analysis assuming gene independence. *Stat. Methods Med. Res.* **2016**. [CrossRef]

137. Dinu, I.; Potter, J.D.; Mueller, T.; Liu, Q.; Adewale, A.J.; Jhangri, G.S.; Einecke, G.; Famulski, K.S.; Halloran, P.; Yasui, Y. Gene-set analysis and reduction. *Brief Bioinform.* **2009**, *10*, 24–34. [CrossRef]

138. Boca, S.M.; Kinzler, K.W.; Velculescu, V.E.; Vogelstein, B.; Parmigiani, G. Patient-oriented gene set analysis for cancer mutation data. *Genome Biol.* **2010**. [CrossRef]

*Article*

# FASTENER Feature Selection for Inference from Earth Observation Data

**Filip Koprivec** [1,2,3,†], **Klemen Kenda** [1,4,*,†] **and Beno Šircelj** [1]

[1]   Jožef Stefan Institute, 1000 Ljubljana, Slovenia; filip.koprivec@ijs.si (F.K.); beno.sircelj@ijs.si (B.Š.)
[2]   Faculty of Mathemathics and Physics, University of Ljubljana, 1000 Ljubljana, Slovenia
[3]   Institute of Mathematics, Physics, and Mechanics, 1000 Ljubljana, Slovenia
[4]   Jozef Stefan International Postgraduate School, 1000 Ljubljana, Slovenia
*   Correspondence: klemen.kenda@ijs.si
†   These authors contributed equally to this work.

Received: 19 May 2020; Accepted: 21 October 2020; Published: 23 October 2020

**Abstract:**  In this paper, a novel feature selection algorithm for inference from high-dimensional data (FASTENER) is presented. With its multi-objective approach, the algorithm tries to maximize the accuracy of a machine learning algorithm with as few features as possible. The algorithm exploits entropy-based measures, such as mutual information in the crossover phase of the iterative genetic approach. FASTENER converges to a (near) optimal subset of features faster than other multi-objective wrapper methods, such as POSS, DT-forward and FS-SDS, and achieves better classification accuracy than similarity and information theory-based methods currently utilized in earth observation scenarios. The approach was primarily evaluated using the earth observation data set for land-cover classification from ESA's Sentinel-2 mission, the digital elevation model and the ground truth data of the Land Parcel Identification System from Slovenia. For land cover classification, the algorithm gives state-of-the-art results. Additionally, FASTENER was tested on open feature selection data sets and compared to the state-of-the-art methods. With fewer model evaluations, the algorithm yields comparable results to DT-forward and is superior to FS-SDS. FASTENER can be used in any supervised machine learning scenario.

**Keywords:**  feature selection; machine learning; earth observation; genetic algorithm; information theory

---

## 1. Introduction

Land cover classification based on satellite imagery is one of the most common and well-researched machine learning (ML) tasks in the Earth Observation (EO) community. In Europe, the Copernicus Sentinel-2 missions provide large amounts of global coverage EO data. With 5-day revisit times, Sentinel-2 generated a total of 9.4 PB of satellite imagery by April 2020 [1]. The total amount of EO data available through the Copernicus services is estimated to exceed 150 PB. The data represent an opportunity for building solutions in various domains (agriculture, water management, biology or law enforcement); however, computationally efficient methods that yield sufficient results are needed to deal with this big data source. The most important issue when trying to maximize the accuracy of statistical learning methods is effective feature engineering. With many features, the algorithms become slower and consume more resources. In this paper, we present FASTENER (FeAture SelecTion ENabled by EntRopy, https://github.com/JozefStefanInstitute/FASTENER) genetic algorithm for efficient feature selection, especially in EO tasks.

EO data sets are characterized by a large number of instances (millions or even billions) and a fair number of features (hundreds). These characteristics differ from the usual feature selection data sets that often contain up to several hundred instances and usually thousands of features. FASTENER exploits entropy-based measures to converge to a (near) optimal set of features for statistical learning faster than other algorithms. The algorithm can be used in many scenarios and can, for example, complement multiple-source data fusion and extensive feature generation in the fields of natural language processing, DNA microarray analysis or the Internet of Things [2]. FASTENER has been tested on a land cover classification EO data set from Slovenia. Additionally, we have tested the algorithm against 25 general feature selection data sets.

The main evaluation use case presented in this paper is based on our previous contributions [3,4] regarding crop (land cover) classification using Sentinel-2 data from the European Space Agency (ESA). The scientific contributions of our work are as follows:

1.  A novel genetic algorithm for feature selection based on entropy (FASTENER). Such an algorithm reduces the number of features while preserving (or even improving) the accuracy of the classification. The algorithm is particularly useful for data sets containing a large number of instances and hundreds of features. A reduced number of features reduces learning and inference times of classification algorithms as well as the time needed to derive the features. By using an entropy based approach, the algorithm converges to the (near) optimal subset faster than competing methods.
2.  Improvement of the state-of-the-art in feature selection in the field of Earth Observation. FASTENER yields better result than current state-of-the-art algorithms in remote sensing. The algorithm has been tested and compared to other methods within the scope of the land-cover classification problem.
3.  Usage of pre-trained models for information gain calculation for feature selection in Earth Observation scenarios. Usually, it is computationally expensive to train a machine learning model. The inference phase is much faster. FASTENER exploits the pre-trained models in order to estimate the information gain of certain features by inferring from data sets with randomly permuted values of the evaluated feature. This approach improves the convergence speed of the algorithm.

The paper is structured as follows. In the next section, related work regarding feature selection algorithms as well as state-of-the-art in feature selection in Earth Observation is described. Section 4 contains a thorough description of the data used in the experiments and describes our algorithm. Section 5 describes while Section 6 discusses the experimental results and compares FASTENER to other algorithms based on NSGA-II. Finally, our conclusions are presented in Section 7.

## 2. Related Work

When tackling problems related to EO, quite often the older and simpler algorithms (ReliefF) have given the best results. This section presents the current state-of-the-art in feature selection and particularly its applications within the EO community. It also describes supporting fields of the work presented in this paper, such as the current status in land-cover classification and the multi-objective optimization approach to feature selection.

### *2.1. Multi-Objective Optimization and Its Usage in Feature Selection*

Data science often reduces its evaluation to a single score (or fitness function) that explains the accuracy or efficiency of a particular methodology. There are other limitations that should be considered. On the one hand, there are the available resources (e.g., memory, processors, space), while on the other hand the response time should be considered (i.e., the preferred response time for a particular methodology). Quite often, the most appropriate approach should minimize the need for resources and response time while maintaining sufficient accuracy. The multi-objective optimization [5] strives to find a set (or Pareto front) of solutions (in our case feature sets) that achieve the best possible classification accuracy with a limited number of features.

Multi-objective optimization methods have already been used for feature selection purposes, although their usage in the field is limited due to their computational complexity [6]. The classical

NSGA-II algorithm [5] was used for feature subset optimization in [7]. NSGA-II uses fast non-dominated sorting for selecting viable candidates for the next generation parent population. The methodology has been improved by the usage of Reduced Pareto Set Genetic Algorithm with Elitism (RPSGAe) [8] for achieving a more efficient directed search in the feature space [9]. The elite-preserving operator suppresses the deterioration of the population fitness along with the successive generations. RPSGAe reduces the number of solutions on the efficient (Pareto) front while maintaining its characteristics intact, similarly to the purge function in the FASTENER algorithm. NSGA-II feature selection approach has also been extended in [10] with six different importance measures, including representation entropy which is based on information theory. The research shows that there is no preferred importance measure that would be a good fit for all feature selection problems. FASTENER uses importance measures that are based on real model evaluations and are therefore more reliable than estimates based on the features themselves.

Pareto-optimal subset selection (POSS) [11] is also a genetic algorithm, but it only uses mutation techniques (while NSGA-II also uses cross-over methods). Other methods explore several variants of directed searches within the feature space to converge to the sub-optimal data set faster (e.g., FS-SDS [12] uses stochastic diffusion search and extreme learning machine [13] as an embedded classifier). Other methods, such as forward selection and backward elimination use an exhaustive step-by-step search to find the most optimal feature set [14]. These methods are much slower (and quite often do not converge to a small-enough data set within a reasonable time) and could even converge to a local minimum.

Our approach extends POSS and NSGA-II algorithms and suggests the usage of entropy-based indices based on previously trained models to steer the iterative feature selection process. The parameterized mutation strategy introduces additional features to the candidate feature sets with respect to changes in the Pareto front. The Pareto front is periodically purged in order to eliminate non-suitable candidates from the loop. With these improvements, FASTENER converges to a (nearly) optimal solution faster than the compared methods.

## 2.2. Feature Selection and Dimensionality Reduction for EO Tasks

The automatic spectro-temporal feature selection (ASTFS) [15] workflow provides a search over the feature space by incrementally extending the feature set with effective features (those that improve classification scores) ordered by their global separability index [16]. A similar approach is implemented in [17], where the authors use mutual information (MI) [18] and Fisher's criterion to select the $k$ most important features. Mutual information based on entropy estimates from $k$-nearest neighbour distances [18] is data-efficient and has a minimal bias. The problem of estimating MI between discrete and continuous features has been solved in [19], which can be applied to classification tasks (discrete) with continuous features. Although the separability index and mutual information provide good heuristics for adding different features, the non-iterative (non-wrapping) algorithms do not necessarily yield the optimal feature set. Our algorithm offers a more thorough search, which is controlled by the genetic nature of the algorithm. The approach has been tested using timeless features for land cover classification [20,21].

A revision and test of multiple feature selection algorithms (similarity-based, statistical, sparse learning, information theoretical, and wrapper methods) [22] showed that ReliefF (a similarity-based approach) [23], although not among the recent favourites, yielded the best results in a particular scenario of parthenium weed infestation detection.

Our algorithm can be applied to any feature selection scenario as it is expected to converge to the (near) optimal subset selection. FASTENER is expected to perform at least as good as any other similarity-based approaches; however, several iterations would be required during the learning process to find the (near) optimal subset of features.

Genetic algorithms have been used in EO scenarios for feature selection of hyper-spectral EO images [24]. In contrast to multi-spectral (as in our case), hyper-spectral images contain a significantly larger number of bands. The approach in [24] does not use any entropy-based features to optimize the search and is, in that respect, similar to POSS. The latter has been tested in land cover classification [4] and yielded good results.

In an EO environment, where data indices are abundant, the calculation of some features is computationally very intensive and external data are difficult to acquire (weather data, soil samples). Reducing the number of features has a significant positive impact on the overall algorithm performance in a real-world scenario. As the production of global EO products sometimes take thousands of computing hours, efficient feature selection algorithms could reduce the costs by 50–95%.

*2.3. Land Cover Classification and Feature Engineering*

There are essentially three approaches to land cover classification based on EO data [25]. The first approach is object-based and focused on a single image, while the other two approaches are pixel-based. From the latter, a simpler approach is based on single-image data, while the other is based on a time-series of images. Single image approaches depend on the current situation (e.g., cloud cover), whereas time series based approaches inherently overcome this shortcoming. Sentinel-2 mission from the Copernicus programme is dedicated to agriculture and their satellites re-visit times are 5 days. It is possible to construct an interpolated time-series of different EO sensors throughout the entire vegetation period [3]. A simple feature extraction involves taking particular sensor values and derived features at pre-selected points in time and combining them into a feature vector. A more intelligent approach involves the extraction of timeless features such as the maximum value of a particular time series, length of the maximum interval, steepness of the steepest slope and others [20,21]. Our evaluation of the FASTENER algorithm is based on the latter approach [3].

The best accuracy scores in land cover classification were achieved with deep learning [26]. These approaches reduce the value of feature engineering that originates mostly from domain experts and/or extensive experimental knowledge. Deep neural networks are able to derive the important features exclusively from raw data. Of course, the data must be extensive, which is the case in EO. However, this comes with a price; namely, extreme computational requirements reduce the usability of these methods.

Our goal is to develop lean EO machine learning models that are capable of capturing most of the information with limited resources. Intelligent feature selection, together with smart feature engineering and a pragmatic choice of the learning models is one of the three pillars of smart EO.

## 3. FASTENER—A Genetic Algorithm for Feature Selection

The presented algorithm is based on the POSS genetic algorithm and the Pareto ensemble pruning proposed in [27,28]. The main objective of the algorithm is to select an optimal subset of available features from a large feature set ($N > 100$). The problem could be represented mathematically as finding the optimal (or near-optimal) binary vector $x$ of an indexed set $\{0,1\}^N$. Finding the optimal subset is NP-complete. Each element in $x$ corresponds to a particular feature that is included or excluded in our decision model. The optimality of such a set is measured by a scoring function, which we will denote by $A(M; x)$ (accuracy measure of model $M$ on a subset of features $x$). Some standard examples of such measures in machine learning classification problems include precision, recall, accuracy and $F_1$ score. In our experiments, the $F_1$ score was used. We also adopt the notation $|x|$ to denote the number of bits set in $x$ that directly corresponds to the number of selected features. Mathematically, for a fixed classification model $M$ and the number of features $k$, we search for such an $x$ where the following maximum is reached:

$$\max_{\substack{x \in \{0,1\}^N \\ |x|=k}} A(M; x).$$

As in POSS, the function $x \mapsto (|x|, A(M; x))$ creates a two-dimensional Pareto front. In one dimension, we evaluate $k$, the number of selected features, and in the other dimension, we present $A$ score, which represents the optimality (e.g., accuracy) of such a subset. In this space, the ordering of instances is defined as follows. The pair $(k_1, s_1)$ dominates the pair $(k_2, s_2)$ if and only if $k_1 \leq k_2$ and $s_1 \geq s_2$. Intuitively, the smaller set (by cardinality) of features provides classification results, which are at least as good as for the larger set. We denote such a relation by $(k_2, s_2) \preceq (k_1, s_1)$. The relation is obviously transitive but does not induce linear ordering. Pairs that are not comparable through such a relation are of particular interest, as they lie on the Pareto front. Feature subsets that lie on the Pareto front (not strongly dominated by other subsets) are the desired subsets. They provide the best classification power using the smallest number of features. The final result of a feature selection algorithm is the final Pareto front. The final Pareto front can be seen as an accumulation of the best results for a fixed $k$. The features with the best performance can then be selected from the front automatically (with a certain cut-off threshold), manually or by using additional information (e.g., time to calculate the features). The results clearly show that the Pareto front "plateaus" after the inclusion of some features and this fact can easily be used to automatically select the best number of features.

Our modification of the POSS algorithm combines Pareto front searching with a genetic algorithm to incorporate additional statistical information when recombining genes. The main ingredient in the algorithm is the concept of an *item*; a subset of features and *scored item*, which is an item with an assigned score corresponding to such a subset of features. The size of an item denoted by $|\text{ITEM}|$ corresponds to the number of features selected. The algorithm works by successively evaluating items, combining them, and updating the Pareto front.

Each subset of features can be represented directly as a binary number, where set bits correspond to the included (selected) features and unset bits correspond to excluded ones. This allows for a natural representation of genes for such an item with a binary string or bit-set and also gives a natural human-readable representation as an arbitrary sized integer. Apart from this, many later needed operations can be quickly implemented as simple operations on a binary string. The size of an item corresponds to the number of set bits, the bit-wise AND operation between two items corresponds to the genes that are common to both of them, while the bit-wise XOR returns all different selected features.

In our implementation, the integer representation of item genes is treated in a little-endian way, as this allows easier extension of feature sets and keeps integer representations valid even when adding new features. For example, in a set of six features with 32 possible feature combinations, binary string 11010 represents a subset containing features with indices 0, 1 and 3 and is represented by a decimal number 11. Even if the number of features is increased, such a subset would still be represented by the number 11, albeit its binary string representation would be extended by zeros on the right side.

An example of the incremental Pareto front update is shown in Figure 1. Each subsequent generation of the FASTENER algorithm produces a Pareto front that dominates the Pareto fronts from previous generations (in rare cases, where no improvement is made, the Pareto fronts of two subsequent generations may be the same). The main objective of the algorithm is to converge as quickly as possible to a theoretically limiting front (in terms of fitness function evaluations).

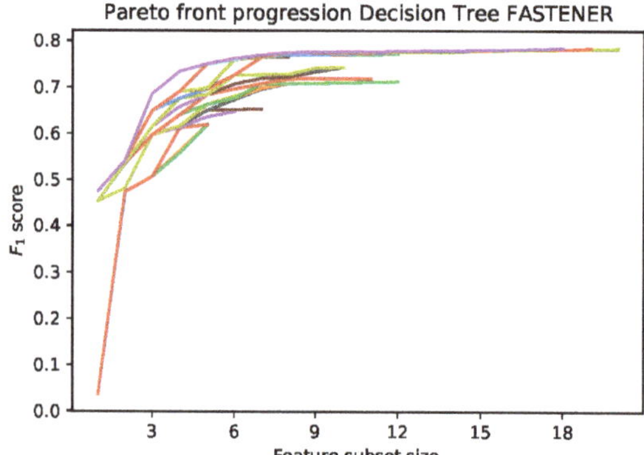


**Figure 1.** Iterative improvements of Pareto front from the first generation. Each next generation's Pareto front achieves better $F_1$ score.

### 3.1. Algorithm

The algorithm stores the current Pareto-optimal items and progresses in generations. For each $k \leq N$ it keeps an item with the best score $A$. If there is no such item this means either that it is strictly dominated by another item or that the algorithm has not yet evaluated an item this size. In our implementation, the Pareto front is implemented as a simple dictionary. Apart from the Pareto front, the current population is kept in a separate set. The decoupling of the current population from the Pareto front allows the algorithm to "explore" different combinations more easily, while the Pareto front continues to be updated at each iteration.

As usual, in each generation, the current population of items is mated, a random genetic mutation is introduced and newly acquired items are evaluated. For each evaluation, the Pareto front is updated. If a new item is strictly better than an item on the front with the same size, the newly evaluated item is placed on the Pareto front. After all new items have been evaluated, the front is purged by removing items that are strictly dominated by other items. The pseudo-code for the basic algorithm is presented in Algorithm 1. The algorithm's flow diagram is depicted in Figure 2.

To prevent the search phase of the algorithm from diverging (the population as a whole is not controlled by the scoring function), the population is periodically reset to the current Pareto optimal items. In the presented algorithm, this decision is based only on the generation number; however, our implementation also considers the rate with which the running Pareto front is being updated between generations.

---

**Algorithm 1** Basic algorithm

---

**Require:**
 Initial population POP
 Evaluation function EVALF
 Number of iterations K
 Mating pool selection strategy SELECTPOOL
 Crossover strategy CROSSOVER
 Mutation strategy MUTATE
 Predicate RESETTOFRONT
 Predicate PURGEFRONTPREDICATE
 Front purging procedure PURGEFRONT
1:  **function** FEATURESUBSETSELECTION
2:   front ← pop
3:   **for** gen = 1 to K **do**
4:    pool ← SELECTPOOL(pop)
5:    pool ← CROSSOVER(pool)
6:    pop ← MUTATE(pop ∪ pool)
7:    pop ← EVALUATE(pop, EVALF)
8:    front ← front ∪ pop       ▷ Add evaluated population to Pareto front
9:    front ← REMOVEDOMINATED(front)  ▷ Remove unoptimal items from Pareto front
10:    **if** PURGEFRONTPREDICATE(gen) **then**
11:     front ← PURGEFRONT(front)
12:    **end if**
13:    **if** RESETTOFRONT(gen) **then**
14:     pop ← front
15:    **end if**
16:   **end for**
17:   **return** front
18: **end function**

---

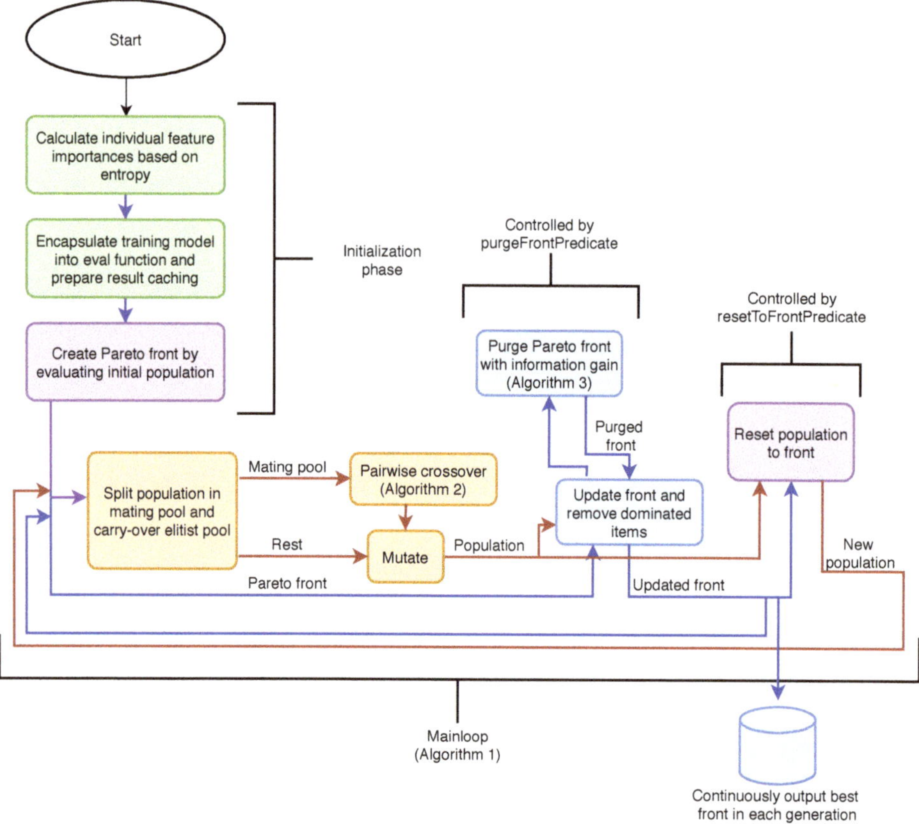



**Figure 2.** The flow diagram of the FASTENER graphically represents Algorithm 1 and puts Algorithms 2 and 3 into context. In the figure, red/orange colour objects represent population-based operations, while blue colour objects represent Pareto front related operations. The violet colour represents both, the population and the Pareto front. Green boxes refer to technical details of the algorithm. The algorithm includes the initialization phase and the main loop. In the initialization phase, the initial population is prepared and evaluated and the technical prerequisites for the algorithm are created. In the main loop, the population is first split into a mating pool and a (gene) carry-over elitist pool. The mating pool first enters the crossover phase based on Algorithm 2 and is being mutated together with the elitist pool. Then the new Pareto front is updated and purged (Algorithm 3). Finally, the main loop is closed by registering the new population as the next generation Pareto front.

### 3.2. Mutation

An important part when considering gene manipulation and modification is to keep the overall objective in mind. We want to find the optimal subset of features with good accuracy characteristics. The ultimate goal is to obtain a subset with a size much smaller than $N$. In preliminary experiments, the number of features with a sufficiently predictive power was around $\sqrt{N}$, which means a significant reduction in both training time and in data preparation time (which is true especially for EO data sets). Mutation and crossover methods should therefore be tailored to help maintain, reduce or only slightly increase the size of items.

The most interesting parts of the algorithm are the CROSSOVER and MUTATE procedures. The procedure for the mutation is adopted from the POSS algorithm and is a simple random bit mutation. Each random bit in the gene is flipped independently with a probability $\frac{1}{N}$. Thus a new item is generated from each of the items selected for mating and, most importantly, the expected value of set bits is kept approximately the same. Formally, the expected value of the number of flips is 1, and therefore the mutations gravitate slowly toward $x$-s where $|x| = \frac{N}{2}$. Since the desired range of features in which we can expect satisfactory score results is much lower than the asymptotic behaviour of mutations, this type of mutation serves as a size increaser. Mutation procedures therefore slowly increase and add new features to be considered. Furthermore, varying the probability of flipping is an easy way to control the speed of the introduction of new features. Setting the mutation rate to $\frac{a}{N}$ for $a > 1$ and varying $a$ with generation number and changes in the Pareto front updates can significantly increase the search space. In addition, all further analyses are still valid even if the mutation is set as an arbitrary function $f : \mathbb{N}^+ \to (0, 1]$ only dependent on $N$ and heuristic data about mutations (independent of bit-index and bit status), called *mutation function*.

*3.3. Crossover*

The crossover procedure is designed to reduce the size of the newly produced item. The mating pool selection used for the algorithm is a simple random mating pool with a fixed size. Two different items from the mating pool are combined using the CROSSOVERPAIR procedure. The CROSSOVERPAIR procedure itself is presented in Algorithm 2 and requires 4 input parameters: two items to be mated, a scaling function for the number of non-intersected genes, and a set of entropy information for each individual feature. Without loss of generality, we may assume that $|\text{ITEM1}| \geq |\text{ITEM2}|$.

**Lemma 1.** *Let ITEM1 and ITEM2 be a random independent subset of $\{0, 1\}^N$ where the probability of each gene being set is some positive function $f : \mathbb{N}^+ \to (0, 1]$. The expected size of an intermediate gene conditional on the sizes of its parent genes is*

$$\mathbb{E}\left[|\text{ITEM1} \cap \text{ITEM2}| \mid |\text{ITEM1}| = a, |\text{ITEM2}| = b\right] = \frac{ab}{N} = \frac{|\text{ITEM1}||\text{ITEM2}|}{N}$$

We can similarly derive conditional expectation for size change.

**Lemma 2.** *The probability of an ITEM having size $k$, assuming that each feature is selected independently with probability $f : \mathbb{N}^+ \to (0, 1]$ is*

$$P(|\text{ITEM}| = k) = \frac{\binom{N}{k}(1 - f(N))^{N-k}f(N)^k}{Nf(N)}.$$

**Lemma 3.** *The probability of an intermediate item having size $k$ conditional on sizes of its parents is similarly*

$$P(|A \cap B| = k \mid |A| = a, |B| = b) = \frac{\binom{N}{k}\binom{N-k}{a-k}\binom{N-a}{b-k}}{\binom{N}{a}\binom{N}{b}}.$$

*Since all features are equally likely to be selected, conditional probability is independent of scaling function $f$.*

**Lemma 4.** *By linearity of expectation and assumption that $|\text{ITEM1}| = a$ and $|\text{ITEM2}| = b$ we can easily derive*

$$\mathbb{E}\left[max\{|\text{ITEM1}|, |\text{ITEM2}|\} - |\text{ITEM1} \cap \text{ITEM2}| \mid |\text{ITEM1}| = a, |\text{ITEM2}| = b\right] = |\text{ITEM1}|\left(1 - \frac{|\text{ITEM2}|}{N}\right).$$

**Remark 1.** *The assumption that each feature is selected independently with probability $f(N)$ should not be confused with each subset of features selected with equal probability. In our case, this is beneficial because smaller sets are intrinsically better, which are easily controlled by the mutation function $f$.*

---

**Algorithm 2** Randomized crossover with information gain weighting

---

**Require:**
    First item ITEM1
    Second item ITEM2
    Scaling function for number of genes ONGENESCALING
    Individual feature entropy INFORMATIONENTROPY
1:  **function** CROSSOOVERPAIR
2:     **if** |ITEM1| < |ITEM2| **then**
3:        SWAP(ITEM1, ITEM2)
4:     **end if**
5:     intermediate ← ITEM1 & ITEM2                  ▷ Intersection of genes
6:     rest ← ITEM1 ⊕ ITEM2         ▷ Bitwise XOR, features present in exactly one item
7:     addNum ← ONGENESCALING(ABS(|ITEM1| - |ITEM2|))
8:                 ▷ Scale the number of new features according to provided function
9:     additional ← RSELECT(addNum, informationEntropy, rest)
10:         ▷ Randomly select addNum features, weighted by precalculated entropy
11:     mated ← intermediate | additional            ▷ Add selected features
12:     **return** mated
13: **end function**

---

After crossover, the resulting genes initially contain features that were selected in both parents. The intersection of features is a good starting point, as it seems to produce good results during the running of the algorithm. Since the implementation of the genes is a simple bit string, the intermediate result is simply bitwise logical AND of its parents' genes. Another way to combine the parents would be to produce offspring with the union of genes. The latter approach produces offspring that is too large (contains too many selected features) and is thus unsuitable for our objective, where we want to select a sufficiently small feature set. The size of the intermediate result obtained by the intersection (line 5) is smaller than or equal to the size of ITEM2 (smallest of the parents), but can be much smaller, depending on the input sets (see Lemma 3). With the intersection, the number of genes decreases too rapidly (see Lemma 4) and disposes of useful information. We therefore use additional genes presented in exactly one of the parents (generated by element-wise exclusive OR) to construct an additional set of features with good information gain. The visualization for the first part of CROSSOVERPAIR procedure on an example case is shown in Figure 3. The genes of two items to be mated are split into two indexed sets: an intermediate set containing genes from both parents and a set called *rest* containing genes to be used for the enrichment of intersected genes.

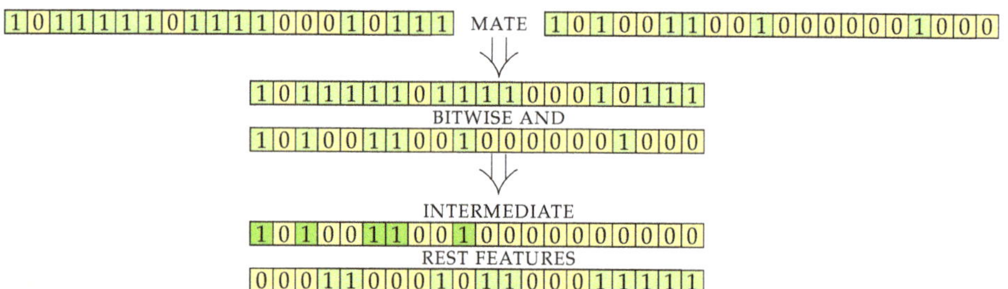


**Figure 3.** Schema of mating of 2 genes. With bit-wise AND operation we produce the intermediate set and with bit-wise exclusive or XOR we produce the rest (features) set.

Enrichment of the intermediate set of genes is done as follows. First, the number of enrichment genes is calculated using the ONGENESCALING function. This function is used to control the size of the final offspring. From Lemma 4, it follows that the expected size of the rest (features) set, with the size of the largest parent being constant, is linear to the size of the smaller parent (ITEM2). Adding all genes from the rest set will result in an offspring too large (and diverge toward the whole feature space being selected), selecting none of the genes will diverge towards very small sets. In our experiments, we used the function $x \mapsto \lfloor \frac{x}{2} \rfloor + 1$ (for ONGENESCALING) that scaled the amount of features appropriately for our data set. For a larger number of features, the functions $\sqrt{x}$ and $\log_a x$ with a sufficiently small base $a$ are a good choice. As with most parameters, ONGENESCALING can be swapped during the run-time and made dependent on the conditions of front updates and statistics.

The second part of the mating algorithm selects the best performing features using heuristics based on information gain. From all features in the rest set, *addNum* are randomly selected (using a weighted random selection, where the weight of each feature is its individual precalculated information gain). The final result is an item with genes from the intersection set and selected genes from rest set. In the implementation, the results are simply combined with element-wise OR. The heuristics for the information gain in our algorithm is based on mutual information gain. The `scikit-learn` implementation [29] of mutual information gain is used for each feature in the train set.

Figure 4a shows the continuation of the example mating algorithm from the previous figure. Each of the initial features is represented with its information gain, which is shown by the height and intensity of the shade of the column above it. Columns for features in the rest set are shown in red (and the corresponding genes are green in the bottom row). Of the eight features in the rest features row, five are selected using weighted random function. Figure 4b shows the visualization of the result (offspring) of the mating procedure. The genes selected by intersection are marked with dark green shading, while the genes selected by information enrichment are shown in light green. The corresponding information gain statistics in shown with column color intensity. In this example, two genes of size 14 and 6 were combined and they produced a result with size 10, which may be further increased by the mutation strategy. The resulting item contains genes that are present in both parents and were therefore rated as good in the previous evaluation. It is enriched with a limited number of genes that are present in either parent, weighted according to their information gain.

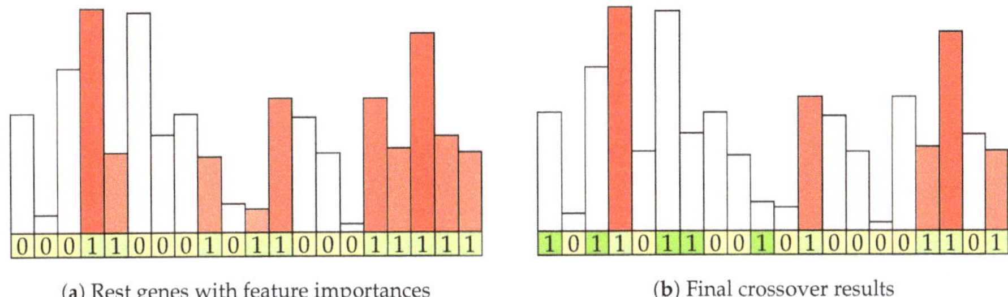


(a) Rest genes with feature importances

(b) Final crossover results

**Figure 4.** Visualization of the rest set and final mating result. Information gain of a particular feature is depicted by the height of the column above a particular feature.

### 3.4. Purging Features from Pareto Front Using Information Gain

An initial version of the algorithm presented in Algorithm 1 has been further improved to exploit pre-trained models for the calculation of the information gain and to purge non-relevant features from items on the Pareto front. For each item added to the front, a model trained on its genes is stored. Since the number of elements is small compared to the learning data, storing the models results in virtually no additional effort in run-time and is a useful tool for evaluating the optimization over time. In the same way as when the population is reset to the Pareto front, another predicate is introduced to indicate when the Pareto front should be purged using the information criterion. The gene purging procedure is presented in Algorithm 3.

---

**Algorithm 3** Gene purging algorithm

---

**Require:**

    Current Pareto front FRONT

    Evaluation function with ability to randomly shuffle features EVALF

1: **function** PURGEPARETOFRONT
2:     newFront = ← Dict.empty         ▷ Empty dictionary representing new Pareto Front
3:     **for** item in front **do**
4:         **if** |ITEM| == 1 **then**
5:             **continue**         ▷ Items with only one feature cannot be reduced
6:         **end if**
7:         baseResult ← item.result         ▷ Score acquired by using all features from item
8:         scoreDecrease ← []         ▷ Score decrease for each feature
9:         **for** geneInd in item.genes **WHERE** ISSET(item.genes[geneInd]) **do** ▷ Check only set genes
10:             newScore ← EVALF(item.model, item.genes, geneInd)

        ▷ Evaluate existing trained model on test data set, where values of feature geneInd are randomly shuffled

11:             scoreDecrease[geneInd] ← baseResult - newScore
12:         **end for**
13:         bestInd ← ARGMIN(scoreDecrease)

        ▷ Take the gene which showed smallest score decrease when shuffled

14:         newItem ← ITEM(genes=UNSETBIT(item.genes, bestInd))

        ▷ Create new item where the gene with the smallest score decrease is unset

15:         newFront[|newItem|] ← EVALUATE(newItem, EVALF)

        ▷ Evaluate newly created item (train and test model)

16:     **end for**
17:     front ← front ∪ newFront         ▷ Update current front with newly calculated items
18:     front ← REMOVEDOMINATED(front)         ▷ Remove unoptimal items from Pareto front
19:     **return** front
20: **end function**

---

The main idea of purging the items on the Pareto front with information criterion is to discard unneeded features from items that already perform well. For each item on the Pareto front, we try to discard a feature that makes the least contribution to the item's score. The obvious way to do this would be to transverse through all the features used in the prediction model, remove superfluous features one by one, and train and re-evaluate such a model. For each item, a new model must be trained and re-evaluated. This is time-consuming and offers only minor improvements. Our proposal is to use a heuristic approach to select a feature to be removed. For each such feature, an existing model corresponding to the item is used (saving time by avoiding the training of a new model) and evaluated according to the previously used test data set. For each selected feature, its values are randomly shuffled across the test data set. The motivation for such heuristics is that the shuffling values of a feature with strong predictive power has a much stronger effect on the resulting score than shuffling values of a feature with only low predictive power. In this way, we can provide heuristics for estimating the predictive power of features in a model-agnostic way using only a linear number of model calls. Such heuristics provide a noticeable acceleration and can be used for any type of model. The approach targets black box models with a high ratio between training and inference time/resource consumption—e.g., gradient boosting or complex neural networks. For each newly shuffled feature in an item, a change in the base score is recorded (line 11). Interestingly, sometimes the change can even be negative if the included features are detrimental to the overall prediction procedure. This is proving to be an effective method for quickly sorting out bad features introduced by mutations at the end of simulations when the exploration is limited to the feature space previously explored.

The feature whose shuffling has the most detrimental effect is taken and a new item is constructed whose genes are the same as those of the original element, except that the resulting feature is omitted (lines 13 and 14). This procedure can be further generalized. For example, first *scoreDecrease* could be scaled with some function (for example, $x^3$) and then the feature to be removed could be sampled by these new *scoreDecrease* weights from the set. Special care should be taken that the scaling function is monotonous (and preferably a bijection) since negative values can occur and should not be bundled with positive ones. If a random selection is used, the final probabilities must also be re-scaled correctly due to possible negative weights. After the feature to be removed is selected, the new item is evaluated and all new items are brought to the front. The front should also be purged to take into account possible new features on the front.

The most important heuristic optimization used by the front purging functions is based on the fact that only a new model is built with the features with the most promising importance score (as inferred by a model when feature values are shuffled). The time required for this operation is linear in the number of features. For each feature, the model requires one evaluation (inference) on the training set. The implementation of the gene purging can be further optimized so that each feature is selected only once. Under the reasonable assumption of sufficiently deterministic training and evaluation, it is easy to see that evaluating an item and (potential) addition of a feature to the front is an idempotent operation. Removing the same feature from the same item on the front several times is not a reasonable change and therefore another feature could be considered for removal. An appropriate threshold for reducing the score may also be introduced so that features with very high predictive power are never removed.

## 4. Data

FASTENER is focused on improving the state-of-the-art in EO land-cover classification. The description of the EO data set is given in Sections 4.1 and 4.2. Apart from that, the algorithm was tested on 25 additional feature selection benchmark data sets with a varying number of features, label classes and instances, in order to be compared against the existing state-of-the-art methods.

### 4.1. EO Data

Earth Observation data were provided by the EU Copernicus program's Sentinel-2 mission, whose main objectives are land observation, land use and change detection, support for generating land cover, disaster relief support and climate change monitoring [30]. The data comprise 13 multi-spectral channels in the visible/near-infrared (VNIR) and short wave infrared (SWIR) spectral range with a temporal resolution of 5 days and spatial resolutions of 10 m, 20 m and 60 m (the latter is used for diagnostics only) [3]. Sentinel's Level-2A products (surface reflectances in cartographic geometry) were retrieved via the SentinelHub (https://www.sentinel-hub.com/) services and processed using the `eo-learn` (https://github.com/sentinel-hub/eo-learn) library. In addition, a digital elevation model for Slovenia (EU-DEM) with 30 m resolution was used.

The ground truth data in our experiments were collected from the Slovenian land parcel identification system (LPIS). The original LPIS data consist of 177 different vegetation classes. These classes were grouped into 23 more general classes proposed by domain experts. The final data set includes 23 separate classes describing the type of farmland and one class describing all non-agricultural surfaces. Data have been collected for the year 2017.

A classification scenario is depicted in Figure 5. The figure shows a subset (true colour) of input EO data, the manually acquired ground truth data and the final automatic classification result of our algorithm. Based on the (easily obtainable EO data), the classification algorithm predicts ground-truth label (which is difficult to obtain and often contains incorrect data). When comparing Figure 5b,c, we observe the similarity between ground truth and the classification result.

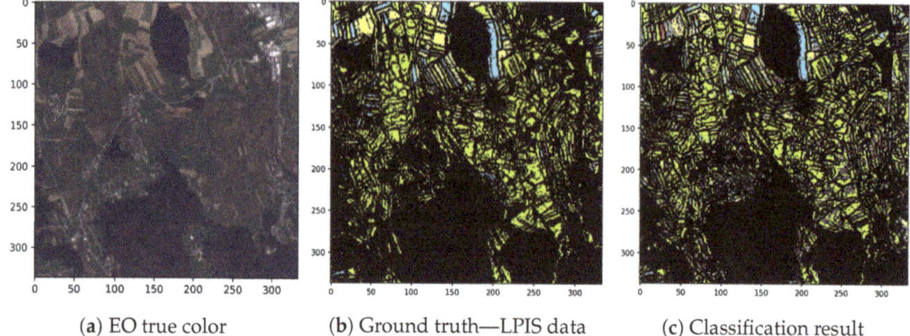

(**a**) EO true color  (**b**) Ground truth—LPIS data  (**c**) Classification result

**Figure 5.** Partial input data, ground truth data and classification results.

*4.2. Feature Engineering and Sampling*

The EO data were collected for the entire year. Four raw band measurements (red, green, blue—RGB and near-infrared —NIR) and six relevant vegetation-related derived indices (normalized differential vegetation index—NDVI, normalized differential water index—NDWI, enhanced vegetation index—EVI, soil-adjusted vegetation index—SAVI, structure intensive pigment index—SIPI and atmospherically resistant vegetation index—ARVI) were considered. The derived indices are based on extensive domain knowledge and are used for assessing vegetation properties. In Figure 6d, an example of NDVI index is depicted, which is an indicator of vegetation health and biomass. Its value changes during the growing season of the plants and differs significantly from other non-planted areas. The NDVI is calculated as:

$$NDVI = \frac{NIR - red}{NIR + red}$$

Timeless features were extracted based on Valero et al. [20]. These features can describe the three most important crop stages: the beginning of greenness, the ripening period and the beginning of senescence [20,21]. Annual time series have different shapes due to the phenological cycle of a crop and characterize the evolution of a crop. With timeless features, they can be represented in a condensed form.

For each pixel, 18 features per each of the 10 time-series were generated. The raw value and maximum inclination for a given pixel were calculated from the elevation data as 2 additional features. In total, 182 features were used in the experiments.

The examples of learning features are depicted in Figure 6: EVI minimal value (Figure 6e), EVI standard deviation (Figure 6f), NDVI maximum mean value in a sliding temporal neighborhood of size 2 (Figure 6g), SIPI mean value (Figure 6h) and SAVI mean value (Figure 6i).

Prior to the experiments, edge detection [31] was performed on EO data, excluding the pixels at the borders of land plots. These pixels are potential mixed-class instances that can have a negative effect on the learning process. An example of an edge detection mask is depicted in Figure 6a. A balanced learning set was sampled from the entire data set with $20 \times 10^3$ data points (pixels) representing each class. The classes with the lowest frequency were oversampled in the vicinity of sampled instances. The entire learning data set [32] consists of $480 \times 10^3$ samples.

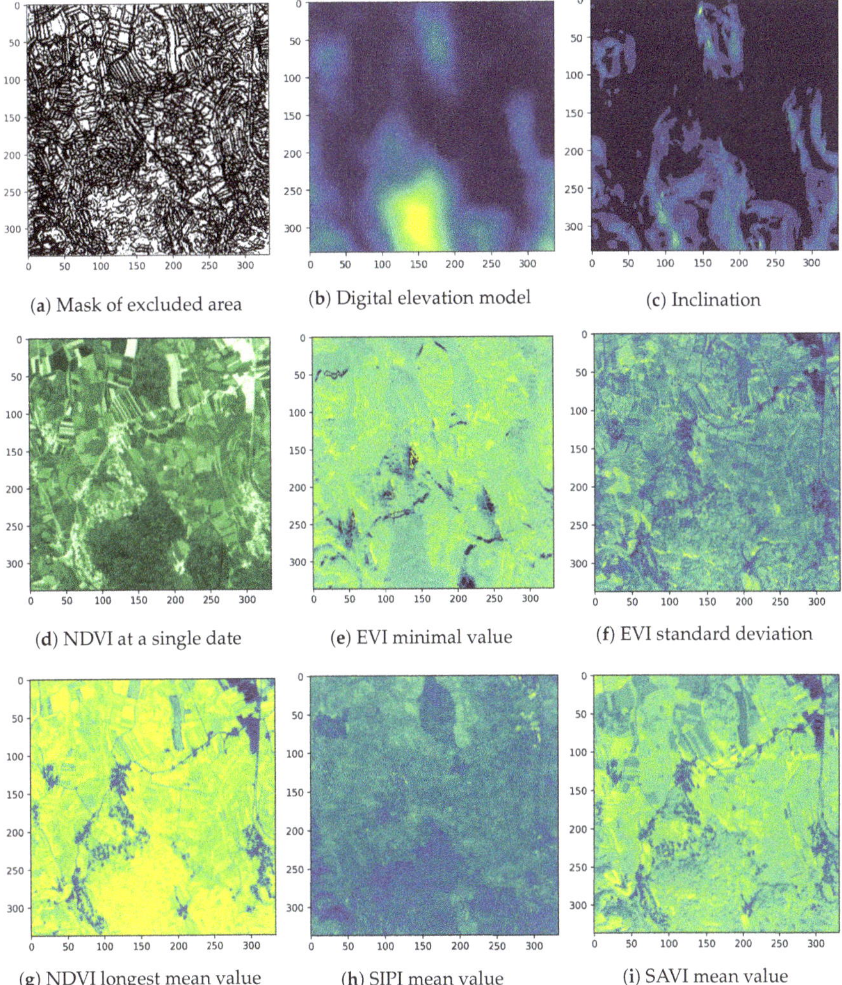

(**a**) Mask of excluded area   (**b**) Digital elevation model   (**c**) Inclination

(**d**) NDVI at a single date   (**e**) EVI minimal value   (**f**) EVI standard deviation

(**g**) NDVI longest mean value   (**h**) SIPI mean value   (**i**) SAVI mean value

**Figure 6.** Examples of features derived from EO time series. Each feature represents a potentially significant parameter for land cover classification [20].

## 4.3. Feature Selection Benchmark Data Sets

Benchmark data sets for feature selection were taken from [6]. The description of all the data sets is given in Table 1. It is apparent from the table that our main data set, EOData [32] in the last row, differs from other data sets with its very high number of instances and relatively low number of features.

**Table 1.** Description of the data sets used for testing the FASTENER with number of instances, features and classes. $I_{C\downarrow}$ and $I_{C\uparrow}$ represent the percentage of instances representing the smallest and the largest class, respectively.

| Data Set | Instances | Features | Classes | $I_{C\downarrow}$ | $I_{C\uparrow}$ |
|---|---|---|---|---|---|
| ALLAML | 72 | 7129 | 2 | 35 | 65 |
| arcene | 200 | 10,000 | 2 | 44 | 56 |
| BASEHOCK | 1993 | 4862 | 2 | 50 | 50 |
| CLL_SUB_111 | 111 | 11,340 | 3 | 10 | 46 |
| COIL20 | 1440 | 1024 | 20 | 5 | 5 |
| colon | 62 | 2000 | 2 | 35 | 65 |
| gisette | 7000 | 5000 | 2 | 50 | 50 |
| GLIOMA | 50 | 4434 | 4 | 14 | 30 |
| GLI_85 | 85 | 22,283 | 2 | 30 | 70 |
| Isolet | 1560 | 617 | 26 | 4 | 4 |
| leukemia | 72 | 7070 | 2 | 35 | 65 |
| lung | 203 | 3312 | 5 | 3 | 68 |
| lung_discrete | 73 | 325 | 7 | 7 | 29 |
| lymphoma | 96 | 4026 | 9 | 2 | 48 |
| nci9 | 60 | 9712 | 9 | 3 | 15 |
| ORL | 400 | 1024 | 40 | 3 | 3 |
| orlraws10P | 100 | 10,304 | 10 | 10 | 10 |
| PCMAC | 1943 | 3289 | 2 | 50 | 50 |
| Prostate_GE | 102 | 5966 | 2 | 49 | 51 |
| RELATHE | 1427 | 4322 | 2 | 45 | 54 |
| TOX_171 | 171 | 5748 | 4 | 23 | 26 |
| USPS | 9298 | 256 | 10 | 8 | 17 |
| warpAR10P | 130 | 2400 | 10 | 10 | 10 |
| warpPIE10P | 210 | 2420 | 10 | 10 | 10 |
| Yale | 165 | 1024 | 15 | 7 | 7 |
| EOData | 480,000 | 182 | 24 | 4 | 4 |

## 5. Results

### 5.1. Experimental Setup

To assess the quality of the selected feature subsets, a *frontSurface* metric was introduced to measure the surface under the Pareto front produced by the resulting subsets. With the analogy to the *area under a curve* (AUC), which is often used in the evaluation of classification methods, the *frontSurface* measure can also be called the *area under a front* (AUF). The area under a front can be interpreted as a measure of the efficiency of the feature selection algorithm. The surface is calculated as a simple sum of the best scores for each number of features $k$. In order to simplify the presentation and later evaluation, the surface is calculated only for a fixed maximum number of features $K$ and then normalized. This makes it easier to compare fronts of different sizes and keeps them resistant to outliers with a large number of features. The measure is defined as

$$AUF := \frac{\sum_{k=1}^{K} opt(k)}{K} \qquad opt(k) := \max_{|\text{ITEM}| \leq k} \{item.result\},$$

where $opt(k)$ represents the item with the highest score on the Pareto front, with a size of, at most, $k$ (i.e., the best performing subset of features for a fixed number of features). In our case, the maximum number of features $K = 20$ was selected because the selected feature subsets rarely have more than 20 features. Intuitively, the $AUF$ measures the average of the best scoring items with a size of less than $K$.

All experiments on benchmarking data sets were performed with the same setup. The data for feature selection were first split into training and test subset (80:20) with stratified random split (`sklearn.model_selection.train_test_split` with a random state 2020 was used). The selected algorithm was run on a training subset, where the accuracy score was calculated as the $F_1$ score of the resulting model on a test set. For the FS-SDS algorithm, which uses internal data set splitting, the full data set was used. EO-data set was further split into two equal parts. The first part was used as a test set in the evaluation phase. The second part was used in the later analysis of the algorithm's generalization in Section 6.2.

The resulting statistics were calculated using the best reported feature subset for a data set and feature selection algorithm. To partially avoid possible feature selection bias [33] (test data set is not separate from the data set used for feature selection), a set of 80:20 training/test balanced splits were done using random seeds from 20 to 50. For wrapping algorithms (which typically use a significant amount of time in the learning phase), it is often impossible to validate according the nested cross-validation loop strategy.

The final results were obtained as the *AUF* of $F_1$ score, produced by the models trained on training data sets using the selected feature subset and evaluated on corresponding test subsets. The results are used for comparative purposes between the different algorithms.

For FASTENER algorithm, the decision tree classifier provided by the `scikit-learn` library was used as the base classification algorithm (using Gini index for measuring quality of a split and a requirement of keeping, at a minimum, two examples in a split). A random mating pool of size 3 was selected and all item pairs were mated using information gain crossover. All items from the Pareto front were also carried over onto the next generation and mutated along with the newly created ones. For DT-forward, we used the same decision tree hyper parameters as with FASTENER. FS-SDS uses Extreme Learning Machine (ELM) for classification. The latter was configured with 160 hidden neurons and the sigmoid activation function.

### 5.2. Comparison with Similarity-Based Methods on EO Data

FASTENER results were compared with the results of the KBEST algorithm and the RELIEFF algorithm. Although outdated, they are presented in the literature as the currently best-performing methods in the field of land cover classification [22]. To no surprise, FASTENER achieves better results. Similar to FASTENER, the best feature subset for a fixed number of features was represented as an item in a Pareto front. The Pareto front visualization for reported optimal feature subsets is shown in Figure 7. The optimal feature subsets reported by the FASTENER algorithm outperform KBEST and RELIEFF. Since the tested implementation of RELIEFF uses Python's native KD tree, only 20% of the original training data was used to speed up the feature selection process in RELIEFF and compared with the subset selected by FASTENER on the same reduced training data.

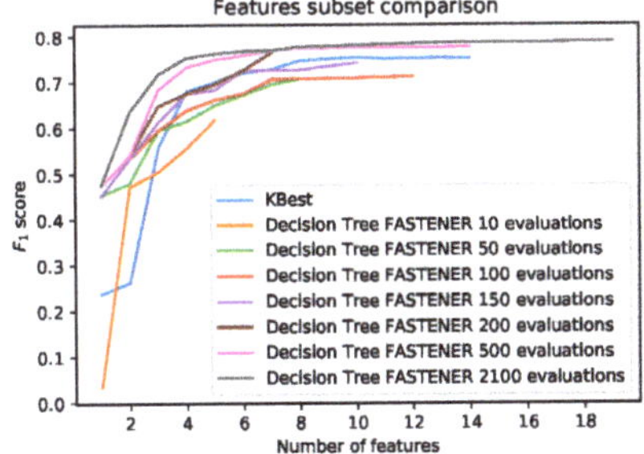


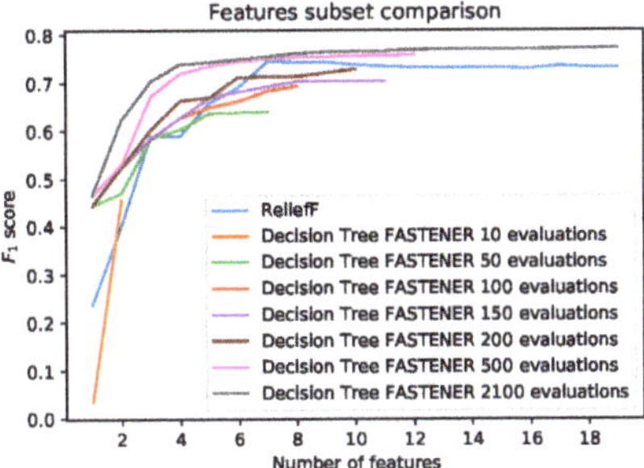


**Figure 7.** Pareto front comparison.

*5.3. Detailed Comparison with Wrapper Methods for EO Validation*

Due to the nature of the EO data set (high number of instances), the evaluation of DT-forward, DT-backward, SVM-forward and SVM-backward [14] was not possible due to extremely long computation times. Along with FASTENER and POSS, we were also able to test the EO data set using FS-SDS [12]. The following evaluation presents a detailed comparison of POSS and FASTENER algorithms, and FS-SDS results are given at the end of the subsection.

Figure 8 shows the progression of the Pareto front produced by POSS and FASTENER algorithms with the same number of model evaluations. It can be clearly seen that the convergence speed of the FASTENER algorithm outperforms the convergence speed of a comparable POSS implementation. Further analysis shows that the FASTENER algorithm converges about three-times faster than the POSS algorithm.

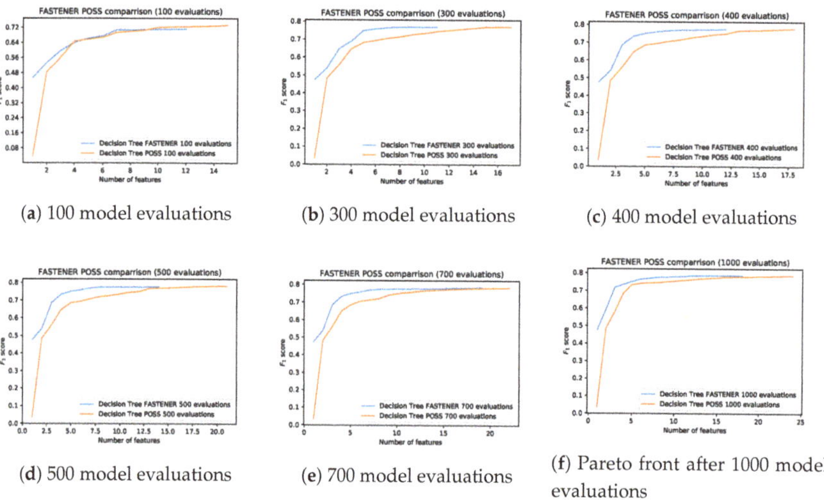

(**a**) 100 model evaluations     (**b**) 300 model evaluations     (**c**) 400 model evaluations

(**d**) 500 model evaluations     (**e**) 700 model evaluations     (**f**) Pareto front after 1000 model evaluations

**Figure 8.** Comparison of the Pareto fronts produced by FASTENER and POSS algorithms after different numbers of iterations.

Comparisons of Pareto *AUF* and *AUF* changes for POSS and FASTENER algorithms during the first 800 iterations are shown in Figure 9. A quick visual comparison of the *AUF* graphs between POSS and the FASTENER results shows that the surface with the FASTENER algorithm is on average of 5% larger. The FASTENER *AUF* also exhibits much larger jumps. Larger jumps indicate a strong improvement in the $F_1$ score. This can either be a large improvement in the score for an existing $k$ number of features or a slightly smaller improvement in the score for many smaller feature sets that dominate a large part of an existing Pareto front. Periodic larger jumps correspond to the invocation of PURGEPARETOFRONT routine, which tries to improve the results with additional information gain.

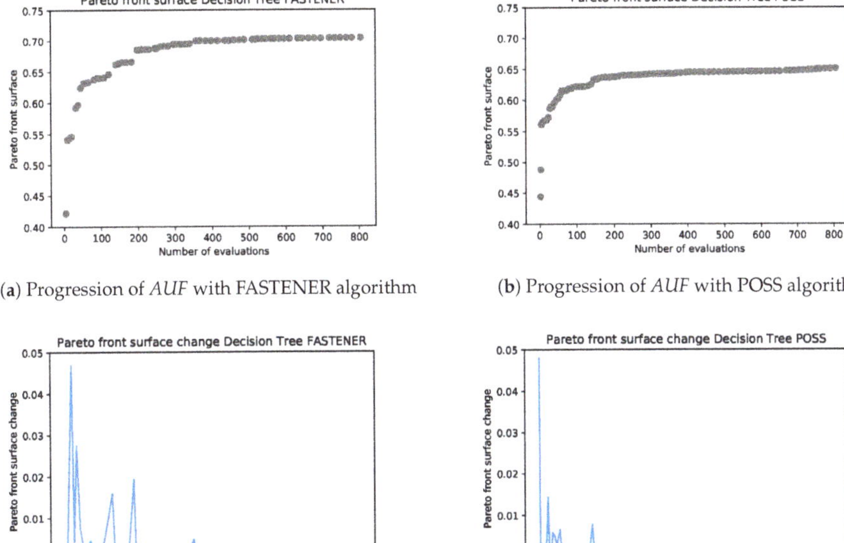

(**a**) Progression of *AUF* with FASTENER algorithm

(**b**) Progression of *AUF* with POSS algorithm

(**c**) *AUF* change with FASTENER algorithm

(**d**) *AUF* change with POSS algorithm

**Figure 9.** The comparison of the Pareto fronts generated by FASTENER and POSS algorithms after a different number of iterations. FASTENER exhibits better *AUF* scores and the discovery of several jumps, indicating the discovery of a distinct new best combination of features.

Another useful tool in convergence analysis is to compare the change in *AUF* presented in Figure 9c,d. The changes in the FASTENER algorithm are an indication of a higher convergence ratio. The computational complexity of *AUF* is minimal (linear in the size of the front) and can be easily computed while purging the Pareto front after each generation. The continuous calculation of the *AUF* (and its change) can be used as a helpful marker to measure the course of convergence in several of the above methods. In particular, the change in the area under a front can be used to modify the probabilities in the crossover and mutate methods to control the feature space search.

Detailed analysis of wrapper methods tested on EO-data is presented in Table 2. All the tested methods produce the area under a front with low variation, which can be attributed to a data set size. It can clearly be seen that FASTENER outperforms both POSS and FS-SDS.

**Table 2.** Results of wrapper feature selection algorithms on EO data set. FASTENER yields the highest average *AUF* and lowest number of model evaluations. Standard deviation of the *AUF* is small for all methods. Best values accross different methods are bolded (in all tables).

| Method | avg. *AUF* | sd | eval_n |
|---|---|---|---|
| POSS | 0.66 | 0.0008 | **1000** |
| FS-SDS | 0.62 | 0.0008 | 120,000 |
| FASTENER | **0.71** | 0.0008 | **1000** |

*5.4. Benchmark Data Set with Wrapper Methods*

Along with FASTENER, forward decision tree selection method (DT-forward) [14] and FS-SDS algorithm [12] were tested for comparison on an open feature selection data sets [6] used for benchmarking.

Tables 3 and 4 present comparisons of the FS-SDS algorithm, FASTENER and forward selection with decision tree (DT-forward). For each of the algorithms, 10 different data splits were used for feature selection and each feature subset was tested on 30 different test/train data splits. FASTENER was run with an exploration setting (larger generations) for 200 generations and the number of evaluations is a lot higher than in the EO setting. Each resulting Pareto front was purged, since adding more features might hurt performance on new data and the area under the Pareto front was calculated. Tables 3 and 4 present mean, max, median and standard deviation of thus obtained surfaces under Pareto fronts, along with the number of model evaluations.

**Table 3.** Mean, max, median value, standard deviation *AUF* and number of model evaluations. FS-SDS uses 30,000 model evaluations per each $k$ (whereas for FASTENER the number of iterations produces the whole Pareto front) in the first iteration. Number of evaluations is reduced with the next iterations and stays in the range between (10,000–30,000).

| Data Set | FS-SDS | | | | | FASTENER | | | | |
|---|---|---|---|---|---|---|---|---|---|---|
| | Mean | Max | Medi | sd | eval_n* | Mean | Max | Medi | sd | eval_n |
| ALLAML | 0.643 | 0.717 | 0.65 | 0.039 | 30,000 | **0.776** | **0.894** | **0.78** | 0.062 | 11,238 |
| arcene | 0.608 | 0.673 | 0.607 | 0.037 | 30,000 | **0.707** | **0.785** | **0.713** | 0.034 | 12,083 |
| BASEHOCK | 0.58 | 0.618 | 0.58 | 0.018 | 30,000 | **0.742** | **0.78** | **0.744** | 0.019 | 13,847 |
| CLL_SUB_111 | 0.537 | 0.671 | 0.534 | 0.053 | 30,000 | **0.626** | **0.79** | **0.623** | 0.067 | 11,911 |
| COIL20 | 0.655 | 0.673 | 0.657 | 0.01 | 30,000 | **0.687** | **0.714** | **0.687** | 0.012 | 14,214 |
| colon | 0.658 | 0.77 | 0.652 | 0.069 | 30,000 | **0.763** | **0.9** | **0.757** | 0.076 | 11,256 |
| gisette | 0.672 | 0.682 | 0.671 | 0.005 | 30,000 | **0.786** | **0.802** | **0.786** | 0.006 | 13,213 |
| GLIOMA | 0.549 | 0.687 | 0.561 | 0.067 | 30,000 | **0.618** | **0.827** | **0.617** | 0.095 | 11,674 |
| GLI_85 | 0.659 | 0.746 | 0.665 | 0.063 | 30,000 | **0.755** | **0.894** | **0.77** | 0.065 | 11,238 |
| Isolet | 0.365 | 0.392 | 0.364 | 0.012 | 30,000 | **0.396** | **0.432** | **0.395** | 0.012 | 14,572 |
| leukemia | 0.62 | 0.77 | 0.618 | 0.06 | 30,000 | **0.824** | **0.9** | **0.833** | 0.057 | 11,248 |
| lung | 0.663 | 0.75 | 0.656 | 0.031 | 30,000 | **0.727** | **0.828** | **0.731** | 0.038 | 11,802 |
| lung_discrete | **0.533** | 0.666 | **0.532** | 0.068 | 30,000 | 0.511 | **0.716** | 0.516 | 0.086 | 11,753 |
| lymphoma | 0.539 | 0.619 | 0.548 | 0.058 | 30,000 | **0.59** | **0.757** | **0.58** | 0.07 | 11,719 |
| nci9 | 0.331 | 0.513 | 0.311 | 0.097 | 30,000 | **0.405** | **0.61** | **0.406** | 0.088 | 12,081 |
| ORL | 0.347 | 0.395 | 0.351 | 0.029 | 30,000 | **0.39** | **0.488** | **0.389** | 0.04 | 13,540 |
| orlraws10P | 0.675 | 0.77 | 0.686 | 0.053 | 30,000 | **0.694** | **0.821** | **0.695** | 0.062 | 13,571 |
| PCMAC | 0.596 | 0.624 | 0.594 | 0.012 | 30,000 | **0.71** | **0.736** | **0.708** | 0.013 | 15,768 |
| Prostate_GE | 0.676 | 0.734 | 0.691 | 0.041 | 30,000 | **0.748** | **0.837** | **0.749** | 0.05 | 14,002 |
| RELATHE | 0.52 | 0.554 | 0.519 | 0.017 | 30,000 | **0.667** | **0.707** | **0.667** | 0.018 | 15,541 |
| TOX_171 | 0.421 | 0.509 | 0.427 | 0.046 | 30,000 | **0.553** | **0.647** | **0.55** | 0.039 | 14,759 |
| USPS | 0.561 | 0.569 | 0.561 | 0.005 | 30,000 | **0.583** | **0.593** | **0.583** | 0.005 | 16,670 |
| warpAR10P | 0.502 | 0.588 | **0.525** | 0.052 | 30,000 | **0.523** | **0.672** | 0.518 | 0.065 | 13,840 |
| warpPIE10P | 0.582 | 0.646 | 0.589 | 0.037 | 30,000 | **0.633** | **0.741** | **0.634** | 0.041 | 13,630 |
| Yale | 0.38 | 0.454 | 0.376 | 0.035 | 30,000 | **0.436** | **0.594** | **0.438** | 0.052 | 15,307 |

Table 3 shows that FASTENER outperforms FS-SDS on almost all data sets, except on the lung_discrete data set, where the mean and median *AUF* score are better with FS-SDS. Apart from outperforming score-wise, FASTENER also produces results faster (with fewer model evaluations) due to additional statistical data, while producing only marginal overhead due to initial entropy calculation.

**Table 4.** Mean, max, median value, standard deviation of *AUF* and number of model evaluations.

| Data Set | DT Forward | | | | | FASTENER | | | | |
|---|---|---|---|---|---|---|---|---|---|---|
| | Mean | Max | Medi | sd | eval_n | Mean | Max | Medi | sd | eval_n |
| ALLAML | **0.826** | 0.893 | **0.833** | 0.044 | 106,830 | 0.776 | **0.894** | 0.78 | 0.062 | 11,238 |
| arcene | 0.659 | 0.76 | 0.665 | 0.054 | 149,895 | **0.707** | **0.785** | **0.713** | 0.034 | 12,083 |
| BASEHOCK | 0.753 | **0.781** | **0.751** | 0.015 | 72,825 | 0.742 | 0.78 | 0.744 | 0.019 | 13,847 |
| CLL_SUB_111 | 0.54 | 0.655 | 0.54 | 0.049 | 169,995 | **0.626** | **0.79** | **0.623** | 0.067 | 11,911 |
| COIL20 | **0.707** | **0.729** | **0.706** | 0.011 | 15,255 | 0.687 | 0.714 | 0.687 | 0.012 | 14,214 |
| colon | 0.751 | 0.888 | 0.745 | 0.066 | 29,895 | **0.763** | **0.9** | **0.757** | 0.076 | 11,256 |
| gisette | **0.8** | **0.815** | **0.799** | 0.007 | 74,895 | 0.786 | 0.802 | 0.786 | 0.006 | 13,213 |
| GLIOMA | 0.592 | 0.743 | **0.626** | 0.103 | 66,405 | **0.618** | **0.827** | 0.617 | 0.095 | 11,674 |
| GLI_85 | 0.724 | 0.82 | 0.733 | 0.066 | 334,140 | **0.755** | **0.894** | **0.77** | 0.065 | 11,238 |
| Isolet | **0.409** | **0.445** | **0.408** | 0.014 | 9150 | 0.396 | 0.432 | 0.395 | 0.012 | 14,572 |
| leukemia | **0.866** | **0.9** | **0.873** | 0.03 | 105,945 | 0.824 | 0.9 | 0.833 | 0.057 | 11,248 |
| lung | **0.76** | 0.827 | **0.761** | 0.038 | 49,575 | 0.727 | **0.828** | 0.731 | 0.038 | 11,802 |
| lung_discrete | **0.526** | 0.689 | **0.532** | 0.08 | 4770 | 0.511 | **0.716** | 0.516 | 0.086 | 11,753 |
| lymphoma | 0.586 | **0.785** | **0.581** | 0.077 | 60,285 | **0.59** | 0.757 | 0.58 | 0.07 | 11,719 |
| nci9 | **0.419** | **0.64** | **0.414** | 0.115 | 145,575 | 0.405 | 0.61 | 0.406 | 0.088 | 12,081 |
| ORL | **0.407** | 0.464 | **0.41** | 0.032 | 15,255 | 0.39 | **0.488** | 0.389 | 0.04 | 13,540 |
| orlraws10P | **0.76** | **0.835** | **0.765** | 0.055 | 154,455 | 0.694 | 0.821 | 0.695 | 0.062 | 13,571 |
| PCMAC | **0.722** | **0.75** | **0.721** | 0.013 | 49,230 | 0.71 | 0.736 | 0.708 | 0.013 | 15,768 |
| Prostate_GE | **0.787** | **0.851** | **0.793** | 0.041 | 89,385 | 0.748 | 0.837 | 0.749 | 0.05 | 14,002 |
| RELATHE | **0.684** | **0.724** | **0.681** | 0.016 | 64,725 | 0.667 | 0.707 | 0.667 | 0.018 | 15,541 |
| TOX_171 | 0.469 | 0.536 | 0.466 | 0.035 | 86,115 | **0.553** | **0.647** | **0.55** | 0.039 | 14,759 |
| USPS | 0.564 | 0.574 | 0.565 | 0.005 | 3735 | **0.583** | **0.593** | **0.583** | 0.005 | 16,670 |
| warpAR10P | **0.533** | 0.659 | **0.527** | 0.06 | 35,895 | 0.523 | **0.672** | 0.518 | 0.065 | 13,840 |
| warpPIE10P | **0.669** | **0.753** | **0.676** | 0.033 | 36,195 | 0.633 | 0.741 | 0.634 | 0.041 | 13,630 |
| Yale | 0.407 | 0.505 | 0.406 | 0.048 | 15,255 | **0.436** | **0.594** | **0.438** | 0.052 | 15,307 |

Table 4 presents a comparison of features selected by FASTENER with features selected by forward selection with decision tree (DT-forward). DT-forward works by successively adding the best performing features, which can be suitable for small data sets, but the computational complexity explodes, as the number of features increases. As can be seen from Table 4 that FASTENER outperforms the DT-forward in a slightly less than half of the data sets. The difference between the maximum in DT-forward selection and FASTENER data in most of other cases is a few percentage points. The important improvement brought in by FASTENER is the efficiency of obtaining better (or just marginally worse) *AUF* with a lot fewer evaluations in comparisons to the forward selection. On average, the FASTENER uses between 2–5-times fewer model evaluations, depending on the number of features in the data set, while obtaining comparable or sometimes even better results. Additional insight presented by the experiment is the standard deviation when using different data splits. Standard deviation in smaller data sets is larger than in EO-data set, and a bit larger than the standard deviation using DT-forward since FASTENER includes additional random part during the algorithm run, while forward feature selection is deterministic.

Another FASTENER strength compared to the FS-SDS or DT-forward is the non-parametricity concerning the number of features. Both FS-SDS and DT-forward are rigid when selecting the feature subset size and increasing (in the case of forward selection) or changing (FS-SDS) subset size is computationally expensive. FASTENER automatically explores available search space unconstrained by the number of features, but they can be additionally constrained when reporting the results.

## 6. Discussion

### 6.1. Comparisons with other Methods

FASTENER builds on the idea of having multiple non-optimal items simultaneously considered for selection. The front in the main loop of the algorithm is a list of items for a fixed number of selected features. The additional parameter controls the size of the buckets, but initial testing concluded that increasing the bucket size to more than 2 decreased the performance and a more elitist parent selection (bucket sizes 1 and 2) was used. The exploration phase, where the population is allowed to deviate from the currently optimal front and only the items in the currently selected population are used as possible parents for a few rounds can be seen as a generalization of multiple fronts used in NSGA-II. The front purging after a selected number of rounds (controlled by a parameter) then returns the front to the current best and maintains an elitist gene selection.

Due to a large number of features and data points in the EO domain, smaller population sizes are more suitable for feature selection. In this case, the increase in bucket sizes (and non-optimal fronts) greatly increases the population size, which produces slower convergence as the elitism part does not kick in quickly enough.

Another specific thing about feature selection is the fact that the feature subset size dimension is discrete and mostly fully dense. Since the optimal sizes of subsets quickly converge even for a small number of selected features, the clustering optimization performed by NSGAA-II and RPSGAe does not improve the efficiency of convergence.

*6.2. Generalization*

An important aspect of feature subset optimization is further generalization to unseen data. As mentioned at the beginning of this section, an additional part of the data was kept unseen to the FASTENER algorithm during the phase of selecting the optimal subset of features. This hidden data set was used to analyze the overfitting of the FASTENER algorithm to the combination of training/test data. For each generation of the FASTENER algorithm, the features from the optimal items were used as features to train the model using the training data. However, this time the models were evaluated using the 10% of data not seen during FASTENER iterations. The overfitting was minimal and ranged from 0.5 to approximately 0.3 percentage points compared to the results on the test data set reported by the FASTENER algorithm. An important piece of information obtained from additional testing on the unseen data was the fact that some items from the reported optimal Pareto fronts were strictly dominated by feature subsets with fewer features. This was not surprising for subsets with a larger number of features, as the Pareto front "levels off" after a larger number of features. Results of the generalization analysis are presented in Figure 10.

The generalization on unseen data is quite good, which is at least partly due to a large data set compared to the number of algorithm iterations. The algorithm did not converge to some local optimum imposed by training data. Another way to analyze training performance on unseen data is to check the difference between *AUF* values on the reported front and unseen data. The plot of the *AUF* difference is shown in Figure 11. The difference levels off after approximately 60 iterations of the algorithm and appears to be constant with some variation.

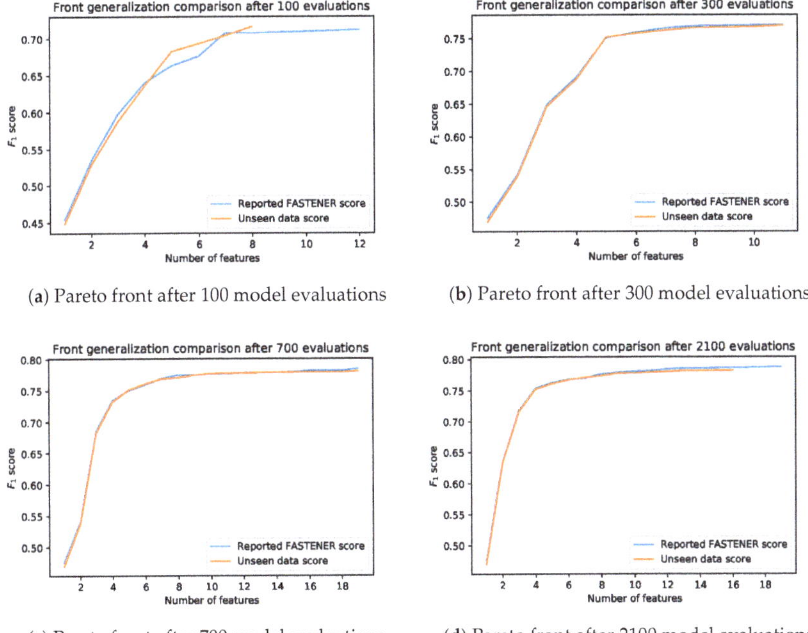

(**a**) Pareto front after 100 model evaluations

(**b**) Pareto front after 300 model evaluations

(**c**) Pareto front after 700 model evaluations

(**d**) Pareto front after 2100 model evaluations

**Figure 10.** The generalization of results on unseen data presents small performance discrepancies between test data and previously unseen data.

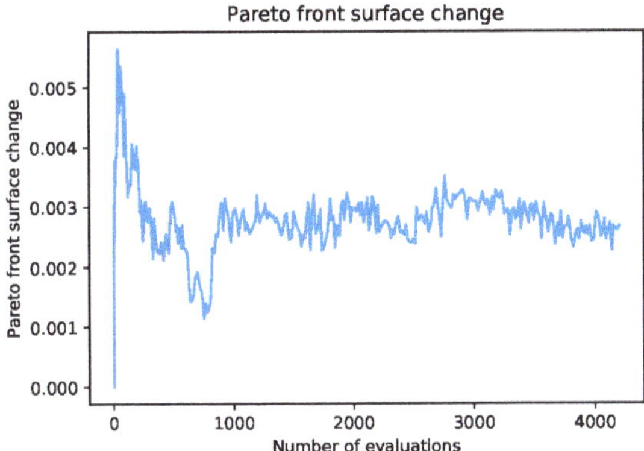

**Figure 11.** Low *AUF* difference between the test and unseen data shows good generalization abilities of the FASTENER algorithm.

Apart from the generalization, it is interesting to observe the statistics for selected items and the statistics of individual genes during the execution of the algorithm. Figure 12 visualizes optimal selected features after the termination of the FASTENER algorithm. The yellow squares represent selected optimal features as genes. With $y$ axis progression, new features are added. Interestingly, the algorithm does not simply extend the feature sets of $k$ for the higher $k$'s, but rather examines previously excluded feature combinations that, with additional features, perform better.

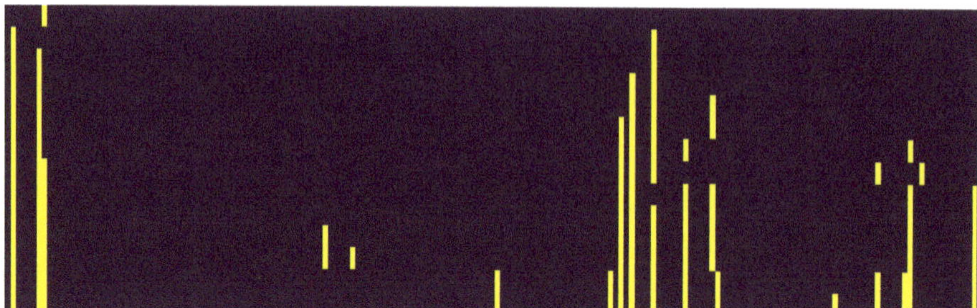

**Figure 12.** Visualization of optimal features for different feature subset sizes. The $x$ axis represents the feature index, while the $y$ axis depicts the number of features $k$ (increasing with each row). FASTENER does not simply add new features as $k$ is increased, but rather finds the best possible combination of features that gives the best possible classification result for a given $k$.

Apart from selecting the optimal subsets, one should also look at the density of the different features, as they were involved during the iterations. The diagrams in Figure 13 show the number of evaluations of features during the run time of the algorithm. If the feature subset [1, 5, 7, 100] was evaluated, each of the features shown is considered to be evaluated once. Figure 13a shows the number of feature evaluations indexed by feature indices. The most evaluated feature is 0, since this was the only item in the starting Pareto front for the algorithm. Looking at other features, there are some areas of higher activity, but interestingly, while features between indices 50 and 75 seem to be fairly highly valued, they are very rarely included in the final optimal Pareto fronts. It seems that they are favored by heuristics. On the contrary, towards the end, features are fairly highly valued and are often present in optimal subsets.

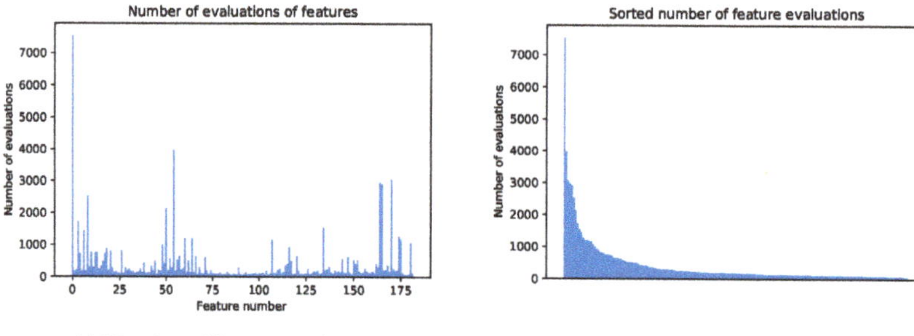

(a) Number of feature evaluations       (b) Sorted number of feature evaluations

**Figure 13.** Number of feature evaluations. Although some features are not represented in Figure 12, they have often been evaluated by the algorithm.

## 7. Conclusions and Future Work

With the FASTENER algorithm, we have combined a wrapping methodology with information-theoretical measures based on entropy. The resulting feature selection method has proven to be very promising (superior in EO scenario), with its fast convergence (less learning iterations needed) and better accuracy (higher $F_1$ score and surface under the Pareto front) than the currently tested methods in the field. In the EO literature, feature selection methodologies have not been thoroughly explored and thus very basic algorithms, such as ReliefF and POSS have been reported to give the best results so far. We have also repeated experiments using FS-SDS, which FASTENER consistently outperformed. FASTENER was compared to FS-SDS and DT-forward algorithms on 25 open feature selection data sets, which are significantly different from the EO data set (a few instances and a lot of features). In terms of accuracy, FASTENER is comparable with DT-forward but generally achieves the same result with far fewer evaluations. Although the method was originally developed for applications within the EO scenarios, its usage in other domains seems promising. Any supervised machine learning problem (classification or regression) that requires optimization of the accuracy measure (either $F_1$ score or *RMSE*) with respect to the number of used features would benefit from the implementation of FASTENER.

Several aspects will be addressed in future work. Better theoretical justification of the algorithm is needed as well as the analysis of its convergence and other relevant properties. Possibilities of parallelization of FASTENER (similar as in [34] for POSS) should be examined to speed up the convergence.

Finally, for EO or similar problems, where the calculation of features themselves is computationally challenging, optimization of the final time of the learning process (including feature engineering and data acquisition) need to be performed.

**Author Contributions:** Conceptualization and methodology, F.K. and K.K.; software, F.K., B.Š.; validation, F.K. and K.K.; writing and editing, K.K., F.K. and B.Š.; visualization, B.Š., F.K.; supervision, project administration, and funding acquisition, K.K. All authors have read and agreed to the published version of the manuscript.

**Funding:** This research was funded by European Union's Horizon 2020 programme project EnviroLENS (Innovation Action) grant number 821918 and project PerceptiveSentinel (Research and Innovation) grant number 776115.

**Acknowledgments:** The authors would like to thank the H2020 Perceptive Sentinel project team members from JSI and Sinergise, who have created an effective and easy to use earth observation Python library `eo-learn` which is based on SentinelHub data access.

**Conflicts of Interest:** The authors declare no conflict of interest. The funders had no role in the design of the study; in the collection, analyses, or interpretation of data; in the writing of the manuscript, or in the decision to publish the results.

## Abbreviations

The following abbreviations are used in this manuscript:

| | |
|---|---|
| ARVI | Atmoshperically adjusted vegetation index |
| ASTFS | Automatic spectro-temporal feature selection |
| AUF | Area Under (Pareto) Front |
| DT | Decision Tree |
| EO | Earth Observation |
| ESA | European Space Agency |
| EVI | Enhanced Vegetation Index |
| FASTENER | Feature Selection Enabled by Entropy |
| LDA | Latent Dirichlet Allocation |
| LPIS | Land Parcel Identification System |
| NIR | Near Infra Red |
| NDVI | Normalized differential vegetation index |
| NDWI | Normalized differential water index |
| NLPCA | Non-linear Principal Components Analysis |
| PB | Petabyte |
| PCA | Principal Components Analysis |
| POSS | Pareto Optimization for Subset Selection |
| PPOSS | Parallel Pareto Optimization for Subset Selection |
| SAVI | Soil-adjusted Vegetation Index |
| FS-SDS | Feature Selection using Stohastic Diffucion Search |
| SIPI | Structure Insensitive Pigment Index |

## References


1. European Space Agency. Mission Status Report 158. Available online: https://sentinel.esa.int/documents/247904/4114743/Sentinel-2-Mission-Status-Report-158-25-Jan-3-Apr-2020.pdf (accessed on 15 May 2020).
2. Kenda, K.; Kažič, B.; Novak, E.; Mladenić, D. Streaming Data Fusion for the Internet of Things. *Sensors* **2019**, *19*, 1955. [PubMed]
3. Koprivec, F.; Čerin, M.; Kenda, K. Crop Classification using Perceptive Sentinel. In Proceedings of the 21th International Multiconference, Ljubljana, Slovenia, 24–28 September 2018; Volume C, pp. 37–40.
4. Koprivec, F.; Peternelj, J.; Kenda, K. Feature Selection in Land-Cover Classification Using EO-learn. In Proceedings of the 22nd International Multiconference, Ljubljana, Slovenia, 7–11 October 2019; Volume C, pp. 37–50.
5. Deb, K. *Multi-Objective Optimization Using Evolutionary Algorithms*; John Wiley & Sons: Hoboken, NJ, USA, 2001; Volume 16.
6. Li, J.; Cheng, K.; Wang, S.; Morstatter, F.; Trevino, R.P.; Tang, J.; Liu, H. Feature selection: A data perspective. *ACM Comput. Surv. (CSUR)* **2018**, *50*, 94. [CrossRef]
7. Khan, A.; Baig, A.R. Multi-objective feature subset selection using non-dominated sorting genetic algorithm. *J. Appl. Res. Technol.* **2015**, *13*, 145–159. [CrossRef]
8. Gaspar-Cunha, A.; Covas, J.A. RPSGAe—reduced Pareto set genetic algorithm: Application to polymer extrusion. In *Metaheuristics for Multiobjective Optimisation*; Springer: Berlin/Heidelberg, Germany, 2004; pp. 221–249.
9. Gaspar-Cunha, A. Feature selection using multi-objective evolutionary algorithms: Application to cardiac SPECT diagnosis. In *Advances in Bioinformatics*; Springer: Berlin/Heidelberg, Germany, 2010; pp. 85–92.
10. Spolaôr, N.; Lorena, A.C.; Diana Lee, H. Feature selection via pareto multi-objective genetic algorithms. *Appl. Artif. Intell.* **2017**, *31*, 764–791. [CrossRef]
11. Qian, C.; Yu, Y.; Zhou, Z.H. Subset Selection by Pareto Optimization. In *Advances in Neural Information Processing Systems 28*; Cortes, C., Lawrence, N.D., Lee, D.D., Sugiyama, M., Garnett, R., Eds.; Curran Associates, Inc.: Red Hook, NY, USA, 2015; pp. 1774–1782.
12. Alhakbani, H.; al Rifaie, M.M. Feature Selection Using Stochastic Diffusion Search. In Proceedings of the Genetic and Evolutionary Computation Conference, Berlin, Germany, 15–19 July 2017; pp. 385–392. [CrossRef]
13. Huang, G.B.; Zhu, Q.Y.; Siew, C.K. Extreme learning machine: Theory and applications. *Neurocomputing* **2006**, *70*, 489–501. [CrossRef]
14. Guyon, I.; Elisseeff, A. An Introduction to Variable and Feature Selection. *J. Mach. Learn. Res.* **2003**, *3*, 1157–1182.
15. Yin, L.; You, N.; Zhang, G.; Huang, J.; Dong, J. Optimizing Feature Selection of Individual Crop Types for Improved Crop Mapping. *Remote Sens.* **2020**, *12*, 162. [CrossRef]
16. Somers, B.; Asner, G.P. Multi-temporal hyperspectral mixture analysis and feature selection for invasive species mapping in rainforests. *Remote Sens. Environ.* **2013**, *136*, 14–27. [CrossRef]
17. Stromann, O.; Nascetti, A.; Yousif, O.; Ban, Y. Dimensionality Reduction and Feature Selection for Object-Based Land Cover Classification based on Sentinel-1 and Sentinel-2 Time Series Using Google Earth Engine. *Remote Sens.* **2019**, *12*, 76. [CrossRef]
18. Kraskov, A.; Stögbauer, H.; Grassberger, P. Estimating mutual information. *Phys. Rev. E* **2004**, *69*, 066138. [CrossRef]
19. Ross, B.C. Mutual Information between Discrete and Continuous Data Sets. *PLoS ONE* **2014**, *9*, e0087357. [CrossRef][PubMed]
20. Valero, S.; Morin, D.; Inglada, J.; Sepulcre, G.; Arias, M.; Hagolle, O.; Dedieu, G.; Bontemps, S.; Defourny, P.; Koetz, B. Production of a Dynamic Cropland Mask by Processing Remote Sensing Image Series at High Temporal and Spatial Resolutions. *Remote Sens.* **2016**, *8*, 55. [CrossRef]
21. Waldner, F.; Canto, G.S.; Defourny, P. Automated annual cropland mapping using knowledge-based temporal features. *ISPRS J. Photogramm. Remote Sens.* **2015**, *110*, 1–13. [CrossRef]
22. Kiala, Z.; Mutanga, O.; Odindi, J.; Peerbhay, K. Feature Selection on Sentinel-2 Multispectral Imagery for Mapping a Landscape Infested by Parthenium Weed. *Remote Sens.* **2019**, *11*, 1892. [CrossRef]

23.  Robnik-Šikonja, M.; Kononenko, I. Theoretical and empirical analysis of ReliefF and RReliefF. *Mach. Learn.* **2003**, *53*, 23–69. [CrossRef]
24.  Walton, N.S.; Sheppard, J.W.; Shaw, J.A. Using a Genetic Algorithm with Histogram-Based Feature Selection in Hyperspectral Image Classification. In Proceedings of the Genetic and Evolutionary Computation Conference, Prague, Czech Republic, 13–17 July 2019; pp. 1364–1372. [CrossRef]
25.  Gómez, C.; White, J.C.; Wulder, M.A. Optical remotely sensed time series data for land cover classification: A review. *ISPRS J. Photogramm. Remote Sens.* **2016**, *116*, 55–72. [CrossRef]
26.  Zhu, X.X.; Tuia, D.; Mou, L.; Xia, G.S.; Zhang, L.; Xu, F.; Fraundorfer, F. Deep learning in remote sensing: A comprehensive review and list of resources. *IEEE Trans. Geosci. Remote Sens.* **2017**, *5*, 8–36. [CrossRef]
27.  Qian, C.; Yu, Y.; Zhou, Z.H. Pareto Ensemble Pruning. In Proceedings of the Twenty-Ninth AAAI Conference on Artificial Intelligence, Austin, TX, USA, 25–30 January 2015.
28.  Qian, C.; Shi, J.C.; Yu, Y.; Tang, K.; Zhou, Z.H. Parallel Pareto Optimization for Subset Selection. In Proceedings of the Twenty-Fifth International Joint Conference on Artificial Intelligence, New York, NY, USA, 9–15 July 2016; pp. 1939–1945.
29.  Pedregosa, F.; Varoquaux, G.; Gramfort, A.; Michel, V.; Thirion, B.; Grisel, O.; Blondel, M.; Prettenhofer, P.; Weiss, R.; Dubourg, V.; et al. Scikit-learn: Machine Learning in Python. *J. Mach. Learn. Res.* **2011**, *12*, 2825–2830.
30.  Drusch, M.; Del Bello, U.; Carlier, S.; Colin, O.; Fernandez, V.; Gascon, F.; Hoersch, B.; Isola, C.; Laberinti, P.; Martimort, P.; et al. Sentinel-2: ESA's optical high-resolution mission for GMES operational services. *Remote Sens. Environ.* **2012**, *120*, 25–36. [CrossRef]
31.  Canny, J. A computational approach to edge detection. *IEEE Trans. Pattern Anal. Mach. Intell.* **1986**, *6*, 679–698. [CrossRef]
32.  Šircelj, B.; Kenda, K.; Koprivec, F. Land Patch Samples. *PANGAEA* **2020**. [CrossRef]
33.  Teschendorff, A.E. Avoiding common pitfalls in machine learning omic data science. *Nat. Mater.* **2019**, *18*, 422–427. [CrossRef]
34.  Zhou, Z.H.; Yu, Y.; Qian, C. Subset Selection: Acceleration. In *Evolutionary Learning: Advances in Theories and Algorithms*; Springer: Singapore, 2019; pp. 285–293. [CrossRef]

*Article*

# Approximate Learning of High Dimensional Bayesian Network Structures via Pruning of Candidate Parent Sets

Zhigao Guo [1,*] and Anthony C. Constantinou [1,2,*]

1   Bayesian Artificial Intelligence Research Lab, School of Electronic Engineering and Computer Science, Queen Mary University of London, London E1 4NS, UK
2   The Alan Turing Institute, British Library, 96 Euston Road, London NW1 2DB, UK
*   Correspondence: zhigao.guo@qmul.ac.uk (Z.G.); a.constantinou@qmul.ac.uk (A.C.C.)

Received: 10 September 2020; Accepted: 7 October 2020; Published: 10 October 2020

**Abstract:** Score-based algorithms that learn Bayesian Network (BN) structures provide solutions ranging from different levels of approximate learning to exact learning. Approximate solutions exist because exact learning is generally not applicable to networks of moderate or higher complexity. In general, approximate solutions tend to sacrifice accuracy for speed, where the aim is to minimise the loss in accuracy and maximise the gain in speed. While some approximate algorithms are optimised to handle thousands of variables, these algorithms may still be unable to learn such high dimensional structures. Some of the most efficient score-based algorithms cast the structure learning problem as a combinatorial optimisation of candidate parent sets. This paper explores a strategy towards pruning the size of candidate parent sets, and which could form part of existing score-based algorithms as an additional pruning phase aimed at high dimensionality problems. The results illustrate how different levels of pruning affect the learning speed relative to the loss in accuracy in terms of model fitting, and show that aggressive pruning may be required to produce approximate solutions for high complexity problems.

**Keywords:** structure learning; probabilistic graphical models; pruning

---

## 1. Introduction

A Bayesian Network (BN) [1] is a probabilistic graphical model represented by a Directed Acyclic Graph (DAG). The structure of a BN captures the relationships between nodes, whereas the conditional parameters capture the type and magnitude of those relationships. A BN differs from other graphical models, such as neural networks, in that it offers a transparent representation of a problem where the relationships between variables (i.e., arcs) represent conditional or causal relationships. Moreover, the uncertain conditional distributions in BNs can be used for both predictive and diagnostic (i.e., inverse) inference, providing the potential for a higher level of artificial intelligence. For example, knowledge-based BNs are often assumed to represent causal or influential networks and enable decision makers to reason about intervention [2]. On the other hand, structure of BNs, and especially those based on search-and-score solutions on which this paper focuses, are generally assumed to represent networks with conditional—rather than causal—relationships, although the class of constraint-based learning (which we cover below) is often used to discover relationships that could, under various assumptions, be interpreted causally [3].

Formally, a BN model represents a factorisation of the joint distribution of random variables $X = (X_1, X_2, \cdots, X_n)$. Each BN has two elements, structure $G$ and parameters $\theta$. Constructing a BN involves both structure learning and parameter learning, and both learning approaches may involve combination of data with knowledge [4–7]. Given observational data $D$, a complete BN $\{G, \theta\}$ can be learnt by maximising the likelihood:

$$P(G, \theta|D) = P(G|D)P(\theta|G, D). \tag{1}$$

and the parameters of the network can be learnt by maximising

$$P(\theta|G, D) = \prod_{i=1}^{n} P(\theta_i|\Pi_i, D) \tag{2}$$

where $\Pi_i$ denotes the parents of node $X_i$ indicated by structure $G$. Depending on the prior assumption, the parameter learning process of each node can be solved using Maximum Likelihood estimation given data $D$, or Maximum A Posteriori estimation given data $D$ and a subjective prior.

On the other hand, the BN structure learning (BNSL) problem represents a more challenging task in that it cannot be solved by simply maximising the fitting of the local networks to the data. The structure learning process must take into consideration the complexity of the model in order to avoid overfitting. In fact, the problem of BNSL is NP-hard, which means that it is generally not possible to perform exhaustive search in the search space of possible graphs, and this is because the number of possible structures grows super-exponentially with the number of nodes $n$ [8]:

$$f(n) = \sum_{i=1}^{n} (-1)^{i+1} \frac{n!}{(n-i)!n!} 2^{i(n-i)} f(n-1) \tag{3}$$

Algorithms that learn BN structures are typically classified into three categories: (a) score-based learning that searches over the space of possible graphs and returns the graph that maximises a scoring function, (b) constraint-based learning that prunes and orientates edges using conditional independence tests, and (c) hybrid learning that combines the above two strategies. In this paper, we focus on the problem of score-based learning.

A score-based algorithm is generally composed of two parts: (a) a search strategy that transverses the search space, and (b) a scoring function that evaluates a given graph with respect to the observed data. Well-known search strategies in the field of BNSL include the greedy hill-climbing search, tabu search, simulated annealing, genetic algorithms, dynamic programming, A* algorithm, and branch-and-bound strategies. The objective functions are typically based on either the Bayesian score or other model selection scores. Bayesian scores evaluate the posterior probability of the candidate structures and include variations of the Bayesian Dirichlet score, such as the Bayesian Dirichlet with equivalence and uniform priors (BDeu) [9,10], and the Bayesian Dirichlet sparse (BDs) [11]. Well-established scores for model selection include the Akaike Information Criterion (AIC) [12] and the Bayesian Informatic Criterion (BIC), often also referred to as the Minimum Description Length (MDL) [13]. Other less popular model selection scores include the Mutual Information Test (MIT) [14], the factorized Normalized Maximum Likelihood (fNML) [15], and the qutotient Normalized Maximum Likelihood (qNML) [16].

Score-based approaches further operate in two different ways. The first approach involves scoring a graph only when the graph is visited by the search method, which typically involves exploring neighbouring graphs and following the search path that maximises the fitting score via arc reversals, additions and removals. The second approach involves generating scores for local networks (i.e., a node and its parents) in advance, and searching over combinations of local networks given the pre-generated scores, thereby formulating a combinatorial optimisation problem. The algorithms that fall in the former category are generally based on efficient heuristics such as hill-climbing, but tend to stuck in local optimum solutions, thereby offering an approximate solution to the problem of BNSL. While algorithms of the latter category are also generally approximate, they can be more easily adjusted to offer exact learning solutions that guarantee to return a graph with score not lower than the global maximum score. This paper focuses on this latter subcategory of score-based learning.

Algorithms such as the Integer Linear Programming (ILP) [17,18] explore local networks in the form of the Candidate Parent Sets (CPSs), usually up to a bounded maximum in-degree, and offer an exact solution. Other exact learning algorithms which cast the BNSL problem as a combinatorial optimisation problem include the Dynamic Programming (DP) [19,20], the A* algorithm [21] and the Branch-and-Bound (B&B) [22,23]. However, exact learning is generally restricted to problems of low complexity. Evidently, the efficiency of these algorithms is determined by the number of CPSs. For example, the ILP algorithm is restricted to CPSs of size up to one million. However, order-based algorithms such as OBS [24], ASOBS [25,26] and MINOBS [27] explore in the node ordering space, in which the number of structures consistent with the orderings that consist of $n$ nodes is [28]

$$f(n) = \prod_{i=1}^{n} 2^{n-1} = 2^{n(n-i)/2}. \tag{4}$$

Compared with Equation (3), the number of structures is substantially reduced. For example, for 10 nodes, there are $3.5 \times 10^{13}$ different structures instead of $4.2 \times 10^{18}$ computed by Equation (3). Therefore, order-based algorithms are able to search for approximate solutions in problems that involve hundreds of variables.

In this paper we focus on score-based algorithms that learn BN graphs via local learning of CPSs. Specifically, we investigate the relationship between different levels of pruning on CPSs and the loss in accuracy in terms of the BDeu score. The remainder of this paper is organised as follows: Section 2 provides the problem statement and methodology, Section 3 provides the results, and we provide our concluding remarks and directions for future research in Section 4.

## 2. Problem Statement and Methodology

Different pruning rules have been proposed to improve the scalability and the efficiency of score-based algorithms that operate on CPSs. Pruning approaches in this context generally aim to reduce the number of CPSs [22,29–32]. The efficacy of a pruning strategy can be significant in the reduction of computational complexity, and this depends on the number of the variables, the level of maximum in-degree and the number of observations in the data. For example, Table 1 presents a sample of the CPSs of node "0" in the Audio-train data set (https://github.com/arranger1044/awesome-spn#dataset), which consists of 100 variables and 15,000 observations. If we assume maximum-in-degree of 1, with no pruning, this produces $100 \cdot (C_{99}^{0} + C_{99}^{1}) = 10,000$ CPSs, whereas for maximum-in-degree of 2 increases to $100 \cdot (C_{99}^{0} + C_{99}^{1} + C_{99}^{2}) = 495,100$ CPSs, and for maximum in-degree of 3 increases to $100 \cdot (C_{99}^{0} + C_{99}^{1} + C_{99}^{2} + C_{99}^{3}) = 16,180,000$ CPSs.

**Table 1.** Sample CPSs of node "0", ordered by BDeu score with max in-degree 3. The example is based on Audio-train dataset which incorporates 100 variables.

| Child Node | Local BDeu Score | CPS Size | CPS |
|:---:|:---:|:---:|:---:|
| 0 | −5149.19 | 3 | {9, 85, 95} |
| 0 | −5150.47 | 3 | {9, 94, 95} |
| 0 | −5174.53 | 3 | {85, 94, 95} |
| 0 | −5207.08 | 3 | {80, 85, 95} |
| 0 | −5208.28 | 3 | {9, 80, 95} |
| ... | ... | ... | ... |
| 0 | −6886.30 | 2 | {48, 67} |
| 0 | −6886.74 | 1 | {67} |
| 0 | −5174.53 | 1 | {81} |
| 0 | −5174.53 | 1 | {75} |
| 0 | −6889.11 | 0 | {} |

Pruning rules can be used to discard CPSs that are impossible to exist in optimal structures. Based on the observation that a parent set cannot be optimal if its subsets have higher scores [24], the CPSs that are left after applying these pruning rules are called "legal CPSs". Table 2 presents an example where, in a network with four nodes $\{1, 2, 3, 4\}$, part of the CPSs (those in bold) are pruned with the remaining CPSs representing the so-called legal CPSs. Figure 1 also illustrates all the possible CPSs of node 1, given a maximum in-degree of 3. As shown in Table 2, the CPSs, $\{2, 3\}$, $\{2, 4\}$ and $\{2, 3, 4\}$ are pruned and do not form part of the legal CPSs of node 1 because a) the score of CPS $\{2, 3\}$ is lower than its subset $\{3\}$, the score of CPS $\{2, 4\}$ is lower than its subset $\{4\}$, and the score of CPS $\{2, 3, 4\}$ is lower than its subset $\{3, 4\}$. In total, six CPSs are pruned out of the 32 possible CPSs. Figure 2 presents the optimal structure and shows that none of the pruned CPSs (or that only the legal CPSs) are part of the optimal structure.

**Table 2.** An example BN with four nodes and the legal CPSs that remain after pruning the CPSs that are impossible to exist (highlighted in bold), as determined by the BDeu score. The example assumes four nodes, a maximum in-degree (ID) of 3, and a sample size of 5000.

| Node | ID = 0 | ID = 1 | ID = 1 | ID = 1 | ID = 2 | ID = 2 | ID = 2 | ID = 3 |
|------|--------|--------|--------|--------|--------|--------|--------|--------|
| 1 | $\{\}$<br>$-2288.7$ | $\{2\}$<br>$-2274.6$ | $\{3\}$<br>$-2196.2$ | $\{4\}$<br>$-2240.7$ | $\mathbf{\{2,3\}}$<br>$\mathbf{-2252.8}$ | $\mathbf{\{2,4\}}$<br>$\mathbf{-2256.1}$ | $\{3,4\}$<br>$-2171.3$ | $\mathbf{\{2,3,4\}}$<br>$\mathbf{-2173.5}$ |
| 2 | $\{\}$<br>$-2003.7$ | $\{1\}$<br>$-1989.6$ | $\{3\}$<br>$-1900.7$ | $\{4\}$<br>$-1915.1$ | $\mathbf{\{1,3\}}$<br>$\mathbf{-1903.8}$ | $\mathbf{\{1,4\}}$<br>$\mathbf{-1918.3}$ | $\{3,4\}$<br>$-1849.2$ | $\mathbf{\{1,3,4\}}$<br>$\mathbf{-1851.4}$ |
| 3 | $\{\}$<br>$-2891.5$ | $\{1\}$<br>$-2799.0$ | $\{2\}$<br>$-2788.5$ | $\{4\}$<br>$-2811.3$ | $\{1,2\}$<br>$-2714.5$ | $\{1,4\}$<br>$-2741.9$ | $\{2,4\}$<br>$-2745.5$ | $\{1,2,4\}$<br>$-2692.6$ |
| 4 | $\{\}$<br>$-1951.6$ | $\{1\}$<br>$-1903.6$ | $\{2\}$<br>$-1862.9$ | $\{3\}$<br>$-1871.4$ | $\{1,2\}$<br>$-1829.5$ | $\{1,3\}$<br>$-1846.5$ | $\{2,3\}$<br>$-1819.9$ | $\{1,2,3\}$<br>$-1807.6$ |

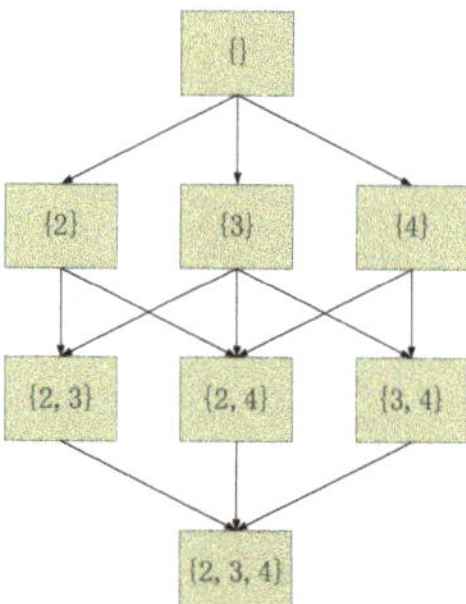


**Figure 1.** All possible CPSs of node "1" (not shown in the diagram) under the assumption the maximum in-degree is 3.

We use the GOBNILP software (https://www.cs.york.ac.uk/aig/sw/gobnilp/) to obtain the legal CPSs. For example, the legal CPSs for the Audio-train data set, under maximum in-degree of 3, is 7,343,077 which represents a 45.4% of the total number of all possible CPSs. Table 3 presents the number and rates of legal CPSs, in relation to the number of all possible CPSs, over different sample sizes of the Audio-train data set and varied maximum in-degrees. The results show that the level of pruning decreases with sample size and increases with maximum in-degree. This is because a higher sample size leads to the detection of more dependencies whereas a higher maximum in-degree produces a higher number of possible dependencies that need to be explored. Still, this level of pruning is insufficient for high dimensionality problems. In general, the combinatorial optimisation problem of CPSs remains unsolvable for data sets that are large in terms of the number of the variables, with higher levels of maximum in-degree and data sample size further increasing complexity.

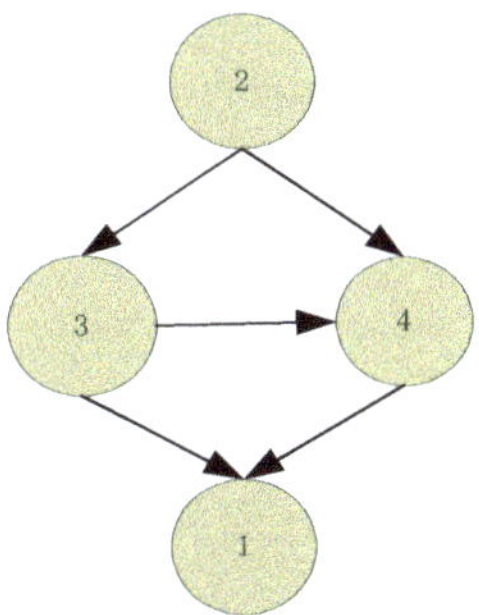

**Figure 2.** The optimal structure learnt from the CPSs presented in Table 2.

**Table 3.** The number and rates of legal CPSs in relation to the all possible CPSs for subsets of Audio-train data over varying samples sizes and maximum in-degrees.

| Maximum In-Degree | Number of All Possible CPSs | Sample Size | | | | |
|---|---|---|---|---|---|---|
| | | 3000 | 6000 | 9000 | 12,000 | 15,000 |
| 1 | 10,000 | 8398 84.0% | 8926 89.3% | 9163 91.6% | 9320 93.2% | 9394 93.9% |
| 2 | 495,100 | 228,197 46.1% | 306,263 61.9% | 349,587 70.6% | 374,007 75.5% | 388,621 78.5% |
| 3 | 16,180,000 | 1,200,429 7.42% | 3,260,399 20.2% | 5,130,502 31.7% | 6,405,394 39.6% | 7,343,077 45.4% |

In this paper, we investigate the effect of different levels of pruning on legal CPSs. The effect is investigated both in terms of the gain in speed and the loss in accuracy, where the loss in accuracy is measured as a discrepancy $\Delta$ between the fitting scores of two learnt graphs defined as

$$\Delta = (S^* - S)/S^* \tag{5}$$

where $S$ denotes the BDeu score of a graph generated frompruning, and $S^*$ denotes the BDeu score of the baseline graph generated without pruning. This approach is based on the B&B algorithm proposed by Cassio de Campos [22], where the construction of the graph iterates over possible CPSs and starts from the most likely CPS per node. Specifically, we explore the CPSs expressed in the format shown in Table 1, where legal CPSs for each node are sorted in a descending order as determined by the local BDeu score. Different levels of pruning are explored by pruning different percentages of legal CPSs for each node, starting from the bottom-ranked CPSs of each node in terms of BDeu score. This means that the search in the space of possible DAGs starts from the most promising parent sets of each node. A valid DAG is ensured by skipping CPSs that lead to cycles. A valid DAG is achieved by ensuring that the learnt DAG incorporates at least one root node, which is a node having no parents [33]. This means that the exploration of CPSs must also include 'empty' CPSs (i.e., no parent), as shown in Table 1.

Three different levels of complexity are investigated, where the pruned result is compared to the unpruned result. The unpruned result represents the exact result for moderate complexity problems, although we cannot guarantee this will be the case for higher complexity problems. We define the three different levels of complexity as follows:

(a) Moderate complexity, which assumes less than 1 million legal CPSs per network;
(b) High complexity, which assumes more than 1 million and less than 10 million legal CPSs per network;
(c) Very high complexity, which assumes more than 10 million legal CPSs per network.

We generate BDeu scores using the GOBNILP software and search for the optimal CPSs using different algorithms. Experiments on networks of moderate and high complexity were carried on an Intel Core i7-8750H CPU at 2.2 GHz with 16 GB of RAM, where each optimisation is assigned a maximum 9.2 GB of memory. Experiments on networks of very high complexity were carried on an Intel Core i7-8700 CPU at 3.2 GHz with 32 GB of RAM, where each optimisation is assigned a maximum 25 GB of memory. All experiments are restricted to 24 h of structure learning runtime.

## 3. Results

The experiments presented in this section are based on BNs and relevant data that are available on the GOBNILP website; link (https://www.cs.york.ac.uk/aig/sw/gobnilp/#benchmarks) for the networks used in Section 3.1, and link (https://www.cs.york.ac.uk/aig/sw/gobnilp/ for the networks used in Sections 3.1 and 3.3.)

### 3.1. Pruning Legal CPSs of Moderate Complexity

We start the investigation by focusing on BNSL problems of moderate complexity, which we restrict to networks with up to 1 million legal CPSs. Table 4 lists the six networks with their corresponding number of legal CPSs for each case study and sample size combination, assuming a maximum in-degree of 3.

**Table 4.** Moderate complexity case studies (nodes|max in-degree in true networks), depicting the total number of legal CPSs per network, as well as the average number of CPSs per node in that network, for network and sample size combination. The number of legal CPSs assume a maximum in-degree of 3.

| | Sample Size | Asia (8\|2) | Insurance (27\|3) | Water (32\|5) | Alarm (37\|4) | Hailfinder (56\|4) | Carpo (61\|5) |
|---|---|---|---|---|---|---|---|
| | 100 | 41 | 279 | 482 | 907 | 244 | 5068 |
| CPSs (graph) | 1000 | 107 | 774 | 573 | 1928 | 761 | 3827 |
| | 10,000 | 161 | 3652 | 961 | 6473 | 3768 | 16,391 |
| | 100 | 5.13 | 10.33 | 15.06 | 24.51 | 4.36 | 84.47 |
| CPSs (per node) | 1000 | 13.38 | 28.67 | 17.91 | 52.11 | 13.59 | 63.78 |
| | 10,000 | 20.12 | 135.26 | 30.03 | 174.95 | 67.29 | 273.18 |

Tables 5 and 6 present the loss in accuracy from different levels of CPSs pruning. The effect is measured as a discrepancy $\Delta$ defined in Section 2. For example, in Table 5, 90% pruning of the CPSs on Asia with sample size 100 leads to a graph that deviates 6.7‰, or 0.67%, in terms of BDeu score from the unpruned baseline graph. Note that 90% pruning of CPSs implies that, for each node, only the 10% most relevant CPSs are considered in searching for the optimal graph.

Overall, the results suggest that higher sample sizes encourage more aggressive pruning. This is reasonable because a higher sample size implies that the ordering of legal CPSs is more accurate and hence, the pruning also becomes more accurate in terms of pruning the least relevant CPSs. The results show that, in most cases, the loss in accuracy increases faster when pruning exceeds the level of 30%. However, the results from the Hailfinder and Carpo networks suggest that even minor levels of pruning can have a negative impact on the BDeu score, however small this impact may be. All the moderate complexity experiments completed search within four minutes, except the case of Carpo-10000 which took approximately twelve minutes to complete. From this, we can conclude that unless the intention is to save seconds or minutes of structure learning runtime, pruning of legal CPSs is less desirable in problems of moderate complexity.

**Table 5.** Loss in accuracy for different levels of pruning, as a discrepancy $\Delta$ in BDeu score from the unpruned score, based on the three different sample sizes for case studies Asia, Insurance and Water.

| Pruning | Asia (100) | Asia (1000) | Asia (10,000) | Insurance (100) | Insurance (1000) | Insurance (10,000) | Water (100) | Water (1000) | Water (10,000) |
|---|---|---|---|---|---|---|---|---|---|
| 90% | −6.70‰ | −1.26‰ | −1.33‰ | −30.74‰ | −62.92‰ | −35.26‰ | −11.84‰ | −28.11‰ | −15.50‰ |
| 80% | −6.70‰ | −1.26‰ | −1.06‰ | −30.74‰ | −37.77‰ | −7.99‰ | −11.15‰ | −19.37‰ | −8.12‰ |
| 70% | −6.70‰ | −1.26‰ | −1.06‰ | −10.50‰ | −13.80‰ | −7.13‰ | −8.21‰ | −2.99‰ | −0.68‰ |
| 60% | −6.70‰ | −1.26‰ | −0.72‰ | −8.32‰ | −6.73‰ | −5.32‰ | −6.70‰ | −2.81‰ | −0.44‰ |
| 50% | −6.68‰ | −1.26‰ | −0.72‰ | −7.94‰ | −4.14‰ | −2.83‰ | −1.24‰ | −1.02‰ | −0.27‰ |
| 40% | −0.04‰ | −1.26‰ | −0.72‰ | −2.33‰ | −1.28‰ | −2.07‰ | −0.64‰ | 0‰ | −0.18‰ |
| 30% | 0‰ | −0.9‰ | −0.25‰ | −2.23‰ | 0‰ | −1.22‰ | −0.32‰ | 0‰ | −0.02‰ |
| 20% | 0‰ | 0‰ | 0‰ | 0‰ | 0‰ | −0.25‰ | −0.32‰ | 0‰ | 0‰ |
| 10% | 0‰ | 0‰ | 0‰ | 0‰ | 0‰ | 0‰ | 0‰ | 0‰ | 0‰ |
| 0% | 0‰ | 0‰ | 0‰ | 0‰ | 0‰ | 0‰ | 0‰ | 0‰ | 0‰ |

**Table 6.** Loss in accuracy for different levels of pruning, as a discrepancy $\Delta$ in BDeu score from the unpruned score, based on the three different sample sizes for case studies Alarm, Hailfinder and Carpo.

| Pruning | Alarm (100) | Alarm (1000) | Alarm (10,000) | Hailfinder (100) | Hailfinder (1000) | Hailfinder (10,000) | Carpo (100) | Carpo (1000) | Carpo (10,000) |
|---|---|---|---|---|---|---|---|---|---|
| 90% | −78.39‰ | −46.86‰ | −23.13‰ | −34.15‰ | −8.21‰ | −8.82‰ | −7.88‰ | −3.84‰ | −2.90‰ |
| 80% | −30.04‰ | −38.71‰ | −14.44‰ | −25.02‰ | −4.17‰ | −6.05‰ | −5.29‰ | −3.13‰ | −1.99‰ |
| 70% | −18.87‰ | −22.93‰ | −3.88‰ | −10.03‰ | −4.17‰ | −4.23‰ | −4.33‰ | −2.02‰ | −1.94‰ |
| 60% | −13.55‰ | −14.33‰ | −1.99‰ | −2.23‰ | −2.16‰ | −1.38‰ | −4.33‰ | −1.78‰ | −1.85‰ |
| 50% | −4.27‰ | −5.23‰ | −1.79‰ | −1.57‰ | −1.60‰ | −0.57‰ | −3.97‰ | −1.73‰ | −1.10‰ |
| 40% | −3.69‰ | −1.82‰ | −0.20‰ | −1.57‰ | −1.03‰ | −0.57‰ | −2.33‰ | −1.54‰ | −1.06‰ |
| 30% | −1.06‰ | −0.30‰ | 0‰ | −1.27‰ | −0.20‰ | −0.06‰ | −1.51‰ | −1.17‰ | −0.93‰ |
| 20% | −1.06‰ | −0.30‰ | 0‰ | −0.07‰ | −0.19‰ | −0.06‰ | −1.01‰ | −0.25‰ | −0.35‰ |
| 10% | 0‰ | −0.15‰ | 0‰ | 0‰ | −0.19‰ | −0.06‰ | −1.01‰ | −0.18‰ | −0.02‰ |
| 0% | 0‰ | 0‰ | 0‰ | 0‰ | 0‰ | 0‰ | 0‰ | 0‰ | 0‰ |

## 3.2. Pruning Legal CPSs of High Complexity

This subsection reports the results from pruning based on high complexity case studies that involve problems where the number of legal CPSs ranges between 1 million and 10 million. For this scenario, we used the real-world data sets called Audio-train which consists of 100 variables and 15,000 observations, and Kosarek-test which consists of 190 variables and 6675 observations. We used GOBNILP to generate the BDeu scores for CPSs. This process took 90 and 1398 seconds to complete respectively, with GOBNILP returning 7,343,077 and 5,748,931 legal CPSs, respectively.

Because GOBNILP's ILP algorithm is restricted to CPSs of size less than 1 million, we replaced ILP with the approximate algorithm called MINOBS (https://github.com/kkourin/mobs). The change from exact to approximate learning was inevitable since exact solutions are only applicable to problems of relatively low complexity. In fact, the results show that even an approximate algorithm such as MINOBS fails to complete search within the 24-hour runtime limit in the absence of pruning. At 24 h of runtime, we stop the search and obtain the highest scoring graph discovered up to that point.

Table 7 presents the results from this set of experiments. The results suggest that problems of high complexity may benefit considerably from pruning compared to problems of moderate complexity. In fact, the results show that it may be safe to perform aggressive pruning on legal CPSs without, or with limited, loss in accuracy, in exchange for a considerable reduction in runtime. For example, 90% pruning on the CPSs of the Audio-train data set is found to reduce runtime needed to first discover the highest scoring graph by approximately 21 folds without, or in exchange for a trivial, reduction in the BDeu score. Note that while the time needed to first discover the highest scoring graph is generally expected to decrease with higher levels of pruning, Table 7 shows that this is generally, although not always, the case.

**Table 7.** Loss in accuracy for different levels of pruning, as a discrepancy Δ in BDeu score from the unpruned score, based on the three different sample sizes for case studies Audio-train and Kosarek-test. Time (secs) represents the time needed by the MINOBS algorithm to first discover the highest scoring graph within the 24 h of search.

| | Audio-Train | | | | Kosarek-Test | | | |
|---|---|---|---|---|---|---|---|---|
| Pruning | CPSs Graph | per Node | Δ | Time (secs) | CPSs Graph | per Node | Δ | Time (secs) |
| 99% | 73,535 | 735 | $-4.352‰$ | 1473 | 58,249 | 307 | $-7.468‰$ | 4260 |
| 95% | 367,258 | 3673 | $-0.669‰$ | 682 | 287,641 | 1514 | $-0.271‰$ | 3265 |
| 90% | 734,414 | 7344 | $-0.002‰$ | 1035 | 575,096 | 3027 | 0‰ | 1803 |
| 80% | 1,468,717 | 14,687 | 0‰ | 2952 | 1,149,980 | 6053 | 0‰ | 16,378 |
| 70% | 2,203,033 | 22,030 | 0‰ | 3908 | 1,724,881 | 9078 | 0‰ | 16,010 |
| 60% | 2,937,329 | 29,373 | 0‰ | 5344 | 2,299,767 | 12,104 | 0‰ | 20,637 |
| 50% | 3,671,663 | 36,717 | 0‰ | 4334 | 2,874,708 | 15,130 | 0‰ | 9033 |
| 40% | 4,405,948 | 44,059 | 0‰ | 4587 | 3,449,544 | 18,155 | 0‰ | 14,903 |
| 30% | 5,140,257 | 51,403 | 0‰ | 10,028 | 4,024,450 | 21,181 | 0‰ | 9288 |
| 20% | 5,874,560 | 58,746 | 0‰ | 10,442 | 4,599,334 | 24,207 | 0‰ | 29,603 |
| 10% | 6,608,876 | 66,089 | 0‰ | 11,385 | 5,174,238 | 27,233 | 0‰ | 42,493 |
| 0% | 7,343,077 | 73,431 | 0‰ | 21,643 | 5,748,931 | 30,258 | 0‰ | 82,758 |

The increased benefit from pruning, observed in the case of the Audio-train and Kosarek-test data sets, relative to the data sets of moderate complexity investigated in Section 3.1, can be explained by the higher number of CPSs in the network. This is because when working with higher numbers of CPSs and a low bounded maximum in-degree (in this case, 3), even 90% pruning of legal CPSs makes it likely that the top three most relevant variables (out of hundreds of variables) will not be part of those pruned.

### 3.3. Pruning Legal CPSs of Very High Complexity

Lastly, we investigate the effect of pruning in case studies that incorporate more than 10 million legal CPSs. For this purpose, we used the EachMovie-train and Reuters-52-train data sets taken from the same repository. EachMovie-train consists of 500 variables and 4524 observations, whereas Reuters-52-train consists of 889 variables and 6532 observations. As with the high complexity cases, we perform the experiments using the MINOBS algorithm. The GOBNILP software generated a total of 21,985,307 and 37,479,789 legal CPSs, in 134 and 616 seconds respectively. However, in these experiments we had to reduce the maximum in-degree from 3 to 2. This was necessary to avoid running out of memory.

The results in Table 8 suggest that we can derive conclusions that are similar to those derived for the high complexity experiments in Section 3.2. Specifically, the pruning strategy appears to have a minor impact on accuracy of very high complexity network scores, in exchange for potentially large reductions in runtime. Note that while higher levels of pruning always reduce the time required to find the highest scoring graphs from those explored within the 24h runtime limit in the case of EachMovie-train, the effectiveness of pruning is not as consistent in the case of Reuters-52-train. This suggests that the pruning effectiveness also depends on the data set and may not be solely due to randomness as discussed in Section 3.2.

**Table 8.** Loss in accuracy for different levels of pruning, as a discrepancy Δ in BDeu score from the unpruned score, based on the three different sample sizes for case studies EachMovie-train and Reuters-52. Time (secs) represents the time needed by the MINOBS algorithm to first discover the highest scoring graph within the 24 h of search.

| | EachMovie-Train | | | | Reuters-52-Train | | | |
| Pruning | CPSs | | Δ | Time (secs) | CPSs | | Δ | Time (secs) |
| | Graph | per Node | | | Graph | per Node | | |
|---|---|---|---|---|---|---|---|---|
| 99% | 220,378 | 441 | −0.671‰ | 1711 | 375,700 | 423 | −1.269‰ | 3368 |
| 95% | 1,099,782 | 2200 | −0.158‰ | 6471 | 1,874,897 | 2109 | −0.051‰ | 6430 |
| 90% | 2,199,065 | 4398 | 0‰ | 9049 | 3,748,921 | 4217 | 0‰ | 10,002 |
| 80% | 4,397,558 | 8795 | 0‰ | 15,273 | 7,496,843 | 8433 | 0‰ | 34,537 |
| 70% | 6,596,133 | 13,192 | 0‰ | 23,133 | 11,244,877 | 12,649 | 0‰ | 41,554 |
| 60% | 8,795,121 | 17,589 | 0‰ | 9195 | 14,992,798 | 16,865 | 0‰ | 12,925 |
| 50% | 10,993,281 | 21,943 | 0‰ | 15,812 | 18,741,002 | 21,081 | 0‰ | 27,914 |
| 40% | 13,191,681 | 26,383 | 0‰ | 37,244 | 22,488,769 | 25,297 | 0‰ | 17,276 |
| 30% | 15,390,238 | 30,780 | 0‰ | 74,312 | 26,236,772 | 29,513 | 0‰ | 72,208 |
| 20% | 17,588,746 | 35,107 | 0‰ | 24,576 | 29,984,724 | 33,729 | 0‰ | 16,969 |
| 10% | 19,787,306 | 39,575 | 0‰ | 35,952 | 33,732,728 | 37,945 | 0‰ | 69,315 |
| 0% | 21,985,307 | 43,971 | 0‰ | 82,758 | 37,479,789 | 42,159 | 0‰ | 48,704 |

## 4. Conclusions

This study investigated the effectiveness of different levels of CPS pruning across problems of varied complexity. The results suggest that it is generally not beneficial to perform pruning of legal CPSs on problems of moderate or lower complexity. This is because the risk of pruning relevant CPSs increases in low complexity case studies that tend to incorporate a lower number of variables, in exchange for relatively minor improvements in speed. On the other hand, the results from problems of higher complexity show potential for major benefit from this type of pruning. This is because these problems tend to incorporate hundreds or thousands of variables, and such a high number of variables makes it easier to determine and prune irrelevant parent-sets, thereby minorly impacting accuracy in exchange for considerable gains in speed. Importantly, problems of very high complexity are often unsolvable and could benefit enormously from any form of effective pruning. The pruning strategy investigated in this paper applies to any type of score-based learning, including the traditional greedy hill-climbing heuristics where pruned legal CPSs could be used to restrict the path of arc additions.

Future work can be extended in various directions. Firstly, more experiments are needed to derive stronger conclusions about the effect of this type of pruning across different algorithms and hyperparameter settings (e.g., over different bounded maximum in-degree). Other research directions include investigating this type of pruning on ordered-based algorithms, where such a pruning strategy could be used to restrict the search space of ordered-based graphs. Lastly, other studies have shown that maximising an objective function does not necessarily imply a more accurate causal graph, especially when the data incorporate noise [34]. This diminishes the importance of exact learning and invites future work where the effect of pruning is judged in terms of graphical structure, in addition to its impact on a fitting score.

**Author Contributions:** Conceptualization, Z.G.; Methodology, Z.G.; Data curation, Z.G.; Validation, Z.G.; Formal analysis, Z.G. and A.C.C.; Investigation, Z.G. and A.C.C.; Writing–original draft preparation, Z.G.; Writing–review and editing, A.C.C.; Supervision A.C.C.; Project administration, A.C.C.; Funding acquisition, A.C.C. All authors have read and agreed to the published version of the manuscript.

**Funding:** This research was funded by the ERSRC Fellowship project EP/S001646/1 on Bayesian Artificial Intelligence for Decision Making under Uncertainty [35], and by the Alan Turing Institute in the UK under the EPSRC grant EP/N510129/1.

**Conflicts of Interest:** The authors declare no conflict of interest.

## References


1. Pearl, J. *Probabilistic Reasoning in Intelligent Systems: Networks of Plausible Inference*; Morgan Kaufmann: Burlington, MA, USA, 1988.
2. Constantinou, A.; Fenton, N. Things to Know about Bayesian Networks. *Significance* **2018**, *15*, 19–23. [CrossRef]
3. Spirtes, P.; Glymour, C.; Scheines, R.; Heckerman, D. *Causation, Prediction, and Search*; MIT Press: Cambridge, MA, USA, 2000.
4. Amirkhani, H.; Rahmati, M.; Lucas, P.; Hommersom, A. Exploiting Experts' Knowledge for Structure Learning of Bayesian Networks. *IEEE Trans. Pattern Anal. Mach. Intell.* **2017**, *39*, 2154–2170. [CrossRef] [PubMed]
5. Guo, Z.; Gao, X.; Ren, H.; Yang, Y.; Di, R.; Chen, D. Learning Bayesian Network Parameters from Small Data Sets: A Further Constrained Qualitatively Maximum a Posteriori Method. *Int. J. Approx. Reason.* **2017**, *91*, 22–35. [CrossRef]
6. Guo, Z.; Gao, X.; Di, R. Learning Bayesian Network Parameters with Domain Knowledge and Insufficient Data. In Proceedings of the 3rd Workshop on Advanced Methodologies for Bayesian Networks, Kyoto, Japan, 20–22 September 2017; pp. 93–104.
7. Yang, Y.; Gao, X.; Guo, Z.; Chen, D. Learning Bayesian Networks using the Constrained Maximum a Posteriori Probability Method. *Pattern Recognit.* **2019**, *91*, 123–134. [CrossRef]
8. Robinson, R. Counting labeled acyclic digraphs. In *New Directions in the Theory of Graphs*; Academic Press: New York, NY, USA, 1973; pp. 239–273.
9. Buntine, W. Theory Refinement on Bayesian Networks. In Proceedings of the 7th Conference on Uncertainty in Artificial Intelligence, Los Angeles, CA, USA, 13–15 July 1991; pp. 52–60.
10. Heckerman, D.; Geiger, D.; Chickering, D. Learning Bayesian Networks: The Combination of Knowledge and Statistical Data. *Mach. Learn.* **1995**, *20*, 197–243. [CrossRef]
11. Scutari, M. An Empirical-Bayes Score for Discrete Bayesian Networks. Available online: www.jmlr.org/proceedings/papers/v52/scutari16.pdf (accessed on 31 May 2020)
12. Akaike, H. Information Theory and an Extension of the Maximum Likelihood Principle. In Proceedings of the 2nd International Symposium on Information Theory, Tsahkadsor, Armenia, 2–8 September 1971; pp. 267–281.
13. Suzuki, J. A Construction of Bayesian Networks from Databases based on an MDL Principle. In Proceedings of the 9th Conference on Uncertainty in Artificial Intelligence, Washington, DC, USA, 9–11 July 1993; pp. 266–273.
14. de Campos, L. A Scoring Function for Learning Bayesian Networks based on Mutual Information and Conditional Independence Tests. *J. Mach. Learn. Res.* **2006**, *7*, 2149–2187.
15. Silander, T.; Roos, T.; Kontkanen, P.; Myllymäki, P. Factorized Normalized Maximum Likelihood Criterion for Learning Bayesian Network Structures. In Proceedings of the 4th European Workshop on Probabilistic Graphical Models, Hirtshals, Denmark, 17–19 September 2008; pp. 257–264.
16. Silander, T.; Leppa-aho, J.; Jaasaari, E.; Roos, T. Quotient Normalized Maximum Likelihood Criterion for Learning Bayesian Network Structures. In Proceedings of the 21st International Conference on Artificial Intelligence and Statistics, Canary Islands, Spain, 9–11 April 2018; pp. 948–957.
17. Jaakkola, T.; Sontag, D.; Globerson, A.; Meila, M. Learning Bayesian Network Structure using LP Relaxations. In Proceedings of the 13th International Conference on Artificial Intelligence and Statistics, Sardinia, Italy, 13–15 May 2010; pp. 358–365.
18. Bartlett, M.; Cussens, J. Integer Linear Programming for the Bayesian Netowork Structure Learning Problem. *Artif. Intell.* **2015**, *244*, 258–271. [CrossRef]
19. Koivisto, M.; Sood, K. Exact Bayesian Structure Discovery in Bayesian Networks. *J. Mach. Learn. Res.* **2004**, *5*, 549–573.
20. Silander, T.; Myllymäki, P. A Simple Approach for Finding the Globally Optimal Bayesian Network Structure. In Proceedings of the 22nd Conference on Uncertainty in Artificial Intelligence, Cambridge, MA, USA, 13–16 July 2006; pp. 445–452.
21. Yuan, C.; Malone, B. Learning Optimal Bayesian Networks: A Shortest Path Perspective. *J. Artif. Intell. Res.* **2013**, *48*, 23–65. [CrossRef]

22. de Campos, C.; Ji, Q. Efficient Structure Learning of Bayesian Networks using Constraints. *J. Mach. Learn. Res.* **2011**, *12*, 663–689.

23. van Beek, P.; Hoffmann, H.F. Machine Learning of Bayesian Networks using Constraint Programming. In Proceedings of the 21st International Conference on Principles and Practice of Constraint Programming, Cork, Ireland, 31 August–4 September 2015; pp. 429–445.

24. Teyssier, M.; Koller, D. Ordering-Based Search: A Simple and Effective Algorithm for Learning Bayesian Networks. In Proceedings of the 21st Conference on Uncertainty in Artificial Intelligence, Edinburgh, UK, 26–29 July 2005; pp. 584–590.

25. Scanagatta, M.; de Campos, C.; Corani, G.; Zaffalon, M. Learning Bayesian Networks with Thousands of Variables. In Proceedings of the 29th Conference on Neural Information Processing Systems, Montreal, QC, Canada, 7–12 December 2015; pp. 1864–1872.

26. Scanagatta, M.; Corani, G.; de Campos, C.; Zaffalon, M. Approximate Structure Learning for Large Bayesian Networks. *Mach. Learn.* **2018**, *107*, 1209–1227. [CrossRef]

27. Lee, C.; van Beek, P. Metaheuristics for Score-and-Search Bayesian Network Structure Learning. In Proceedings of the 30th Canadian Conference on Artificial Intelligence, Edmonton, AB, Canada, 16–19 May 2017; pp. 129–141.

28. Jensen, F.; Nielsen, T. *Bayesian Networks and Decision Graphs*; Springer: Berlin, Germany, 2007.

29. de Campos, C.; Ji, Q. Properties of Bayesian Dirichlet Scores to Learn Bayesian Network Structures. In Proceedings of the 24th AAAI Conference on Artificial Intelligence, Atlanta, GA, USA, 11–15 July 2010; pp. 431–436.

30. Cussens, J. An Upper Bound for BDeu Local Scores. Available online: https://miat.inrae.fr/site/images/6/69/CussensAIGM12_final.pdf (accessed on 31 May 2020).

31. Suzuki, J. An Efficient Bayesian Network Structure Learning Strategy. *New Gener. Comput.* **2017**, *35*, 105–124. [CrossRef]

32. Correia, A.; Cussens, J.; de Campos, C.P. On Pruning for Score-Based Bayesian Network Structure Learning. *arXiv* **2019**, arXiv:1905.09943.

33. Pearl, J. *Causality*; Cambridge University Press: Cambridge, UK, 2009.

34. Constantinou, A.; Liu, Y.; Chobtham, K.; Guo, Z.; Kitson, N. Large-scale Empirical Validation of Bayesian Network Structure Learning Algorithms with Noisy Data. *arXiv* **2020**, arXiv:2005.09020.

35. Constantinou, A. Bayesian Artificial Intelligence for Decision Making under Uncertainty. Available online: https://www.researchgate.net/profile/Anthony_Constantinou/publication/325848089_Bayesian_Artificial_Intelligence_for_Decision_Making_under_Uncertainty/links/5b28e595aca27209f314c4a8/Bayesian-Artificial-Intelligence-for-Decision-Making-under-Uncertainty.pdf (accessed on 31 May 2020).

*Article*

# A Blended Artificial Intelligence Approach for Spectral Classification of Stars in Massive Astronomical Surveys

Carlos Dafonte [1,*], Alejandra Rodríguez [1], Minia Manteiga [2], Ángel Gómez [1] and Bernardino Arcay [1]

1   CITIC—Department of Computer Science and IT, University of A Coruna, 15071 A Coruña, Spain; alejandra.rodriguez@udc.es (A.R.); angel.gomez@udc.es (Á.G.); bernardino.arcay@udc.es (B.A.)
2   CITIC—Department of Navigation and Earth Sciences, University of A Coruna, 15071 A Coruña, Spain; minia.manteiga@udc.es
*   Correspondence: dafonte@udc.es; Tel.: +34-881-011-329

Received: 25 March 2020; Accepted: 29 April 2020; Published: 1 May 2020

**Abstract:** This paper analyzes and compares the sensitivity and suitability of several artificial intelligence techniques applied to the Morgan–Keenan (MK) system for the classification of stars. The MK system is based on a sequence of spectral prototypes that allows classifying stars according to their effective temperature and luminosity through the study of their optical stellar spectra. Here, we include the method description and the results achieved by the different intelligent models developed thus far in our ongoing stellar classification project: fuzzy knowledge-based systems, backpropagation, radial basis function (RBF) and Kohonen artificial neural networks. Since one of today's major challenges in this area of astrophysics is the exploitation of large terrestrial and space databases, we propose a final hybrid system that integrates the best intelligent techniques, automatically collects the most important spectral features, and determines the spectral type and luminosity level of the stars according to the MK standard system. This hybrid approach truly emulates the behavior of human experts in this area, resulting in higher success rates than any of the individual implemented techniques. In the final classification system, the most suitable methods are selected for each individual spectrum, which implies a remarkable contribution to the automatic classification process.

**Keywords:** hybrid systems; MK classification; spectral features; astronomical databases; artificial neural networks

## 1. Introduction

Today's astrophysicists are frequently dealing with the analysis of complex data from one or more astronomical surveys, which typically contain millions or even hundreds of millions of sources, from which they want to determine attributes such as their membership to a given class of astronomical objects (star, galaxy, and quasar), or their main physical parameters. To analyze the information in these enormous volumes of data, it is necessary to resort to automatic processing techniques. Many of these files are open to the international scientific community for study, but their analysis is a challenge for astronomers of the 21st century, since it requires mastery of advanced computing techniques, based on statistics and the use of methodologies such as those derived from Artificial Intelligence (AI), in what has come to be called "data mining in astronomy". In this paper, we present the application of a variety of AI-based techniques to the retrieval of information present in stellar spectra obtained from telescopes, with the goal of providing a reliable hybrid system that makes it easier for the astronomer to classify stars in the MK system, a standard in stellar astrophysics.

The radiation spectrum of a star (i.e., the energy distribution as a function of the wavelength) is a black body-like curve that shows the temperature of the outer regions of stars, their photosphere, while the occurrence of spectral lines and bands in the light distribution is a consequence of the energy transitions of the elements and molecules that compose the stellar plasma. The relative intensities of these spectral characteristics are strongly dependent on the physical features (temperature, pressure, etc.) and on the presence and quantities of chemical elements in the stellar atmosphere, in such a way that stellar spectroscopy has become one of the most important tools to study those properties in stars.

In the 1880s, Williamina P. Fleming, Antonia C. Maury, and Ann Jump Cannon, among others, developed a pioneering work of massive classification of stellar spectra: The Henry Draper catalog of stars [1]. Stellar types were originally arranged in alphabetic order, beginning at A for stars with the strongest hydrogen lines. Soon it became clear that the spectral types were related to the star temperature, thus the final series was obtained after some subsequent modifications: O, B, A, F, G, K, and M, from hot to cool stars (additional letters were used to designate nova and less common types of stars). Numbers 0–9 are used to subdivide the types, applying the higher numbers to cooler stars. The hottest stars are known as early stars and the coldest as late stars. This scheme was created at the Harvard College Observatory, thus it was called the Harvard stellar classification system.

In 1940, the American astronomers Morgan, Keenan, and Kellman carried out an expansion of the Harvard sequence to include luminosity [2]. In the so-called MKK, or simply MK, system (following the authors' names or first two authors' names) a roman numeral is appended to the Harvard type to indicate the luminosity class of the star: I for super-giants, II for bright giants, III for giants, IV for sub-giants, and V for dwarfs or main sequence stars. The different levels of luminosity refer to the absolute magnitude of the star, which is a measure of its luminosity on a logarithmic scale with a negative factor, whereby smaller magnitudes correspond to brighter stars.

Human experts often classify stellar spectra with the support of a guiding catalog of prototypical spectra with a reliable classification in the MK system, which have been previously selected to be used as a complete reference in the manual classification process. The classical spectral classification process mainly focuses on the substantial information provided by certain lines and spectral areas, so that, to directly compare the non-classified stars with those of the reference catalog, it is necessary to adapt all the spectra to the same scale and then normalize them and isolate their continuous component (affected by interstellar reddening).

Once human classifiers have scaled and normalized target spectra, they try to guess their spectral type and luminosity in the MK system. The morphological differences between spectral types are found in the intensity of the absorption lines of hydrogen and helium, and in the presence or absence of certain metals and molecular bands (Ca, Mg, Fe, C, Ti, etc. bands). However, the luminosity is related to the width of some specific lines in the spectrum. Therefore, the expert classifiers measure and analyze the relationship between some absorption lines (H, He, Ca, etc.) and the depth of certain relevant molecular bands (TiO, CH, etc.), obtaining an initial classification that usually includes a first approximation to the spectral type and luminosity class [3]. This tentative classification is completed by assigning the spectral subtype that best matches when the unclassified spectra and the catalog templates are superposed. Sometimes it is impossible to decide which spectral subtype a star belongs to, so that they finally adopt a mixed classification (e.g., types B02, F79, etc.).

The described manual classification technique is quite subjective and strongly depends on the criteria and experience of the expert classifiers. Moreover, their practical application is often not feasible, especially when the number of spectra is very high, as it would require a great deal of time and human resources. In fact, nowadays it is possible to collect spectral information of several hundreds of stars in one night of telescope observation, using modern techniques of multi-object spectroscopy, or many more objects in the case of spatial missions. In these situations, the classical manual technique is no longer operative, and it would be desirable to replace it with automatic and non-subjective classification schemes.

In previous works, we described an initial knowledge-based system for the classification of the low-resolution optical spectra of super-giant, giant, and dwarf stars [4]. Then, we refined this system with a first hybrid approach based on the combination of the most effective neural networks [5]. The obtained results encouraged us to adapt such a system to manage stars belonging to other evolutionary stages (with different luminosity classes), while adding new intelligent techniques, such as Takagi–Sugeno fuzzy reasoning, to complete the MK computerized classification process.

Certain well-known previous works have also applied AI techniques and deep learning to the stellar classification field [6–12], obtaining different performance in the classification. The big data and AI processing on massive catalogs (that do not stop growing in number) go hand in hand. From the first works that explicitly mention the datamining concept in 2000 [13], to publications that propose robust machine learning techniques for this task (also using the MK system) on observations gathered in the Dark Sky observatory [14], they are also currently working with collaborative concepts innovators such as Galaxy Zoo in the LSST survey [15,16] and applying streaming techniques to characterize variable stars in this same survey [17]. In all these cases, and many others, ML techniques such as the one proposed here are the core of the classification or parameterization, using and combining fuzzy logic, artificial neural networks, and rule-based systems.

Of course, it is not our objective to re-test methods that have already proved their suitability, but to determine our best approach for the problem of automatic spectral classification, to design the system while mainly using a first catalog, to optimize and above all validate its operation based on another well-known standard catalog.

Thus, this paper presents the implementation of several intelligent models to carry out an analysis of the sensitivity and the adaptation capability of different AI techniques to the classification of stellar spectra. This exhaustive study aims to integrate all the techniques into a single intelligent system that is able to guide the process and apply the most appropriate classification method for each situation. Considering the subjective human classification process, we believe that the combination of fuzzy expert systems and neural networks in a single intelligent system could be more flexible and suitable than one that uses only a specific AI technique, and might even mean a better adjustment to the peculiarities of stellar classification. Furthermore, we consider that such a system could also be very helpful in the training of new stellar spectroscopists.

The following section describes the materials and methods that have been used in the development of this research. Firstly, we specify the different groups of spectra that were chosen for the implementation and evaluation of the models, and all the preprocessing stages that are applied to the data before they are presented to the different techniques. Subsequently, we include some of the morphological algorithms that were developed for knowledge-based systems but are used here to extract the set of parameters that characterizes each spectrum, which are then used as an input for most of the intelligent models. After that, we describe in detail the different AI techniques that were implemented, to finally compare their results for the same sets of unclassified spectra. To conclude, we present the option of a hybrid solution that combines the best tested AI techniques.

## 2. Materials and Methods

### 2.1. Spectral Catalogs

At the beginning of our research, we mainly worked in the study of a large set of observed sources, candidates to be classified as stars in the post-AGB phase of evolution [18]. In this first approach, we chose as templates the spectra of luminosity level I from the reference catalog used at that time in the manual MK classification method [19].

The results of this initial development showed that, to deal with the whole problem of MK classification of stars, it is highly advisable to compile a solid and robust catalog that provides good resolution for all target spectral types and as many classes of luminosity as possible. Consequently, to design and develop intelligent tools for the rationalization of the manual classification process, we have built a digital database that includes a large sample of optical spectra from public online catalogs: The D. Silva and M. Cornell optical spectral library, now extended to all luminosity levels [19], the catalog of observations of G. Jacoby, D. Hunter and C. Christian [20], and the digital spectra collected by A. Pickles [21].

We decided to add a broader set of spectra to facilitate the optimal convergence and generalization of the artificial neural networks, since the training patterns sets notably influence the performance of the models. For this purpose, we selected an additional database of 908 spectra obtained with the ELODIE spectrograph at the French observatory in Haute-Provence; these spectra present greater coverage of MK spectral subtypes and luminosity levels for metallicities ([Fe/H]) from −3.0 to 0.8 [22].

Before applying this final database to the design of automatic classification systems, the classification experts who collaborate in this research visually studied, morphologically analyzed, and contrasted the values of the spectral parameters for the four above-mentioned catalogs; for that purpose, they used the information available in public databases such as SIMBAD [23]. In this way, those spectra that did not satisfy our quality criteria were eliminated: a lot of noise, significant gaps in regions with an abundance of classification indices (e.g., some early types from Pickles), very high or very low levels of metallicity (e.g., some spectra from Silva library with values of [Fe/H] inferior to −1), spectra where MK classifications widely differ from one source to another (e.g., several Jacoby's spectra), and the spectral subtypes and/or luminosity levels with very few representatives (e.g., the entire spectral type O, unfortunately).

Afterwards, the remaining spectra were analyzed by means of statistical clustering techniques, also discarding those that could not be placed into one of the groups obtained with the different implemented algorithms (K-means, ISODATA, etc.). Because of this complete selection process, we obtained a main catalog with 258 spectra from the libraries of Silva (27), Pickles (97) and Jacoby (134), and a secondary catalog with 500 spectra extracted from the Prugniel database.

Figure 1 shows the final distribution of spectral types and luminosity classes of the reference catalog. The main catalog (Silva, Pickles, and Jacoby) is quite homogeneous in relation to the number of spectra of each type, but it presents some deficiencies in the luminosity classes (II and IV); in the secondary final catalog (Prugniel), most spectra belong to intermediate spectral types (F–K) and to luminosity class V.

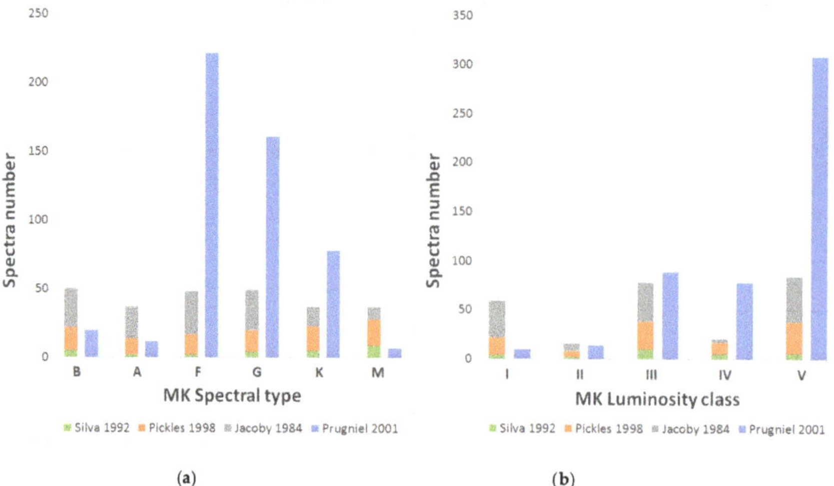



(a)  (b)

**Figure 1.** Composition of our definitive reference catalog: (**a**) number of spectra per MK spectral type; and (**b**) number of spectra per MK luminosity class.

All the spectra of our final database were analyzed, obtaining the measurement of some representative parameters such as spectral lines of absorption or molecular bands, so that for each spectral type the typical values were delimited. In most of our implementations, the unclassified spectra that are supplied to the system will be compared with this numerical characterization of the reference spectra; moreover, they are useful to determine the rules of the expert systems as well as to train and validate the neural models.

More than ten years ago, our research group started working on the Gaia Project, the astrometric cornerstone mission of the European Space Agency (ESA) that was successfully launched and set into orbit in December 2013. Gaia is an astrometric mission that measures parallaxes and movements of the stars in the Milky Way, and also includes a spectrophotometric survey of all objects in the sky up to a visible magnitude of approximately 20.5. The scientific work to prepare the mission archive is organized around the Data Processing and Analysis Consortium (DPAC) where we lead the Outlier Analysis Working Package (WP) and collaborate in others [24–26]. In the context of DPAC, we have been testing the performance of different ANNs, both unsupervised and supervised, for classification and parameterization of astrophysical properties of the sources. In particular, we have gained experience in the problem of parameterization of stellar atmospheric properties using the spectra of Gaia RVS instrument [27]. This instrument was mainly built to measure the radial velocity of stars in the near infrared CaII spectral region, but it is also a most helpful tool to estimate the most important stellar APs: effective temperature (Teff), logarithm of surface gravity (log g), abundance of metal elements with respect to hydrogen ([Fe/H]), and abundance of alpha elements with respect to iron ([alpha/Fe]). The results obtained have encouraged us to try those same techniques by applying them to the stellar classification problem in the classic MK system.

*2.2. Spectral Indices and Sensitivity Analysis*

The different types of stars have different observable peculiarities in their spectrum, which are commonly studied to determine the MK classification and also to collect relevant information about stellar properties such us temperature, pressure, density, or radial velocity. Most of these typical spectral features can be isolated by the definition of spectral classification indices [3].

Although there are some classical indices included in the most important classification bibliographical sources, the set of spectral parameters used in this project corresponds to the main features that the spectroscopic experts of our group visually study when trying to obtain the manual classification of the spectra. Most of these spectral indices are grouped into three general types, Intensity (I), Equivalent Width (EW), or Full Width at Half Maximum (FWHM) of absorption and emission spectral lines (He, H, Fe, Ca, K, Mg, etc.); depth (B) of molecular absorption bands (TiO, CH, etc.); and relations between the different measures of the absorption/emission lines (H/He, CaII/ H ratios, etc.).

Once the set of spectral indices for each particular classification scenario is determined, their values are obtained by applying specific morphological processing algorithms primarily based on the calculation of the molecular bands energy and on the estimation of the local spectral continuum for the absorption/emission lines. These algorithms allow for the extraction of a large number of relevant spectral markers that can be included as classification criteria, although only a small subset of them will be in the final selection of indices. To carry out a study as exhaustive as possible, we initially measured some peculiar parameters that are not normally included in the rules used in the traditional classification process. Likewise, the equivalent width, intensity, and width at half maximum are measured for each spectral line, even when they are not included in the typical classification criteria for a particular context.

Among all the classification criteria collected in the preliminary study of the problem (those indicated by the classification experts as well as those found in the bibliography), a set of 33 morphological features was initially selected. We have contrasted our own classification criteria with those included in the different bibliographic sources, with special emphasis on the study of Lick indices [28].

The suitability of this selected subset of spectral indices was verified by performing a complete sensitivity analysis through the spectra included in the chosen reference catalogs. Thus, the value of each chosen morphological feature was estimated for each spectrum of the reference main catalog (Silva 1992, Pickles 1998, Jacoby 1984); then, these values were ordered so that we can examine their variation with the spectral type and luminosity class. Thanks to this meticulous study, it was possible to establish the real resolution of each index, while delimiting the different types of spectra that it is able to discriminate.

Figure 2 shows two examples of the methodology followed to evaluate the behavior of some of the most significant spectral indices. First, a spectral index (intensity of line Hiβ at 4861 Å) is included in the final selection of parameters for the MK classification because of its excellent performance in the template spectra. Second, we also include an index that was discarded (depth of molecular band TiO at 5805 Å) because its real behavior was not completely congruent with the theoretical one (in fact, it presents similar values for very different spectral types, such as B and M). A more detailed description of this analysis procedure can be found in [29].

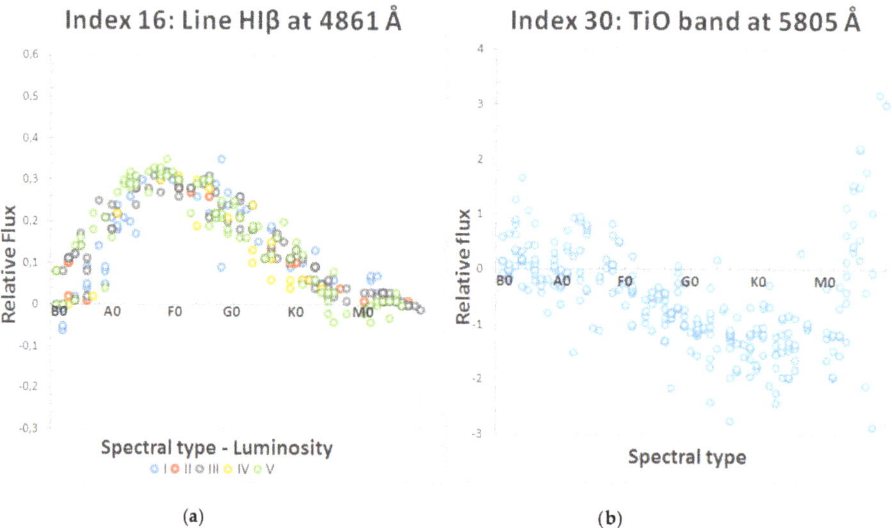


**Figure 2.** Spectral indices sensitivity analysis: (**a**) hydrogen line at 4861 Å; and (**b**) band of TiO at 5805 Å.

Table 1 lists the most relevant results of our sensitivity study, contrasting them with the theoretical classification capability indicated by the experts. Parameters 1–25 constitute the final set of proper indices that are considered to be able to address the MK classification, as they present a clear and reproducible pattern of discrimination between spectral types and/or luminosity classes in the analysis on the 258 spectra from the main reference catalog. The first column contains the definition of each index (an asterisk indicates parameters that are not usually contemplated in the manual classification); the second column shows the types, subtypes, and levels of luminosity that each parameter could potentially discriminate; and the third column specifies the types/classes that it is actually able to delimit, according to the performance achieved in the reference catalogs. The indices highlighted in blue have necessarily been excluded in the processing of the spectra of the secondary database (Prugniel 2001), since they correspond to morphological characteristics that are out of the spectral range of this catalog (4100–6800 Å) and are, therefore, impossible to estimate; unfortunately, this limitation has an unavoidable impact on the performance of some of the implemented techniques.

**Table 1.** Sensitivity analysis results for the selected classification indices.

| Index | Theoretical Resolution | Actual Resolution |
|---|---|---|
| 1. B (4950 Å) | Global [1] | K-M types<br>Luminosity I for F-G, Luminosity III for M |
| 2. B (6225 Å) | Global [1] | G-M types, K-M subtypes<br>Luminosity I for G-K, Luminosity V for K-M |
| 3. B (5160 Å) | K-M types | G-M types |
| 4. B (5840 Å) | K-M types | K-M subtypes |
| 5. B (5940 Å) | K-M types | M subtypes |
| 6. B (6245 Å) | K-M types | K-M types |
| 7. B (6262 Å)* | K-M types | K-M types |
| 8. B (6745 Å) | K-M types | M subtypes<br>Luminosity III-V for M |
| 9. B (7100 Å)* | K-M types | M subtypes |
| 10. I (3933 Å) | A-G types | F-G subtypes<br>Luminosity I for B-F |
| 11. I (3968 Å) | A-G types | A-M types [2] Luminosity I for B-F |
| 12. I (4340 Å) | B-G types<br>Luminosity I for early | A-G subtypes [2]<br>Luminosity I for B-A |
| 13. I (4102 Å) | B-G types<br>Luminosity I for early | A-F subtypes |
| 14. I (4026 Å) | O-B types | B type |
| 15. I (4471 Å) | O-B types | B type |
| 16. I (4861 Å) | B-G types<br>Luminosity for early | B-K types, A-G subtypes [3]<br>Luminosity for B-A |
| 17. I (6563 Å) | B-G types<br>Luminosity I for early<br>Spectra with emission | B-K types[1], A subtypes |
| 18. EW (4300 Å) | A-G types | A-G types, F-G subtypes<br>Luminosity I for K-M |
| 19. I (3933 Å)/I (3968 Å) | A-G types | A-F types |
| 20. I (4102 Å)/I (4026 Å) | B-A types | B type |
| 21. I (4102 Å)/I (4471 Å) | O-A types | B subtypes<br>Luminosity for B |
| 22. EW (4300Å)/I (4340Å) | F- K types | G-K types |
| 23. $\Sigma_{1,2}$Bi* | Not included | Global [1]<br>F-M types, G-M subtypes<br>Luminosity I for F-K, Luminosity V for K-M |
| 24. $\Sigma_{3..9}$Bi* | Not included | K-M subtypes |
| 25. $\int I(\lambda_i)\,d\lambda$* | Not included | Early, intermediate |
| 26. B (4953 Å) | K-M types | Not valid (value only from M5) |
| 27. B (6140 Å) | K-M types | Not valid (value only from M4) |
| 28. B (4435 Å) | K-M types | Not valid (includes other indices) |
| 29. B (5622 Å) | K-M types | Not valid (arbitrary behavior for B-M) |
| 30. B (5805 Å) | K-M types | Not valid (includes other indices) |
| 31. I (4144 Å) | O-B types | Not valid (wrong behavior for M) |
| 32. I (4481 Å)/I (4385 Å) | F-G types | Not valid (arbitrary behavior for B, F) |
| 33. I (4045 Å)/I (4173 Å) | A-G types | Not valid (arbitrary behavior for F-M) |

[1] Early, Intermediate, Late. [2] With previous global classification. [3] With previous type classification. *B* means the depth of the molecular bands, *I* is the intensity, and *EW* is the equivalent width of the absorption/emission lines

During the successive tuning processes carried out at this stage, eight indices were excluded from the 33 morphological characteristics of the initial selection. In some cases, as shown in Table 1, the actual found resolution and the theoretical one differed significantly, since they cannot separate the spectral types that, according to experts and bibliographic sources, they should be able to.

The final set of selected spectral indices is included in the classification criteria implemented by means of rules for the expert systems, and in the elaboration of the training and validation patterns for neural models. However, in some specific cases, we have also used certain interesting spectral regions or even the full range of wavelengths for that purpose. Besides, in some implementations, this set of parameters has been reduced by means of another adjustment process based on techniques that allow us to minimize and optimize the number of indices necessary to obtain the MK classification, such as principal component analysis (PCA) or clustering algorithms (K-means, Max-Min, etc.).

### 2.3. Knowledge-Based Systems

In our initial classification project, we designed and developed an automatic system that integrated signal processing, knowledge-based techniques, and a very rudimentary fuzzy logic implementation in order to process the spectra of three luminosity levels (specifically I, III, and V). A complete description of the different versions of this expert system can be found in [4,5,29].

In the current phase of our research, we have modified the first expert systems to adapt them to all the spectral types and luminosity levels of the selected spectra of the new reference catalogs. Therefore, the present stellar classifier consists of a knowledge-based system that manages uncertainty and imprecision—characteristic features of human reasoning—by combining traditional production rules, fuzzy logic, and credibility factors. We adapted the methodology of Shortliffe and Buchanan [30] with the development of a fuzzy reasoning scheme by taking the Takagi–Sugeno model [31] as a starting point, and appropriately adapting the Max-Min inference method to include the singularities of the spectral classification process.

Morphological Algorithms

As mentioned in previous sections, the spectral indices are usually defined by the depth of a molecular absorption band (B), the measurement of an absorption/emission line (I, EW, or FWHM), or by the relationships between the different values of the absorption/emission lines.

A molecular band is a spectral zone where the flux suddenly decreases from the local continuum during a wide wavelength interval. This implies that we can decide whether a molecular band is sufficiently significant by simply measuring its energy. In the present case, after experts have defined the interval for each band by means of the main reference catalog spectra, we calculate the upper threshold line in the limit of that interval by making use of the linear interpolation between the fluxes. Then, we apply a discrete integral to calculate the area between this line and the abscissa axis; and we integrate the flux signal between the band extremes to measure the area around each band. We then subtract the resulting energies and hereby obtain the band flux. When the band deepens and widens, its value turns more negative: this is why positive as well as negative values that are close to zero are not considered bands.

Absorption/emission lines are dark or bright lines that appear in a uniform and continuous spectrum because of a lack (absorption) or excess (emission) of photons in a specific region of wavelengths, in comparison with their closer regions. From a morphological point of view, an absorption line is a descending (ascending for emission) deep peak that appears in an established wavelength zone. The intensity of a line (in $erg^{-1}cm^{-2}s^{-1}Å^{-1}$) is the measurement of its energy flux at the wavelength of the peak with respect to the local spectral continuum of the zone; the equivalent width (usually measured in Å) corresponds to the base of a rectangle whose height is the measurement of the local spectral continuum and has the area of the profile of the line; the width at half maximum (also in Å) will be the width of the line measured at half the maximum intensity. Therefore, it is clearly necessary to calculate the spectral continuum of the surrounding area of each line in order to establish the value of these three parameters.

In the analysis module, we designed an algorithm that estimates the local continuum for each spectral line and obtains a pseudo-continuum that is valid to calculate the value of the spectral parameters of absorption/emission lines (I, EW, and FWHM). During the different development phases of our automatic classification system [4,5,29], the calculation method for this local continuum was modified and adjusted, as the measurements obtained in each option were compared with the actual estimations manually made by the experts.

In the final algorithm, we combined the procedures that proved to be more efficient during the previous implementations with a new method inspired by the technique developed by Cardiel and Gorgas for their INDEX program [32]. Thus, we defined two adjacent intervals to each line (one towards the red and one towards the blue), and then we obtained a value for each lateral continuum by smoothing the flux signal with a low pass filter that excludes the samples with greater standard deviation in each interval (central moving-average of five points). Both estimations (left and right local continuum) are interpolated with polynomial adjustment to obtain the final pseudo-continuum measurement at the wavelength of each line. Through a manual calibration process performed on the spectra belonging to the main catalog, the algorithm established the side intervals, achieving as such a proper adaptation to the peculiarities of each specific absorption/emission line. That is:

$$
\left.
\begin{array}{l}
Cl = \frac{\sum_{j=ll}^{rl} X_j \, F(\lambda_j)}{N} \\[2mm]
Cr = \frac{\sum_{j=ll}^{rl} X_j \, F(\lambda_j)}{N}
\end{array}
\right| X_j = \left\{ \begin{array}{l} 1, \ \ if \ \sigma(\lambda_j) < \sigma_n \\[1mm] 0, \ \ if \ \sigma(\lambda_j) > \sigma_n \end{array} \right. ,
\tag{1}
$$

where $Cl$ and $Cr$ are the partial estimations of the continuum in the selected range to the left and to the right of the sample where the peak has been detected ($\lambda_p$); $F(\lambda_j)$ is the flux in sample $j$; $N$ the number of samples used in the computation of the left and right partial continuum; $X$ is a binary vector that indicates the representative fluxes of the local continuum in the zone; $\sigma(\lambda_j)$ is the local standard deviation in sample $j$; and $\sigma_n$ is a standard deviation threshold used to decide if the sample $j$ really represents the local continuum and not another morphological feature in that area (band, absorption line, etc.).

Figure 3 shows the estimation of the local continuum using the final method for an intermediate spectrum. The automatic adjustment is shown in blue, while the continuum estimated by the experts in each case is shown in red. As can be seen, the estimation is generally quite correct, but in some areas (marked in green) there are deviations from the estimation made manually by experts; in some regions where there is a lot noise or a large profusion of spectral lines, the signal does not smooth properly, since the computational algorithm includes more peaks than desired, resulting in higher values than the actual value estimated by the experts.

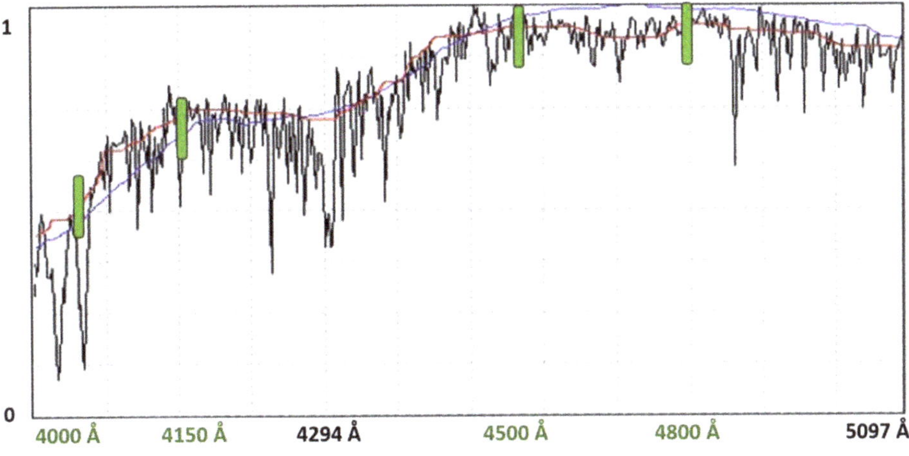


**Figure 3.** Continuum estimation for a G8 type star.

### 2.4. Artificial Neural Networks

Neural networks are an ideal technique for solving classification problems, especially for their ability to learn by induction and their capability to discover the intrinsic relationships that underlie the data processed during their learning (training phase). One of the major advantages of this AI technique is its generalization capability, which means that a trained network could classify data of a similar nature to those in the training set without any previous presentation.

At this point in our development, we again used spectra from the main guiding catalog, originating from the same libraries that were applied to design and implement the knowledge-based systems' reasoning rules [4,5,29]. By following this design strategy, the applied techniques could easily be compared, since their design principles and bases are the same.

Although it is possible to use all the available data for the learning phase, our experience in previous works advise us to split the available spectra into three sets, dedicating approximately 50% of them to the training, 10% to validating the learning, and 40% to testing the implemented networks. In the design of some of the neural networks, the spectra of the secondary catalog have also been used, because they allow us to evaluate the behavior of the networks in a sample with a greater variety of spectral features; these spectra are real, not estimations, and are affected by factors such as interstellar reddening or atmosphere distortion effects. The detailed distribution of the spectra sets is shown in Table 2.

**Table 2.** Datasets for artificial neural networks, detailed by spectral type and luminosity class.

| MK Spectral Type | Training Set | Validation Set | Test Set |
| --- | --- | --- | --- |
| B | 25 | 5 | 20 |
| A | 20 | 3 | 14 |
| F | 24 | 5 | 19 |
| G | 25 | 5 | 19 |
| K | 20 | 3 | 14 |
| M | 20 | 3 | 14 |
| TOTAL | 134 | 24 | 100 |
| **MK Luminosity** | **Training Set** | **Validation Set** | **Test Set** |
| I | 30 | 6 | 23 |
| II | 10 | 1 | 5 |
| III | 40 | 7 | 31 |
| IV | 11 | 2 | 8 |
| V | 43 | 8 | 33 |
| TOTAL | 134 | 24 | 100 |

The spectral analyzer, designed according to the knowledge-based system approach, is equipped with functions that allow it to obtain automatically the training, testing, and validation patterns presented to the neural networks. In most cases, neural networks designed for spectral classification have been trained and tested primarily with input patterns that include the measurement of the 25 spectral features obtained after performing the sensitivity analysis (Indices 1–25 in Table 1).

Since we did not previously select the relevant parameters per spectral type or luminosity class, most neural models that were implemented count 26 units in the input layer (25 for the selected indices and 1 additional neuron, the so-called teaching input, for the supervised training models). On the other hand, all networks that use spectra from the secondary catalog (Prugniel 2001) only have 16 input patterns, since this database does not cover the entire spectral range in which the 25 selected indices are located. However, in some cases, we have also implemented networks that use a subset of spectral characteristics, with the objective of analyzing the influence of different sets of parameters in obtaining the MK classification. Furthermore, in some particular designs, we have used input patterns representing the real spectral flux in specific regions of wavelengths, which means that full spectral areas are provided to the neural network so that it is capable of extracting the relevant information on its own.

We started our development by scaling the spectra of every catalog to 100 at wavelength 5450 Å to obtain normalized values that are adapted to fluxes of the reference catalogs. We then calculated the value of the 25 spectral parameters in order to build the pattern sets. There was no need to cover the full spectral range, merely the range 3900–7150 Å, which is why the applied template spectra cover several spectral ranges. It is perfectly possible for the sampling frequency to differ according to the catalog. What happens is that the analyzer searches the lines and bands in a determined spectral area and then calculates the catalog resolution by means of a computational algorithm.

As soon as the spectral analyzer has gathered the input values, these values should be normalized and presented to the neural networks. The inputs were standardized by means of a contextualized normalization, which allows us to normalize the values in the [0, 1] interval and adequately scale and center each parameter's distribution function. We assigned a lowest ($X_1$) and a highest value ($X_2$) to each classification index, establishing that 95% of the values lie between these two values; this provides us with constants $a$ and $b$, which can be determined for each spectral parameter, according to values $X_1$ and $X_2$:

$$
\begin{aligned}
0.025 &= \frac{1}{1+e^{-(aX_1+b)}} \\
0.975 &= \frac{1}{1+e^{-(aX_2+b)}}
\end{aligned}
\qquad (2)
$$

At the present day, there are many models of artificial neural networks with different design, learning rules, and outputs. We have selected different architectures to address the automatic classification of optical spectra, carrying out the training process with both supervised and unsupervised learning algorithms.

In particular, we have designed backpropagation networks (multi-layer architecture with supervised learning), SOM networks (self-organized maps with grid architecture and unsupervised learning) [24,25,33,34], and RBF networks (architecture in layers and hybrid learning) [35]. We have also analyzed some variants of these general network types (e.g., BP momentum), to decide the networks with a best performance in determining the MK classification. In the same way, we have studied each network behavior when using the different possible topologies (number of layers, nodes of the intermediate layers, etc.). As a preliminary step, we analyzed the ability of each of these three models to discriminate between consecutive types individually, corresponding to early, intermediate, and late stars. This initial experimentation certainly makes it easier to determine which network best fits each couple of successive spectral types.

We have used the SNNS v4.3 simulator (Stuttgart Neural Network Simulator) [36] for the design and training of classification networks. This software tool incorporates an option to convert directly the trained networks to C-code, which has been used in our work to incorporate the implemented networks into the global analysis and classification system for stars.

### 2.4.1. Learning Algorithms

Our research has applied three different backpropagation (BP) learning algorithms for its initial developments: standard backpropagation, enhanced backpropagation (with a momentum term and flat spot elimination), and batch backpropagation (the weight changes are summed over each full presentation of all the training patterns).

In a first phase of experimentation, we have designed neural networks using the above-mentioned BP algorithms for the spectral types from B to M (type O was previously discarded due to its lack of presence in the reference catalogs). As for topology, classification networks have an input layer with 25 neurons (1 per each spectral index), one or more hidden layers, and an output layer with six units (1 for each MK spectral type). Since the backpropagation learning algorithm is a supervised one, it is necessary to provide the network with an extra unit (the teaching input) for each input pattern so that it can calculate the error vector and appropriately update the synaptic weights of its connections.

In simulated systems, the most commonly used design strategy for configuring intermediate layers is to try to simplify them as much as possible, including the smallest number of neurons in each hidden layer, since each extra layer would involve a higher processing load in the software simulation. Thus, following this strategy, we have designed networks with different topologies, starting with simple structures and progressively increasing both the number of hidden layers and the number of units included in them ($25 \times 3 \times 6$, $25 \times 5 \times 6$, $25 \times 10 \times 6$, $25 \times 2 \times 2 \times 6$, $25 \times 5 \times 3 \times 6$, $25 \times 5 \times 5 \times 6$, $25 \times 10 \times 10 \times 6$, $25 \times 3 \times 2 \times 1 \times 6$, $25 \times 5 \times 3 \times 2 \times 6$, and $25 \times 10 \times 5 \times 3 \times 6$). During the training, we updated the weights in topological order (input, hidden, and output layers). We opted for a random initiation of the weights with values from $-1$ to $1$. We also modified the total training cycles, the validation frequency, and the learning parameters values ($\eta$, $\mu$ and flat) through the learning stage of the various implemented topologies.

Kohonen self-organized map unsupervised networks are unique in that they build preservation maps of the training data topology, where semantic information is transmitted by the location of a unit [33].

With a similar strategy to that of feed-forward nets, we have tested several Kohonen maps by grouping the MK spectral types into the same three sets (B-A, F-G, K-M). The SOM neural models do not validate the learning progress, so the validation patterns were added to the training dataset. In these networks, we have deliberately introduced another spectral type for every pair of consecutive types, in order to make it easier for the network to cluster the data more efficiently, contrasting the target spectra with very different ones in terms of morphology. Hence, this extra spectral type was selected to be as separate as possible from the spectra of each group, that is, we chose type M for early types and type B spectra for late and intermediate ones.

The input patterns of the Kohonen networks were also built by using the 25 essential spectral features shown in Table 1. In the learning phase of the designed networks, the total cycles and the learning parameters were adjusted (h(t), r(t), decrease factor for h(t), and decrease factor for r(t)).

The calculation of the chemical compositions or abundances of stellar spectra, and by extension MK classifications, could be also considered as a nonlinear approximation problem in which there are data affected by noise, and even the absence of data in the series occurs in some situations. The classical resolution strategy consists in performing a functional regression, which implies the determination of the specific function that best expresses the relation between the dependent and independent variables. In the case of RBF neural networks, it is not necessary to make any assumptions about the functions that relate outputs to inputs, since their basic principle is to center radial-based functions around the data that need to be approximated. Unlike MLP networks, the typical architecture of RBF networks is based on a feed-forward model composed strictly of three layers: an input layer with n neurons, a single hidden layer with k processing units, and a layer of output neurons [35].

We have designed RBF networks to obtain the different levels of classification in the MK system and, using again the simplification strategy of the network structure, we have considered the configurations $25 \times 2 \times 6$, $25 \times 4 \times 6$, $25 \times 6 \times 6$, and $25 \times 8 \times 6$.

After repeatedly trying the three described neural schemes for all the spectral types, we carefully chose the best network of each type so as to contrast their performance and determine which would be the best choice at the different scenarios of classification. The best implemented networks were studied by analyzing their results for the spectra of the test set.

The backpropagation (trained with enhanced learning algorithm) and RBF networks obtain a similar overall performance, 96% ($0.96 \pm 0.050$) and 95% ($0.95 \pm 0.056$), respectively (both intervals calculated for a statistical confidence level of 0.99). The BP network solves the problem better for some spectral types (F, K, M), whereas for other types (A) the RBF is a preferable option. However, the SOM maps obtained an inferior performance, about 68% ($0.68 \pm 0.117$), in the best situations (networks trained with small decrease factors). These results are probably a consequence of the size and distribution of the training spectra, as these are not supervised networks and they need to make the clustering autonomously, so that they require a training dataset that is large enough to be able to abstract similarities and properly group the inputs.

As for the validation set, BP and RBF networks trained with a validation set of spectra present a higher success rate than the same networks trained without it. It is clear that the cross-validation technique influences the weights and allows the network to correct the weights dynamically during the training process and, therefore, improve the learning process.

Finally, the different topologies that were tested for each supervised learning algorithm (feed-forward and RBF) do not significantly affect the final performance of the network. However, we obtained the best results with a $25 \times 5 \times 3 \times 6$ backpropagation momentum network and with a $25 \times 4 \times 6$ RBF network. SOM networks with a higher number of units in their competitive layer present a higher success rate; in fact, the networks designed with few units in the competitive layer (fewer than 12) are not capable of clustering the data properly.

Once we know the network with the best performance for each pair of spectral types, we are better able to propose neural models to undertake the complete MK classification process.

### 2.4.2. ANN Models for Two-Dimensional MK Classification

The whole MK spectral classification includes the estimation of the spectral subtype and the luminosity class of the stars. Therefore, to carry it out completely, we here propose the implementation of several neural models based on two different design philosophies: classification based on a global network and classification structured in levels.

The first method proposed to obtain the whole MK classification is based on the design of two global artificial neural networks that estimate simultaneously and independently the spectral subtype and the luminosity class of the stars.

In this way, this global networks include only BP and RBF neural structures, since SOM architectures were discarded because they obtained more modest results during previous experimentation. In the design of the networks, we implemented models that only support the 16 spectral indices present in the range 4100–6800 Å of the secondary catalog, and another alternative networks that accept as input the 25 spectral parameters of the full range (3510–7427 Å, corresponding to the main guiding catalog). Figure 4 illustrates the conceptual design of this first proposed model.

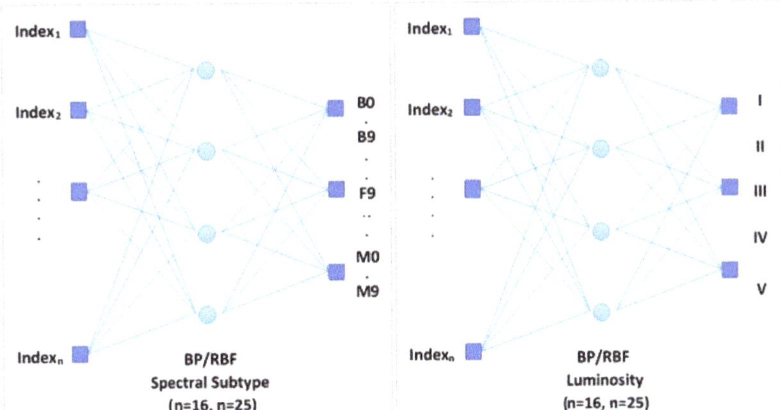


**Figure 4.** Conceptual scheme of the global networks designed for the complete MK classification.

Our second strategy to accomplish the whole MK classification is based on a combination of the networks described in the previous sections, properly interconnected; specifically, we use those that have obtained a higher performance for each level of spectral classification (global, spectral type, spectral subtype, and luminosity) during the previous evaluation phase of the different neural configurations. These networks are now organized in a tree-like composition similar to the reasoning scheme of the previous designed knowledge-based systems [29].

In this way, BP (specifically networks trained with the backpropagation momentum algorithm) and RBF are implemented again, and they also accept as input both the 25 spectral indices belonging to the range of the main reference catalog and the 16 included in the most restrictive range of the secondary catalog.

Regardless of the learning algorithm and the spectral range, the MK classification net of networks is formed by a first level BP/RBF network that decides the global classification, that is early, intermediate, or late star; a second level made up by three different networks that classify each previous pair of spectral types; and a third level formed by twelve networks, two for each spectral type (B, A, F, G, K, and M), so that one of them is responsible for establishing the specific subtype (0–9) and the other one appropriately determines the luminosity class (I–V) depending on the specific estimated spectral type.

When implementing this tree structure in C++, the communication between the different classification levels was arranged appropriately, so that the networks at each level can decide which network in the next level send the spectra to and direct them towards it accompanied by a partial classification and a probability estimation. In some specific cases, it will be necessary to send them to two different networks, especially with spectra at the boundary between two consecutive spectral types, for example B9A0 or G9K0. Figure 5 shows a conceptual scheme of this second strategy.

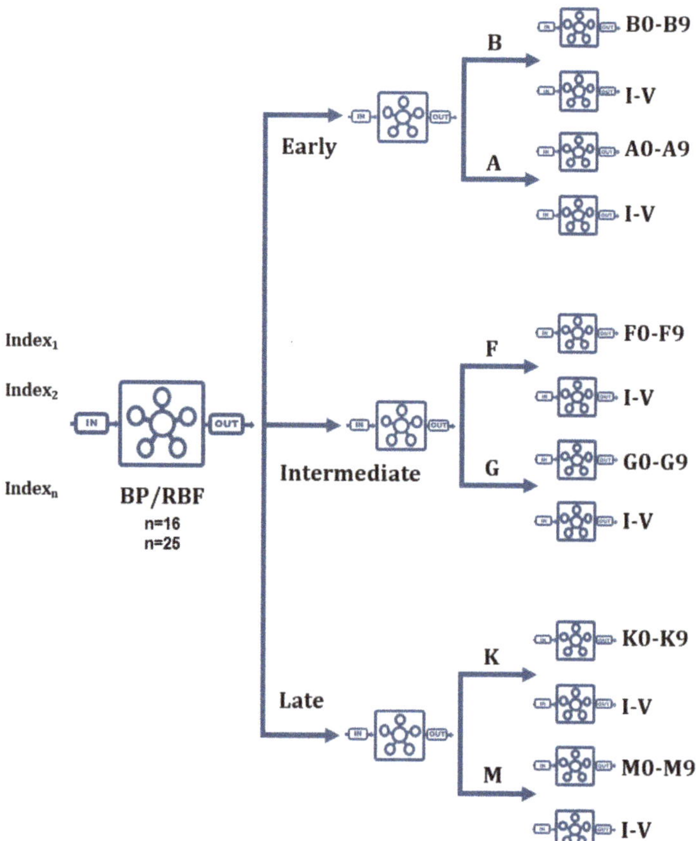


**Figure 5.** Conceptual scheme of the net of networks designed for the complete MK classification.

Table 3 lists the different topologies selected in the two alternative strategies formulated to obtain the complete MK classification, as well as their performance for test spectra.

**Table 3.** Implemented neural architectures for complete MK classification.

| Network Type | Spectral Range | Topology (Success Rate) |
|---|---|---|
| Net of networks BP Momentum | 3510 Å–7427 Å | Global: $25 \times 10 \times 3$ (98%); Type: $25 \times 5 \times 3 \times 2$ (94%)<br>Sub.: $25 \times 20 \times 20 \times 10$ (65%); Lum.: $25 \times 10 \times 10 \times 5$ (60%) |
| Net of networks BP Momentum | 4100 Å–6800 Å | Global: $16 \times 10 \times 3$ (95%); Type: $16 \times 10 \times 5 \times 2$ (93%)<br>Sub.: $16 \times 50 \times 20 \times 10$ (76%); Lum.: $16 \times 20 \times 10 \times 5$ (79%) |
| Net of networks RBF | 3510 Å–7427 Å | Global: $25 \times 15 \times 3$ (93%); Type: $25 \times 4 \times 2$ (92%);<br>Sub.:$25 \times 20 \times 10$ (62%); Lum.: $25 \times 20 \times 5$ (60%) |
| Net of networks RBF | 4100 Å–6800 Å | Global: $16 \times 15 \times 3$ (92%); Type: $16 \times 4 \times 2$<br>Sub.: $16 \times 25 \times 10$ (75%); Lum.: $16 \times 30 \times 5$ (75%) |
| Global network BP Momentum | 3510 Å–7427 Å | Sub.: $25 \times 30 \times 20 \times 10$ (67%)<br>Lum.: $25 \times 10 \times 10 \times 5$ (68%) |
| Global network BP Momentum | 4100 Å–6800 Å | Sub.: $16 \times 50 \times 20 \times 10$ (74%)<br>Lum.: $16 \times 20 \times 10 \times 5$ (75%) |
| Global network RBF | 3510 Å–7427 Å | Sub.: $25 \times 30 \times 10$ (65%)<br>Lum.: $25 \times 20 \times 5$ (65%) |
| Global network RBF | 4100 Å–6800 Å | Sub.: $16 \times 30 \times 10$ (75%)<br>Lum.: $16 \times 30 \times 5$ (73%) |

The classification results obtained with the different implemented methods will be now shown through the interface of the spectral analysis application. In the case of neural networks, besides the two-dimensional MK classification in spectral subtype and luminosity class, we also provide a probability that somehow informs about the quality of the offered answers, i.e., the network confidence in its own conclusions. In those cases, where the network output is ambiguous, falling within the range of non-resolution ([0.45–0.55]), two alternative classifications will be provided, each of them supported by their corresponding probability.

*2.5. Final Approach: Hybrid System*

The increasing complexity and sophistication of modern information systems clearly indicates that it is necessary to study the suitability of each technique at our disposal to develop adequate and efficient solutions, as we have done in the analysis presented in this paper.

The hybrid strategies are mainly based on the integration of different intelligent methods (fuzzy logic, evolutionary computation, artificial neural networks, knowledge-based systems, genetic algorithms, probabilistic computation, chaos theory, automatic learning, etc.), in a single robust system that explores their complementarities with the objective of achieving an optimal result in the resolution of a specific problem [37]. In this way, hybrid systems combine different methods of knowledge representation and reasoning strategies in order to overcome the possible limitations of their integrated individual AI techniques, and increase their performance in terms of exploiting all available knowledge in an application field (symbolic, numerical, inaccurate, inexact, etc.) and, ultimately, offer a more powerful search for answers and/or solutions.

Although there are several possibilities for the integration and cooperation between the different components of the hybrid systems, the conclusions of our previous experience with the different individual techniques, naturally point to a hybrid modular architecture for the automation of the MK classification. Our proposal would be made up of both symbolic (knowledge-based systems) and connectionist (neural networks) components, combined in a single functional hybrid system that operates in co-processing mode.

The methods with a more satisfactory performance during the previous experimentation stage were directly transferred from their respective development environments to a single hybrid system developed in language C ++.

In this global system, it will now be possible to perform the spectral classification from two different perspectives: separately with each of the individual techniques, obtaining a set of isolated predictions that the user will be responsible for analyzing to arrive at a compromise solution; or initiating a combined classification process that automatically invokes the most efficient methods of each implemented technique (expert systems, backpropagation net of networks, RBF global network, etc.) and finally presents the user a set of answers accompanied by an associated probability, making possible a quick evaluation of the conclusions quality. In this second classification strategy, the system itself is responsible for analyzing the results of each technique, grouping them together and providing a summary, so that it is possible to offer better conclusions at the same time that we grant the automatic procedure of some degree of self-explanation.

Based on the experience acquired during the study phase, the experts assigned to each different technique a set of probability values according to its resolution capacity in the classification of the spectra belonging to each spectral type or luminosity class. Then, when the hybrid strategy is launched, the different optimal classifiers must combine the probability of their respective conclusions (derived from the stellar classification process) with the numerical value that they have associated by default, and which is indicative of the grade of confidence of their estimations for a specific spectral type or luminosity class.

Figure 6 shows in the hybrid system interface the results obtained in a specific classification assumption when selecting the option based on the described hybrid approach.

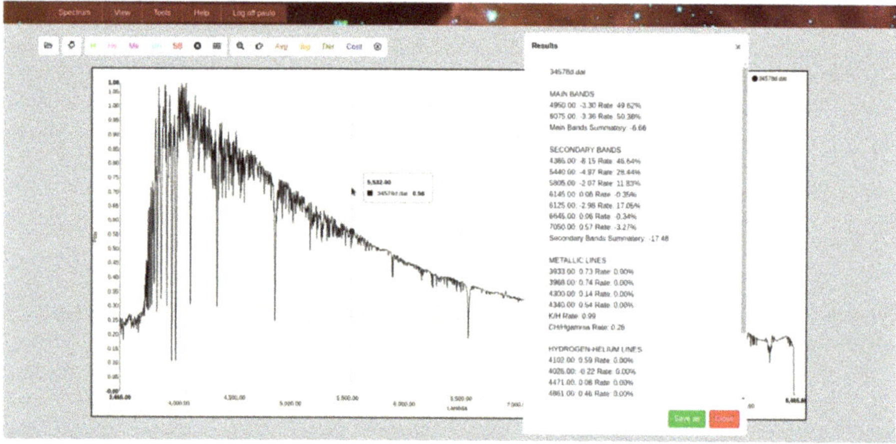

**Figure 6.** Hybrid system interface showing the combined classification with optimal methods.

The integration of the two described types of classification into a single hybrid system allows this computational implementation to emulate, in a more satisfactory way, the classical methods of analysis and classification of stellar spectra that we describe in previous sections. The option that combines the optimal implementations of each technique obtains a classification based on solid foundations that result in clear and resumed conclusions for each classification level. At the same time, it is possible to carry out a detailed analysis of each individual spectrum with all the methods implemented and to study the discrepancies or coincidences between the estimations provided by each of them.

## 3. Results

To accomplish a complete analysis of the performance achieved with the different computational techniques, we now study their results in the spectra of the test datasets from both the main and the secondary reference catalogs. It is important to remark that these sets had not been previously used in the design of the expert systems or in the training of the different neural networks.

Every method processes and analyzes the spectra and obtains a two-dimensional classification in the MK system (spectral subtype, indicative of the surface temperature, and stellar luminosity), both accompanied by an estimation of the confidence in the classification (in form of probability). In the analysis of results, we do not consider as correct classifications the conclusions whose probability or confidence does not exceed 70%. In addition, in the study of the outputs of the subtype networks, the procedure for evaluating the successes is similar to that carried out by most experts when they compare their classifications, since we accept a resolution of ±1 spectral subtypes. In this sense, if the network places the subtype within the correct group, the result is computed as valid; otherwise, the two adjacent subtypes are examined (higher and lower temperatures), and the result would be considered correct if the output belongs to one of these subtypes. Otherwise, it would be computed as a real error. Regarding luminosity, as most of expert spectroscopists do, we consider that the valid outputs are only those where the class estimated by the model exactly matches the one specified in the reference catalog.

Figure 7 shows the comparative performance analysis of the three neural architectures finally selected (BP momentum networks, RBF networks, and SOM maps) for each classification level (global, spectral type, spectral subtype, and luminosity class) and each spectral range.

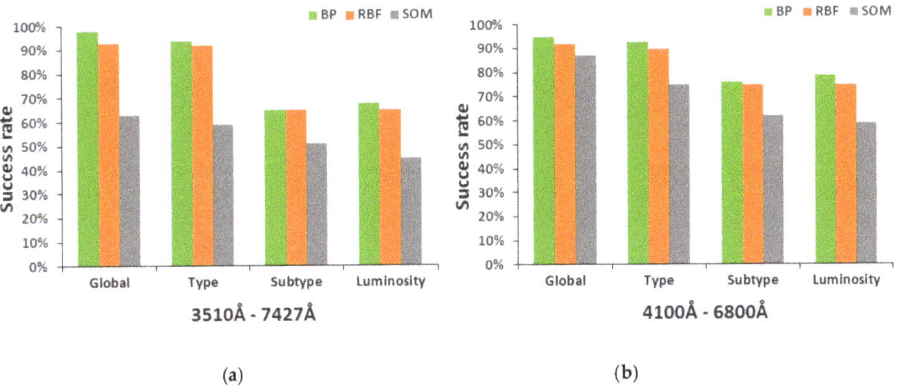


**Figure 7.** Performance analysis of classification neural network architectures: (**a**) spectral range 3510–7427 Å; and (**b**) spectral range 4100–6800 Å.

Table 4 shows the confusion matrix [38] for the spectral type corresponding to the analysis of the outputs of the neural model with optimal results, that is, the net of networks trained with the backpropagation momentum learning algorithm. In the rows of this useful graphical tool, we include the actual number of spectra of each spectral type, following the catalog classifications, and, in the columns, we show the number of predictions obtained for each of them with our neural implementation.

**Table 4.** Initial and final confusion matrix for the spectral type in the optimal neural model.

| Real/Estimated | B | A | F | G | K | M |
|---|---|---|---|---|---|---|
| B | 19/20 | 1/0 | 0/0 | 0/0 | 0/0 | 0/0 |
| A | 1/1 | 10/11 | 3/2 | 0/0 | 0/0 | 0/0 |
| F | 0/0 | **4/2** | 14/17 | 1/0 | 0/0 | 0/0 |
| G | 0/0 | 0/0 | 1/1 | 16/18 | 2/0 | 0/0 |
| K | 0/0 | 0/0 | 0/0 | 1/0 | 12/14 | 1/0 |
| M | 0/0 | 0/0 | 0/0 | 1/1 | 1/0 | 12/13 |

By examining the resulting matrix, it becomes clear that errors generally occur between contiguous types, that is, which presumably is due to the edge problem already mentioned in previous sections (errors in subtypes that fall on the boundaries between two different spectral types, such as B9A0 or K9M0). Therefore, if we use the classical resolution of ±1 spectral subtypes in the analysis of the outputs, the resulting confusion matrix (second numbers of each cell included in Table 4) shows a considerable decrease in classification errors.

Figure 8 shows the performance comparison of the two proposed strategies for the entire MK classification, detailed by spectral type, again using test spectra in both catalog spectral ranges. Figure 9 presents the homologous performance comparison graphs for luminosity level.

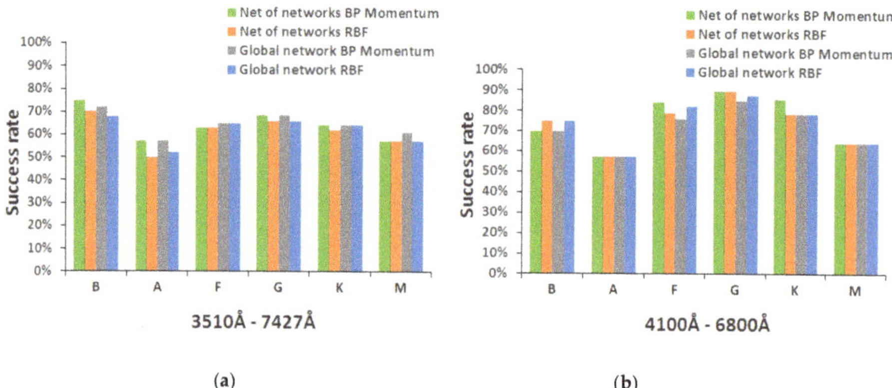

**Figure 8.** Performance analysis of neural models for final MK spectral type classification: (**a**) spectral range 3510–7427 Å; and (**b**) spectral range 4100–6800 Å.

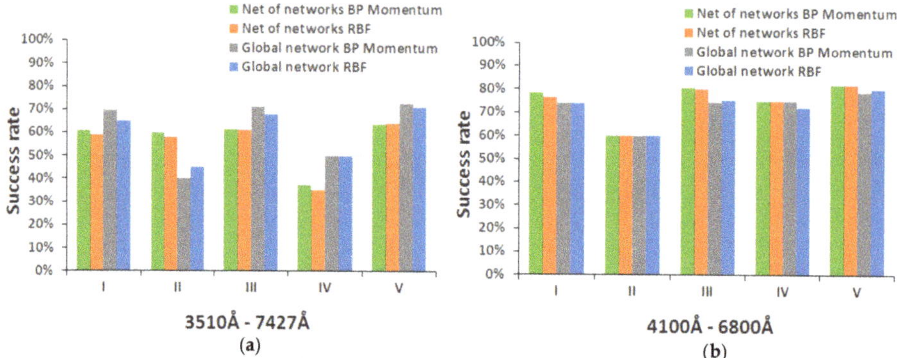

**Figure 9.** Performance analysis of neural models for final MK luminosity classification: (**a**) spectral range 3510–7427 Å; and (**b**) spectral range 4100–6800 Å.

According to the results obtained with the different neural models implemented for the complete MK classification, we considered that it would be convenient to include in our final system the net of neural networks for the determination of both the spectral subtype and the luminosity class, since the success rates are satisfactory in the two dimensions (around 75%, $0.75 \pm 0.049$ for a statistical confidence level of 0.99) and slightly higher than those achieved by the specific global network (about 73%, $0.73 \pm 0.051$ for a statistical confidence level of 0.99).

As mentioned in Section 2.5, the final classifier that we propose is a hybrid system that combines the methods that resulted in better performance during our previous experimentation stages. Figure 10 shows the results for both MK system dimensions for the whole test set of 600 spectra, versus the actual classification provided by the guiding catalogs for these spectra.

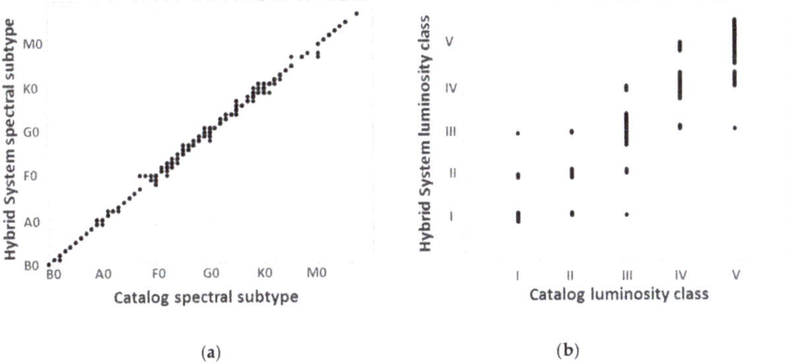

**Figure 10.** Comparison of the MK classifications obtained by the hybrid final system with the reference catalog stellar classifications: (**a**) spectral type; and (**b**) luminosity class.

Table 5 shows the mean absolute errors (in black color) and standard deviations (in blue) for the complete MK classification, which were obtained with our hybrid system using again the 600 test spectra.

**Table 5.** Mean absolute errors and standard deviations for system classifications by spectral type.

|  | All | B | A | F | G | K | M |
|---|---|---|---|---|---|---|---|
| Spectral type | 0.25/0.54 | 0.13/0.33 | 0.50/0.91 | 0.29/0.56 | 0.23/0.50 | 0.15/0.42 | 0.24/0.77 |
| Luminosity | 0.24/0.44 | 0.10/0.30 | 0.35/0.56 | 0.31/0.47 | 0.24/0.44 | 0.10/0.30 | 0.05/0.22 |
| Number of spectra | 600 | 40 | 26 | 241 | 180 | 92 | 21 |

To complement the evaluation of the results, we carried out an additional validation of the optimal individual models and the proposed hybrid system, also contrasting their conclusions with the classifications obtained by the group of expert spectroscopists of our project. In this final study, we again consider the two sets of test spectra, which correspond to 600 spectra in both the spectral range 3510–7427 Å and the spectral range 4100–6800 Å.

Figure 11 shows the success rates obtained by the global hybrid system and each of the individual artificial intelligence methods in the complete classification of the 600 test spectra, for both spectral ranges.

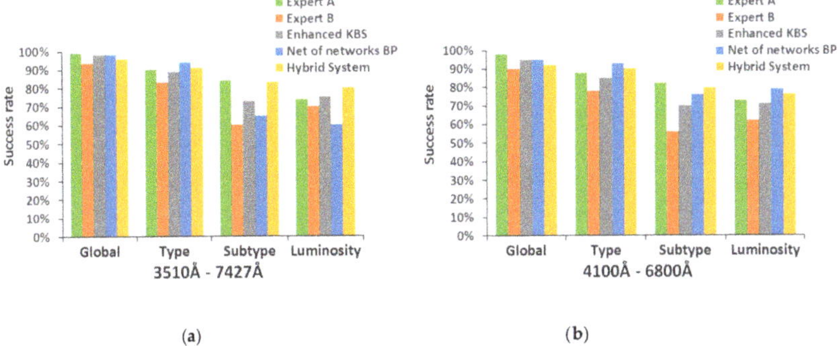

**Figure 11.** Final performance analysis comparing hybrid model with each individual technique: (**a**) spectral range 3510–7427 Å; and (**b**) spectral range 4100–6800 Å.

Table 6 shows, as a percentage of agreement, the results of a double-blind study that consists of evaluating the correctness of the classifications assigned to the spectra without knowing their origin. On the one hand, the estimations obtained with the different computational techniques were compared with those of the human experts who collaborate in the development of this system. On the other hand, the spectroscopists of our group analyzed again the spectra already classified by the different automatic techniques, expressing their agreement or disagreement with the classifications concluded by each one (highlighted in Table 6).

**Table 6.** Double blind study comparing human experts and computational sources.

|  | Expert A | Expert B | Fuzzy KBS | BP Networks | Hybrid System |
|---|---|---|---|---|---|
| **Expert A** | 100% | 78% | 86% | 85% | 89% |
| **Expert B** | 87% | 100% | 86% | 86% | 83% |
| **Fuzzy KBS** | 84% | 76% | 100% | 79% | 78% |
| **BP networks** | 84% | 78% | 79% | 100% | 75% |
| **Hybrid system** | 85% | 72% | 78% | 75% | 100% |

Finally, Figure 12 shows the performance of the best techniques when applied to the classification of larger and heterogeneous spectral sets, such as the one formed by 49 real post-AGB stars from our own observation campaigns [39] or the set composed by the 486 spectra initially discarded from all the reference catalogs for reasons such as an atypical metallicity level, lack of spectral fluxes in some areas, or an incomplete classification in the available sources.

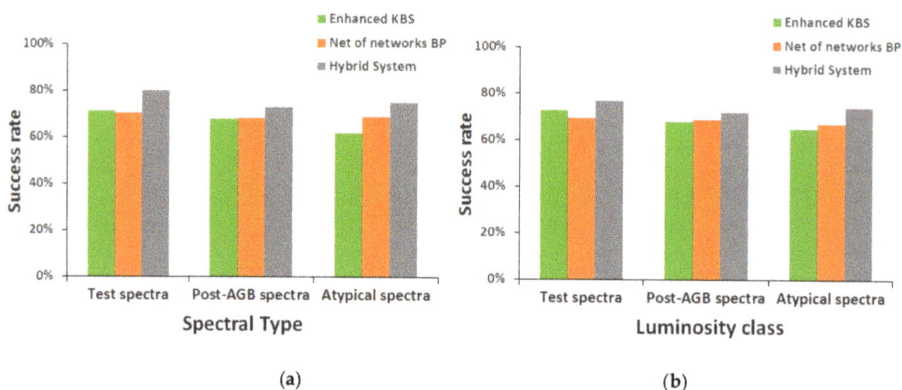


**Figure 12.** Results with different sets of peculiar spectra (**a**) spectral type; and (**b**) luminosity class.

## 4. Discussion

In our preliminary studies of the different selected neural models, we found (as the graphs in Figure 7 show), that the backpropagation and RBF networks obtain a very similar results (with some slight differences), whereas the Kohonen networks present significantly lower success rates in all implementations. As mentioned above, self-organized maps are mainly characterized by the use of a non-supervised and competitive learning algorithm, in which it is essential that the training set is large enough to allow the network to extract the similarities between the input data and cluster them appropriately. For this reason, the SOM networks trained with the spectra from the secondary catalog (Figure 7b) show a more adequate behavior, with success rates more similar to those obtained with the other neural models. It is clear that the addition of more spectra to the training set allows the creation of two-dimensional output maps more adapted to the problem of spectral classification.

Analyzing the results obtained during the previous experimentation phases, we can conclude that the neural architectures based on the multi-layer perceptron (BP) and on radial basis functions (RBF) have a more congruent behavior, being able to achieve a very acceptable final performance for the different classification levels. That is, for both spectral ranges, they reach success rates above 90% ($0.9 \pm 0.077$ for a statistical confidence level of 0.99) in the preliminary discrimination and in the determination of the spectral type, and around 70% ($0.7 \pm 0.012$, for a statistical confidence level of 0.99) in obtaining the subtype and luminosity class. For this reason, these architectures have been included in the implementation of our final developed system of spectral analysis and classification.

The detailed analysis of the performance achieved with the two design strategies for the full MK classification (shown in the comparative charts of Figure 8) indicates that, taking into account the results as a whole, the model based on a tree structure presents a more satisfactory success rate, since it is able to obtain correctly and unequivocally (probability greater than 70%) the spectral subtype of about 76% of the spectra in networks trained with the backpropagation momentum algorithm ($0.76 \pm 0.049$, for a statistical confidence level of 0.99) and around 74% when they are trained with the RBF algorithm ($0.74 \pm 0.05$, for a statistical confidence level of 0.99). By contrast, in implementations based on the direct determination of the subtype by means of global networks, this performance is slightly lower (72–73%, $0.72 \pm 0.052$ for a statistical confidence level of 0.99). The inclusion of spectra from the secondary catalog (Figure 8b) generally produces a significant improvement in the performance of each of the models implemented (between 7% and 13%).

Thus, for the first MK dimension (the spectral type), the study of the results denotes that there could be a certain correlation between the final success rate and the composition and size of the training set, since the networks performance decreases for the types in which that set includes less spectra (A and M, mainly), regardless of the model and the spectral range. When the spectra of the secondary catalog are introduced, the results improve significantly (by more than 20% in some cases) for the intermediate types (F, G, and K), because in these types the catalog presents a greater increase in both the number and the variety of spectra. On the other hand, for the other spectral types (B, A, and M), the performance does not increase noticeably and sometimes even decreases, since the stars of these types are more sensitive to the spectral indices that have been excluded from the secondary catalog range (mainly some important lines of hydrogen and helium around 4000 Å and some late bands of TiO in the region of 7000 Å).

To continue the study of the results for the MK spectral type, we built the confusion matrix of the BP net of networks, which is the architecture that achieved the best performance. In this case, the neural model correctly assigns the spectral type to a total of 83 test spectra (shown in the main diagonal in Table 4), which implies a success rate of 83% ($0.83 \pm 0.096$, for a statistical confidence level of 0.99) with a mean error of 0.2 types. A detailed observation of the matrix shows that most errors takes place between adjacent types, which is consistent with the typical problem concerning to the subtypes that are located in the limits between spectral types, already mentioned in previous sections. For this reason and with the aim of getting closer to the real analysis carried out by the experts, we decided to rebuild the matrix using the resolution of ±1 spectral subtypes that human experts normally use (shown in the second numbers of each cell in Table 4). As expected, the percentage of errors decrease with this adaptation, since the network is now capable of assigning a spectral type congruent with the one that gives their own catalog to 93 of the 100 evaluation spectra ($0.93 \pm 0.065$, for a statistical confidence level of 0.99), reducing the mean error to 0.08. Although the success rate increases by 10%, in this new matrix, there is still some confusion between spectral types A and F (highlighted in Table 4). These errors in classification are in some way predictable, since these spectral types are usually affected by the classic degeneracy problem between the surface temperature of the star and the metallicity of its spectrum.

The exploration of the results of the other MK dimension, luminosity (Figure 9), shows that both implemented final models reach a similar performance (60–68%) for the spectral range of the main catalog, although the option based on specific networks of luminosity shows a slight improvement (global network approach). This can be due to the additional errors that are included in the net of networks model when the spectral type is determined by following a wrong path in the tree, since it adds an extra and prior error that conditions the further estimation of the luminosity class. However, this trend is reversed by adding the spectra from the secondary catalog, since the model based on a net of networks (composed of several consecutive neural networks that obtain the luminosity after having determined the spectral type) achieves an optimal performance. They estimate adequately (probability greater than 70%) the luminosity class of 79% ($0.79 \pm 0.047$, for a statistical confidence level of 0.99) of the spectra for backpropagation networks and around 75% ($0.75 \pm 0.05$, for a statistical confidence level of 0.99) for RBF networks.

As in the spectral type, if we analyze the results considering the level of luminosity, we can note again a clear dependence on the size of the training set, since the performance decreases significantly in those luminosity levels where this set is scarcer (II and IV), independently of the neural model and the spectral range. Obviously, the more spectra are available for each class, the better is the delimitation of the spectral features that define it, so that the obtained results would be perfectly consistent with the design of the neural models. For this reason, when secondary spectra are added, performance tends to improve for all classes but especially for the less favored ones, where the sample size was significantly smaller, reaching an increase of more than 25% in some implementations. Similarly, the highest success rate is achieved for the luminosity V, because the number of patterns that belong to this class is significantly higher, with more than 200 spectra of difference in some cases.

The detailed analysis of the performance achieved in the two dimensions of the MK classification, led us to finally use the strategy based on a net of neural networks for both the spectral subtype and the luminosity class, since it presents sufficiently adequate success rates in both dimensions (about 75%, $0.75 \pm 0.05$ for a statistical confidence level of 0.99) and somewhat better than those obtained with the strategy of global networks (73%, $0.73 \pm 0.051$ for a statistical confidence level of 0.99). As we could already see in the sensitivity analysis carried out on spectral classification indices (Table 1), many of the parameters used by experts affect temperature and stellar luminosity at the same time, and it is extremely difficult to separate both influences. Therefore, a neural model based on a multi-level classification that determines the luminosity class as a function of the previous spectral type estimation will also be more consistent with the deduction process followed in the classical stellar classification techniques.

The next phase of the results analysis was the evaluation of the hybrid model proposed as a final implementation of the classification system (Figure 10 and Table 5), also analyzing its behavior compared to the other previously implemented classification methods (Figure 11). We also carried out a double-blind study to validate the correction of the classifications without knowing their source; in this way, the classifications assigned by the human experts who collaborate in this project were compared with those obtained with the different computational techniques (Table 6). As an additional test, we studied the performance of the implemented methods when applied to larger, heterogeneous, and unusual spectral sets (Figure 12), i.e., real stars of our observation campaigns, spectra with peculiarities, etc.

As previously, the errors in some spectral types such as A and F appear again with the hybrid system, which can be clearly seen in the graph in Figure 10, where a greater dispersion with respect to the diagonal (perfect classification) can be observed in the regions corresponding to the subtypes of these spectral types. Likewise, these types present a greater mean absolute error and a greater standard deviation. As for the other dimension, the luminosity classes II and IV are where the greatest errors continue to be concentrated.

The results obtained in these final comparative studies confirm the hypothesis that the hybrid approach proposed emulates the behavior of human experts in this field more satisfactorily than any of the individual computational techniques. The hybrid system's success rate in discrimination of the spectral subtype, 80.16% ($0.8016 \pm 0.042$, for a statistical confidence level of 0.99) with an absolute mean error of $\pm 0.25$ subtypes and a standard deviation of 0.54, and the level of luminosity, 76.83%

(0.7683 ± 0.044, for a statistical confidence level of 0.99) with an absolute mean error of ±0.24 classes and a standard deviation of 0.44, is superior (even for the most refined versions of individual methods) and it is maintained at a higher level when determining the classification of broader sets of spectra that have special characteristics, such as a high noise level or an insufficient resolution (shown in graphs in Figure 12). In addition, the agreement between this hybrid system and the experts who participated in the blind study is also superior to the one presented with any other methods.

This propensity towards performance improvement was somewhat expected, especially in view of the fact that the hybrid approach is essentially based on a classification strategy that selects the best conclusions from the optimal methods according to a heuristic estimation of the confidence level of the predictions. Thus, it is possible to mitigate some of the problems that individual methods have proved unable to manage, such as the confusion between stars B and M in cases where emission occurs instead of absorption. However, other problems do not find solution in this alternative either, for example the classical confusion between spectra A and F of low metallicity level.

## 5. Conclusions

In this paper, we propose a detailed analysis of how knowledge-based systems and ANNs can be applied to the MK classification of stellar spectra. If knowledge-based systems are found most appropriate for global and generic classifications, neural networks are shown to adequately establish the stellar spectral types and luminosity.

Following the classic method of manual classification, we built two different catalogs of spectra covering the whole range of MK spectral types and luminosity classes. The spectra of these databases were used to define the rules of expert systems and to train and validate the artificial neural networks.

Since human experts visually study certain spectral features, we defined an initial set of 33 spectral indices that characterizes each spectrum and reduces its processing complexity. This first set was refined by performing an exhaustive sensitivity analysis that helped us to determine the actual classification capacity of each of the proposed indices. As a result, we obtained a final set of 25 spectral features that were used both to define the fuzzy sets in TKS expert systems and to build the input patterns of neural networks.

The paper describes our complete experimentation with various ANN models and analyzes their performance and results. Thus, we designed and tested BP, SOM, and RBF networks, and checked a variety of topologies, which can provide the stars' MK classification in three levels (global, spectral type and subtype, and luminosity class). We established that the finest networks could determine the spectral type in a sample of 100 testing spectra with an accuracy of almost 95% (0.95 ± 0.056, for a statistical confidence level of 0.99).

To accomplish the whole two-dimensional MK classification of stars, we have adopted two neural strategies that essentially differ in their structure: global approach and nets of networks. According to the results obtained, we can conclude that it is more efficient to use a net of neural networks for the determination of the spectral subtype and the luminosity class, since the success rates are satisfactory enough in both dimensions (around 75%, 0.75 ± 0.05 for a statistical confidence level of 0.99) and also slightly higher than those obtained with the specific global network (around 73%, 0.73 ± 0.051 for a statistical confidence level of 0.99).

In our final proposal, all the computational techniques were combined into a hybrid system that decides the proper procedure for the different classification contexts, therefore reaching greater adaptation to the problem than a classification tool based on a single technique. The obtained results suggest that the hybrid strategy is a more adequate and versatile approach, since it achieves higher success rates than any other individual method (around 80%, 0.80 ± 0.042 for a statistical confidence level of 0.99).

Our most recent research work consists in the analysis of functional networks, to establish whether they are apt for stellar classification, and in the completion of our stellar database and its adaptation to the Internet. We want to make our database accessible so that users all over the world may store and classify spectra and as such contribute to making our automatic analysis and classification system more adaptable and accurate.

Gaia's DPAC, to which our research group belongs, is currently developing several software modules to approach stellar parameterization by the use of RVS instrument data from different perspectives [40]. A second step will be the inclusion of a MK classifier such as the one presented here, only for late and intermediate type stars, adapted to the Gaia RVS spectral region. We started the work based on simulations, although in recent months we dispose of real data for algorithms validation (under NDA, not publishable results). Some data and public results of the WP will be publicly available in Data Release 3 (DR3) at the end of 2021 and as such will constitute a hard big data environment (more than 1 Petabyte of data) for some of the techniques presented here.

**Author Contributions:** Conceptualization, C.D., A.R. and M.M.; Methodology, A.R., M.M. and B.A.; Software, C.D., A.R. and Á.G.; Validation, A.R., M.M. and B.A.; Investigation, C.D., A.R. and M.M.; Resources, A.R. and M.M.; Data Curation, A.R. and Á.G.; Writing—Original Draft Preparation, C.D., A.R. and M.M.; Writing-Review & Editing, C.D., A.R. and M.M.; Visualization, A.R. and Á.G.; Supervision, C.D., M.M. and B.A.; Funding Acquisition, C.D., M.M. and B.A. All authors have read and agreed to the published version of the manuscript.

**Funding:** This work was supported by Ministry of Science, Innovation and Universities (FEDER RTI2018-095076-B-C22) and Xunta de Galicia (ED431B 2018/42).

**Acknowledgments:** The authors acknowledge the infrastructure and support of the Centre for Information and Communications Technology Research (CITIC)

**Conflicts of Interest:** The authors declare no conflict of interest.

## References


1. Cannon, A.J.; Pickering, E.C. The Henry Draper catalogue 0h, 1h, 2h, and 3h. *Ann. Harv. Coll. Obs.* **1918**, *91*, 1–290.
2. Morgan, W.W.; Keenan, P.C.; Kellman, E. *An Atlas of Stellar Spectra, with an Outline of Spectral Classification*, 1st ed.; The University of Chicago Press: Chicago, IL, USA, 1943; ISBN 978-0-598-92469-8.
3. Gray, R.O.; Corbally, C.J. *Stellar Spectral Classification*, 1st ed.; Princeton University Press: Princenton, NJ, USA, 2009; ISBN 978-0-691-12511-4.
4. Rodríguez, A.; Arcay, B.; Dafonte, C.; Manteiga, M.; Carricajo, I. Automated Knowledge-based Analysis and Classification of Stellar Spectra using Fuzzy Reasoning. *Expert Syst. Appl.* **2004**, *27*, 237–244. [CrossRef]
5. Rodríguez, A.; Arcay, B.; Dafonte, C.; Manteiga, M.; Carricajo, I. Hybrid Approach to MK Classification of Stars. In Neural Networks and Knowledge-based Systems. In Proceedings of the 6th WSEAS International Conference on Artificial Intelligence, Knowledge Engineering and Data Bases, Corfu Island, Greece, 16–19 February 2007.
6. Weaver, W.B.; Torres-Dodgen, A.V. Accurate two-dimensional Classification of Stellar Spectra with Artificial Neural Networks. *Astrophys. J.* **1997**, *487*, 847–857. [CrossRef]
7. Von Hippel, T.; Storrie-Lombardi, L.J.; Storrie-Lombardi, M.C.; Irwin, M.J. Automated Classification of Stellar Spectra—I. Initial Results with Artificial Neural Networks. *R.A.S. Mon. Not.* **1994**, *269*, 97–104. [CrossRef]
8. Snider, S.; Allende, C.; von Hippel, T.; Beers, T.C.; Sneden, C.; Qu, Y.; Rossi, S. Three-Dimensional Spectral Classification of Low-Metallicity Stars using Artificial Neural Networks. *Astrophys. J.* **2001**, *562*, 528–548. [CrossRef]
9. Manteiga, M.; Ordoñez, D.; Dafonte, C.; Arcay, B. ANNs and Wavelets: A Strategy for Gaia RVS Low S/N Stellar Spectra Parameterization. *Publ. Astron. Soc. Pac.* **2010**, *122*, 608–617. [CrossRef]
10. Sharma, K.; Kembhavi, A.; Kembhavi, A.; Sivarani, T.; Abraham, S.; Vaghmare, K. Application of convolutional neural networks for stellar spectral classification. *Mon. Not. R. Astron. Soc.* **2020**, *491*, 2280–2300. [CrossRef]
11. Brice, M.J.; Andonie, R. Automated Morgan Keenan Classification of Observed Stellar Spectra Collected by the Sloan Digital Sky Survey Using a Single Classifier. *Astron. J.* **2019**, *158*, 188. [CrossRef]
12. Liu, W.; Zhu, M.; Dai, C.; He, D.Y.; Yao, J.; Tian, H.F.; Wang, B.Y.; Wu, K.; Zhan, Y.; Chen, B.Q. Classification of large-scale stellar spectra based on deep convolutional neural network. *Mon. Not. R. Astron. Soc.* **2019**, *483*, 4774–4783. [CrossRef]
13. Paczynski, B. Monitoring All Sky for Variability. *Publ. Astron. Soc. Pac.* **2000**, *112*, 1281–1283. [CrossRef]

14. Gray, O.; Corbally, C.J. An expert computer program for classifying stars on the MK spectral classification system. *Astron. J.* **2014**, 147. [CrossRef]

15. Baehr, S.; Vedachalam, A.; Borne, K.; Sponseller, D. Data Mining the Galaxy Zoo Mergers. In Proceedings of the 2010 Conference on Intelligent Data Understanding, Mountain View, CA, USA, 5–6 October 2010; pp. 133–144.

16. Beck, M.R.; Scarlata, C.; Fortson, L.F.; Lintott, C.J.; Simmons, B.D.; Galloway, M.A.; Willett, K.W.; Dickinson, H.; Masters, K.L.; Marshall, P.J.; et al. Integrating human and machine intelligence in galaxy morphology classification tasks. *Mon. Not. R. Astron. Soc.* **2018**, *476*, 5516–5534. [CrossRef]

17. Zorich, L.; Pichara, K.; Protopapas, P. Streaming classification of variable stars. *Mon. Not. R. Astron. Soc.* **2020**, *492*, 2897–2909. [CrossRef]

18. Suárez, O.; Manteiga, M.; Rodríguez, A.; Dafonte, C.; Arcay, B.; Ulla, A.; García-Lario, P.; Manchado, A. Automatic classification of optical sources candidates to be in the post-AGB stage. In *Highlights of Spanish Astrophysics II, Proceedings of the 4th Scientific Meeting of the Spanish Astronomical Society, Santiago de Compostela, Spain, 11–14 September 2000*; Kluwer Academic Publishers: Dordrecht, The Netherlands, 2001.

19. Silva, D.R.; Cornell, M.E. A New Library of Stellar Optical Spectra. *Astrophys. J. Suppl. Ser.* **1992**, *81*, 865–881. [CrossRef]

20. Jacoby, G.H.; Hunter, D.A.; Christian, C.A. A Library of Stellar Spectra. *Astrophys. J. Suppl. Ser.* **1984**, *56*, 257–281. [CrossRef]

21. Pickles, A.J. A Stellar Spectral Flux Library: 1150-25000 Å. *Publ. Astron. Soc. Pac.* **1998**, *110*, 863–878. [CrossRef]

22. Prugniel, P.; Soubiran, C. A database of high and medium-resolution stellar spectra. *Astron. Astrophys.* **2001**, *369*, 1048–1057. [CrossRef]

23. SIMBAD Astronomical Database. Available online: http://simbad.u-strasbg.fr/simbad/ (accessed on 29 January 2020).

24. Ordoñez, D.; Dafonte, C.; Arcay, B.; Manteiga, M. HSC: A multi-resolution clustering strategy in Self-Organizing Maps applied to astronomical observations. *Appl. Soft Comput.* **2012**, *12*, 204–215. [CrossRef]

25. Fustes, D.; Manteiga, M.; Dafonte, C.; Arcay, B.; Ulla, A.; Smith, K.; Borrachero, R.; Sordo, R. An approach to the analysis of SDSS spectroscopic outliers based on self-organizing maps. *Astron. Astrophys.* **2013**, *559*, 1–10. [CrossRef]

26. Dafonte, C.; Fustes, D.; Manteiga, M.; Garabato, D.; Álvarez, M.A.; Ulla, A.; Allende Prieto, C. On the estimation of stellar parameters with uncertainty prediction from Generative Artificial Neural Networks: Application to Gaia RVS simulated spectra. *Astron. Astrophys.* **2016**, *594*, 1–10. [CrossRef]

27. Ordoñez, D.; Dafonte, C.; Manteiga, M.; Arcay, B. Parameterization of RVS synthetic stellar spectra for the ESA Gaia mission: Study of the optimal domain for ANN training. *Expert Syst. Appl.* **2010**, *37*, 1719–1727. [CrossRef]

28. Worthey, G.; Faber, S.M.; González, J.J.; Burstein, D. Old stellar populations. 5: Absorption feature indices for the complete LICK/IDS sample of stars. *Astrophys. J. Suppl. Ser.* **1994**, *94*, 687–722. [CrossRef]

29. Manteiga, M.; Carricajo, I.; Rodríguez, A.; Dafonte, C.; Arcay, B. STARMIND: A Fuzzy Logic Knowledge-based System for the Automated Classification of Stars in the MK System. *Astron. J.* **2009**, *137*, 3245–3253. [CrossRef]

30. Shorliffe, E.H.; Buchanan, B.G. *Rule-Based Expert Systems: The MYCIN Experiments of the Stanford Heuristic Programming Project*, 1st ed.; Addison Wesley: Boston, MA, USA, 1984; ISBN 978-0201101720.

31. Takagi, T.; Sugeno, M. Fuzzy identification of systems and its applications of modeling and control. *Ieee Trans. Syst. Man Cybern.* **1985**, *15*, 116–132. [CrossRef]

32. Cardiel, N. Formación Estelar en Galaxias Dominantes de Cúmulos. Ph.D. Thesis, Universidad Complutense de Madrid, Madrid, Spain, 1999. Available online: http://reduceme.readthedocs.io/en/latest/ (accessed on 29 February 2020).

33. Kohonen, T. Self-organized formation of topologically correct feature maps. *Biol. Cybern.* **1982**, *43*, 59–69. [CrossRef]

34. Fustes, D.; Dafonte, C.; Arcay, B.; Vallenari, A.; Luri, X. SOM ensemble for unsupervised outlier analysis. Application to outlier identification in the Gaia astronomical survey. *Expert Syst. Appl.* **2013**, *40*, 1530–1541. [CrossRef]

35. Howlett, R.J.; Jain, L.C. *Radial Basis Function Networks 1: Recent Developments in Theory and Applications*, 1st ed.; Physica-Verlag: Heilderberg, Germany, 2001; ISBN 978-3-7908-1367-8.

36. SNNS. Stuttgart Neural Network Simulator. Available online: http://www.ra.cs.uni-tuebingen.de/SNNS/welcome.html (accessed on 20 January 2020).

37. Kandel, A.; Langholz, G. *Hybrid Architectures for Intelligent Systems*, 1st ed.; CRC Press: Boca Ratón, FL, USA, 1992; ISBN 978-0-8493-4229-5.

38. Kohavi, R.; Provost, F. Glossary of Terms. *Spec. Issue Appl. Mach. Learn. Knowl. Discov. Process* **1998**, *30*, 271–274.

39. Suárez, O.; García-Lario, P.; Manchado, A.; Manteiga, M.; Ulla, A.; Pottasch, S.R. A spectroscopic atlas of post-AGB stars and planetary nebulae selected from the IRAS point source catalogue. *Astron. Astrophys.* **2006**, *458*, 173–180. [CrossRef]

40. Recio-Blanco, A.; de Laverny, P.; Allende Prieto, C.; Fustes, D.; Manteiga, M.; Arcay, B.; Bijaoui, A.; Dafonte, C.; Ordenovic, C.; Ordoñez Blanco, D. Stellar parametrization from Gaia RVS spectra. *Astron. Astrophys.* **2016**, *585*, 1–22. [CrossRef]

*Article*

# Improved Practical Vulnerability Analysis of Mouse Data According to Offensive Security based on Machine Learning in Image-Based User Authentication

**Kyungroul Lee [1] and Sun-Young Lee [2],***

[1] R&BD Center for Security and Safety Industries (SSI), Soonchunhyang University, Asan-si, Chungnam 31538, Korea; carpedm@sch.ac.kr

[2] Department of Information Security Engineering, Soonchunhyang University, Asan-si, Chungnam 31538, Korea

* Correspondence: sunlee@sch.ac.kr; Tel.: +82-41-530-1357

Received: 9 February 2020; Accepted: 17 March 2020; Published: 18 March 2020

**Abstract:** The objective of this study was to verify the feasibility of mouse data exposure by deriving features to improve the accuracy of a mouse data attack technique using machine learning models. To improve the accuracy, the feature appearing between the mouse coordinates input from the user was analyzed, which is defined as a feature for machine learning models to derive a method of improving the accuracy. As a result, we found a feature where the distance between the coordinates is concentrated in a specific range. We verified that the mouse data is apt to being stolen more accurately when the distance is used as a feature. An accuracy of over 99% was achieved, which means that the proposed method almost completely classifies the mouse data input from the user and the mouse data generated by the defender.

**Keywords:** practical security; offensive security; user authentication; machine learning; vulnerability analysis

---

## 1. Introduction

Due to the emergence of an online society, the need for technologies for online user authentication has been on the rise. A representative online user authentication technology is a password-based authentication technology where the information, ID, and password that necessary for authentication is input from the keyboard [1]. However, attackers have come up with keyboard data attack techniques that they use to expose keyboard data input from a user [2–4]. To counteract these security threats of keyboard data, researchers have come up with an image-based authentication technology [5–7]. This technology uses a specific location chosen by the user as a password in the image displayed on the screen. Since in this image-based authentication, the password is input from the mouse, security threats of the password-based user authentication methods that result from the exposure of the keyboard data are avoided. Nevertheless, just like with the keyboard data, a security threat has been found that does not ensure the security of image-based authentication because of the exposure of mouse data [8–11].

Specifically, in image-based authentication, the authentication information that must be protected is image data to be displayed and mouse-click data to be input. In this study, we focused on mouse-click data. The operating system manages the mouse position for the interaction with the user and provides API, GetCursorPos() function, to obtain the current mouse position on the screen. Thus, an attacker can call the API periodically at short intervals to track the movement of the mouse data that is input from the user, i.e., the attacker can steal the mouse position that the user inputs, which means that image-based authentication is neutralized by exposing the password. To counteract this vulnerability, the mouse data protection technology has emerged. The key idea of this technology is not to prevent the exposure of mouse data, but to confuse the attacker so that the actual mouse position input from the user is unknown. Namely, by using the SerCursorPos() function to randomly generate the mouse position known only to the defender, the attacker cannot differentiate the coordinates input from the mouse from the coordinates generated by the defense tool [10]. As a result, the attacker collects both the random mouse data and the real data input by the user, which means that the attacker does not steal the user's password in the image-based authentication, because the attacker does not know the random location generated by the defender.

Accordingly, attackers have come up with attack techniques to track mouse movements by neutralizing the mouse data protection technique. Among them, an attack technique based on machine learning that effectively classifies the actual mouse position input by the user among all collected mouse data positions has emerged [12]. This technique not only requires the attacker to have a high level of attack techniques, but has also a high accuracy rate. Specifically, the maximum accuracy is 98%, and in order to classify the mouse position with high accuracy, the feature of the datasets for machine learning models is composed of the collectable X and Y coordinates as well as the elapsed time of the collected mouse data. Although this technique has a high accuracy demand, attackers still require a technique to completely steal mouse data. Therefore, in this study, we examined the feasibility of an attack technique that improves the accuracy by constructing features to the existing datasets.

The contributions of this paper are as follows:

- Mouse data attack technique using the existing machine learning has defined the elapsed time and collected coordinates as features and has a high attack success rate. In this study, we improved the accuracy of the attack success rate by constructing the distance of the collected coordinates to the features. Hence, attackers can more effectively steal passwords by tracking the mouse movement input from the users in image-based authentication;
- The accuracy of the existing attack technique is 98%; that is, attackers have a 98% attack success rate. Although this attack has a high accuracy, 2 of 100 do not track the mouse position. Moreover, when the user moves the mouse once, attackers may be unable to track the mouse movement almost completely, considering that many mouse coordinates are generated quickly. On the other hand, the attack technique proposed in this study has more than 99% accuracy of many machine learning models in multiple datasets, which means that the attack success rate is higher than that of the existing attack techniques. Overall, the proposed attack technique tracks the mouse movement input by the user more effectively in image-based authentication and can steal passwords with high probability.

This paper is organized as follows: Section 2 describes the mouse data transfer process and the existing mouse data attack and defense techniques that are necessary for understanding the proposed mouse data attack technique. Section 3 presents the dataset configuration and features for the proposed attack technique, and Section 4 shows the experiment results. Finally, we conclude this paper in Section 5.

## 2. Related Studies

This section describes the process of transferring mouse data from a mouse device to an application program, which is part of the important information in the image-based authentication technology.

### 2.1. Mouse Data Transfer Process

The operating system supports the interaction between the mouse device and the user and manages the cursor position on the screen based on the data transmitted from the mouse device. Data transferred from the mouse device is collected by a device driver for the mouse device of the operating system, and the collected mouse data are transferred to the application program to support the interaction with the user. Figure 1 shows the mouse data transfer process.

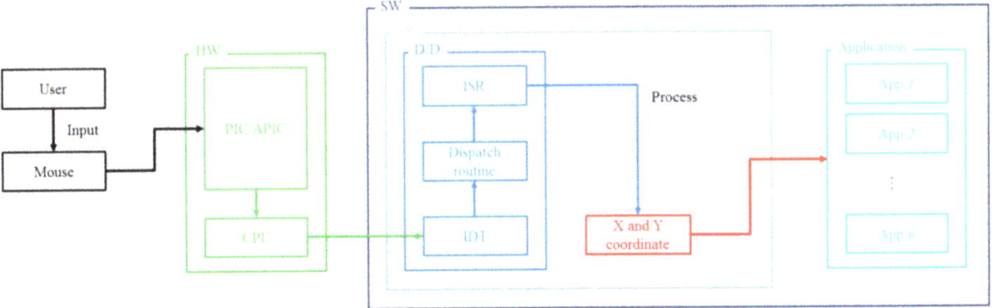


**Figure 1.** Mouse data transfer process.

The host PC uses interrupts to handle requests for input and output from the devices, i.e., when the user moves the mouse device, the mouse device generates an interrupt to the host PC (CPU), which calls the interrupt handler in the device driver of the operating system to handle it, and the mouse data is processed in that handler and passed to the application program.

## 2.2. Mouse Data Attack and Defense Techniques

The operating system provides a mouse cursor for interacting with the user and positions the mouse cursor based on the mouse data input by the user. For the interaction to occur, the operating system must manage the position of the mouse cursor and provide APIs related to the mouse, GetCursorPos() function, for supporting requests from application programs. Therefore, the attacker collects the mouse position on the screen, which represents the movement of the mouse input by the user, by periodically calling this GetCursorPos() function at short intervals.

To protect the mouse movement from being exposed by the attack technique, a defender has been proposed to prevent stealing the mouse data input from the actual user rather than preempting the mouse data [10]. This technique prevents exposure of the actual mouse data by generating random mouse positions in a short period of random times by the defender. Attackers collect both the random mouse positions generated by the defender and the actual mouse positions input by the user, however, they cannot classify the mouse positions since the random mouse positions generated by the defender are not known. Therefore, the defender effectively prevents mouse data from being exposed.

An attacker requires a technology to completely steal mouse data, and the feasibility of data stealing has been studied when a mouse data protection technology is applied [12]. In this study, both random mouse data generated by the defender and actual mouse data input by the user were collected, and the security of mouse data was verified by classifying mouse data using machine learning models by configuring datasets from the collected data. As a result, this technique verified that the mouse data was stolen with a high attack success rate of 98%, by applying the machine learning.

In this study, we proposed a method to improve the accuracy by the defining and analyzing the feature appearing between mouse coordinates input by the user.

## 3. Feature Extraction and Dataset Configuration

### 3.1. Feature Extraction

The proposed method uses an existing attack technique that collects all mouse positions by periodically calling the GetCursorPos() function. As described in Section 2.2, simply using this attack technique does not guarantee mouse data attack success. However, in order to effectively steal mouse data, we analyze the characteristics of mouse coordinates and configure features based on the analyzed results. The previous attack technique based on machine learning used the elapsed time and the collected coordinates as a feature. To analyze the characteristics of this feature, we derived the distribution of coordinates with respect to elapsed time, X coordinates, Y coordinates, distance between X coordinates, and distance between Y coordinates. Figure 2 shows the distribution of the coordinates according to the elapsed time.

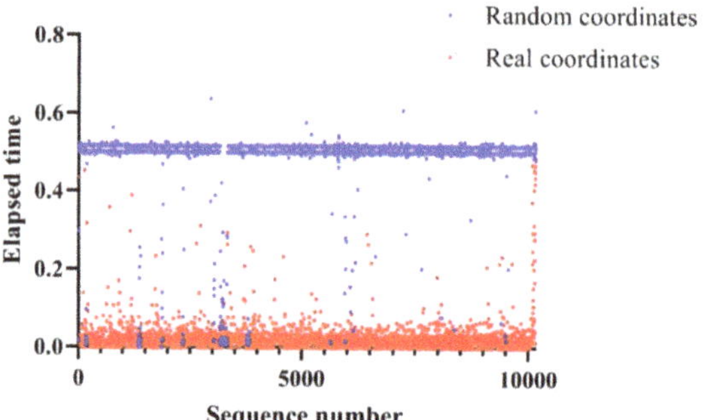

**Figure 2.** Distribution of the coordinates according to the elapsed time.

In Figure 2, all points are a distribution of the coordinates according to the elapsed time. The red dots are the real coordinates while the blue dots are the random coordinates. The distribution of random coordinates is relatively periodic and has a distribution close to 0.5, because random coordinates are generated by the defense tool. Conversely, the distribution of real coordinates is aperiodic and has a distribution near zero, because real coordinates are input from the user. Therefore, we hypothesized that it is possible to classify random coordinates and real coordinates based on the elapsed time. Figure 3 shows the distribution of the coordinates with respect to the X coordinate.

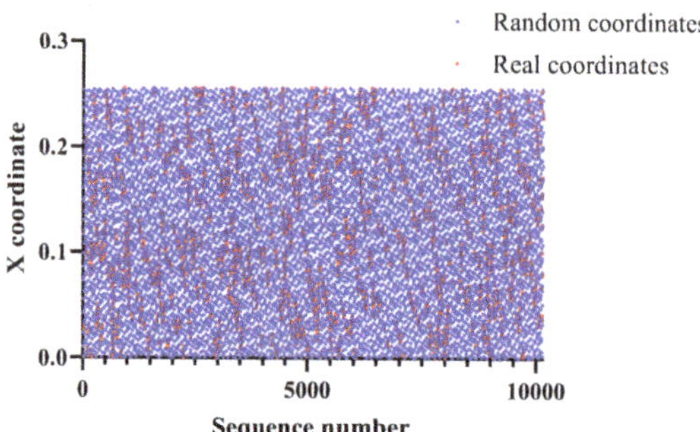

**Figure 3.** Distribution of the coordinates according to the X coordinate.

Specifically, all points are a distribution of the coordinates according to the X coordinate. The red dots are the real coordinates while the blue dots are the random coordinates. Random coordinates have an overall wide distribution ranging from 0 to 0.24, while real coordinates have an intensive distribution with a relatively arbitrary range. Therefore, we hypothesized that it is possible to classify random coordinates and real coordinates based on X coordinates. Figure 4 shows the distribution of the coordinates according to the Y coordinate.

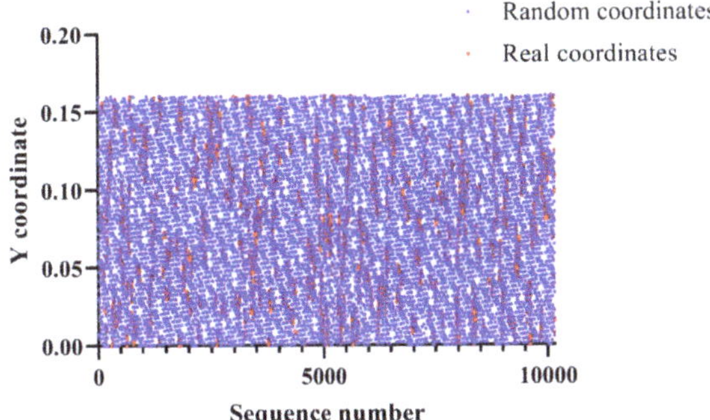

**Figure 4.** Distribution of the coordinates according to the Y coordinate.

All points are a distribution of the coordinates with respect to the Y coordinate. The red dots are the real coordinates while the blue dots are the random coordinates Random coordinates have an overall wide distribution ranging from 0 to 0.16, while real coordinates have an intensive distribution with a relatively arbitrary range. Therefore, we hypothesized that it is possible to classify random coordinates and real coordinates based on Y coordinates. Figure 5 shows the distribution of the coordinates with respect to the distance between previous X coordinates and current X coordinates.

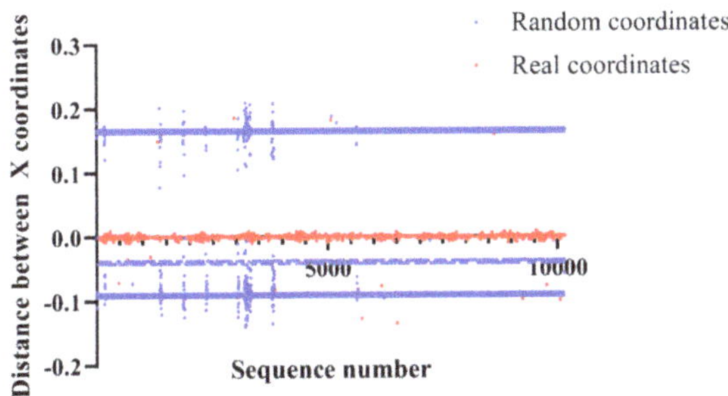

**Figure 5.** Distribution of the coordinates according to the distance between previous X coordinates and current X coordinates.

All points are a distribution of the coordinates with respect to the distance between previous X coordinates and current X coordinates. The red dots are the real coordinates while the blue dots are the random coordinates. Random coordinates have intensive distributions at 0.16, −0.04, and −0.08, while real coordinates have an intensive distribution range from −0.01 to 0.01. Therefore, we hypothesized that it is possible to classify random coordinates and real coordinates based on the distance between previous X coordinates and current X coordinates. Figure 6 shows the distribution of the coordinates with respect to the distance between previous Y coordinates and current Y coordinates.

All points are a distribution of the coordinates with respect to the distance between previous Y coordinates and current Y coordinates. The red dots are the real coordinates while the blue dots are the random coordinates. Random coordinates have intensive distributions at 0.12, 0.04, 0.02, −0.03, −0.11, and −0.13, while real coordinates have an intensive distribution range from −0.01 to 0.01. Therefore, we hypothesized that it is possible to classify random coordinates and real coordinates based on the distance between previous Y coordinates and current Y coordinates. We analyzed the distribution of coordinates based on the elapsed time, X coordinates, Y coordinates, distance between X coordinates, and distance between Y coordinates. After a comprehensive analysis, we derived that each distribution has a distribution that classifies random and real coordinates. Consequently, defining these characteristics as features is expected to classify mouse data more precisely.

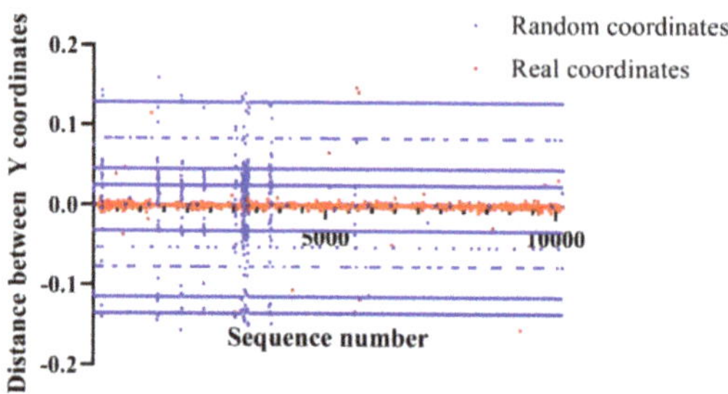



**Figure 6.** Distribution of the coordinates according to the distance between previous Y coordinates and current Y coordinates.

To classify the actual mouse data input by the mouse, the mouse data collected from the attack tool is composed as a dataset, with data collected by setting four periods, 50 ms, 100 ms, 250 ms, and 500 ms. As in a previous study [12], datasets collected using the existing attack technique were used in the same way, and the distance between the previous coordinates and the current coordinates was added as a feature to configure datasets for the experiment. Specifically, the collected X and Y coordinates, the elapsed time, and the distance between the previous coordinates and the current coordinates are defined as features.

*3.2. Feature Definition*

In this subsection, we describe the features defined in this paper. The previous research defined time difference and coordinates as features, while this paper defined five features for machine learning models. The defined features are the elapsed time, X coordinates, Y coordinates, distance between X coordinates, and distance between Y coordinates.

3.2.1. Elapsed Time

Elapsed time is the time when the X and Y coordinates are acquired in ns unit. We defined the difference between the time when the previous mouse data was collected and the time when the current mouse data was collected as a feature as shown in Equation (1), and used it as the elapsed time feature.

$$\text{elapsed time} = T_C - T_P. \tag{1}$$

3.2.2. X Coordinates and Y Coordinates

X coordinate features and Y coordinate features are the X and Y coordinates data of all the random and real coordinates, respectively. Each coordinate is represented as the X and Y coordinate on the screen, so it has a range equal to the screen size. Therefore, we preprocess these values to range from 0 to 1 in order to facilitate learning of data for machine learning models. The real mouse data input from the user collected arbitrary coordinates by manually moving the mouse device by human hand, while the random mouse data generated by the defense tool generated random coordinates obtained by calling a random function. Specifically, random coordinates are generated periodically by a timer function called at 50 ms, 100 ms, 250 ms, and 500 ms intervals.

3.2.3. Distance between X Coordinates and Distance between Y Coordinates

Distance between the X coordinates feature and distance between the Y coordinates feature are the distance from the previously collected X and Y coordinates to the currently collected X and Y coordinates. To measure the distance, the previous coordinate value is subtracted from the current coordinate value, and the final distance value ranges from 0 to 1, because the coordinates have values between 0 and 1 after preprocessing. Nevertheless, if the value is subtracted, it is expressed as a negative number, therefore, to facilitate the data learning for machine learning models, the absolute value is obtained to have a positive value as shown in Equations (2) and (3). These distances are used as features because it is easy to classify the data as the coordinates have sequential characteristics when the user moves the mouse device.

$$\text{Distance between X coordinates} = |\, C_{Xcor} - P_{Xcor}\,|, \tag{2}$$

$$\text{Distance between Y coordinates} = |\, C_{Ycor} - P_{Ycor}\,|. \tag{3}$$

*3.3. Machine Learning Models and Performance Evaluation*

The machine learning models utilized in this study are KNN [13], Logistic Regression [14], Decision Tree [15], Random Forest [16], Gradient Boosting Regression Tree [16], Support Vector Machine (SVM) [17], and MLP [18]. KNN classifies data based on the decision boundary, and Logistic Regression classifies data using linear function as shown in Equation (4). Decision Tree learns data by repeating yes and no questions until it reaches a decision, and SVM classifies the data based on the distance from the data points located at the decision boundary, as shown in Equation (5). Finally, MLP classifies data through deep learning by configuring hidden units.

$$\hat{y} = w[0] \times x[0] + w[1] \times x[1] + \ldots + w[p] \times x[p] + b > 0, \tag{4}$$

$$k_{rbf}(x_1, x_2) = \exp(-\gamma \,\|\, x_1 - x_2 \,\|\, 2). \tag{5}$$

To evaluate the performance of machine learning models, we used accuracy, precision, recall, and F1-score for each model. Accuracy refers to an attack success rate as shown in Equation (6), and precision refers to the ratio of the results that are actually true to the total results as shown in Equation (7). Recall is the ratio of the results classified by the model as true to the total result as shown in Equation (8), and the F1-score is the harmonic mean of precision and recall as shown in Equation (9).

$$ACCURACY = \frac{TP + TN}{TP + TN + FP + FN} \tag{6}$$

$$PRECISION = \frac{TP}{TP + FP} \tag{7}$$

$$RECALL = \frac{TP}{TP + FN} \tag{8}$$

$$F1 - score = 2 \times \frac{PRECISION\,RECALL}{PRECISION + RECALL} \tag{9}$$

## 4. Experiment Results

This section describes the experiment results of model validation, accuracy, precision, recall, F1-score, and AUC to evaluate the performance of the proposed attack technique in order to prove its contribution. The machine learning models used in this paper are KNN, Logistic Regression, Decision Tree, Random Forest, Gradient Boosting Regression Tree, SVM, and MLP. For the experiments, the training sets, validation sets, and test sets were classified into any number to overcome overfitting and underfitting. Table 1 shows the results of each set based on dataset 1-1. For learning, we divided the training set and the test set into a ratio of 3:1 for all the data, and for the verification, the training set divided the training set and the validation set into a ratio of 3:1. Moreover, we set the seed values to include data randomly in each set when dividing the data for reasonable learning.

As shown in the table, the training set has the highest score with a Random Forest of 1.0, and the validation set and the test set have a score of 0.98 for most of the models except for Logistic Regression. To evaluate the performance of the proposed attack technique, Figure 7 shows the evaluations of accuracy, precision, recall, F1-score, and AUC for datasets 1-1 to 1-4.

**Table 1.** Training set, validation set, and test set scores of dataset 1-1 with optimal parameters.

| Model | Parameters | Training Set Score | Validation Set Score | Test Set Score |
|---|---|---|---|---|
| KNN | n_neighbors = 4 | 0.98 | 0.98 | 0.98 |
| Logistic regression | C = 10, L2 regularization | 0.75 | 0.76 | 0.75 |
| Decision tree | max_depth = 6 | 0.98 | 0.98 | 0.98 |
| Random forest | n_estimators = 12 | 1.00 | 0.98 | 0.98 |
| Gradient boosting | learning_rate = 0.01, max_depth = 9 | 0.99 | 0.98 | 0.98 |
| SVM | C = 100,000 | 0.98 | 0.98 | 0.98 |
| MLP | max_iter = 1000, alpha = 0.01 | 0.97 | 0.98 | 0.98 |

As shown in the figure, most of the datasets work well except for the Logistic Regression. The lowest performances are 0.75, 0.88, 0.96, and 0.97 for the Logistic Regression in datasets 1-1, 1-2, 1-3, and 1-4, respectively. The highest performances are 0.981 for the Random Forest in dataset 1-1, 0.995 for the Decision Tree, Random forest, and Gradient Boosting Regression Tree in dataset 1-2, 0.998 for the KNN, Decision Tree, Random Forest, Gradient Boosting Regression Tree, and SVM in dataset 1-3, and 0.998 for KNN, Decision Tree, Gradient Boosting Regression Tree, and SVM in dataset 1-4. Comparing between the proposed method and the existing method, the performances are improved in most of the models except for Logistic Regression and Decision Tree in dataset 1-1, and performances are improved in most of the models except for Logistic Regression in dataset 1-2. In dataset 1-3, Random Forest and SVM improved performance, while only Random Forest improved performance in dataset 1-4. Experiment results do not show an improved performance in many models. However, in most recall and AUC results, there were performance degradations of only 0.0018 on average (17 items, a total decrease of 0.031), while all other items increased by an average of 0.073 (141 items, a total increase of 10.351); hence, the increase rate is 4.055% higher than the decrease rate.

To evaluate the performance of the proposed attack technique, Figure 8 shows the results of each set of accuracy, precision, recall, F1-score, and AUC based on datasets 2-1 to 2-4.

As shown in the figure, most of the datasets show a high performance except for Logistic Regression. The lowest performances were 0.82, 0.91, 0.97, and 0.98 for Logistic Regression in datasets 2-1, 2-2, 2-4, and 2-4, respectively. The highest performances were 0.986 for the Random Forest and Gradient Boosting Regression Tree in dataset 2-1, 0.997 for the Random Forest in dataset 2-2, 0.998 for the KNN, Decision Tree, Random Forest, Gradient Boosting Regression Tree, SVM, and MLP in dataset 2-3, and 0.998 for the Decision Tree, Random Forest, Gradient Boosting Regression Tree, and SVM in dataset 2-4. Comparing between the proposed method and the existing method, performances are improved in all models in dataset 2-1 and in most of the models except for Logistic Regression in dataset 2-2. In dataset 2-3, performances are improved in Logistic Regression, Random Forest, and Gradient Boosting Regression Tree, while only Random Forest and MLP have an improved performance in dataset 2-4. Just as in datasets 1-1 to 1-4, the results do not seem to improve the performance in many models, however, most result in only a 0.001 performance degradation on recall and AUC (13 items, a total decrease of 0.129) and all other items were increased by an average of 0.054 (147 items, a total increase of 8.015); hence, the increase rate was 540%, which is higher than the decrease rate.

As described above, in this study, we evaluated the performance of datasets including the elapsed time and the X and Y coordinates, which is the existing attack technique, and the datasets including the distance between the previous and current coordinates. Consequently, we verified through experiments that the performance of datasets with distance is higher than that of datasets with elapsed time only. Therefore, the results of the performance evaluation are changed by defining features that can be included in the datasets, and we proved the contribution of the proposed method by comparing the performance evaluations. We compared the results of the training set, the validation set, and the test set, and also compared the results of performance, accuracy, precision, recall, F1-score, and AUC. Figure 9 shows the comparison of the training set, the validation set, and the test set.

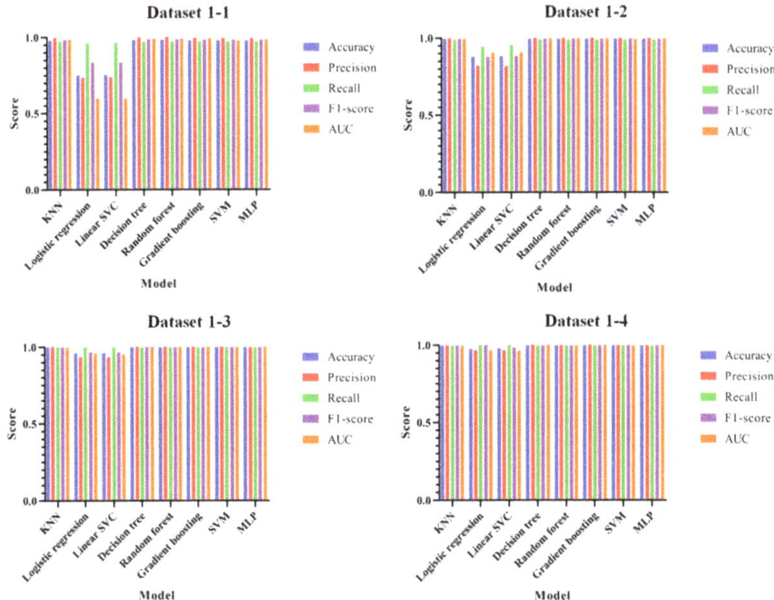

**Figure 7.** Performance evaluations of accuracy, precision, recall, F1-score, and AUC for datasets 1-1 to 1-4.

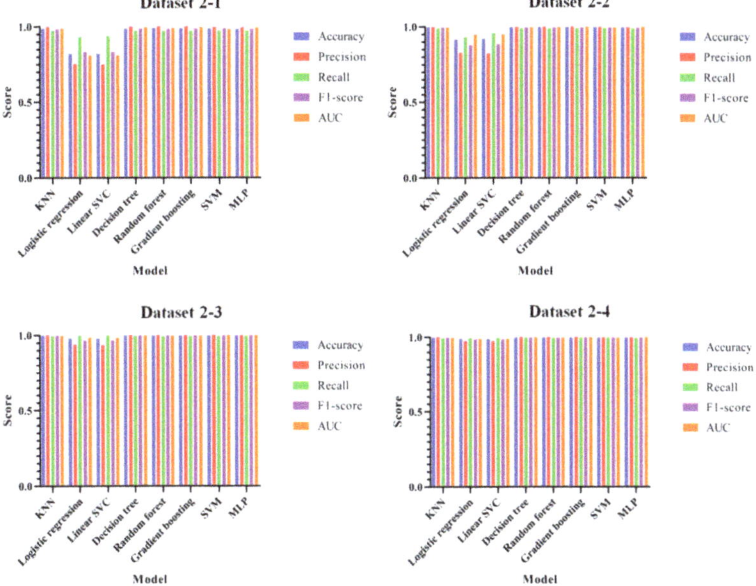

**Figure 8.** Performance evaluations of accuracy, precision, recall, F1-score, and AUC for datasets 2-1 to 2-4.

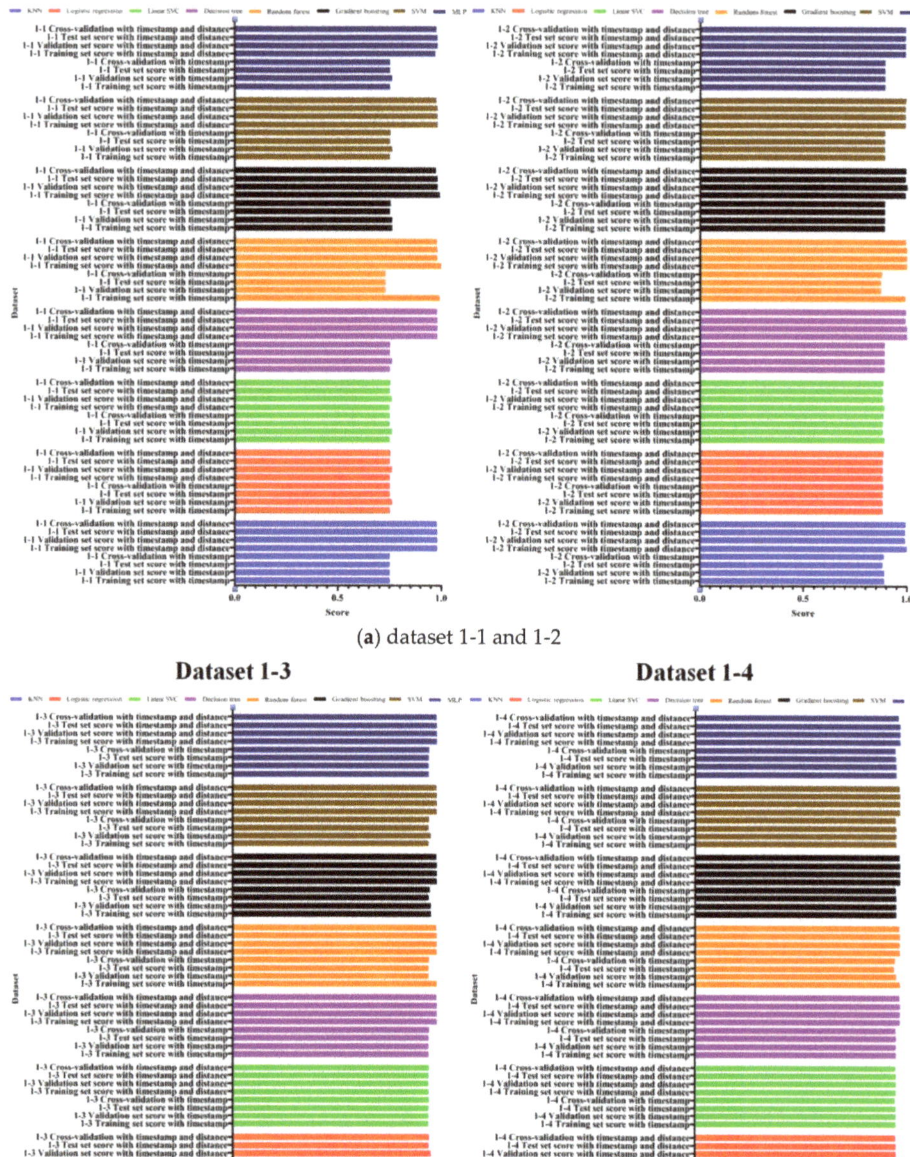

(**a**) dataset 1-1 and 1-2

(**b**) dataset 1-3 and 1-4

**Figure 9.** *Cont.*

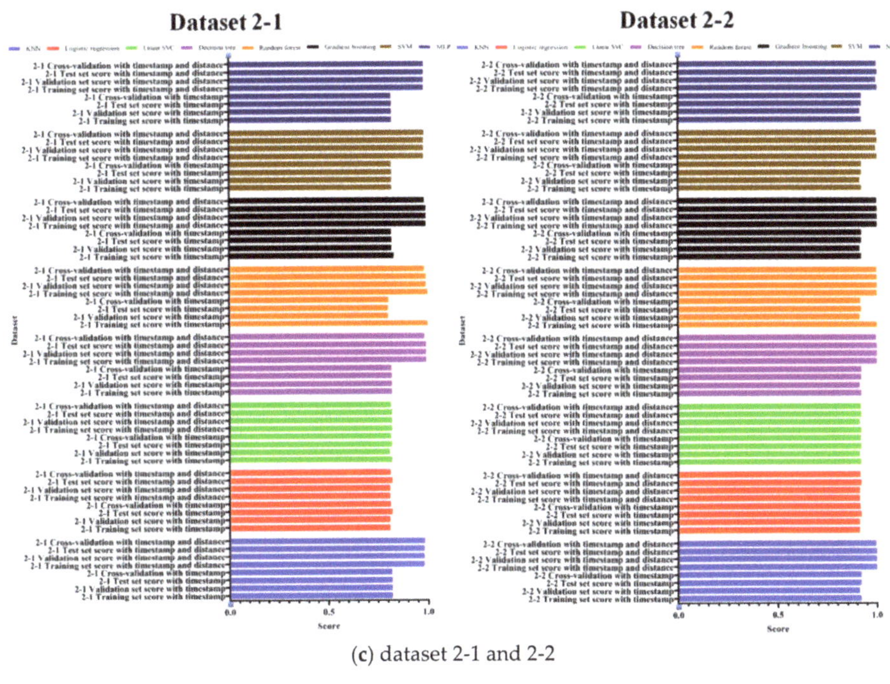

(**c**) dataset 2-1 and 2-2

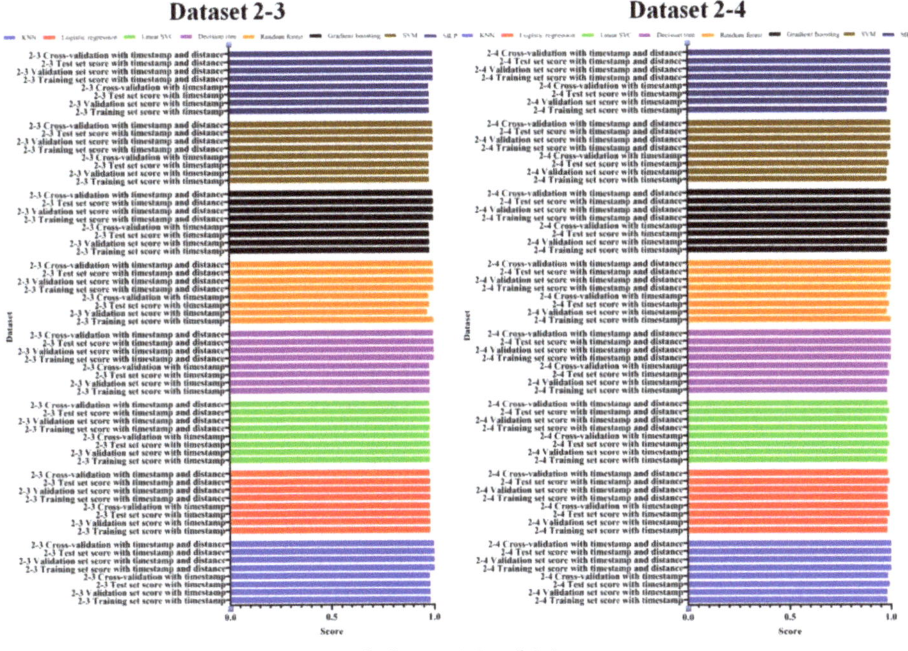

(**d**) dataset 2-3 and 2-4

**Figure 9.** Performance evaluation of the proposed and existing methods in the training set, validation set, and test set. (**a**) dataset 1-1 and 1-2; (**b**) dataset 1-3 and 1-4; (**c**) dataset 2-1 and 2-2; (**d**) dataset 2-3 and 2-4.

As shown in the figures, the performances of each model are represented by colors. The top four in each model are the results of the datasets that include the distance between the previous and current coordinates proposed in this study, while the bottom four are the results of the datasets that do not include the distance and were used in the previous study. Therefore, the performance of the proposed method is improved in most datasets and machine learning models. An increase in the performance trend was observed from dataset 1 to 4, and on average, all datasets including distance outperformed all the datasets that do not include distance. Moreover, all scores in datasets 1-4 and 2-4 have a near-perfect performance of over 99%, which results in a higher performance for constructing features with datasets containing distances. Regarding the changes in the training set, the validation set, and the test set, the results of most models of datasets that do not include distance have significant changes compared to models of datasets including distance that have relatively small changes.

As described above, the results of the training set, the validation set, and the test set showed significant differences, which means that the datasets including the distance are verified to have a high performance. Figure 10 compares the accuracy, precision, recall, F1-score, and AUC to evaluate the more practical performance.

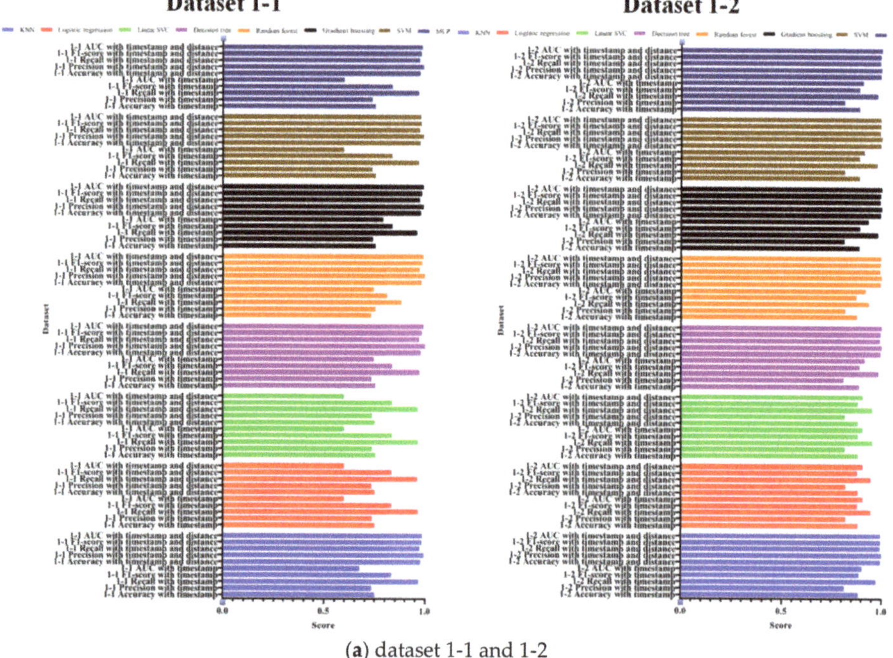

(**a**) dataset 1-1 and 1-2

**Figure 10.** *Cont.*

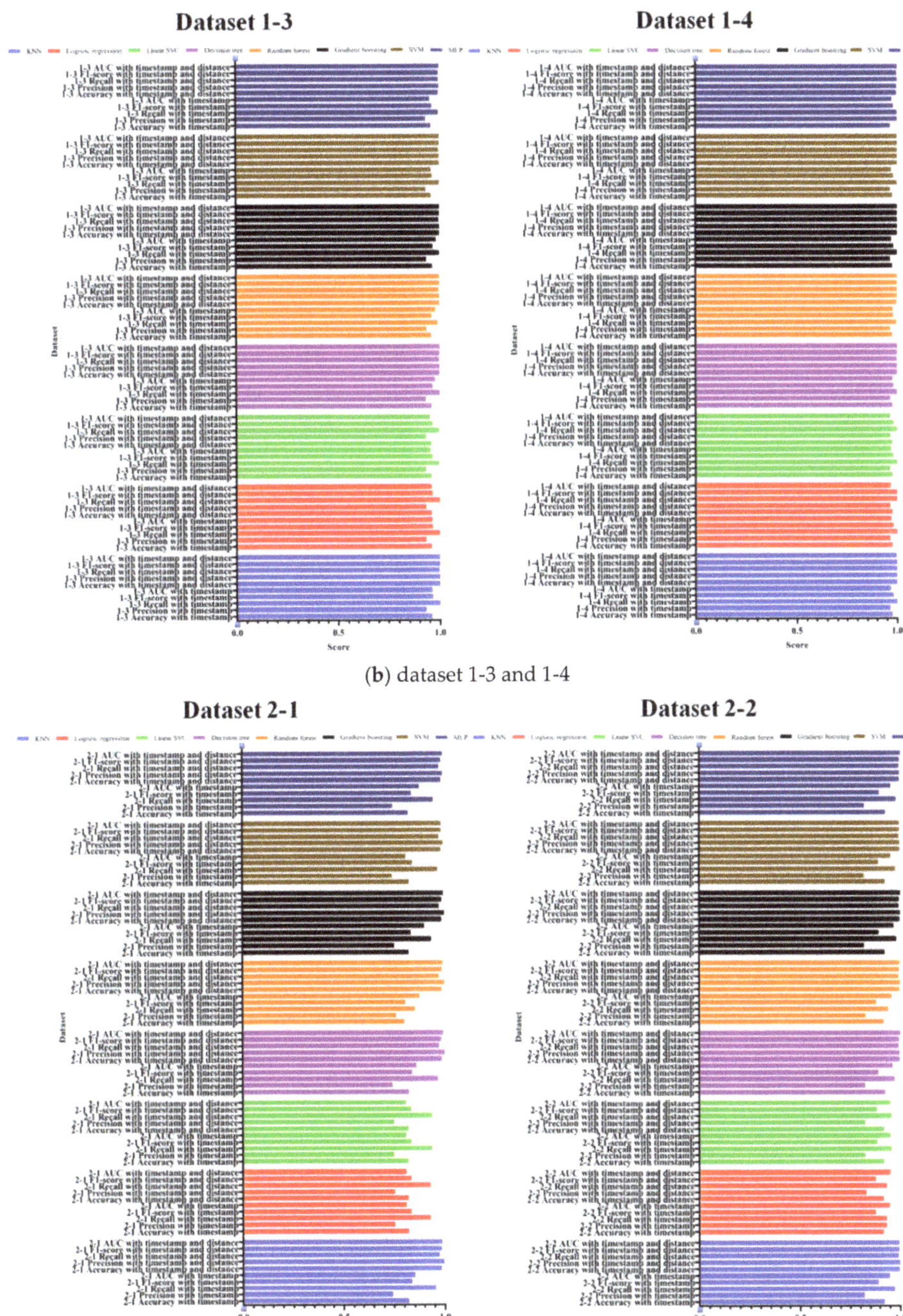

(**b**) dataset 1-3 and 1-4

(**c**) dataset 2-1 and 2-2

**Figure 10.** *Cont.*

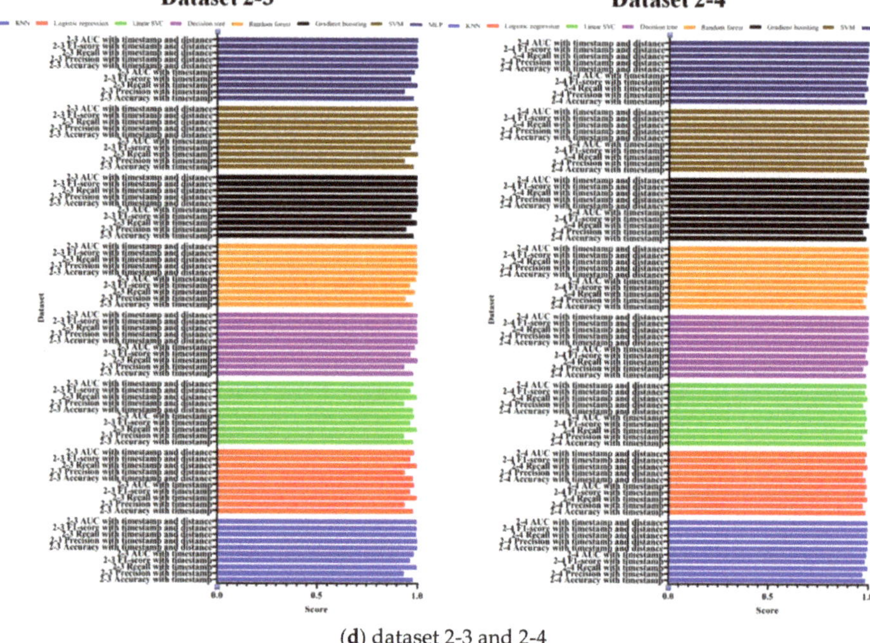



(**d**) dataset 2-3 and 2-4

**Figure 10.** Performance evaluation of the proposed and existing methods with respect to accuracy, precision, recall, F1-score, and AUC. (**a**) dataset 1-1 and 1-2; (**b**) dataset 1-3 and 1-4; (**c**) dataset 2-1 and 2-2; (**d**) dataset 2-3 and 2-4.

As shown in the figures, the performances of each model are shown by colors. The top four in each model are the results of the datasets that include the distance between the previous and current coordinates as proposed in this study, while the bottom four are the results of the datasets that do not include the distance. The performance of the proposed method is improved in most datasets and machine learning models. A trend of better performance was observed as datasets went from 1 to 4, and datasets with distance have a significantly higher performance than datasets without distance. Moreover, most models of all datasets except for datasets 1-1 and 2-1 have more than 99% accuracy. In other words, most datasets have 99% accuracy, which indicates that the proposed technique effectively classifies random coordinates generated by the defender in the image-based authentication method, compared to 98%, which is the accuracy of the previous attack technique, by simply collecting mouse coordinates. Hence, the proposed method more effectively steals passwords, which are the actual mouse data input by the user.

## 5. Conclusions

This study analyzed the security of mouse data by deriving features to further improve the accuracy of existing attack techniques using machine learning based on mouse data in image-based authentication. The existing attack technique defines elapsed time and X and Y coordinates as features, classifies random mouse data generated by defender and actual mouse data input by user, with a maximum accuracy of 98%. In this study, we analyzed the distribution of coordinates based on the distance between coordinates and the elapsed time. The distance between the mouse coordinates input by the user is concentrated in a specific range. Therefore, we configured datasets that defined the distance between the previous coordinates and the current coordinates as a feature. With experiment results based on the configured datasets, we verified that the actual mouse data input by the user was classified more effectively than it was by the existing attack techniques, with an accuracy of 99%, which means that the mouse data input by the user is classified almost completely. In conclusion, the proposed attack technique can effectively steal the mouse data input by the user and the passwords as the actual mouse data input by the user.

**Author Contributions:** Conceptualization, K.L. and S.-Y.L.; methodology, S.-Y.L.; software, K.L.; data curation, K.L.; writing—original draft preparation, K.L.; writing—review and editing, S.-Y.L.; project administration, S.-Y.L. All authors have read and agreed to the published version of the manuscript.

**Funding:** This research was partially supported by the National Research Foundation of Korea (NRF) grant funded by the Korea government (MSIT) (No. 2018R1A4A1025632) and the Basic Science Research Program through the National Research Foundation of Korea (NRF) funded by the Ministry of Education (NRF-2018R1D1A1B07047656), and the Soonchunhyang University Research Fund.

**Acknowledgments:** A part of this paper was presented in a conference at the 5th International Congress on Information and Communication Technology (ICICT), 20 and 21 February 2020, London, United Kingdom.

**Conflicts of Interest:** The authors declare no conflicts of interest.

## References

1. Conklin, A.; Dietrich, G.; Walz, D. Password-based authentication: A system perspective. In Proceedings of the 37th Annual Hawaii International Conference on System Sciences, Big Island, HI, USA, 5–8 January 2004.
2. Lee, K.; Yim, K. Password sniff by forcing the keyboard to replay scan codes. In Proceedings of the Joint Workshop Information Security, Guangzhou, China, 5–6 August 2010.
3. Lee, K.; Yim, K. Keyboard security: A technological review. In Proceedings of the International Conference on Innovative Mobile and Internet Services in Ubiquitous Computing, Seoul, Korea, 30 June–2 July 2011.
4. Oh, I.; Lee, K.; Lee, S.; Do, K.; Ahn, H.; Yim, K. Vulnerability Analysis on the Image-Based Authentication Through the PS/2 Interface. In Proceedings of the International Conference on Innovative Mobile and Internet Services in Ubiquitous Computing, Matsue, Japan, 4–6 July 2018.
5. Takada, T.; Koike, H. Awase-E: Image-based authentication for mobile phones using user's favorite images. In Proceedings of the International Conference on Mobile Human-Computer Interaction, Udine, Italy, 8–11 September 2003.
6. Parekh, A.; Pawar, A.; Munot, P.; Mantri, P. Secure authentication using anti-screenshot virtual keyboard. *Int. J. Comput. Sci. Issues* **2011**, *8*, 534–537.
7. Newman, R.E.; Harsh, P.; Jayaraman, P. Security analysis of and proposal for image-based authentication. In Proceedings of the 39th Annual 2005 International Carnahan Conference on Security Technology, Las Palmas, Spain, 11–14 October 2005.
8. Lee, H.; Lee, Y.; Lee, K.; Yim, K. Security assessment on the mouse data using mouse loggers. In Proceedings of the International Conference on Broad-Band Wireless Computing, Communication and Applications, Asan, Korea, 5–7 November 2016.
9. Oh, I.; Lee, Y.; Lee, H.; Lee, K.; Yim, K. Security Assessment of the Image-based Authentication using Screen-capture Tools. In Proceedings of the International Conference on Innovative Mobile and Internet Services in Ubiquitous Computing, Torino, Italy, 10–12 July 2017.
10. Lee, K.; Oh, I.; Yim, K. A Protection Technique for Screen Image-based Authentication Protocols Utilizing the SetCursorPos function. In Proceedings of the World conference on Information Security Applications, Jeju Island, Korea, 24–26 August 2017.
11. Lee, K.; Yim, K. Vulnerability Analysis on the Image-based Authentication: Through the WM_INPUT message. In Proceedings of the International Workshop on Convergence Information Technology, Busan, Korea, 21–23 December 2017.
12. Lee, K.; Esposito, C.; Lee, S. Vulnerability Analysis Challenges of the Mouse Data based on Machine Learning for Image-based User Authentication. *IEEE Access* **2019**, *7*, 177241–177253. [CrossRef]
13. Zhang, M.; Zhou, Z. ML-KNN: A lazy learning approach to multi-label learning. *Pattern Recognit.* **2007**, *40*, 2037–2048. [CrossRef]
14. Cheng, W.; Hüllermeier, E. Combining instance-based learning and logistic regression for multilabel classification. *Mach. Learn.* **2009**, *76*, 211–225. [CrossRef]

15. Sinclair, C.; Pierce, L.; Matzner, S. An application of machine learning to network intrusion detection. In Proceedings of the 15th Annual Computer Security Applications Conference, Phoenix, AZ, USA, 6–10 December 1999.

16. Rovert, E.; Lawrence, O.; Kevin, W.; Kegelmeyer, W. A Comparison of Decision Tree Ensemble Creation Techniques. *IEEE Trans. Pattern Anal. Mach. Intell.* **2007**, *29*, 173–180.

17. Jan, S.U.; Lee, Y.; Shin, J.; Koo, I. Sensor Fault Classification Based on Support Vector Machine and Statistical Time-Domain Features. *IEEE Access* **2017**, *5*, 8682–8690. [CrossRef]

18. Yin, C.; Zhu, Y.; Fei, J.; He, X. A Deep Learning Approach for Intrusion Detection Using Recurrent Neural Networks. *IEEE Access* **2017**, *5*, 21954–21961. [CrossRef]

*Article*

# Improved Parsimonious Topic Modeling Based on the Bayesian Information Criterion

**Hang Wang and David Miller** *

Electrical Engineering and Computer Science Department, The Pennsylvania State University, State College, PA 16802, USA; hzw81@psu.edu
* Correspondence: djm25@psu.edu

Received: 14 February 2020; Accepted: 9 March 2020; Published: 12 March 2020

**Abstract:** In a previous work, a parsimonious topic model (PTM) was proposed for text corpora. In that work, unlike LDA, the modeling determined a subset of salient words for each topic, with topic-specific probabilities, with the rest of the words in the dictionary explained by a universal shared model. Further, in LDA all topics are in principle present in every document. In contrast, PTM gives sparse topic representation, determining the (small) subset of relevant topics for each document. A customized Bayesian information criterion (BIC) was derived, balancing model complexity and goodness of fit, with the BIC minimized to jointly determine the entire model—the topic-specific words, document-specific topics, all model parameter values, and the total number of topics—in a wholly unsupervised fashion. In the present work, several important modeling and algorithm (parameter learning) extensions of PTM are proposed. First, we modify the BIC objective function using a lossless coding scheme with low modeling cost for describing words that are non-salient for *all* topics—such words are essentially identified as wholly noisy/uninformative. This approach increases the PTM's model sparsity, which also allows model selection of more topics and with lower BIC cost than the original PTM. Second, in the original PTM model learning strategy, word switches were updated sequentially, which is myopic and susceptible to finding poor locally optimal solutions. Here, instead, we jointly optimize all the switches that correspond to the same word (across topics). This approach jointly optimizes many more parameters at each step than the original PTM, which in principle should be less susceptible to finding poor local minima. Results on several document data sets show that our proposed method outperformed the original PTM model with respect to multiple performance measures, and gave a sparser topic model representation than the original PTM.

**Keywords:** topic model; Bayesian information criterion; expectation maximization algorithm; medical abstracts

---

## 1. Introduction

Topic modeling [1] is a type of statistical modeling for finding the "topics" that occur in a collection of documents. The latent Dirichlet allocation (LDA) [2] is one of the well-known topic models. The LDA topic model assumes that each topic is a probability mass function defined over the given vocabulary, and for each of the documents, every word follows a document specific mixture over the topics. In an improvement of LDA called parsimonious topic modeling (PTM) [3], two shortcomings of LDA are discussed. First, all words have their own probability parameters under every topic in LDA; this strategy uses a huge number of parameters, with the model potentially prone to overfitting. Second, in LDA, every topic is assumed to be present in every document, with a non-zero probability. However, PTM gives a sparser description in the two aspects mentioned above. First, some words are not topic specific for certain topics—they use a universal shared model. Second, in each document, only some of the topics occur with a non-zero probability.

The Bayesian information criterion (BIC) [4,5] is a widely used criterion for model selection. There are two parts in the negative logarithm of the Bayesian marginal likelihood: the likelihood of the data and the model complexity cost. So, we can use BIC to balance data fitting goodness and model complexity. The BIC cost function derived for PTM [3] improves over "vanilla" BIC in two aspects. First, the proposed BIC has differentiated cost terms based on different effective sample sizes for different types of parameters; second, PTM introduces a shared feature representation to decrease the effective feature dimensionality of a topic. Our contribution here over PTM [3] is that we give a cheaper expression to describe wholly uninformative words, which thus encourages an even sparser model. Inspired by the intuition that in text corpora there are a large proportion of words that are not related to any topics, we implemented an optimization method to jointly optimize all the parameters related to a single word under each topic, to encourage the possibility that all topics choose to use a shared (uninformative) representation of a given word. Our work improves on the PTM model in encouraging a sparser and more reasonable representation of topics.

Our model solves both the unsupervised feature selection and model order (number of topics) selection problems. First we fix the hyper-parameter (number of topics) to a large number, and we get an optimized structure of the model by determining the topic-specific words under each topic and which topics occur in each document. Then, we change (reduce) the number of topics (hyper-parameter) by removing some topics. For each value of this hyper-parameter we train the model and compute BIC, and the optimized hyper-parameter (chosen model order) is the one with the lowest BIC value.

## 2. Notation

Suppose a corpus consists of $D$ documents and $N$ unique words, with $d \in \{1, 2, ..., D\}$ and $n \in \{1, 2, ..., N\}$ the document and word indices, respectively, and with unique topic indexed by $j \in \{1, 2, ..., M\}$, $M$ being the total number of topics (model order). Here are some different definitions used in this paper:

$L_d$ is the number of unique words in document $d$.

$w_{id} \in \{1, 2, ..., N\}$, $i = 1, ..., L_d$ is the $i$-th word in document $d$.

$v_{jd} \in \{0, 1\}$ is the topic "switch"; it indicates whether topic $j$ is present in document $d$.

Topic $j$ is present in document $d$ if $v_{jd} = 1$; otherwise $v_{jd} = 0$.

$M_d \equiv \sum_{j=1}^{M} v_{jd} \in \{1, 2, ...M\}$ is the number of topics present in document $d$.

$\alpha_{jd}$ is the proportion for topic $j$ in document $d$.

In each topic:

$\beta_{jn}$ is the topic-specific probability of word $n$ under topic $j$.

$\beta_{0n}$ is the shared probability of word $n$.

$u_{jn} \in \{0, 1\}$ indicates whether ($u_{jn} = 1$) or not ($u_{jn} = 0$) word $n$ is topic-specific under topic $j$.

$Nj \equiv \sum_{n=1}^{N} u_{jn}$ is the total number of topic-specific words under topic $j$.

$\bar{L}_j \equiv \sum_{n=1}^{N} L_d v_{jd}$ is the sum of the length of the documents for which topic $j$ is present.

## 3. Methodology

### 3.1. "Bag of Words" Model

A bag of words model [6] is commonly used in document classification where the count of each word is used as a feature for class discrimination. Using the bag of words model, we can transform the text corpora into a feature matrix, where each row vector $\mathbf{x} = (x_1, x_2, \ldots, x_D)$ in the matrix is a bag for each document, with the length of the vector the total number of unique words in the whole text corpora. In the row vector, each position represents a single word, and the value in that position is the number of times this word occurs in the document.

### 3.2. Parsimonious Topic Model (PTM)

We first introduce PTM's data generation method.

1. For each document $d = 1, 2, ..., D$
2. For each word $i = 1, 2, ..., L_d$

- Randomly select a topic based on the probability mass function (pmf) $\{\alpha_{jd}v_{jd}, j = 1, 2, ..., M\}$.
- Given the selected topic $j$, randomly generate the $i$-th word based on the topic's pmf over the word space $\{\beta_{jn}^{u_{jn}} \beta_{0n}^{1-u_{jn}}, n = 1, 2, ..., N\}$

Here $v_{jd}$ is the topic switch that indicates whether topic $j$ is present in document $d$. If $v_{jd} = 1$, it means that topic $j$ is present in document $d$ and $\alpha_{jd}$ is treated as a model parameter. $\beta_{jn}$ and $\beta_{0n}$ are the topic-specific probability of word $n$ under topic $j$ and the shared probability of word $n$, respectively.

Based on the above data generation, we can get the data likelihood of a document dataset $\chi$ under our model $(H, \Theta)$:

$$p(\chi|H, \Theta) = \prod_{d=1}^{D} \prod_{i=1}^{L_D} \prod_{j=1}^{M} [\alpha_{jd}v_{jd}\beta_{jn}^{u_{jw_{id}}} \beta_{0n}^{1-u_{jw_{id}}}]. \tag{1}$$

Here $u_{jn}$ is the word switch that indicates if word $n$ is topic-specific under topic $j$. The model *structure* parameters, denoted by $H\{v, u, M\}$, consist of two kinds of switches and the number of topics, $M$ (model order). Likewise, the model parameters, given a fixed model structure, are denoted by $\Theta = \{\{\alpha_j\}, \{\beta_{jn}\}, \{\beta_{0n}\}\}$. The model structure together with the model parameters constitutes the PTM model. In PTM the parameters are constrained by the following two conditions:

First, $\alpha_{jd}$ is the probability that topic $j$ is present in document $d$, and $v_{jd}$ determines whether or not topic $j$ is present. The probability mass function must sum to one. So, we have:

$$\sum_{j=1}^{M} \alpha_{jd}v_{jd} = 1, \forall d. \tag{2}$$

Additionally, the word probability parameters $\{\beta_{jn}, n = 1, ..., N\}$ and $\{\beta_{0n}, n = 1, ..., N\}$ must satisfy a pmf constraint for each topic:

$$\sum_{n=1}^{N} (u_{jn}\beta_{jn} + (1 - u_{jn})\beta_{0n}) = 1, \forall j. \tag{3}$$

Based on the PTM described above, we must determine the model parameters $\Theta$ and the model structure, $H$. Assuming the model structure is known, we can estimate the model parameters using the expectation maximization (EM) algorithm. By introducing, as hidden data, random variables that indicate which topic generates each word in each document, we can compute the expected complete data log likelihood and maximize it subject to the two constraints mentioned above. The model selection is more complicated. We need to derive a BIC cost function to balance the model complexity and the data likelihood. For the PTM model a generalized expectation maximization (GEM) [7,8] algorithm was proposed to update the model parameters ($\Theta$) and the model structure ($H$) iteratively. In the following section, we give a derivation of BIC and the GEM algorithm for our modified PTM model.

### 3.3. Derivation of PTM-Customized Bayesian Information Criterion (BIC)

In this section we derive our BIC objective function, which generalizes the PTM BIC objective [3]. A naive BIC objective has the following form:

$$BIC = K \log(n) - 2\log(\hat{L}). \tag{4}$$

Here $\hat{L}$ is the maximized value of the data likelihood of the model with structure $H$, that is, $\hat{L} = p(D|H,\hat{\Theta}))$, where $\hat{\Theta}$ is the collection of parameters that maximize the likelihood function.

Additionally, $n$ is the number of data points (documents) in the dataset $\chi$ and $K$ is the number of free parameters in the model.

However, the Laplace approximation used in deriving this BIC form is only valid under the assumption that the feature space is far smaller than the sample size, and for our topic model, the feature space (word dictionary) is quite large in practice. Moreover, in the naive BIC form, all the parameters incur the same description length penalty (the log of the sample size), but in PTM different types of parameters contribute unequally to the model complexity. So, a new customized BIC is derived for PTM.

The Bayesian approach to model selection is to maximize the posterior probability of the model $H$ given the dataset $\chi$. When applying Bayes' theorem to calculate the posterior probability, we get:

$$p(H|\chi) = \frac{p(\chi|H)p(H)}{p(\chi)}. \tag{5}$$

Here we define:

$$I = p(\chi|H) = \int p(\chi|H,\Theta)p(\Theta|H)d\Theta, \tag{6}$$

where $p(\Theta|H)$ is the prior distribution of the parameters given the model structure $H$. Then, we need to use Laplace's method to approximate $I$, given the knowledge that for large sample size $p(\chi|H,\Theta)p(\Theta|H)$ peaks around the maximum point (posterior mode $\hat{\Theta}$). We can rewrite $I$ as:

$$I = p(\chi|H) = \int exp(log(p(\chi|H,\Theta)p(\Theta|H)))d\Theta. \tag{7}$$

We can now expand $log(p(\chi|H,\Theta)p(\Theta|H))$ around the posterior mode $\hat{\Theta}$ using a Taylor series expansion.

$$log(p(\chi|H,\Theta)p(\Theta|H)) \approx log(p(\chi|H,\hat{\Theta})p(\hat{\Theta}|H)) + (\Theta - \hat{\Theta})\nabla_{\Theta}Q|_{\hat{\Theta}} - \frac{1}{2}(\Theta - \hat{\Theta})^T\tilde{\Sigma}_{\Theta}(\Theta - \hat{\Theta}), \tag{8}$$

where $Q \equiv log(p(\chi|H,\Theta)p(\Theta|H))$ and $\tilde{\Sigma}_{\Theta} \equiv -\Sigma_{\Theta}$, where $\Sigma_{\Theta}$ is the Hessian matrix (i.e., $\Sigma_{i,j} = \partial^2 \frac{Q}{\partial\Theta_i\partial\Theta_j}|_{\hat{\Theta}}$).

Since $Q$ attains its maximum at $\hat{\Theta}$, $\nabla_{\Theta}Q|_{\hat{\Theta}} = 0$ and $\tilde{\Sigma}_{\Theta} \equiv -\Sigma_{\Theta}$ is negative definite. We can thus approximate $I$ as follows:

$$I = p(\chi|H) \approx p(\chi|H,\hat{\Theta})p(\hat{\Theta}|H)\int e^{\frac{1}{2}(\Theta-\hat{\Theta})^T\tilde{\Sigma}_{\Theta}(\Theta-\hat{\Theta})}. \tag{9}$$

With the above form of the approximation, $e^{\frac{1}{2}(\Theta-\hat{\Theta})^T\tilde{\Sigma}_{\Theta}(\Theta-\hat{\Theta})}$ is a scaled Gaussian distribution with mean $\hat{\Theta}$ and covariance $\tilde{\Sigma}_{\Theta}$. Thus:

$$\int e^{\frac{1}{2}(\Theta-\hat{\Theta})^T\tilde{\Sigma}_{\Theta}(\Theta-\hat{\Theta})} = (2\pi)^{\frac{k}{2}}|\tilde{\Sigma}_{\Theta}|^{-\frac{1}{2}}, \tag{10}$$

where $k$ is the number of parameters in $\Theta$. So we have the approximation of $I$:

$$I = p(\chi|H) \approx p(\chi|H,\hat{\Theta})p(\hat{\Theta}|H)(2\pi)^{\frac{k}{2}}|\tilde{\Sigma}_{\Theta}|^{-\frac{1}{2}}. \tag{11}$$

BIC is the negative log model posterior:

$$BIC = -log(\hat{I}p(H)) \approx \frac{k}{2}log(2\pi) + \frac{1}{2}log(|\tilde{\Sigma}_{\Theta}|) - log(p(\chi|H,\hat{\Theta})) - log(p(\hat{\Theta}|H)) - log(p(H)). \tag{12}$$

Note that $p(\hat{\Theta}|H)$, the prior of the parameters given the structure $H$ can be assumed to be a uniform distribution (i.e., a constant). So, this term can be neglected. The $log(p(\chi|H,\hat{\Theta}))$ term is the data likelihood, and $k$ is the total number of model parameters. Now we need to approximately calculate $\frac{1}{2}log(|\widetilde{\Sigma}_\Theta|)$ and $log(p(H))$.

To do so, we assume that $\widetilde{\Sigma}_\Theta$ is a diagonal matrix. We thus obtain:

$$\frac{1}{2}log(|\widetilde{\Sigma}_\Theta|) \approx \frac{1}{2}\sum_{d=1}^{D}(M_d-1)log(L_d) + \frac{1}{2}\sum_{j=1}^{M}\sum_{d=1}^{D}u_{jd}log(\bar{L}_j) + \frac{1}{2}\sum_{d=1}^{D}log(\sum_{j=1}^{M}\bar{L}_J). \tag{13}$$

The terms on the right represent the cost of the model parameters $\{\alpha_{jd}\}$, $\{\beta_{jn}\}$, and $\{\beta_{0n}\}$, respectively. Note that in the naive BIC form, each parameter pays the same cost $\frac{1}{2}log(samplesize)$. Here we instead use the effective sample size. The effective sample size of $\alpha_{hd}$ is $L_d$, parameter $\beta_{jn}$ has sample size $\bar{L}_j$, and the parameter $\beta_{0n}$ has sample size $\sum_{j=1}^{M}\bar{L}_j$.

Another term to be estimated is $log(p(H)) = log(p(v)) + log(p(u))$. For $log(p(v))$, in each document $d$, suppose the number of topics follows a uniform distribution, and the switch configuration also follows a uniform distribution over all $\binom{M}{M_d}$ switch configurations. We then obtain:

$$-log(p(v)) = Dlog(M) + \sum_{d=1}^{D}log(\tbinom{M}{M_d}). \tag{14}$$

For $log(p(u))$ we propose here a probability model that can jointly estimate $log(p(u))$ and the corresponding parameter cost of $\beta_{jn}$, $\beta_{0n}$. For each word $n$, we define three types of configurations of the word switches $\{u_{jn}, j = 1, ..., M\}$: (1) each word is topic-specific (i.e., $\sum_{j=1}^{M}u_{jn} = M$); (2) all the words are not topic-specific (i.e., $\sum_{j=1}^{M}u_{jn} = 0$); (3) some, but not all, components use the shared distribution (i.e., $0 < \sum_{j=1}^{M}u_{jn} < M$). For cases 1 and 2, there is only one possible configuration of the word switches related to a word (all *open* or all *closed*), so the probability associated with this configuration is 1; for case 3 there are $2^M$ possible configurations. Assuming these are equally likely under this case $log(p[u]) = Mlog2$. We can then estimate $-log(p(u))$ plus the two terms $\sum_{j=1}^{M}\sum_{d=1}^{D}u_{jd}log(\bar{L}_j) + \sum_{d=1}^{D}log(\sum_{j=1}^{J}\bar{L}_J)$ in $log(|\widetilde{\Sigma}_\Theta|)$. That is, we have

$$-log(p(u)) + \frac{1}{2}\sum_{j=1}^{M}\sum_{d=1}^{D}u_{jd}log(\bar{L}_j) + \frac{1}{2}\sum_{d=1}^{D}log(\sum_{j=1}^{M}\bar{L}_j) \approx \sum_{n=1}^{N}(\frac{F_1(u_n)}{2}log(\sum_{j=1}^{M}\bar{L}_j) + \frac{F_2(u_n)}{2}\sum_{j=1}^{M}log(\bar{L}_j)$$
$$+F_3(u_n)(\frac{1}{2}log(\sum_{j=1}^{M}\bar{L}_j) + \frac{1}{2}\sum_{j=1}^{M}u_{jn}log(\bar{L}_j)). \tag{15}$$

Here,
$$F_1(u_n) = \begin{cases} 1, if \sum_{j=1}^{M}u_{jn} = 0 \\ 0, otherwise \end{cases}, \quad F_2(u_n) = \begin{cases} 1, if \sum_{j=1}^{M}u_{jn} = M \\ 0, otherwise \end{cases}, \quad F_3(u_n) = \begin{cases} 1, if 0 < \sum_{j=1}^{M}u_{jn} < M \\ 0, otherwise \end{cases}.$$

Based on the derivation above, the BIC cost function for our modified PTM model is:

$$BIC = Dlog(M) + \sum_{d=1}^{D}log(\tbinom{M}{M_d}) + \frac{1}{2}\sum_{d=1}^{D}(M_d-1)log(\frac{L_d}{2\pi}) - log(p(D|H,\Theta))$$
$$\sum_{n=1}^{N}(\frac{F_1(u_n)}{2}log(\frac{\sum_{j=1}^{M}\bar{L}_j}{2\pi}) + \frac{F_2(u_n)}{2}\sum_{j=1}^{M}log(\frac{\bar{L}_j}{2\pi}) + F_3(u_n)(\frac{1}{2}log(\frac{\sum_{j=1}^{M}\bar{L}_j}{2\pi}) + \frac{1}{2}\sum_{j=1}^{M}u_{jn}log(\frac{\bar{L}_j}{2\pi})). \tag{16}$$

### 3.4. Generalized Expectation Maximization (EM) Algorithm

The EM [9] algorithm is a popular method for maximizing the data log-likelihood. For unsupervised learning tasks, we only have the data points $\chi$, but we do not have any "labels" for those data points, so during the maximum likelihood estimation (MLE) process we introduce "label" random variables which are called latent variables. The EM algorithm can be described as follows:

E-step: With the parameters fixed, we compute the expectation of the latent variables $p(Z|\chi, \Theta)$, which gives the class information for each data point. Using the expectation of the latent variables we can compute the expectation of the complete data log-likelihood:

$$E[L_c] = \sum_Z p(Z|D, \Theta) log(p(D, Z|\Theta). \tag{17}$$

M-step: We update the parameters $\Theta$ to find the maximum value of the expected complete data log-likelihood. By doing the E-step and M-step iteratively, the expected complete data log-likelihood strictly increases and typically converges to a local optimum of the *incomplete* data log-likelihood (or of the expected BIC cost function). However, note that in PTM there are not only the model parameters $\Theta$ but also the model structure $H$ over which we need to optimize. The original EM algorithm cannot be applied here because one cannot get jointly optimal closed form estimates of both the model parameters $\Theta$ and the structure parameters $H$ to maximize $E[L_c]$. However, a generalized expectation maximization (GEM) [7,8] algorithm is proposed, which alternately jointly optimizes $E[L_c]$ over $\Theta$ and then over *subsets* of the structure parameters, $H$ given fixed $\Theta$.

Our GEM algorithm is specified for fixed model order, M. First we introduce the hidden data $Z$: $Z_{id}$ is an M-dimensional binary vector, with a "1" indicating the topic of origin for the word $w_{id}$. For example, if the element $Z_{id}^{(j)} = 1$ and other elements of $Z_{id}$ are all equal to zero, topic $j$ is the topic of origin for word $w_{id}$.

Our GEM algorithm strictly descends in the BIC cost function Equation (16). It consists of an E-step followed by an M-step that minimizes the expected BIC cost function. These steps are given as follows: In the E-step, first we compute the expectation of the hidden data $Z$:

$$\begin{aligned}
p(Z_{id}^{(j)} = 1|w_{id}; \Theta, H) &= \frac{p(w_{id}|Z_{id}^{(j)}; \Theta, H)p(Z_{id}^{(j)}|\Theta, H)}{\sum_{Z_{id}} p(w_{id}|Z_{id}^{(j)}; \Theta, H)p(Z_{id}^{(j)}|\Theta, H)} \\
&= \frac{\alpha_{jd}v_{jd}\beta_{jn}^{u_{jw_{id}}}\beta_{0n}^{1-u_{jw_{id}}}}{\sum_{l=1}^{M} \alpha_{ld}v_{ld}\beta_{ln}^{u_{lw_{id}}}\beta_{0n}^{1-u_{lw_{id}}}}.
\end{aligned} \tag{18}$$

With the expectation of the hidden data, we can compute the expected complete data log-likelihood using Equation (17). By replacing the $log(p(D|H, \Theta))$ term in BIC with the expected complete data log-likelihood($E[L_c]$), we get the expected complete data BIC.

In the generalized M-step, based on the expectation of the complete data BIC we computed in the E-step, we update the model structure $H$ and the model parameters $\Theta$. First we optimize the model parameters given fixed model structure. Then we optimize the model structure given fixed model parameters, both steps taken to minimize the expected BIC. These steps are alternated until convergence.

When updating the model parameters, note that the only term in BIC that is related to the model parameters is the data likelihood term, so we can just maximize the expected complete data log-likelihood computed in the E-step to choose the model parameters. Taking those two constraints (Equations (2) and (3)) into consideration, we have our Lagrangian objective function.

$$
\begin{aligned}
L = &\sum_{d=1}^{D} \sum_{i=1}^{L_d} (v_{jd} p(Z_{id}^{(j)} = 1 | w_{id}; \Theta, H)(log(\alpha_{jd}) + u_{jw_{id}} log(\beta_{jw_{id}}) + (1 - u_{jw_{id}}) log(\beta_{0w_{id}}))) \\
&- \sum_{d=1}^{D} \lambda_d (\sum_{j=1}^{M} \alpha_{jd} v_{jd} = 1) - \sum_{j=1}^{M} \mu_j (\sum_{j=1}^{M} (u_{jn} \beta_{jn} + (1 - u_{jn}) \beta_{0n}) - 1),
\end{aligned}
\tag{19}
$$

where $\mu_j$ and $\lambda_d$ are Lagrange multipliers. By computing the partial derivative of each parameter type and setting those derivatives to zero, we can get the optimized model parameters, satisfying necessary optimality conditions as:

$$
\alpha_{jd} = \frac{\sum_{i=1}^{L_d} p(Z_{id}^{(j)} = 1 | w_{id}; \Theta, H) v_{jd}}{\sum_{l=1}^{M} \sum_{i=1}^{L_d} p(Z_{id}^{(l)} = 1 | w_{id}; \Theta, H) v_{ld}}, j = 1, ..., M, d = 1, ..., D,
\tag{20}
$$

$$
\beta_{jn} = \frac{u_{jn} \sum_{d=1}^{D} \sum_{i=1:w_{id}=n}^{L_d} p(Z_{id}^{(j)} = 1 | w_{id}; \Theta, H) v_{jd}}{\mu_j}, j = 1, ..., M, n = 1, ..., N.
\tag{21}
$$

We compute $\mu_j$ by multiplying both sides of Equation (21) by $u_{jn}$, summing over all $n$, and applying the distribution constrains on topic $j$. This gives:

$$
\mu_j = \frac{\sum_{n=1}^{N} u_{jn} \sum_{d=1}^{D} \sum_{i=1:w_{id}=n}^{L_d} p(Z_{id}^{(j)} = 1 | w_{id}; \Theta, H) v_{jd}}{1 - \sum_{j=1}^{M} (1 - u_{jn}) \beta_{0n}}, \forall j.
\tag{22}
$$

For the shared parameters, we only estimate them once via global frequency counts at initialization and hold them fixed during the GEM algorithm. That is, we set:

$$
\beta_{0n} = \frac{\sum_{d=1}^{D} \sum_{i=1:w_{id}=n} 1}{\sum_{d=1}^{D} L_d}, \forall n.
\tag{23}
$$

When updating the model structure, we implement an iterative loop in which all the topic switches $u$ are visited one by one. If the current change reduces BIC, we accept the change; otherwise we keep the switch unchanged. Note that in updating the word switches $v$, we update all the word switches for a single word jointly to see if it is optimal to choose all the switches related to a single word to be closed (i.e., to specify that this word is completely uninformative). This process is repeated over all the switches until there is no decrease in BIC or until we reach a pre-defined maximum number of iterations. We first update the word switches $u$ until convergence; then we update the topic switches $v$ until convergence. Then we go back to the E-step. Note that when updating the word switches, for each single search of all the switches related to one word, we have three configurations. All switches are closed, all switches are open, or some are closed and some are open. We compute the minimized BIC for each configuration, and then choose the configuration that has the lowest BIC value.

*3.5. Selecting the Model Order*

The optimization process discussed above is under the assumption that the model order is known. Model order selection is based on applying the optimization process for different model orders, in a top-down fashion. We initialize the model with a specific number of topics ($M_{max}$, chosen to upper bound the number of topics expected to be present in the corpus) and reduce the number by a predefined step $\Delta$. For the model trained at each order, we remove the $\Delta$ topics with the smallest mass. This process is applied iteratively until the predefined minimum order is reached. We then retain the model (and model order) with the smallest BIC cost.

For our model, the only "hyper-parameters" are $M_{max}$ and $\Delta$. We can expect the best performance by choosing $\Delta = 1$. For $M_{max}$, if this is set too small, it will underestimate the model order. If it is set very large, the learning and model order selection will require more computation. In principle, choosing any value of $M_{max}$ above the ground truth $M^*$ or the BIC minimizing $\hat{M}$ should be reasonable to find a good solution (the bottom line is that our method requires no true hyper-parameters—the only tradeoff is more computation by choosing $\Delta = 1$ and sufficiently large $M_{max}$).

## 4. Experiments and Results

In this section we compare the original PTM, LDA, and the new PTM method. All methods were used to solve the unsupervised density modeling problem, with no knowledge of class labels. Hence, the comparison of the methods is fair, and we evaluated multiple measures that assess the quality of the learned models as a function of the number of components in the model. None of the methods set hyper-parameters (except LDA, for which M was set to the the maximum)

Performance measurements: BIC was compared between the PTM model and our revised PTM model. For both models, the held-out log-likelihood and the class label purity were also compared. In [3] the performance of the PTM model and the LDA model were compared, and PTM was found to outperform LDA. Here we only include the label purity of the LDA model on different datasets.

When computing the held-out likelihood, we used the method described in [10,11] to compare model fitness on a held-out test set. We divided the documents in the test set into two parts: the observed part and the held-out part. First we computed the topic proportions based on the observed part, then we computed the held-out log-likelihood based on the held-out part:

$$\sum_{d=1}^{D_{test}} \sum_{i=1}^{L_d^{heldout}} log(\sum_{j=1}^{M} E_q[\alpha_{jd}] E_q[\beta_{jw_{id}}]). \tag{24}$$

In our model $E_q[\alpha_{jd}]$ is directly the topic proportions $\alpha_{jd}$. $E_q[\beta_{jw_{id}}]$ is $u_{jn}\beta_{jn} + (1 - u_{jn})\beta_{0n}$.

We evaluated PTM, modified-PTM, and LDA on three datasets as next discussed.

### 4.1. Reuters-21578

The Reuters-21578 dataset is a collection of documents from Reuters news in 1987. There are in total 7674 documents from 35 categories. After stemming and stop word removal, there were 17,387 unique words. There were 5485 documents in the training set and 2189 documents in the test set.

### 4.2. 20-Newsgroups

The 20-Newsgroup dataset is a collection of 18,821 newsgroup documents from 20 classes. There were 53,976 words after stemming and stop word removal. It was split into a 11,293-document training set and a 7528-document test set.

### 4.3. Ohsumed

The Ohsumed dataset includes medical abstracts from the MeSH categories of the year 1991. It consists of 34,389 documents, each assigned to one or multiple labels of the 23 MeSH disease categories. Each document has on average 2.26 labels. The dataset was divided into 24,218 training and 10,171 test documents. There were 12,072 unique words in the corpus after applying standard stop word removal and stemming.

Note that for the Ohsumed dataset, each document may be associated with multiple labels. We computed the label purity for this dataset as follows: We first associated to each topic a multinomial distribution on the class labels. We learned these label distributions for each topic by frequency counting over the ground truth class labels of all documents, weighted by topic proportions:

$$p_j(c) = \frac{\sum_{d=1}^{D} \sum_{i=1:l_{id}=c}^{|C_d|} \alpha_{jd} v_{jd}}{\sum_{d=1}^{D} \sum_{i=1}^{|C_d|} \alpha_{jd} v_{jd}}, \forall j, c, \tag{25}$$

where $p_j(c)$ is the class proportion for class $c$, for topic $j$. Here $l_{id}$ is the $i$-th class label in document $d$ and $|C_d|$ is the number of class labels in document $d$. For labeling a text document, we then computed the probability of each class label based on the topic proportions in that document, that is, $\sum_{j=1}^{M} \alpha_{jd} p_j(c)$, and assigned the labels that had probability higher than a threshold value $T$:

$$\hat{C}_d = \{c : \sum_{j=1}^{M} \alpha_{jd} p_j(c) > T\}. \tag{26}$$

We changed the threshold $T$ and measured the area under the precision/recall curve (AUC). Note that here precision is the number of true discovered labels divided by the total number of ground-truth labels. Recall is the number of correctly classified labels divided by the total number of labels assigned to documents by our classifier.

### 4.4. Discussion

Our results on the three text corpora are shown in Figures 1–3. Here we only include comparison with LDA on label purity. The comparison of LDA with the original PTM on held-out log-likelihood is in [3], and demonstrates that original PTM gives better results. Here, we see that the new PTM method convincingly outperformed original PTM (and hence LDA). Previous work showed that the original PTM method outperformed LDA with respect to two performance measurements: held-out log-likelihood and label purity [3]. This was attributed to the fact that PTM models are sparser than LDA models. However, the modified-PTM method dominated the original PTM (and LDA) with respect to all three of these measures, on all three datasets. Note in particular the large gain in held-out log-likelihood on Reuters-21578, in all measures on 20-Newsgroups, and with respect to BIC and held-out log-likelihood on Ohsumed.

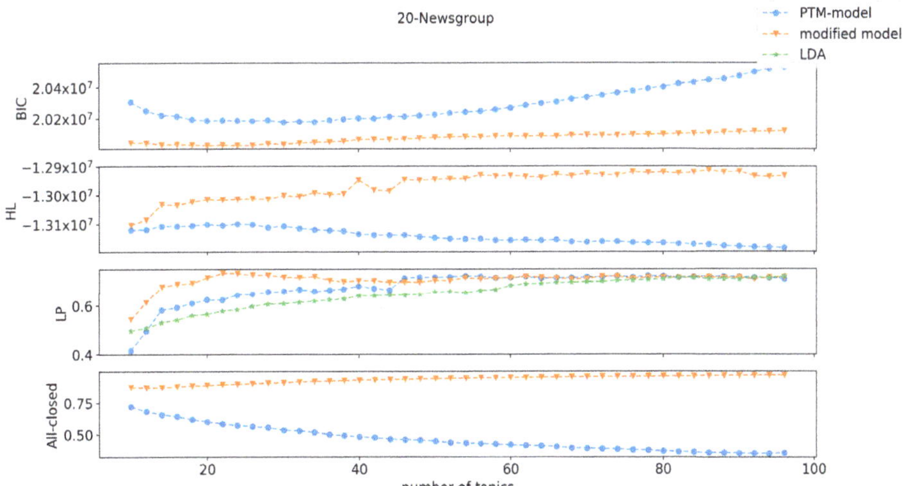



**Figure 1.** Performance comparison between the original parsimonious topic model (PTM), our modified PTM, and latent Dirichlet allocation (LDA) on the Reuters-21578 dataset. The measurements were the Bayesian information criterion (BIC), held-out log-likelihood (HL), label purity (LP), and the proportion of words with all-closed word switches (All-closed). LDA is only shown for label purity.

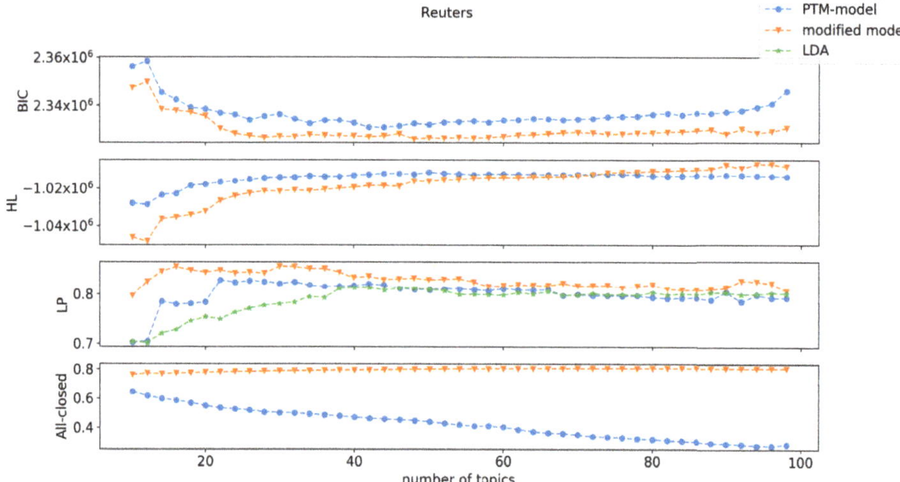

**Figure 2.** Performance comparison between the original PTM, our modified PTM, and LDA on the 20-Newsgroup dataset. The measurements were BIC, held-out log-likelihood (HL), label purity (LP), and the proportion of the words with all-closed word switches (All-closed). LDA is only shown for label purity.

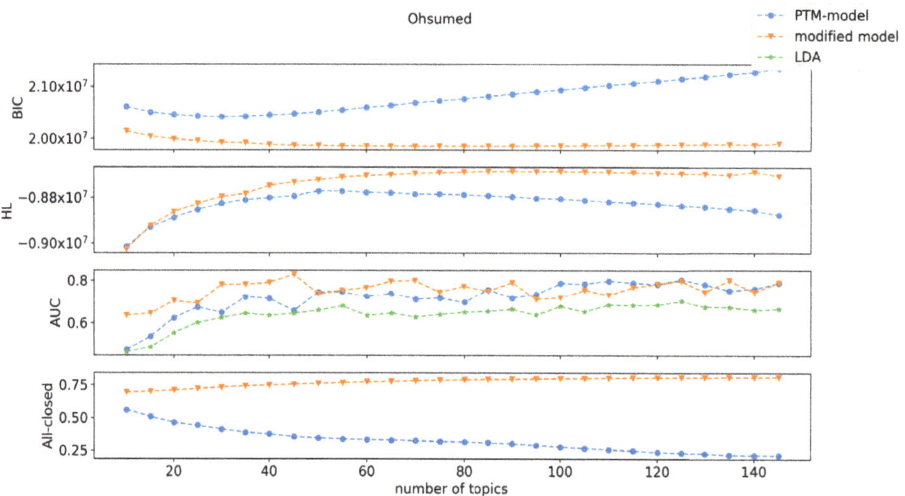

**Figure 3.** Performance comparison between the original PTM, our modified PTM, and LDA on the Ohsumed dataset. The measurements were BIC, held-out log-likelihood (HL), AUC, and the proportion of the words with all-closed word switches (All-closed). LDA is only shown for label purity.

The modified PTM minimizes BIC by selecting a much larger (richer) set of topics than the original PTM. This is achieved by the low description length penalty associated with making words completely uninformative (all closed), which leads to many more words being deemed completely uninformative compared to the original PTM. This is a key advantage of the new PTM method. The original PTM method offers no incentive within the BIC cost function to decide that a word is wholly uninformative about (irrelevant to) the topics that are present in the document. The new method provides substantial incentive for such determination with great reduction in model description complexity gained by such a choice. This allows the model to "afford" having many more topics than in the original PTM method. Especially, close to 80% of words were closed under modified PTM compared to approximately 40% for PTM for Reuters (at the selected model order), approximately 80% compared to approximately 50% are closed for 20-Newsgroups, and approximately 70% compared approximately to 30% were closed for Ohsumed. That is, modified PTM chose higher model orders (number of topics) with fewer topic-specific words and more wholly uninformative words than the original PTM. Choosing more topics was seen to yield better performance for several measures (label purity and held-out log-likelihood).

*4.5. Computational Complexity*

In both the PTM model and our modified model, we need to optimize over the parameters $\Theta$ and the model structure parameters $\{v, u\}$.

Consider learning the parameters $\Theta$ when M is fixed and with $D$ documents. The computational complexity of the PTM model and our modified model are both $O(M_D D L_D)$, where $M_D$ and $L_D$ are the number of topics present in a document and the length of a document, respectively. In LDA all topics are present in all documents, and the computational complexity is $O(MDL_D)$—that is, the total computational complexity of the LDA model. However, in our model and the PTM model we also need to optimize over the model structure parameters $\{v, u\}$.

For the word switches $\{u_{jn}\}$, in the PTM model, updating them involves an iterative loop over all $N$ words in M topics. Thus, the complexity is $O(MN)$. In our modified PTM model for each of the $N$ words, we need to check two more cases—thart is, case 1) and case 2) discussed before Equation (15). So, the computational complexity is $O((M + 2)N) = O(MN)$.

For the topic switches $\{v_{jd}\}$, both our model and the PTM model need computations of order $O(M_D L_D)$ to update each switch $v_{jd}$. We have in total $MD$ switches to be updated. So the computational complexity of updating the topic switches is $O(MM_D DL_D)$ for both PTM and our modified model. The computational complexity comparison is shown in Table 1.

We recorded the execution time for the PTM model and the revised PTM model on each dataset (in Table 1). We ran the experiment on a machine with an Intel Core i5, 2.3 GHz processor. The execution time of our method was slightly less than that of the PTM model, which may be because in our modified model, for the word switches related to a single word, when it comes to the point that all the switches are closed, it is very likely that no update will be done in the future iterations. We did an experiment with the LDA model using the Python package, but our modified PTM model is implemented in C. So, the comparison between our model and LDA may be unfair. A comparison of the execution time between LDA and PTM models is reported in [3].

**Table 1.** Comparison of thecomputational complexity and execution time of different models.

| | Modified PTM | PTM | LDA |
|---|---|---|---|
| Complexity | $O(MM_D DL_D + MN)$ | $O(MM_D DL_D + MN)$ | $O(MDL_D)$ |
| **Execution Time (min)** | | | |
| 20-Newsgroup | 387 | 493 | |
| Reuters-21578 | 39 | 44 | |
| Ohsumed | 424 | 659 | |

### 4.6. Limitations of Our Work

One limitation of our model is that the computational complexity is higher than for the LDA model. Another limitation is the highly non-convex optimization, with discrete and continuous parameters and no guarantee of finding the global minimum.

## 5. Conclusions

In this paper, we proposed two improvements on the PTM method. One is improving the modeling by giving a cheaper expression to describe the wholly uninformative words. The other is improving the learning—we optimized all the parameters related to a single word under each topic, which encourages a sparser model. Our improvements led to consistent and substantial gains in modeling accuracy with respect to multiple performance measures, compared with both the original PTM method and the well-known LDA model. While we demonstrated significant gains on three document corpora, there is increased computational complexity for our method compared to LDA. Future work could aim to develop a fully Bayesian version of our approach, based on posteriors on topic and word switches, rather than on deterministic (binary) switch parameters. Future work could also investigate extensions that model word order dependency, but somehow in a low-complexity fashion.

**Author Contributions:** Conceptualization, D.M.; Data curation, H.W.; Funding acquisition, D.M.; Methodology, D.M.; Project administration, D.M.; Software, H.W.; Supervision, D.M.; Validation, H.W.; Writing – review & editing, D.M. All authors have read and approved the final published version.

**Funding:** This research received no external funding.

**Conflicts of Interest:** The authors declare no conflict of interest.

## Reference


1. Wallach, H.M. Topic modeling: beyond bag-of-words. In Proceedings of the 23rd International Conference on Machine Learning, Pittsburgh, PA, USA, 25–29 June 2006; pp. 977–984. Available online: http://dirichlet.net/pdf/wallach06topic.pdf (accessed on 10 March 2020).
2. Blei, D.M.; Ng, A.Y.; Jordan, M.I. Latent dirichlet allocation. *J. Mach. Learn. Res.* **2003**, *3*, 993–1022.
3. Soleimani, H.; Miller, D.J. Parsimonious topic models with salient word discovery. *IEEE Trans. Knowl. Data Eng.* **2014**, *27*, 824–837. [CrossRef]
4. Schwarz, G. Estimating the dimension of a model. *Ann. Stat.* **1978**, *6*, 461–464. [CrossRef]
5. Bhat, H.S.; Kumar, N. On the derivation of the Bayesian Information Criterion. Available online: https://faculty.ucmerced.edu/hbhat/BICderivation.pdf (accessed on 10 March 2020).
6. Zhang, Y.; Jin, R.; Zhou, Z.H. Understanding bag-of-words model: A statistical framework. *Int. J. Mach. Learn. Cybern.* **2010**, *1*, 43–52. [CrossRef]
7. Dempster, A.P.; Laird, N.M.; Rubin, D.B. Maximum likelihood from incomplete data via the EM algorithm. *J. R. Stat. Soc. Series B. Stat. Methodol.* **1977**, *39*, 1–22.
8. Meng, X.L.; Van Dyk, D. The EM algorithm—An old folk-song sung to a fast new tune. *J. R. Stat. Soc. Ser. B. Stat. Methodol.* **1997**, *59*, 511–567. [CrossRef]
9. Do, C.B.; Batzoglou, S. What is the expectation maximization algorithm? *Nat. Biotechnol.* **2008**, *26*, 897. [CrossRef] [PubMed]
10. Hoffman, M.D.; Blei, D.M.; Wang, C.; Paisley, J. Stochastic variational inference. *J. Mach. Learn. Res.* **2013**, *14*, 1303–1347.
11. Teh, Y.W.; Newman, D.; Welling, M. A collapsed variational Bayesian inference algorithm for latent Dirichlet allocation. In Proceedings of the Advances in Neural Information Processing Systems, Vancouver, BC, Canada, 3–6 December 2007; pp. 1353–1360.

*Article*

# Sub-Graph Regularization on Kernel Regression for Robust Semi-Supervised Dimensionality Reduction

**Jiao Liu [1], Mingbo Zhao [2,*] and Weijian Kong [2,*]**

[1]  School of Management Studies, Shanghai University of Engineering Science, Shanghai 201600, China; liujiaolndx@163.com

[2]  School of Information Science and Technology, Donghua University, Shanghai 201620, China

*   Correspondence: mzhao4@dhu.edu.cn (M.Z.); kongweijian@dhu.edu.cn (W.K.); Tel.: +86-21-67792355 (M.Z.); +86-21-67792315 (W.K.)

Received: 7 October 2019; Accepted: 7 November 2019; Published: 15 November 2019

**Abstract:**  Dimensionality reduction has always been a major problem for handling huge dimensionality datasets. Due to the utilization of labeled data, supervised dimensionality reduction methods such as Linear Discriminant Analysis tend achieve better classification performance compared with unsupervised methods. However, supervised methods need sufficient labeled data in order to achieve satisfying results. Therefore, semi-supervised learning (SSL) methods can be a practical selection rather than utilizing labeled data. In this paper, we develop a novel SSL method by extending anchor graph regularization (AGR) for dimensionality reduction. In detail, the AGR is an accelerating semi-supervised learning method to propagate the class labels to unlabeled data. However, it cannot handle new incoming samples. We thereby improve AGR by adding kernel regression on the basic objective function of AGR. Therefore, the proposed method can not only estimate the class labels of unlabeled data but also achieve dimensionality reduction. Extensive simulations on several benchmark datasets are conducted, and the simulation results verify the effectiveness for the proposed work.

**Keywords:**  kernel regression; semi-supervised learning; dimensionality reduction; anchor graph regularization

---

## 1. Introduction

Dimensionality reduction is an important issue when handing high-dimensional data in many real-world applications, such as image classification, text recognition, etc. In general, dimensionality reduction is achieved by finding a linear or nonlinear projection matrix that casts the original high-dimensional data into a low-dimensional subspace so that the computational complexity can be reduced and the key intrinsic information can be preserved [1–10]. Principal component analysis (PCA) and linear discriminant analysis (LDA) [11] are two of the most widely-used methods for dimensionality reduction. PCA is achieved by finding a projection matrix along the maximum variance of the dataset with the best reconstruction. While LDA is utilized to search for the optimal direction ensuring that the dataset in the reduced subspace can maximize the between-class scatter while minimizing the within-class scatter. As LDA is a supervised approach, it generally outperforms PCA by giving sufficient labeled information.

A key problem is that obtaining a large amount of labeled data is time-consuming and expensive. On the other hand, unlabeled data may be abundant in some real world applications. Therefore, semi-supervised learning (SSL) approaches have become increasingly important in the area of pattern recognition and machine learning [1,2,4,12–14]. Over the past decades, according to the manifold or clustering assumptions—i.e., nearby data likely have the same labels [1,2,4]—graph based SSL is one of the most popular methods in the aspect of SSL, which includes the manifold regularization

(MR) [3], learning with local and global consistency (LGC) [2] and Gaussian fields and harmonic functions (GFHF) [1] methods. All of these utilize labeled and unlabeled sets to formulate a graph for approximating the geometry of data manifolds [5].

The above graph-based SSL can be usually divided into two categorizations: The first is the inductive learning method and the second is the transductive learning one. The transductive learning methods aim to propagate the labeled information via a graph [1,2,4], so that the labels of an unlabeled set are estimated. However, a key problem for transductive learning methods is that they cannot estimate the class labels of new incoming data, therefore suffering from the out-of-sample problem. In contrast, the inductive learning methods, known as MR [3] and Semi-supervised Discriminant Analysis (SDA) [5], aim to study a decision function for classification on the original data space, so that they can reduce the dimensionality as well as naturally solve out-of-sample problems.

It can be noted that the graph in SSL tends to be a $k$ nearest neighborhood ($k$NN) based graph that is first to find the $k$-neighborhoods of each data [15–17] and then define a weight matrix measuring the similarity between any pair-wise data [1,2,4,18–21]. However, $k$NN graph has a key limit in that it cannot be scalable to a large-scale dataset, as the computational complexity for searching the $k$ neighborhoods of data is $O\left(kn^2\right)$, which is not linear with $n$. To solve this problem, Liu et al. [22,23] proposed an efficient anchor graph (AGR), where each data point is first to find the $k$ neighborhoods of anchor points, then the graph is constructed by the inner product of coefficients between the data and anchors, through which the class labels can be inferred from anchors to the whole dataset. As a result, the computational complexity can be greatly reduced. While there are different ways to build the adjacency matrix $S$ in AGR [24–26], we argue that most of them are developed intuitively and lack a probability explanation. In addition, AGR cannot directly infer the class labels of incoming data.

In this paper, we aim to enhance AGR by solving the above problems. From the element concept idea of AGR, we point that the anchors should have the same probability distribution to those of data points, as the anchors refer to the data that can roughly approximate the distribution of data points. Based on this assumption, we then analyze $S$ from the stochastic view and further extend it to be doubly-stochastic. As a result, the distribution of anchors is the same to those of data points, and the updated $S$ can be treated as a transition matrix, where each value in $S$ can be viewed as a transition probability value between any data point and anchor point. Benefiting from $S$, we then develop a sub-graph regularized framework for SSL. The new sub-graph is constructed by $S$ in an efficient way and can preserve the geometry of data structure. Accordingly, an SSL strategy based on such a sub-graph is also developed, which is first to infer the labels of anchors and then to calculate those of the training data. The is quite different from conventional graph-based SSL, which is directly to infer the class labels of datasets on the whole graph and may result in a huge computational cost if the dataset is large-scale. However, this SSL strategy is efficient and suitable for handling a large-scale dataset. The experiments on extensive benchmark datasets show the effectiveness and efficiency of the proposed SSL method.

The main contributions of this paper are given as follows:

(1) We develop a doubly-stochastic $S$ that measures the similarity between data points and anchors. The new updated $S$ has probability means and can be viewed as transition probability between data points and anchors. In addition, the proposed $S$ is also a stochastic extension to the ones in AGR.

(2) We develop a sub-graph regularized framework for SSL. The new sub-graph is constructed by $S$ in an efficient way and can preserve the geometry of the data manifold.

(3) We also adopt a linear predictor for inferring the class labels of new incoming data, which can handle out-of-sample problems. In addition, the computational complexity of this linear predictor is linear with the number of anchors, and hence is efficient.

The organization of the paper is as follows: In Section 2, basic notations and reviews for SSL are provided; in Section 3, the proposed model for graph construction and SSL are developed. In Section 4, we conduct extensive simulations, and give our final conclusions in Section 5.

## 2. Notations and Preliminary Work

### 2.1. Notations

Let $X = [X_l, X_u] \in R^{d \times (l+u)}$ be the data matrix, where $d$ presents the feature number, $l$ and $u$ are the number of labeled and unlabeled sets, respectively, so that $X_l$ and $X_u$ are respectively the labeled and unlabeled sets, $Y = [y_1, y_2, \ldots, y_{l+u}] \in R^{c \times (l+u)}$ be the one hot labels of data, $F = [f_1, f_2, \ldots, f_{l+u}] \in R^{c \times (l+u)}$ is the predicted label matrix satisfying $0 \le f_{ij} \le 1$.

### 2.2. Review of Graph Based Semi-Supervised Learning

We will review the prior graph based SSL methods. Two well-known methods for SSL include LGC [1] and GFHF [2]. The objective of LGC and GFHF can be given as:

$$
\begin{aligned}
g_L(F) &= \tfrac{1}{2} \sum_{i,j=1}^{l+u} \left\| \frac{f_i}{\sqrt{D_{ii}}} - \frac{f_j}{\sqrt{D_{jj}}} \right\|_F^2 W_{ij} + \lambda \sum_{i=1}^{l+u} \| f_i - y_i \|_F^2 \\
g_G(F) &= \tfrac{1}{2} \sum_{i,j=1}^{l+u} \| f_i - f_j \|_F^2 W_{ij} + \lambda_\infty \sum_{i=1}^{l} \| f_i - y_i \|_F^2
\end{aligned}
\tag{1}
$$

where $\lambda$ is a balancing parameter that controls the trade off between the label fitness and the manifold smoothness. $\lambda_\infty$ is a large value such that $\sum_{i=1}^{l} \| f_i - y_i \|_F^2 = 0$, or $f_i = y_i, \forall i = 1, 2, \ldots, l$.

### 2.3. Anchor Graph Regularization

Anchor graph regularization (AGR) is an efficient graph based learning method for large-scale SSL. In detail, let $A = \{a_1, a_2, \ldots a_m\} \in R^{d \times m}$ be the anchor point set, $G = \{g_1, g_2, \ldots g_m\} \in R^{c \times m}$ be the label matrix of $A$, $Z \in R^{m \times n}$ be the weight matrix measuring the similarity between each $x_j$ and $a_i$ with constraints $Z_{ij} \ge 0$ and $\sum_{i=1}^{m} Z_{ij} = 1$, which is usually formulated by the kernel weights or the local reconstructed strategy making the computational complexity for both two strategies linear with the data number. Then, the label matrix $F$ can be estimated as:

$$
f_j = \sum_{i=1}^{m} g_i Z_{ij},
\tag{2}
$$

so that AGR is to minimize the following objective function:

$$
\begin{aligned}
J(G) &= \sum_{j=1}^{l} \| G z_j - y_j \|_F^2 + \tfrac{\gamma}{2} \sum_{i,j=1}^{n} W_{ij}^a \| G z_i - G z_j \|_F^2 \\
&= \| G Z_l - Y_l \|_F^2 + \gamma Tr \left( GZ (I - W^a) Z^T G^T \right) \\
&= \| G Z_l - Y_l \|_F^2 + \gamma Tr \left( G L^r G^T \right)
\end{aligned}
\tag{3}
$$

where the first term is the loss function and the second term is the manifold regularized term, $W^a = Z^T \Delta^{-1} Z \in R^{n \times n}$ is the anchor graph, and $\Delta \in R^{m \times m}$ is a diagonal matrix with each element satisfying $\Delta_{ii} = \sum_{j=1}^{n} Z_{ij}$. It can be easily proven that $W^a$ is doubly-stochastic, hence it has probability meaning. In addition, given two data points $x_i$ and $x_j$ with common anchor points, it follows $W_{ij}^a > 0$; otherwise $W_{ij}^a = 0$. This indicates that the data points with common anchor points have similar semantic concepts hence $W^a$ can characterize the semantic structure of datasets. $L^r = Z (I - W^a) Z^T \in R^{m \times m}$ is the reduced Laplacian matrix, $Z_l \in R^{m \times l}$ is formed by the first $l$ columns of $Z$. Here, we can see that although AGR is performed with a regularization term on all data points, it is equivalent to being regularized on anchor points with a reduced Laplacian matrix $L^r$. Finally, the labels of data points can be inferred from those of anchor points, where the computational complexity can be reduced to $O(n)$. Therefore, both graph construction and the regularized procedure in AGR are efficient and scalable to a large-scale dataset.

## 3. A Sub-Graph Regularized Framework for Efficient Semi-Supervised Learning

### 3.1. Analysis of Anchor Graph Construction

The key point for anchor graph construction is to define the weight matrix for measuring the similarity between each data point and anchor data. A typical way is to use kernel regression [22]:

$$S_{ij} = \frac{K_\delta\left(x_i, b_j\right)}{\sum_{s\in\langle i\rangle} K_\delta\left(x_i, b_s\right)} \forall s \in \langle i\rangle \tag{4}$$

where $\delta$ is the bandwidth of Gaussian function and $\langle i\rangle$ denotes the indices of the $k$ neighborhood anchors of $x_i$. Obviously, we have $S^T 1_q = 1_n$, where $1_n \in R^{n\times 1}$ and $1_q \in R^{q\times 1}$ is the column vectors with $n$ and $q$ ones, respectively, so that the sum of each column of $S$ is equal to 1. This means $S_{ij}$ can be viewed as a probability value $P\left(b_i|x_j\right)$, which represents the transferred probability from $x_j$ to $b_j$. Then, following the Bayes rule, we have:

$$P\left(b_i\right) = \sum_{j=1}^{n} P\left(x_j\right) P\left(b_i|x_j\right) \approx \frac{1}{n} P\left(b_i|x_j\right) \tag{5}$$

where $P\left(x_j\right) \approx 1/n$ follows a uniform distribution based on the strong law of large number $n \to \infty$. In addition, since the anchors are also sampled from the dataset, we can further assume $P\left(b_i\right)$ also follows a uniform distribution, i.e., $P\left(b_i\right) = 1/q$. With these assumptions, we have:

$$\begin{cases} P\left(b_i\right) = 1/q, P\left(x_j\right) = 1/n \\ P\left(b_i\right) = \sum_{j=1}^{n} P\left(x_j\right) P\left(b_i|x_j\right) \end{cases} \tag{6}$$
$$\Rightarrow \sum_{j=1}^{n} P\left(b_i|x_j\right) = \frac{n}{q} \Rightarrow S_i.1_n = \sigma$$

where $S_i$ is the *i-th* row of $S$ and $\sigma = n/q$ is a fixed value so that $S1_n = (n/q)\,1_q = \sigma 1_q$. We thereby have two constraints on $S$, i.e., $S^T 1_q = 1_n$ and $S1_n = \sigma 1_q$ (the advantages will be shown in the next subsection). Our goal is to calculate a weight matrix $S$ that follows the above constraints so that $S$ has clear stochastic meaning.

Fortunately, this can be simply achieved by iteratively normalizing $S$ both in row and column, i.e.,

$$S^0 \xrightarrow{P_r()} S^1 \xrightarrow{P_c()} S^1 \xrightarrow{P_r()} S^2 \xrightarrow{P_c()} S^2 \to \cdots \tag{7}$$

where $P_c(S) = S\Delta_c^{-1}$ and $P_r(S) = \Delta_r^{-1}S$, $\Delta_c = diag(1S) \in R^{(l+u)\times(l+u)}$ and $\Delta_r = diag(S1) \in R^{q\times q}$. Acutally, the above iterative procedure is equivalent to solving the following optimization problem:

$$\min_S \|S - S_0\|_F^2 \quad s.t. \ S \geq 0, \ S^T 1_q = 1_n, \ S1_n = \sigma 1_q \tag{8}$$

where $S_0$ is the initial $S$ as calculated in Equation (4). Equation (8) involves an instance of quadratic programming (QP), which can be divided into two convex sub-problems:

$$\min_S \|S - S_0\|_F^2 \quad s.t. \ S \geq 0, \ S^T 1_q = 1_n \tag{9}$$

$$\min_S \|S - S_0\|_F^2 \quad s.t. \ S \geq 0, \ S1_n = \sigma 1_q. \tag{10}$$

By the above derivations, the initial QP problem in Equation (8) is tackled by successively alternating between two sub-problems in Equations (9) and (10). This alternate optimization procedure will converge due to Von-Neumann's lemma [27,28]. In addition, Von-Neumann's lemma guarantees that alternately solving the sub-problems in Equations (9) and (10) with the current solution is theoretically guaranteed to converge to the global optima of Equation (8).

### 3.2. Sub-Graph Construction

We have now obtained $q$ anchors and the coefficient $s_j$ of each data $x_j$. The weight matrix $S$ reflects the affinities between data points and anchors, i.e., $X \approx BS$. If we further assume such affinities in the original high-dimensional dataset can be preserved in the low-dimensional class labels, then we have $F \approx ZS$, where $Z = [z_1, z_2, \ldots, z_q] \in R^{c \times q}$ represents the class labels of anchors $B$. This indicates that the class labels of the dataset can be easily obtained by $F = ZS$, given that the class labels of anchors have already been inferred. Since the number of anchors is smaller than that of the dataset, the computational cost for calculating $Z$ can be much lower than directly calculating $F$ in certain conventional graph-based SSL methods. We thereby present an efficient method for semi-supervised learning, in which we aim to develop a sub-graph regularized (SGR) framework for semi-supervised learning by utilizing the information of anchors.

Here, in order to develop our proposed sub-graph SSL method, we need to first construct a sub-graph on the set of anchors and define the adjacency matrix to measure the similarity between any two anchors. There are many approaches to construct the graph by utilizing the anchors, such as conventional $k$NN graph [1,18,20,21]. However, intuitively, we will design the adjacency matrix $W^d \in R^{q \times q}$ by using $S$ as follows:

$$W^d = \frac{1}{\sigma} S S^T. \tag{11}$$

It can be easily proven that $W^d 1_q = (1/\sigma) S S^T 1_q = (1/\sigma) S 1_n = 1_q$. This indicates $W_d$ is a doubly-stochastic matrix. Therefore, the above graph construction can be theoretically derived by a probabilistic means. More straightforward, it can be easily noted that $W^d$ in Equation (11) is an inner product of $S$ with each element $W_{ij}^d = s_i^r$, where $s_i^r s_j^{r^T}$ and $s_j^r$ are the $i$-th and $j$-th rows of $S = \{s_1^r, s_2^r, \ldots, s_q^r\}$. This indicates that the rows of $S$ are denoted as the representations of anchors. In addition, given $b_i$ and $b_j$ share more common data points choosing them as anchors, their corresponding $s_i^r$ and $s_j^r$ will be similar and $W_{ij}^d$ will become a large value; To the constrast, $W_{ij}^d$ will be equal to 0, if $b_i$ and $b_j$ do not share any data points. Therefore $W^d$ derived in Equation (11) can be viewed as an adjacency matrix to measure the similarity between any two anchors.

### 3.3. Efficient Semi-Supervised Learning via Sub-Graph Construction

With the above graph construction, we then develop our sub-graph model for efficient semi-supervised learning. Since the number of anchors is much smaller than that of the dataset, our goal is first to estimate the labels of anchors $Z$ from labeled data via the sub-graph model, and then to calculate those of unlabeled samples by the weight matrix. Here, we first give the objective function of the proposed sub-graph regularized framework for calculating the class labels of anchors as follows:

The first term in Equation (12) is to measure the smoothness of estimated labels on the graph, while the second term is to measure how the estimated labels are consistent original labels, and the third one is a Tikhonov regularization term to avoid the singularity of possible solutions. $\eta_A$ and $\eta_I$ are the parameters balancing the tradeoff of the three terms. By conducting the derivation of $J(Z)$ with regard to $Z$, we can calculate the class labels for anchors as follows:

$$Z^* = Y U S^T \left( S U S^T + \eta_A I + \eta_I L^d \right)^{-1} \tag{12}$$

where $U$ is a diagonal matrix where the first $l$ and the remaining $u$ element are 1 and 0, respectively, $L^d$ is the graph Laplacian matrix of $W^d$. Following Equation (13), we can observe that key computations for $Z^*$ are the inverse of $S_l S_l^T + \eta_I L^d + \eta_A I$, where the complexity is $O(q^3)$. Note that $q \ll l + u$, calculating $Z$ can be much smaller than directly calculating $F$ as in LGC and GFHF. Finally, the class labels of the dataset can be calculated by

$$F = Z^* S = Y U S^T \left( S U S^T + \eta_I L^d + \eta_A I \right)^{-1} S. \tag{13}$$

The basic steps of the proposed SGR are in Algorithm 1.

---

**Algorithm 1:** The proposed SGR

---

1 **Input:** Data $X \in R^{D \times (l+u)}$, label matrix $Y \in R^{c \times (l+u)}$, the number of anchors $q$ and other parameters.

2 From $S$ as Equation (8).

3 Form sub-graph weight matrix as $SS^T$ in Equation (11).

4 Estimate the label matrix of anchors $Z^* = YUS^T \left( SUS^T + \eta_I L^d + \eta_A I \right)^{-1}$ as in Equation (12).

5 Estimate the label matrix of dataset by $F = Z^*S$.

6 **Output:** The predicted label matrix of anchors and dataset $Z \in R^{c \times q}$, $F \in R^{c \times (l+u)}$, respectively.

---

### 3.4. Out-of-Sample Extension via Kernel Regression

The proposed SGR can be used to estimate the labels of unlabeled data. It cannot directly infer the labels of new data. One way to handle such problems is to find a linear projective model by regressioning anchors $B$ on $Z$, i.e.,:

$$V = \arg \min_{V,b} \left\| V^T B + b^T e - Z \right\|_F^2 + \gamma \|Z\|_F^2 \tag{14}$$

where $V \in R^{d \times c}$ is the projection and $b$ is the bias term. Though this linearization assumption $Z = V^T B + b^T e$ provides an effective and efficient solution to the out-of-sample problem. However it is not able to fit the nonlinear distribution. Therefore, we solve the above problem in two ways: (1) We combine the objective function of SGR and the regression term to form a unified framework, so that the class labels of $Z$, the projection $V$, and the bias $b$ can be simultaneously calculated; (2) we utilize the kernel trick to search a nonlinear projection. Specifically, we give the objective function as:

$$\begin{aligned} J(V,Z,b) = \min_{V,Z,b} \sum_{j=1}^{l} \|Zs_j - y_j\|_F^2 + \eta_A \|V\|_F^2 \\ + \eta_R \left\| V^T \varphi(X) + b^T e - Z \right\|_F^2 \\ + \eta_I \sum_{i,j=1}^{q} W_{ij}^d \|z_i - z_j\|_F^2. \end{aligned} \tag{15}$$

It should be noted that $\varphi(B)$ is only implicit and not available. To calculate the optimal $V$, we have to involve some restrictions. In detail, let $V$ have a linear combination of $\varphi(B)$, i.e., $V = \varphi(B) A$, where $A \in R^{q \times c}$ is the coefficient for $V$, then:

$$\begin{aligned} J(V,Z,b) = \min_{V,Z,b} \sum_{j=1}^{l} \|Zs_j - y_j\|_F^2 + \eta_A Tr\left(A^T K A\right) \\ + \eta_R \left\| A^T K + b^T e - Z \right\|_F^2 \\ + \eta_I \sum_{i,j=1}^{q} W_{ij}^d \|z_i - z_j\|_F^2 \end{aligned} \tag{16}$$

where $K$ represents the kernel matrix and we can select Gaussian kernel. By setting the derivatives of Equation (16), if follows:

$$\begin{cases} b = \left( 1_q Z^T - 1_q KA \right) \Big/ 1_q 1_q^T \\ A = \left( KL_c K^T + \eta K \right)^{-1} KL_c Z^T \\ Z = YUS^T \left( SUS^T + \eta_I L^d + \eta_R L^r \right)^{-1} \end{cases} \tag{17}$$

where $\eta = \eta_I/\eta_R$, $L_c = I - 1_q{}^T 1_q / 1_q 1_q{}^T$ is to subtract the mean of all data, $L^r = L_c - L_c K^T (K L_c K^T + \eta I)^{-1} K L_c$. Here, denote $x$ as a new coming data and $x_k$ as its kernel representation, its projected data $t$ can be given $t = V^T x_k + b$ and the label of $x$ is estimated as:

$$c_t = \arg\max_i t\,(i) \tag{18}$$

One toy model example for verifying out-of-sample extensions can be given in Figure 1. In this toy example, we annotate two datasets as labeled sets in each class. We then infer the labels in the region $\{(x,y)\,|\,x \in [-2,2], y \in [-2,2]\}$ by out-of-sample extension both in the linear version and kernel version. The experiment results show that the decision boundary learned by the kernel version is satisfied, since they are both consistent with the data manifold. While the linear version fails to handle the task, due to the two-cycle dataset following a nonlinear distribution.

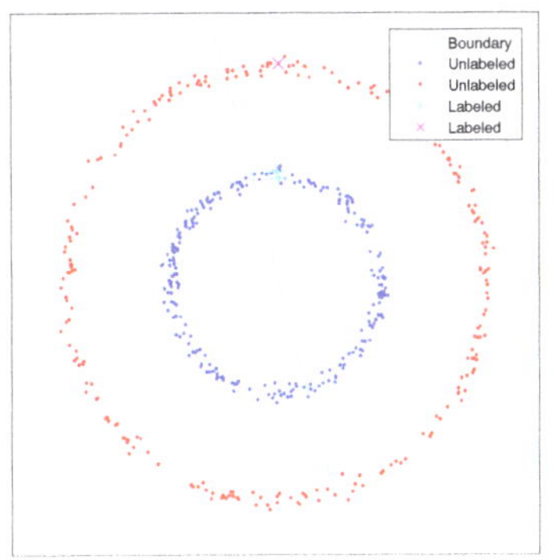

(a) Contour lines of decision boundary for linear version

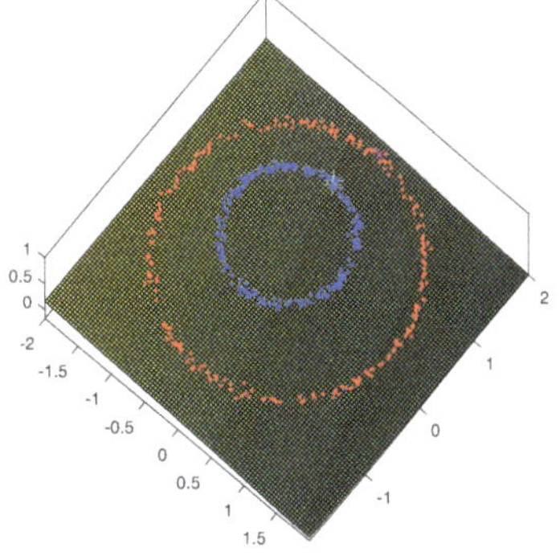

(b) Contour surface of decision region for linear version

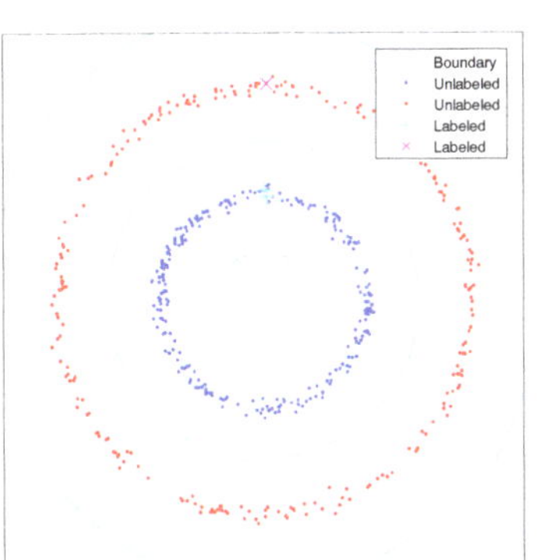

(c) Contour lines of decision boundary for kernel version

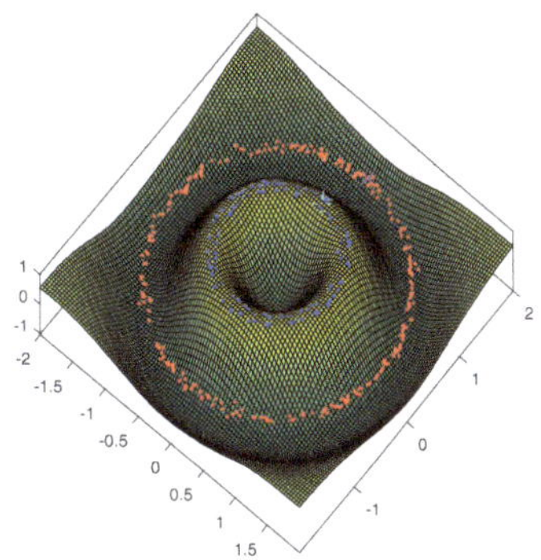

(d) Contour surface of decision region for kernel version

**Figure 1.** Out-of-sample extension: two-cycle dataset in $\{(x,y)\,|\,x \in [-2,2], y \in [-2,2]\}$. (**a,c**) the contour lines of the decision boundary; (**b,d**) the contour surface is the estimated label values in the region. In this experiment, the figures in the upper row represent the results by using a linear prediction model $Z = V^T B + b^T e$, while those on the bottom row represent the results by using a kernel based prediction model $Z = V^T \varphi(B) + b^T e$. Clearly, the kernel prediction model is much better than the linear prediction model since the two-cycle dataset follows a nonlinear distribution.

Note that the proposed method includes three stages of training: (1) initialize the anchors by $k$-means; (2) construct the sub-graph $w^d$; (3) perform SSL. Here, the computational cost of k-means in the first stage is $O\left(q\left(l+u\right)\right)$, while the one for sub-graph construction and SSL strategy in the second and third stage are $W^d$ is $O\left(q\left(l+u\right)\right)$ and $O\left(q^3+\left(l+u\right)q\right)$, respectively. The summary of the computational complexity is in Table 1, from which we can see that if we use a fixed $q$ $(q \ll l+u)$ anchors for large scale dataset, the computational complexity of proposed SGR scales linearly with $l+u$, which indicates the proposed SGR is suitable for handling large-scale data.

**Table 1.** The computational complexity of different stages. Semi-supervised learning (SSL).

| The Proposed Method | The First Stage (Initialization) | The Second Stage (The Proposed Model) | The Third Stage (SSL) | Totals (Considering Large-Scale Data $q \ll l+u$) |
|---|---|---|---|---|
| Computational Complexity | $O\left(q\left(l+u\right)\right)$ | $O\left(q\left(l+u\right)\right)$ | $O\left(q^3+q\left(l+u\right)\right)$ | $O\left(q\left(l+u\right)\right)+O\left(q\left(l+u\right)\right)+$ $O\left(q^3+q\left(l+u\right)\right) \approx O\left(q\left(l+u\right)+q^3\right)$ |

It should be noted a recent work, [29], has proposed another SSL method based coupled graph Laplacian regularization, which is similar to our proposed work. The main advantages for our proposed work compared to [29] can be issued as follows: (1) The proposed constructed graph is doubly-stochastic, so that the constructed graph Laplacian is normalized in each row or column. For the coupled graph Laplacian rigorization, their constructed graph may not be doubly-stochastic; (2) the proposed work can directly handle out-of-sample problems by projecting the newly-coming data on the projection matrix so that the class membership of newly-coming data can be inferred. While for the coupled graph Laplacian regularization, it does not consider this point.

## 4. Experiments

*4.1. Toy Examples for Synthetic Datasets*

We will first show the iterative approach of the proposed method can adaptively reduce the bias of a data manifold, where a dataset of two classes with noises is generated with a half-moon distribution in each class. Here, we use a kernel version of the proposed method to learn the classification model to handle such nonlinear distribution. Figure 2 shows the decision surfaces and boundaries obtained by the proposed method during the iterations. From Figure 2, we can observe that for the two-moon dataset, the results converge fast by only using four iterations. In Figure 2, we can observe that by initially treating each local regression term equal, the boundary learned by the proposed method cannot well separate the two classes as there are many mis-classified data points. However, during the iterative rewrighted process, the converged boundary in Figure 2 after four iterations can be more and more accurate and distinctive due to the reason that the biases caused by the noisy data are seriously reduced.

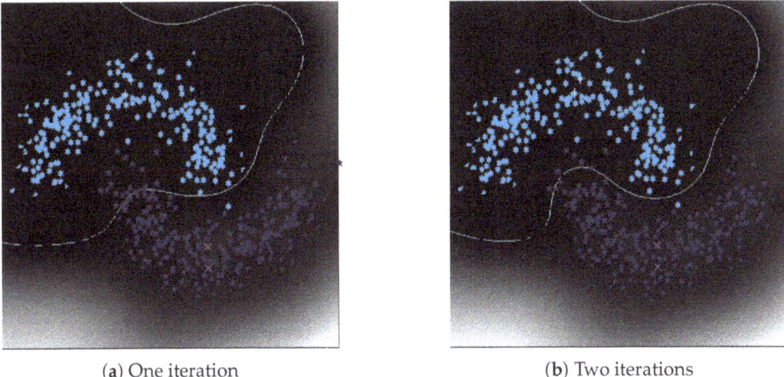

(**a**) One iteration  (**b**) Two iterations

**Figure 2.** *Cont.*

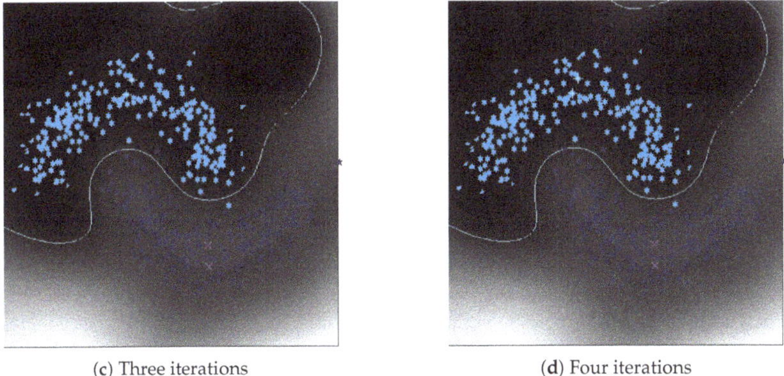

(**c**) Three iterations  (**d**) Four iterations

**Figure 2.** Gray image of reduced space learned by the proposed method: two-moon dataset.

## 4.2. Description of Dataset

In this section, we will utilize six real-world datasets for verification. The six datasets are the Extended Yale-B, Carnegie Mellon University Pose, Illumination and Expression (CMU-PIE), Columbia Object Image Library 100 (COIL-100), Eidgenössische Technische Hochschule 80 (ETH80), U. S. Post Station (USPS) digit image and Chinese Academy of Sciences, Institute of Automation, Hand-Written Digit Base (CASIA-HWDB) datasets. For each dataset, we only select 5%, 10%, 15%, and 20% of the data points to formulate a labeled set randomly, 20% of the data to formulate a test set, and the remaining ones to formulate an unlabeled set. The information of the data and sampled images can be observed in Table 2 and Figure 3, respectively.

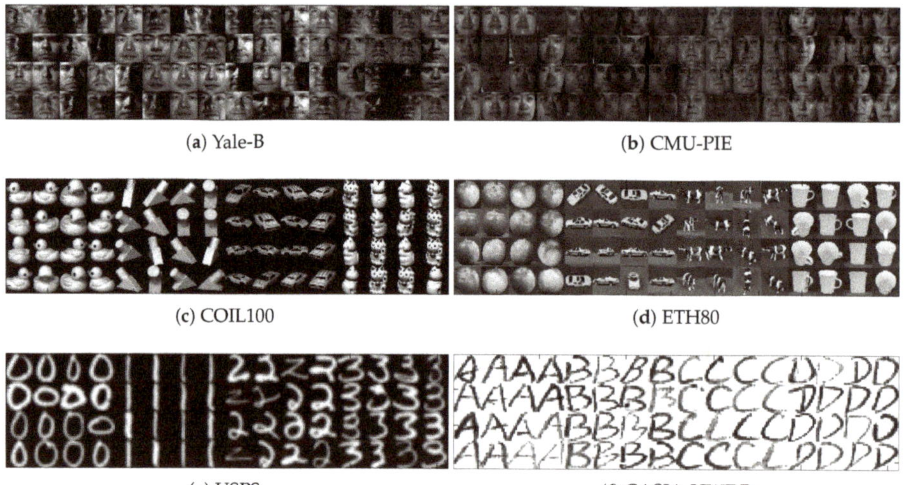

(**a**) Yale-B     (**b**) CMU-PIE

(**c**) COIL100     (**d**) ETH80

(**e**) USPS     (**f**) CASIA-HWDB

**Figure 3.** Sample images of real-world datasets: Yale-B, Carnegie Mellon University Pose, Illumination and Expression (CMU-PIE), Columbia Object Image Library 100 (COIL-100), Eidgenössische Technische Hochschule 80 (ETH80), U. S. Post Station (USPS) digit image and Chinese Academy of Sciences, Institute of Automation, Hand-Written Digit Base (CASIA-HWDB) datasets.

**Table 2.** Information of different datasets.

| Dataset | Database Type | Sample | Dim | Class | Train per Class | Test per Class |
| --- | --- | --- | --- | --- | --- | --- |
| Extended Yale-B [30] | Face | 16123 | 1024 | 38 | 80% | 20% |
| CMU-PIE [31] | Face | 11,000 | 1024 | 68 | 80% | 20% |
| COIL100 [32] | Object | 7200 | 1024 | 100 | 58 | 14 |
| ETH80 [33] | Object | 3280 | 1024 | 80 | 33 | 8 |
| USPS [34] | Hand-written digits | 9298 | 256 | 10 | 800 | remaining |
| CASIA-HWDB [35] | Hand-written letters | 12456 | 256 | 52 | 200 | remaining |

## 4.3. Image Classification

We will show the effectiveness of the proposed SGR for image classification. The experiment settings are as follows [36,37]: For most SSL methods, e.g., LGC, Special Label Propagation (SLP), Linear Neighborhood Propagation (LNP), AGR, Efficient Anchor Graph Regularization (EAGR) and MR, the parameter $k$ for constructing the $k$NN graph is determined by five-fold cross validation, which is chosen from 6 to 20. For LGC, LNP AGR, and EAGR, the regularized parameter is needed to set, which is determined from $\left\{ 10^{-6}, 10^{-3}, 10^{-1}, 1, 10, 10^{3}, 10^{6} \right\}$. The average accuracies of over 50 random splits with changed numbers of labeled data are shown in Tables 3–8. From the classification results, we have:

**Table 3.** Classification accuracies of the Yale-B dataset.

| Methods | 5% Training Labeled | | 10% Training Labeled | | 15% Training Labeled | | 20% Training Labeled | |
|---|---|---|---|---|---|---|---|---|
| | Unlabeled | Test | Unlabeled | Test | Unlabeled | Test | Unlabeled | Test |
| SVM | $53.1 \pm 1.1$ | $52.7 \pm 1.0$ | $68.8 \pm 2.0$ | $67.7 \pm 0.6$ | $75.2 \pm 1.1$ | $73.7 \pm 1.3$ | $80.0 \pm 1.8$ | $78.8 \pm 1.2$ |
| MR | $59.0 \pm 1.2$ | $58.5 \pm 1.3$ | $70.3 \pm 1.1$ | $69.4 \pm 0.5$ | $76.4 \pm 1.3$ | $74.9 \pm 1.5$ | $80.7 \pm 1.3$ | $79.0 \pm 1.1$ |
| LGC | $64.7 \pm 1.0$ | | $71.8 \pm 1.1$ | | $76.4 \pm 4.2$ | | $80.8 \pm 1.0$ | |
| SLP | $65.6 \pm 2.3$ | | $73.9 \pm 1.0$ | | $78.0 \pm 1.8$ | | $81.8 \pm 1.0$ | |
| LNP | $64.9 \pm 1.3$ | $53.8 \pm 2.7$ | $72.0 \pm 1.2$ | $71.2 \pm 0.4$ | $78,0 \pm 2.4$ | $76.6 \pm 2.1$ | $81.6 \pm 1.0$ | $80.0 \pm 1.4$ |
| AGR | $66.6 \pm 1.5$ | $65.8 \pm 1.3$ | $74.3 \pm 1.2$ | $72.2 \pm 0.4$ | $78.1 \pm 1.5$ | $77.3 \pm 1.7$ | $83.0 \pm 1.2$ | $80.0 \pm 4.5$ |
| EAGR | $66.9 \pm 0.8$ | $66.5 \pm 1.8$ | $74.4 \pm 1.1$ | $73.2 \pm 1.5$ | $78.0 \pm 1.5$ | $77.2 \pm 1.9$ | $84.4 \pm 2.4$ | $83.6 \pm 3.1$ |
| SGR | $69.9 \pm 0.4$ | $67.2 \pm 1.0$ | $75.7 \pm 1.1$ | $74.0 \pm 3.3$ | $79.4 \pm 1.0$ | $78.3 \pm 1.1$ | $86.3 \pm 2.5$ | $82.8 \pm 2.4$ |

**Table 4.** Classification accuracies of the CMU-PIE dataset.

| Methods | 5% Training Labeled | | 10% Training Labeled | | 15% Training Labeled | | 20% Training Labeled | |
|---|---|---|---|---|---|---|---|---|
| | Unlabeled | Test | Unlabeled | Test | Unlabeled | Test | Unlabeled | Test |
| SVM | $42.5 \pm 1.3$ | $41.5 \pm 1.1$ | $56.8 \pm 2.2$ | $55.8 \pm 1.5$ | $64.6 \pm 1.2$ | $63.8 \pm 1.8$ | $69.3 \pm 1.7$ | $68.9 \pm 1.2$ |
| MR | $47.8 \pm 1.1$ | $46.7 \pm 1.6$ | $59.3 \pm 1.8$ | $58.8 \pm 1.3$ | $65.6 \pm 1.6$ | $64.5 \pm 1.6$ | $69.9 \pm 1.4$ | $69.1 \pm 1.4$ |
| LGC | $53.5 \pm 1.6$ | | $60.3 \pm 1.7$ | | $66.5 \pm 2.8$ | | $70.5 \pm 1.3$ | |
| SLP | $55.3 \pm 1.9$ | | $63.4 \pm 1.8$ | | $67.2 \pm 1.9$ | | $70.9 \pm 1.3$ | |
| LNP | $55.2 \pm 1.2$ | $54.8 \pm 1.9$ | $62.9 \pm 1.5$ | $61.8 \pm 0.9$ | $68,3 \pm 2.7$ | $67.3 \pm 2.3$ | $71.1 \pm 1.2$ | $71.0 \pm 1.6$ |
| AGR | $56.4 \pm 1.4$ | $55.3 \pm 1.8$ | $64.8 \pm 1.3$ | $64.7 \pm 0.5$ | $68.5 \pm 2.1$ | $66.9 \pm 1.8$ | $72.8 \pm 1.7$ | $71.3 \pm 3.5$ |
| EAGR | $57.2 \pm 1.0$ | $56.4 \pm 1.6$ | $64.4 \pm 1.2$ | $63.7 \pm 1.9$ | $68.4 \pm 1.8$ | $67.7 \pm 2.3$ | $73.1 \pm 2.0$ | $72.4 \pm 2.7$ |
| SGR | $59.0 \pm 0.7$ | $58.4 \pm 1.3$ | $65.6 \pm 1.2$ | $64.6 \pm 1.9$ | $69.8 \pm 1.6$ | $67.9 \pm 1.6$ | $75.0 \pm 2.4$ | $73.9 \pm 2.3$ |

**Table 5.** Classification accuracies of the COIL100 dataset.

| Methods | 5% Training Labeled | | 10% Training Labeled | | 15% Training Labeled | | 20% Training Labeled | |
|---|---|---|---|---|---|---|---|---|
| | Unlabeled | Test | Unlabeled | Test | Unlabeled | Test | Unlabeled | Test |
| SVM | $83.6 \pm 0.9$ | $83.2 \pm 0.8$ | $88.5 \pm 0.8$ | $86.6 \pm 0.8$ | $91.8 \pm 0.8$ | $91.4 \pm 0.7$ | $95.3 \pm 0.8$ | $94.5 \pm 1.6$ |
| MR | $83.7 \pm 1.0$ | $83.4 \pm 0.9$ | $89.0 \pm 0.9$ | $87.3 \pm 0.9$ | $92.1 \pm 0.8$ | $91.6 \pm 0.9$ | $95.3 \pm 0.7$ | $94.7 \pm 1.3$ |
| LGC | $85.5 \pm 0.8$ | | $89.3 \pm 0.9$ | | $92.4 \pm 0.8$ | | $95.5 \pm 0.6$ | |
| SLP | $86.4 \pm 0.7$ | | $89.3 \pm 0.9$ | | $92.8 \pm 0.6$ | | $95.6 \pm 0.8$ | |
| LNP | $86.5 \pm 0.7$ | $85.6 \pm 0.7$ | $89.6 \pm 0.9$ | $88.7 \pm 0.7$ | $92.9 \pm 0.7$ | $92.4 \pm 0.8$ | $95.8 \pm 0.7$ | $95.1 \pm 1.3$ |
| AGR | $86.5 \pm 0.6$ | $85.8 \pm 0.9$ | $90.9 \pm 0.9$ | $88.8 \pm 0.8$ | $93.3 \pm 0.6$ | $92.7 \pm 0.9$ | $95.8 \pm 0.7$ | $95.3 \pm 1.4$ |
| EAGR | $86.6 \pm 0.7$ | $85.7 \pm 1.3$ | $89.9 \pm 0.9$ | $89.0 \pm 1.5$ | $93.2 \pm 0.6$ | $92.7 \pm 1.5$ | $96.0 \pm 0.7$ | $95.2 \pm 0.9$ |
| SGR | $87.0 \pm 0.6$ | $86.7 \pm 1.0$ | $91.8 \pm 0.9$ | $89.7 \pm 0.8$ | $94.7 \pm 0.6$ | $93.2 \pm 0.8$ | $97.0 \pm 0.6$ | $95.6 \pm 0.9$ |

**Table 6.** Classification accuracies of the ETH80 dataset.

| Methods | 5% Training Labeled | | 10% Training Labeled | | 15% Training Labeled | | 20% Training Labeled | |
|---|---|---|---|---|---|---|---|---|
| | Unlabeled | Test | Unlabeled | Test | Unlabeled | Test | Unlabeled | Test |
| SVM | $61.1 \pm 1.3$ | $59.4 \pm 0.3$ | $71.1 \pm 1.9$ | $70.2 \pm 2.0$ | $75.9 \pm 1.5$ | $75.3 \pm 3.1$ | $78.9 \pm 2.0$ | $77.9 \pm 2.5$ |
| MR | $62.3 \pm 0.8$ | $60.0 \pm 0.2$ | $71.7 \pm 2.0$ | $71.0 \pm 2.7$ | $76.2 \pm 1.0$ | $75.3 \pm 2.8$ | $78.9 \pm 1.9$ | $78.3 \pm 2.5$ |
| LGC | $65.7 \pm 1.4$ | | $73.5 \pm 1.4$ | | $76.8 \pm 1.5$ | | $79.0 \pm 1.7$ | |
| SLP | $65.9 \pm 1.5$ | | $73.9 \pm 1.2$ | | $76.9 \pm 1.6$ | | $79.3 \pm 1.8$ | |
| LNP | $64.9 \pm 0.9$ | $62.2 \pm 0.2$ | $73.4 \pm 2.0$ | $71.4 \pm 2.6$ | $76.7 \pm 1.1$ | $76.0 \pm 2.6$ | $79.0 \pm 1.8$ | $78.5 \pm 2.0$ |
| AGR | $66.4 \pm 1.6$ | $65.1 \pm 0.2$ | $75.0 \pm 1.7$ | $72.2 \pm 2.2$ | $76.9 \pm 1.7$ | $76.1 \pm 2.5$ | $79.6 \pm 2.0$ | $78.9 \pm 1.9$ |
| EAGR | $68.2 \pm 1.7$ | $67.7 \pm 2.1$ | $74.9 \pm 1.4$ | $74.2 \pm 1.9$ | $77.3 \pm 1.7$ | $77.0 \pm 1.9$ | $80.0 \pm 2.2$ | $79.4 \pm 2.8$ |
| SGR | $69.4 \pm 1.9$ | $67.2 \pm 0.1$ | $74.0 \pm 1.3$ | $74.2 \pm 2.2$ | $77.5 \pm 1.9$ | $77.3 \pm 1.8$ | $79.8 \pm 22$ | $79.0 \pm 2.2$ |

**Table 7.** Classification accuracies of the ETH80 dataset.

| Methods | 5% Training Labeled | | 10% Training Labeled | | 15% Training Labeled | | 20% Training Labeled | |
|---|---|---|---|---|---|---|---|---|
| | Unlabeled | Test | Unlabeled | Test | Unlabeled | Test | Unlabeled | Test |
| SVM | $71.7 \pm 0.7$ | $70.6 \pm 1.5$ | $77.9 \pm 0.7$ | $77.8 \pm 0.2$ | $91.9 \pm 4.4$ | $90.9 \pm 4.2$ | $96.1 \pm 1.9$ | $95.7 \pm 0.9$ |
| MR | $74.1 \pm 0.7$ | $73.0 \pm 1.5$ | $80.9 \pm 0.8$ | $79.8 \pm 0.1$ | $92.6 \pm 3.4$ | $91.7 \pm 3.4$ | $96.1 \pm 2.2$ | $95.0 \pm 1.0$ |
| LGC | $74.7 \pm 0.7$ | | $87.1 \pm 0.8$ | | $94.6 \pm 3.3$ | | $96.5 \pm 2.3$ | |
| SLP | $75.0 \pm 0.5$ | | $89.7 \pm 0.7$ | | $95.4 \pm 3.0$ | | $96.5 \pm 2.3$ | |
| LNP | $76.5 \pm 0.6$ | $74.8 \pm 0.8$ | $92.0 \pm 0.7$ | $90.8 \pm 0.5$ | $95.5 \pm 3.4$ | $95.0 \pm 3.4$ | $96.9 \pm 2.5$ | $96.5 \pm 0.9$ |
| AGR | $78.7 \pm 0.6$ | $76.1 \pm 0.7$ | $93.6 \pm 0.7$ | $92.6 \pm 0.7$ | $96.0 \pm 2.4$ | $95.8 \pm 2.4$ | $97.1 \pm 2.8$ | $96.7 \pm 0.9$ |
| EAGR | $79.9 \pm 0.6$ | $79.4 \pm 1.2$ | $93.6 \pm 0.7$ | $92.9 \pm 1.1$ | $96.3 \pm 3.6$ | $95.5 \pm 3.5$ | $97.2 \pm 1.7$ | $96.3 \pm 2.2$ |
| SGR | $80.7 \pm 0.5$ | $79.7 \pm 0.7$ | $95.0 \pm 0.5$ | $93.3 \pm 0.8$ | $97.2 \pm 3.1$ | $96.2 \pm 3.1$ | $97.4 \pm 1.5$ | $97.3 \pm 0.7$ |

(1) For almost all methods, the classification results increase given that the number of labeled data increases. For instance, the results of SGR will increase 15% as the number of labeled data is increased from 5% to 20% in most cases. This can almost get 17% increase in CASIA-HWDB dataset. In addition, the classification results will not increase given the number of labeled samples are sufficient especially in the cases of COIL100, USPS, and ETH80 datasets;

(2) The proposed SGR can outperform other methods in all cases. For instance, SGR can achieve 5%–9% superiority over SLP, LNP, and MR in almost all cases. Especially in the CASIA-HWDB dataset, this improvement can even achieve 9%. AGR and EAGR can obtain competitive results as SGR by tuning the parameters. However, the proposed SGR can automatically adjust them while achieving satisfying results;

(3) The accuracies of the unlabeled set outperform those of the test set. This is because the testing data are not utilized for training. However, the accuracies of the test set are still good showing that SGR is able to handling the new incoming data.

**Table 8.** Classification accuracies of the CASIA-HWDB dataset.

| Methods | 5% Training Labeled | | 10% Training Labeled | | 15% Training Labeled | | 20% Training Labeled | |
|---|---|---|---|---|---|---|---|---|
| | Unlabeled | Test | Unlabeled | Test | Unlabeled | Test | Unlabeled | Test |
| SVM | $56.8 \pm 5.4$ | $55.8 \pm 0.6$ | $65.7 \pm 0.6$ | $64.0 \pm 1.7$ | $79.0 \pm 0.5$ | $78.2 \pm 4.0$ | $83.4 \pm 1.8$ | $82.1 \pm 1.9$ |
| MR | $58.7 \pm 3.3$ | $57.3 \pm 0.5$ | $73.0 \pm 0.6$ | $62.0 \pm 1.4$ | $79.4 \pm 0.6$ | $78.4 \pm 2.7$ | $86.6 \pm 1.9$ | $85.5 \pm 1.5$ |
| LGC | $63.1 \pm 2.4$ | | $76.1 \pm 0.4$ | | $80.7 \pm 0.5$ | | $88.1 \pm 1.4$ | |
| SLP | $63.4 \pm 1.6$ | | $77.4 \pm 0.4$ | | $85.3 \pm 0.5$ | | $88.6 \pm 1.7$ | |
| LNP | $66.5 \pm 1.4$ | $64.8 \pm 0.6$ | $78.5 \pm 0.5$ | $77.5 \pm 0.7$ | $85.9 \pm 0.5$ | $84.8 \pm 1.7$ | $89.2 \pm 1.7$ | $90.6 \pm 8.2$ |
| AGR | $72.0 \pm 0.9$ | $71.0 \pm 0.6$ | $80.9 \pm 2.8$ | $77.8 \pm 0.6$ | $87.2 \pm 0.5$ | $86.4 \pm 1.6$ | $91.8 \pm 1.6$ | $90.0 \pm 4.1$ |
| EAGR | $74.9 \pm 0.7$ | $74.4 \pm 1.2$ | $78.6 \pm 3.3$ | $78.0 \pm 3.1$ | $87.6 \pm 0.4$ | $87.2 \pm 1.0$ | $91.6 \pm 1.8$ | $91.2 \pm 2.2$ |
| SGR | $75.3 \pm 0.7$ | $73.6 \pm 0.5$ | $83.6 \pm 2.2$ | $80.3 \pm 0.6$ | $88.7 \pm 0.3$ | $86.5 \pm 1.6$ | $93.2 \pm 1.7$ | $91.7 \pm 3.3$ |

*4.4. Parameter Analysis with Different Numbers of Anchors*

In this subsection, we will verify the accuracies of SGR against different numbers of anchors. In this study, we selected 5% data to formulate a labeled set and the remaining ones to formulate an unlabeled set. Then, in Figure 4, we give the accuracy curve of SGR under different numbers of anchors, where the candidate set is chosen from $\sqrt{n}$ to $10\sqrt{n}$.

From Figure 4, we can see that in ETH80 dataset, the classification results increase when the number of anchors increase. However, the accuracies will not increase anymore given sufficient number of anchors, such as $10\sqrt{n}$. Here, $10\sqrt{n}$ is still much smaller compared with that of original data. For other datasets, the classification accuracies have no change and are less sensitive to the number of anchors.

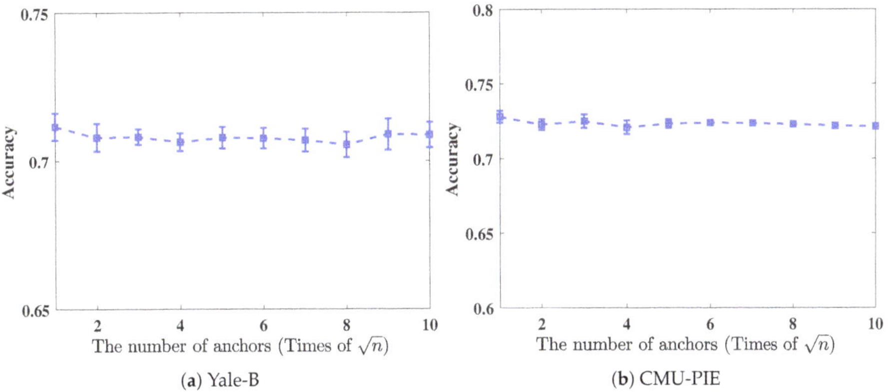

(**a**) Yale-B  (**b**) CMU-PIE

**Figure 4.** *Cont.*

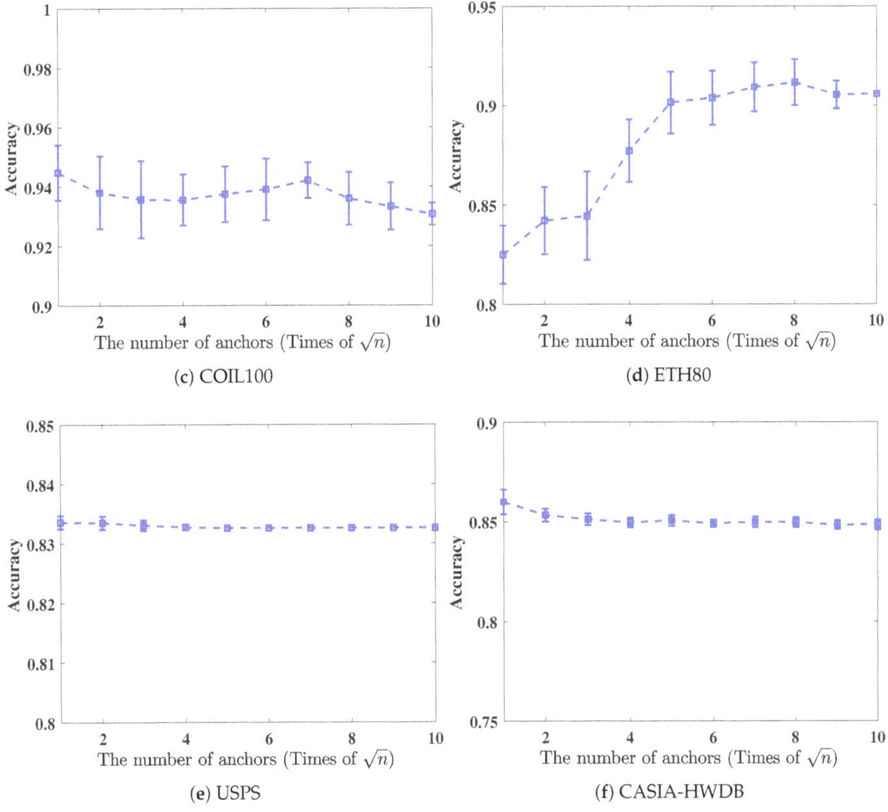

(**c**) COIL100  (**d**) ETH80

(**e**) USPS  (**f**) CASIA-HWDB

**Figure 4.** Classification accuracies over different numbers of anchors.

## 4.5. Image Visualization

In this subsection, we will demonstrate the visualization of the proposed method to show its superiority. In this study, we choose the digit and letter images of the first five classes from

CASIA-HWDB dataset for experiment, where we randomly select 20 data and 80 data in each class to formulate a labeled set and an unlabeled set, respectively. The rest are used to formulate testing data. We then project the test set on the 2D subspace by utilizing a 2D projection matrix for visualization. Since the out-of-sample extension of the proposed SGR and MR are derived from the regression problem, we perform PCA operator on the projection data of $V^T X$ to reduce its dimensionality into two in order to handle the sub-manifold visualization problem. Then, the test data can be visualized on 2D subspace. The experiment results are shown in Figures 5 and 6. From the experiment results, we can observe that SGR can obtain the better performance especially in CASIA-HWDB digit image data.

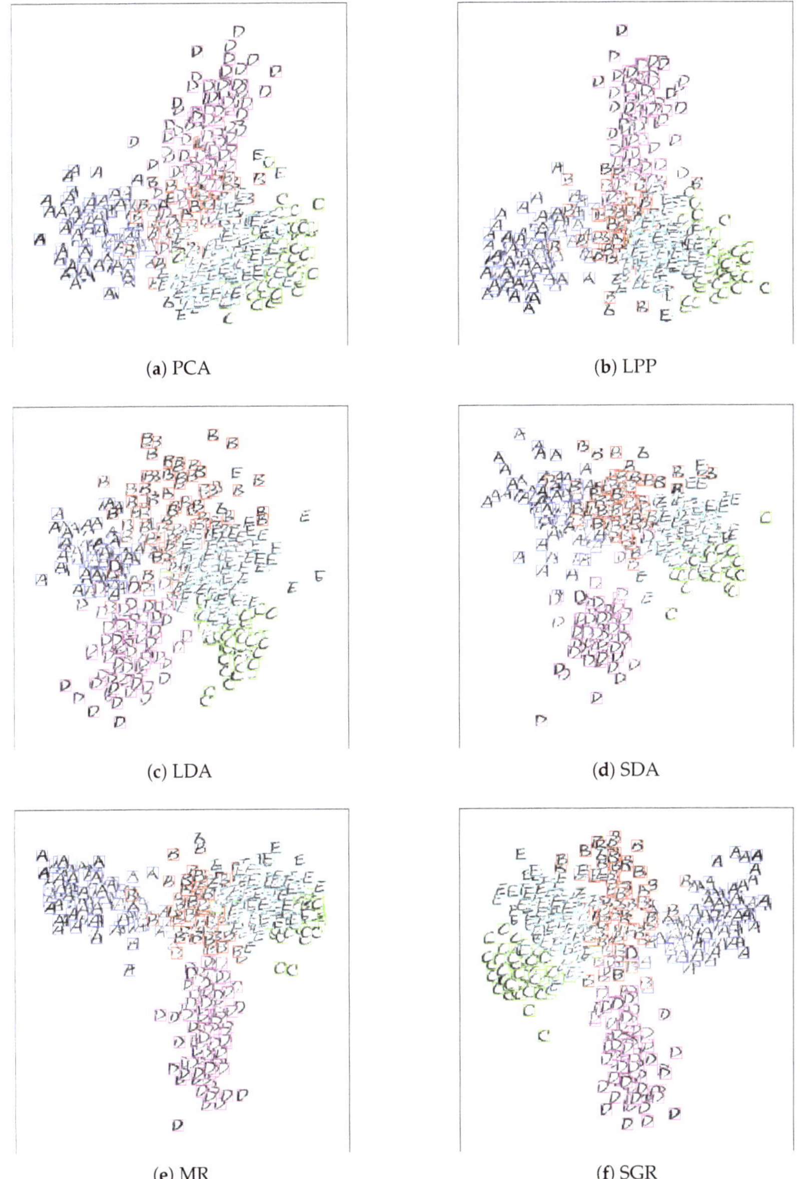

**Figure 5.** Visualization performance of different methods: Five letters images from CASIA-HWDB: Principal Component Analysis (PCA), Locality Preserving Projection (LPP), Linear Discriminant Analysis (LDA), Semi-supervised Discriminant Analysis (SDA), Manifold Regularization (MR) and Sub-Graph Regularization (SGR).

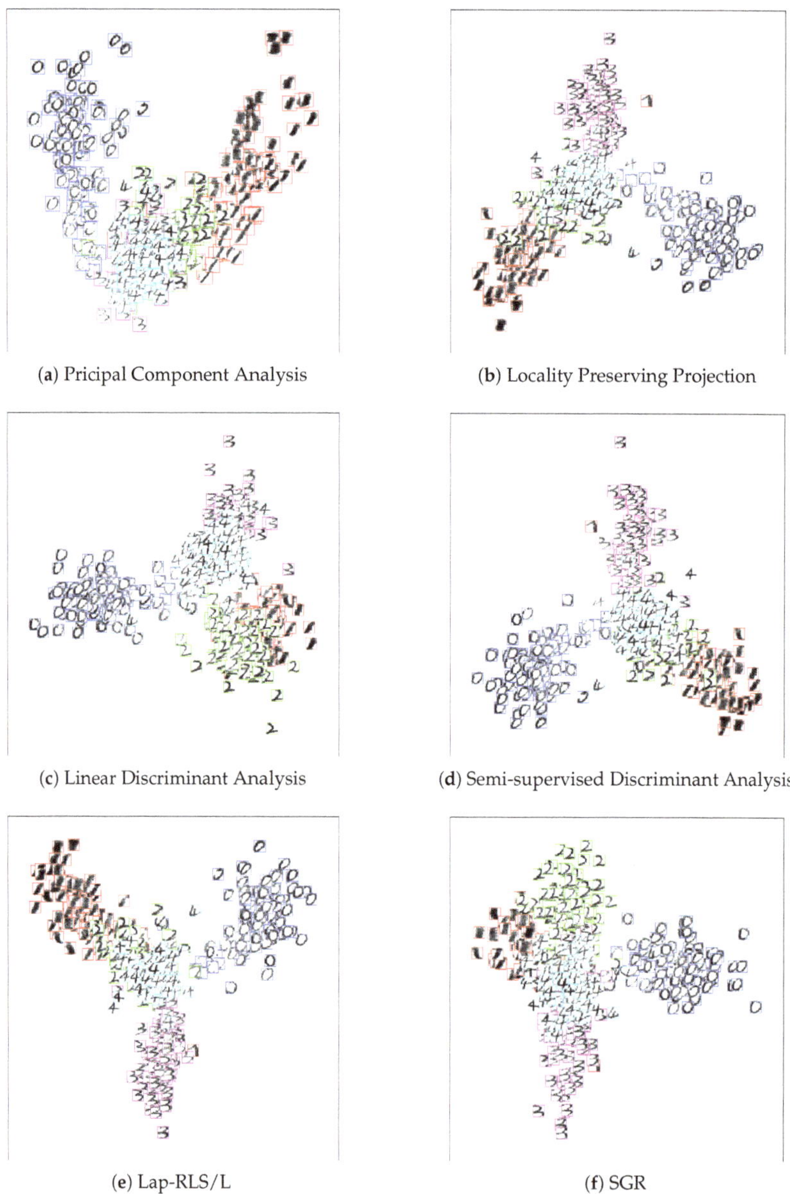

(**a**) Pricipal Component Analysis

(**b**) Locality Preserving Projection

(**c**) Linear Discriminant Analysis

(**d**) Semi-supervised Discriminant Analysis

(**e**) Lap-RLS/L

(**f**) SGR

**Figure 6.** Visualization performance of different methods: five digits images from CASIA-HWDB: Principal Component Analysis (PCA), Locality Preserving Projection (LPP), Linear Discriminant Analysis (LDA), Semi-supervised Discriminant Analysis (SDA), Manifold Regularization (MR) and Sub-Graph Regularization (SGR).

## 5. Conclusions

In this paper, we proposed a sub-graph-based SSL for image classification. The main contributions of the proposed work are as follows:

(1)  We developed a doubly-stochastic $S$ that measures the similarity between data points and anchors. The new updated $S$ has probability means and can be viewed asa transition probability between data points and anchors. In addition, the new sub-graph is constructed by $S$ in an efficient way and can preserve the geometry of data manifold. Simulation results verify the superiority of the proposed SGR;

(2)  We also adopt a linear predictor for inferring the labels of new incoming data, which can handle out-of-sample problems. The computational complexity of this linear predictor is linear with the number of anchors; hence it is efficient. This shows that SGR can handle a large-scale dataset, which is quite practical;

From the above analysis, we can see that the main advantages for the proposed work is the effectiveness for handling the classification problems and that it needs less computational complexity for both graph construction and SSL. It can also handle out-of-sample problems based on a kernel regression on anchors. However, it also suffers the drawback that the parameters are not adaptive. In addition, the graph construction and SSL inference are in two different stages. Our future work can lie in developing a unified framework for optimization with adaptive adjusted parameters.

While the proposed work mainly focuses on image classification, our future work can also lie in handling other state-of-the-art applications, such as image retagging [38], and context classification in the natural language processing field [39,40].

**Author Contributions:** Conceptualization, Software, Methodology, J.L.; Formal analysis, Funding acquisition, Original Draft, M.Z.; Supervision, Validation, Review and editing: W.K.

**Funding:** This work is supported by the National Science Foundation of China under Grant No. 61971121, 61601112 and 61603088, the Fundamental Research Funds for the Central Universities and DHU Distinguished Young Professor Program.

**Conflicts of Interest:** The authors declare no conflict of interest.

## References

1.  Zhu, X.; Ghahramani, Z.; Lafferty, J.D. Semi-supervised learning using gaussian fields and harmonic functions. In Proceedings of the 20th International conference on Machine learning (ICML-03), Washington, DC, USA, 21–24 August 2003.

2.  Zhou, D.; Bousquet, O.; Lal, T.N.; Weston, J.; Scholkopf, B. Learning with local and global consistency. *Advances in Neural Information Processing Systems*; MIT: Cambridge, MA, USA, 2004.

3.  Belkin, M.; Niyogi, P.; Sindhwani, V. Manifold regularization: A geometric framework for learning from labeled and unlabeled samples. *J. Mach. Learn. Res.* **2006**, *7*, 2399–2434.

4.  Nie, F.; Xiang, S.; Liu, Y.; Zhang, C. A general graph based semi-supervised learning with novel class discovery. *Neural Comput. Appl.* **2010**, *19*, 549–555. [CrossRef]

5.  Cai, D.; He, X.; Han, J. Semi-supervised discriminant analysis. In Proceedings of the 2007 IEEE 11th International Conference on Computer Vision, Rio de Janeiro, Brazil, 14–20 October 2007; pp. 1–7.

6.  Zhao, M.; Zhang, Z.; Chow, T.W.; Li, B. Soft label based linear discriminant analysis for image recognition and retrieval. *Comput. Image Underst.* **2014**, *121*, 86–99. [CrossRef]

7.  Zhao, M.; Zhang, Z.; Chow, T.W.; Li, B. A general soft label based linear discriminant analysis for semi-supervised dimensionality reduction. *Neural Netw.* **2014**, *55*, 83–97. [CrossRef]

8.  Zhao, M.; Chow, T.W.; Wu, Z.; Zhang, Z.; Li, B. Learning from normalized local and global discriminative information for semi-supervised regression and dimensionality reduction. *Inf. Sci.* **2015**, *324*, 286–309. [CrossRef]

9.  Zhao, M.; Chow, T.W.; Zhang, Z.; Li, B. Automatic image annotation via compact graph based semi-supervised learning. *Knowl.-Based Syst.* **2015**, *76*, 148–165. [CrossRef]

10.  Zhao, M.; Zhang, Z.; Chow, T.W. Trace ratio criterion based generalized discriminative learning for semi-supervised dimensionality reduction. *Pattern Recognit.* **2012**, *45*, 1482–1499. [CrossRef]

11.  Fukunaga, K. Introduction to statistical pattern classification. *Patt. Recognit.* **1990**, *30*, 1149.

12.  Gao, Y.; Ma, J.; Yuille, A.L. Semi-supervised sparse representation based classification for face recognition with insufficient labeled samples. *arXiv* **2016**, arXiv:1609.03279.

13. Ma, J.; Zhao, J.; Jiang, J.; Zhou, H.; Guo, X. Locality Preserving Matching. *Int. J. Comput. Vis.* **2019**, *127*, 512–531. [CrossRef]

14. Gao, Y.; Yuille, A.L. Estimation of 3D Category-Specific Object Structure: Symmetry, Manhattan and/or Multiple Images. *Int. J. Comput. Vis.* **2019**, *127*, 1501–1526. [CrossRef]

15. Tenenbaum, J.B.; de Silva, V.; Langford, J.C. A global geometric framework for nonlinear dimensionality reduction. *Science* **2000**, *290*, 2319–2323. [CrossRef] [PubMed]

16. Roweis, S.T.; Saul, L.K. Nonlinear dimensionality reduction by locally linear embedding. *Science* **2000**, *290*, 2323–2326. [CrossRef]

17. He, X.; Yan, S.; Hu, Y.; Niyogi, P.; Zhang, H. Face recognition using Laplacianfaces. *IEEE Trans. Pattern Anal. Mach.* **2005**, *27*, 328–340.

18. Wang, F.; Zhang, C. Label propagation through linear neighborhoods. *IEEE Trans. Knowl. Data Eng.* **2008**, *20*, 55–67. [CrossRef]

19. Wang, J.; Wang, F.; Zhang, C.; Shen, H.C.; Quan, L. Linear neighborhood propagation and its applications. *IEEE Trans. Pattern Anal. Machine Intell.* **2009**, *31*, 1600–1615. [CrossRef]

20. Yang, Y.; Nie, F.; Xu, D.; Luo, J.; Zhuang, Y.; Pan, Y. A multimedia retrieval framework based on semi-supervised ranking and relevance feedback. *IEEE Trans. Pattern Anal. Mach. Intell.* **2012**, *34*, 723–742. [CrossRef]

21. Xiang, S.; Nie, F.; Zhang, C. Semi-supervised classification via local spline regression. *IEEE Trans. Pattern Anal. Mach.* **2010**, *32*, 2039–2053. [CrossRef]

22. Liu, W.; He, J.; Chang, S.-F. Large graph construction for scalable semi-supervised learning. In Proceedings of the 27th international conference on machine learning (ICML-10), Haifa, Israel, 21–24 June 2010; pp. 679–686.

23. Liu, W.; Wang, J.; Chang, S.-F. Robust and scalable graph-based semisupervised learning. *Proc. IEEE* **2012**, *100*, 2624–2638. [CrossRef]

24. Wang, M.; Fu, W.; Hao, S.; Tao, D.; Wu, X. Scalable semi-supervised learning by efficient anchor graph regularization. *IEEE Trans. Knowl. Data Eng.* **2016**, *28*, 1864–1877. [CrossRef]

25. Fu, W.; Wang, M.; Hao, S.; Mu, T. Flag: Faster learning on anchor graph with label predictor optimization. *IEEE Trans. Big Data* **2017**. [CrossRef]

26. Wang, M.; Fu, W.; Hao, S.; Liu, H.; Wu, X. Learning on big graph: Label inference and regularization with anchor hierarchy. *IEEE Trans. Knowl. Data Eng.* **2017**, *29*, 1101–1114. [CrossRef]

27. Von Neumann, J. *Functional Operators: Measures and Integrals*; Princeton University Press: Princeton, NJ, USA, 1950; Volume 1.

28. Liu, W.; Chang, S.-F. Robust multi-class transductive learning with graphs. In Proceedings of the 2009 IEEE Conference on Computer Vision and Pattern Recognition, Miami, FL, USA, 20–25 June 2009.

29. Zhao, X.; Wang, D.; Zhang, X.; Gu, N.; Ye, X. Semi-supervised learning based on coupled graph laplacian regularization. *Proceedings of the 2018 Chinese Intelligent Systems Conference*; Springer: Berlin/Heidelberg, Germany, 2019; pp. 131–142.

30. Georghiades, A.S.; Belhumeur, P.N.; Kriegman, D.J. From few to many: Illumination cone models for face recognition under variable lighting and pose. *IEEE Trans. Pattern Anal. Mach.* **2001**, *23*, 643–660. [CrossRef]

31. Baker, S.; Bsat, M. The CMU pose, illumination, and expression database. *IEEE Trans. Pattern Anal. Mach.* **2003**, *25*, 1615.

32. Nene, S.A.; Nayar, S.K.; Murase, H. *Columbia Object Image Library (COIL-100)*; Technical Report CUCS-005-96; Columbia University: New York, NY, USA, 1996.

33. Leibe, B.; Schiele, B. Analyzing appearance and contour based methods for object categorization. In Proceedings of the 2003 IEEE Computer Society Conference on Computer Vision and Pattern Recognition, Madison, WI, USA, 18–20 June 2003; p. II-409.

34. Hull, J.J. A database for handwritten text recognition research. *IEEE Trans. Pattern Anal. Mach. Intell.* **1994**, *16*, 550–554. [CrossRef]

35. Liu, C.-L.; Yin, F.; Wang, D.-H.; Wang, Q.-F. CASIA online and offline chinese handwriting databases. In Proceedings of the 2011 International Conference on Document Analysis and Recognition, 18–21 September 2011; pp. 37–41.

36. Hou, C.; Nie, F.; Wang, F.; Zhang, C.; Wu, Y. Semisupervised learning using negative labels. *IEEE Trans. Neural Netw.* **2011**, *22*, 420–432.

37. Rodriguez, M.Z.; Comin, C.H.; Casanova, D.; Bruno, O.M.; Amancio, D.R.; Costa, L.D.F.; Rodrigues, F.A.; Kestler, H.A. Clustering algorithms: A comparative approach. *PLoS ONE* **2019**, *14*, e0210236. [CrossRef]

38. Tang, J.; Shu, X.; Li, Z.; Jiang, Y.G.; Tian, Q. Social anchor unit graph regularized tensor completion for large scale image retagging. *IEEE Trans. Pattern Anal. Mach. Intell.* **2019**, *41*, 2027–2034. [CrossRef]

39. Amancio, D.R.; Silva, F.N.; Costa, L.d.F. Concentric network symmetry grasps authors' styles in word adjacency networks. *EPL (Europhys. Lett.)* **2015**, *110*, 68001. [CrossRef]

40. Koplenig, A.; Wolfer, S. Studying lexical dynamics and language change via generalized entropies: The problem of sample size. *Entropy* **2019**, *21*, 464. [CrossRef]

*Article*

# Identify Risk Pattern of E-Bike Riders in China Based on Machine Learning Framework

**Chen Wang *, Siyuan Kou and Yanchao Song**

Intelligent Transportation System Research Center, Southeast University, Nanjing 211189, China;
220183009@seu.edu.cn (S.K.); 220173180@seu.edu.cn (Y.S.)
* Correspondence: chen_david_wang@seu.edu.cn; Tel.: +86-150-0517-2397

Received: 28 September 2019; Accepted: 31 October 2019; Published: 6 November 2019

**Abstract:** In this paper, the risk pattern of e-bike riders in China was examined, based on tree-structured machine learning techniques. Three-year crash/violation data were acquired from the Kunshan traffic police department, China. Firstly, high-risk (HR) electric bicycle (e-bike) riders were defined as those with at-fault crash involvement, while others (i.e., non-at-fault or without crash involvement) were considered as non-high-risk (NHR) riders, based on quasi-induced exposure theory. Then, for e-bike riders, their demographics and previous violation-related features were developed based on the crash/violation records. After that, a systematic machine learning (ML) framework was proposed so as to capture the complex risk patterns of those e-bike riders. An ensemble sampling method was selected to deal with the imbalanced datasets. Four tree-structured machine learning methods were compared, and a gradient boost decision tree (GBDT) appeared to be the best. The feature importance and partial dependence were further examined. Interesting findings include the following: (1) tree-structured ML models are able to capture complex risk patterns and interpret them properly; (2) spatial-temporal violation features were found as important indicators of high-risk e-bike riders; and (3) violation behavior features appeared to be more effective than violation punishment-related features, in terms of identifying high-risk e-bike riders. In general, the proposed ML framework is able to identify the complex crash risk pattern of e-bike riders. This paper provides useful insights for policy-makers and traffic practitioners regarding e-bike safety improvement in China.

**Keywords:** e-bike rider; crash risk; machine learning; traffic violation

---

## 1. Introduction

As a convenient, economical, energy-saving, and environmentally-friendly travel tool, electric bicycles (e-bikes) can not only meet the needs of people traveling short or medium distances, but they are also be able to propel the development of the green sustainability concept. At present, China has become the largest e-bike producer and consumer country around the world. According to statistics, e-bike ownership in China has sharply increased from 58 thousand to 29.96 million, from 1998 to 2014 [1,2]. Experts also predict that the ownership of e-bikes may be in a continuous growth condition with the technological innovation and policy improvement of e-bikes in the future [3].

However, increased e-bike ownership also brings a lot of challenges for traffic safety [4]. Yao claimed that the number of crashes reached nearly 56.2 thousand, with an approximately 8.6% annual increase until the year of 2017 in China [5]. Feng et al. revealed the fact that although the total crash rate in China was in a declining state, the crash rate of e-bikes was still a rising trend [6]. Many studies have also shown that the driving behavior of e-bike riders is non-identical to the regular bike riders, that they tend to be more aggressive and careless. Moreover, compared with other drivers, e-bike riders are more prone to violate traffic rules [7]. Those findings suggest that identifying the risk pattern of e-bike riders is important and necessary. Previous literature has revealed many findings regarding the contributory factors of e-bike crash risks, including e-bike violation behaviors, e-bike demographics, and the psychological factors of e-bike riders. However, there is limited literature, to our best knowledge, for identifying the complex risk pattern of e-bike riders.

Thus, in our study, the risk pattern of e-bike riders in Kunshan, China, was deeply examined. At-fault e-bike riders were considered as a high-risk group, while other e-bike riders were used as a non-high-risk group. A number of violation-related features were developed, and four tree-structured machine learning models were utilized to capture the complex risk patterns of e-bike riders. This study is expected to provide policy-makers and traffic practitioners with useful insights into e-bike safety.

## 2. Literature Review

Previous studies have concluded that the demographics, aberrant riding behaviors, and environmental conditions can be selected as predictors of the driving risk of e-bike riders [8]. For example, Haustein et al. found that = age has a negative correlation with e-bike riding safety. They also found that male riders were more likely to encounter riding risk [9]. A study implemented in China indicated that male and aggressive riders were more likely to increase riding risk [10]. Petzoldt et al. claimed that riding e-bikes at a high speed can induce a higher crash risk [11]. Du proved that unregulated riding behavior and riding without a helmet are prone to increase riding risk [12]. Hertach et al. also found that poor roadway conditions would increase riding risk [13]. In addition, some literature has also found that the crash risk of e-bike riders was related to whether they were registered or not [14].

Regarding the crash modeling technique, most previous literature adopted traditional statistical methods to identify crash-related contributing factors. For example, Cherry et al. used simple mathematical statistical indicators to analyze the riding behavior of e-bike riders [15]. Although this method can analyze the driving behavior of e-bike riders, it requires a large amount of historical data and failed to predict the driver's future driving behavior quantitatively [16]. In addition, some mathematical analysis models have also been introduced into the study of the riding behavior of e-bike riders. For example, the logit model was selected to analyze the risk behavior of e-bike riders [17]. Guo et al. adopted a probit model to analyze the factors related to e-bike riders' riding risk [18]. Wang et al. applied the regression model to identify the association between the riding risk and the related factors [19]. Although these studies reasonably identify some risk aspects of e-bike riders, a comprehensive risk pattern recognition of e-bike riders is still lacking. In particular, a complex risk pattern (e.g., non-linear relationships) was not found.

In general, considerable accomplishments have been achieved in e-bike-related safety research. However, few studies have been found to focus on exploring the complex risk pattern of e-bike riders, based on a machine learning framework.

## 3. Study Design

This study was conducted based on e-bike violation records and crash records during the years between 2015 and 2018, throughout the Kunshan City of Jiangsu Province in China, which were collected by the local traffic police department. There were a total of 242,030 e-bike riders identified in both of the two records. The combined dataset contains the demographic, violation-related information, and crash-related information for each e-bike rider, by matching their previous violation and crash records.

Previous studies have shown that the crash rate and fatality rate of e-bike riders were significantly higher than for other vehicle drivers [20]. Once a crash occurs, it is very likely to induce road congestion [21,22]. These findings indicate that when an e-bike accident occurs, it is a significant hazard to both the rider and the traffic condition around him/her. Therefore, it is essential to identify e-bike riders who are prone to crashes. According to previous literature, at fault e-bike riders in crashes were largely considered as a high-risk group [23], while those not at-fault were often used as quasi-induced exposure [24]. There were some arguments on the quasi-induced exposure method about whether not at-fault drivers are representatives of the total population. However, it has been agreed that at-fault drivers are a specific group with a higher crash risk.

Thus, in this study, e-bike riders were divided into two groups, namely: high-risk (HR) riders and non-high-risk (NHR) riders. Those with at-fault crash involvements were determined as HR, while all of the other riders (non-at-fault and no crash involvement) were considered NHR. After cleaning the origin data, 2605 e-bike riders were labeled as HR and 239,425 riders were labeled as NHR.

In order to identify the HR e-bike riders based on the machine learning methods, a number of features were developed. As for the demographics, the age groups were divided into four categories based on exclusive class intervals, namely, teenagers (<18 years old), young-aged riders (18~35 years old), middle-aged riders (35~65 years old), and old-aged riders (>65 years old), according to previous literature [25].

Temporal violation features were developed. For example, if one has two prior violation records during the morning peak-hour traffic in Kunshan (i.e., 07:00–09:00), the feature value for morning peak-hour violation will be coded as two. Spatial violation features were also considered. All of the violation locations were classified into the three groups (high, medium, and low) based on their prior cumulative violation frequency. For determining the high and low locations, 15 and 85 percentiles were used. For example, if one has three violation records at low-violation locations, the low-violation location feature value will be coded as three.

From the original violation dataset, the violation behavior features in the three years were directly extracted for each e-bike rider. For example, if one had two red-light running violation records, the corresponding feature value (i.e., type three: violating traffic signal) would be coded as two. Violation punishment-related features were also developed, including the frequency of violations, frequency of violation type, cumulative penalty, amount of violation penalty, and so on. Table 1 depicts a detailed explanation of the features extracted from the original dataset.

**Table 1.** The features extracted from the original dataset.

| Type | Feature | Variables |
|---|---|---|
| **Demographic Information** | Gender | Male<br>Female |
| | Age | Teenager (<18)<br>Young age (18~35)<br>Middle age (35~65)<br>Old age (>65) |
| **Environmental Information** | Temporal violation features | Late night violation (LNV; 0–6)<br>Morning peak hour violation (MPHV; 7–9)<br>Evening peak hour violation (EPHV; 17–19)<br>Night violation (NV; 20–24)<br>Other time (OT; 10–16) |
| | Spatial violation features | High-frequency location (H_LOC)<br>Middle-frequency location (M_LOC)<br>Low-frequency location (L_LOC) |
| **Violation Information** | Violation behavior features | Not riding on the non-vehicle lane (Type 1)<br>Retrograding (Type 2)<br>Red-light running (Type 3)<br>Aggressive riding (Type 4)<br>Speeding (Type 5)<br>Drunk driving (Type 6)<br>Overloading (Type 7)<br>Other violations (Type 8) |
| | Punishment-related features | Cumulative violation frequency (CVF)<br>Cumulative violation type frequency (CVTF)<br>Cumulative prior penalty (CPP)<br>Average prior penalty (APP)<br>Maximum prior penalty (MPP) |

## 4. Methodology

In order to reasonably identify the risk patterns of e-bike riders, a systematic machine learning-based approach was proposed to deal with a number of data mining issues, including imbalanced datasets, machine learning model selection, hyper-parameter tuning, and model validation. These issues will be discussed in the following sub-sections in Section 4. Figure 1 briefly describes the technical flow of the approach.

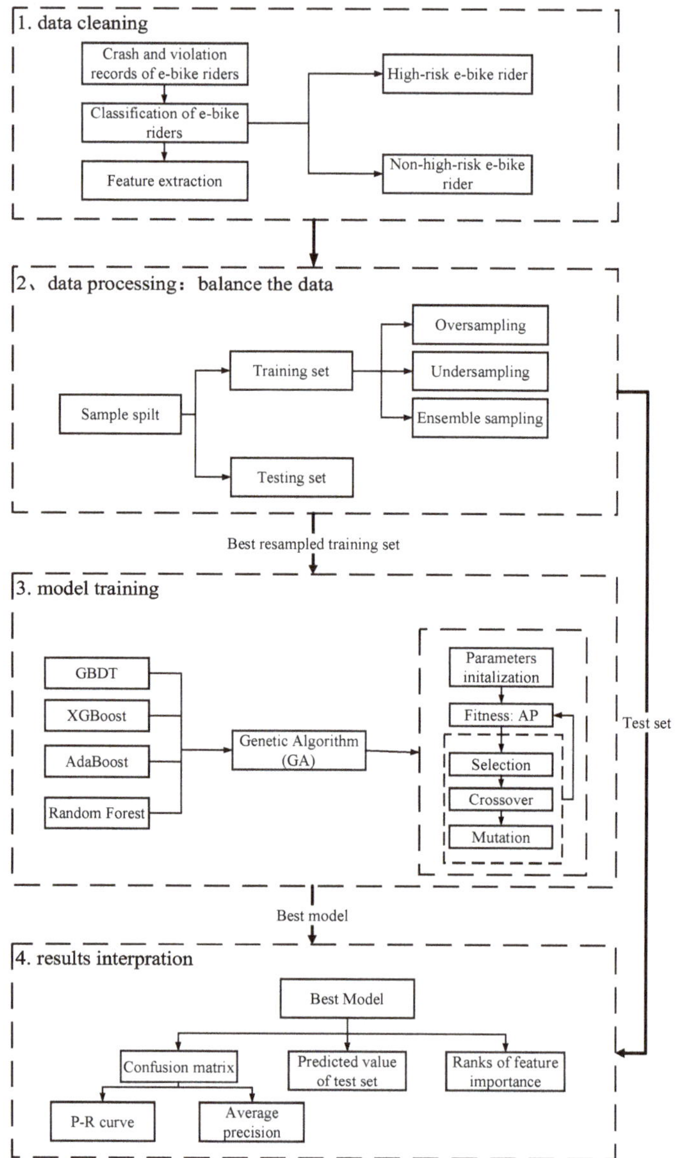


**Figure 1.** Technique flow chart. P–R—precision–recall; GBDT—gradient boosting decision tree; AP—average precision value.

### 4.1. Sampling Techniques

In the dataset, the number of NHR riders greatly exceeds the number of HR riders, which could cause over-fitting issues. Therefore, sampling techniques should be adopted in order to balance the training dataset. In this study, an over-sampling technique (i.e., synthetic minority oversampling technique (SMOTE)), an under-sampling technique (i.e., cluster-centroid method (CC)), and an ensemble method (i.e., balanced bagging classifier (BBC)) were compared. The average precision values (APs) were used as performance measures.

The under-sampling method adopted in this study is the cluster-centroid (CC) method, which makes use of a k-means algorithm to cut down the number of samples in a majority class. The SMOTE method was considered to be an easy and practical over-sampling method [26], the main idea of which is to synthesize new samples based on the K-nearest neighbors of the original samples in the minority class [27]. The ensemble method used in this study is the balanced bagging method, which combines both the sampling and classification techniques. This method first divides the imbalanced dataset into several subsets first. Then, the under-sampling technique and the classifying estimators will be applied to each subset. At last, the final result can be calculated by synthesizing the result of each subset.

After balancing the training set with the above three sampling techniques, a random forest (RF) classifier was provided for the model training. Figure 2 shows the precision–recall (P–R) curves and AP values of the three sampling methods using RF.

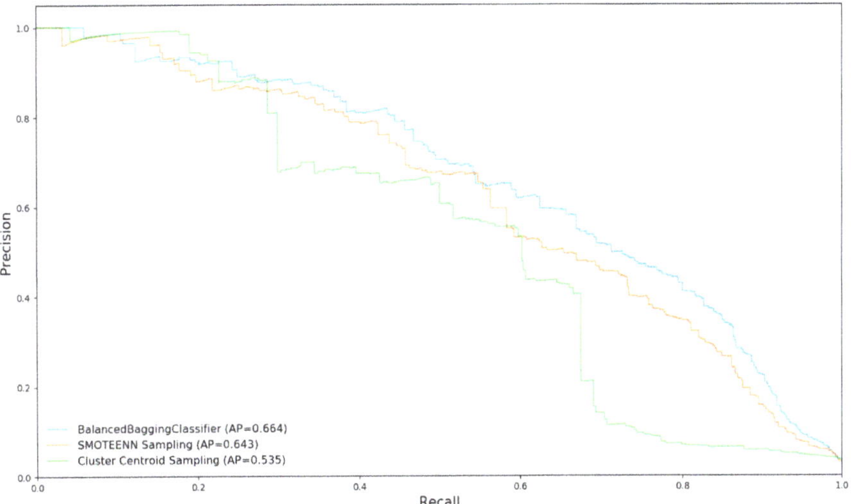


**Figure 2.** P–R curve of each sampling technique.

From the figure, it can be concluded that the balanced bagging method has the best performance. Thus, it was chosen as the standard sampling method, based on which four different tree-structured machine learning models (RF, Adaboost, XGboost, and GBDT) were further trained and compared.

### 4.2. Model Training

In this study, four tree-structured machine learning methods were utilized for the model training, including the RF algorithm, AdaBoost algorithm with a decision tree, XGBoost algorithm, and the gradient boosting decision tree (GBDT). They are all ensemble machine learning algorithms based on decision trees, so that they are more interpretable than other machine learning models.

### 4.2.1. Random Forest

The random forest model is an optimization of the decision tree-based algorithm. The random forest introduces two kinds of randomness, namely: (1) random selected samples and (2) random selected feature variables, which make the algorithm insensitive to noise and difficult to over fit. Therefore, the prediction precision of the random forest algorithm can be improved effectively.

The principle of this algorithm is to randomly extract $n$ subsets from the dataset, and to randomly extract $m$ features for each subset. Then, $n$ incomplete decision trees can be established by the randomly selected samples and features. At last, the output result of each decision tree can be integrated, and then the final classification result of each sample is determined through means of voting.

### 4.2.2. AdaBoost

The AdaBoost classifying technique was first discussed in 1995 by Schapire and Freund [28]. The main idea of AdaBoost is to combine the weak classifiers with suitable weights so as to form a strong classifier [29]. The weighted value can be calculated over multiple iterations. For example, if the sample is classified into the wrong class, its weights will be increased, and vice versa. If the weak classifier has a worse performance than the others, its weight will be reduced, and vice versa. In this study, the selected weak classifier in AdaBoost will also choose a decision tree. The steps of this algorithm can be concluded as follows:

1.  Initialize weights $D_1 = \{w_{11}, w_{12}, w_{13}, \ldots, w_{1n}\}$,

    where, $w_{1i} = \frac{1}{n}$, $i = 1,2,3\ldots n$
2.  For $m = 1$ *to M*:

    (1)  Update training set $\{(x_i, y_i)\}_{i=1}^{n}$ by weighted values;
    (2)  Fit a decision tree $T_m$, namely:

    $$T_m(x) : x \rightarrow \{-1, +1\}$$

    (3)  Compute classification error $e_m$, as follows:

    $$e_m = \sum_{i=1}^{n} w_i I[T(x_i)]$$

    If $T(x_i) \neq y_i$, then, $I[T(x_i)] = 1$; if $T(x_i) = y_i$, then $I[T(x_i)] = 0$;
    (4)  Compute weights $\beta_m$ of decision tree $T_m(x)$, as follows:

    $$\beta_m = \frac{1}{2}log\frac{1 - e_m}{e_m}$$

    (5)  Update the model, as follows:

    $$F_m(x) = sign\left(\sum_{m} \beta_m T_m(x; w_{m,i})\right)$$

    (6)  For $i = 1$ *to n*, update weights $D_{m+1}$ of the training set, as follows:

    $$w_{m+1,i} = \frac{w_{m,i}}{z_m}exp\{-\beta_m y_i I[T(x_i)]\}$$

    Let $z_m = \sum_{i=1}^{N} s_{m,i}exp\{-\beta_m y_i I[T(x_i)]\}$.
3.  Output $F_M(x)$.

### 4.2.3. Gradient Boosting Decision Tree

As for GBDT, it generates a strong classifier by weighting the weak classifier. The purpose of this algorithm is to minimize the loss function. The loss function can be defined as $L(y, F(x)) = \log(1 + e^{-yF(x)})$, in which $F(x)$ is the predicted value. The function of $F(x)$ is $F(x) = \sum_{m=1}^{M} \beta_m h_m(x; s_{m,i})$. To minimize the loss function, a residual term can be added to the prediction function $F(x)$ in each iteration [30,31]. The steps of this algorithm can be concluded as follows:

1. Initialize model as constant $c$:

$$F_0(x) = argmin \sum_{i=1}^{N} L(y, c);$$

2. For $m = 1$ *to* $M$, as follows:

    (1) For $i = i$ to $N$, compute the so-called pseudo-residuals, as follows:

    $$r_{im} = -\left[\frac{\partial L(y_i, F(x_i))}{\partial F(x_i)}\right]_{F(x)=F_{m-1}(x)};$$

    (2) Fit decision tree $h_{mi}$ to $r_{im}$ by training set $\{(x_i, y_i)\}_{i=1}^{n}$;

    $$h_{mi} = \frac{y_i}{1 + \exp(y_i F_{m-1}(x_i))};$$

    (3) Calculate weights $\beta_m$ by minimizing the loss function $L$:

    $$\beta_m = argmin_\beta L(y, F_{m-1}(x) + \beta_m h_{mi});$$

    (4) Update $F_m = F_{m-1}(x) + \beta_m h_{mi}$;

3. Output $F_M(x)$

### 4.2.4. XGBoost

The last adopted algorithm in this study is XGBoost. The main idea of this algorithm is similar to GBDT. Both of them are trying to form a strong classifier by iterating the previous classifier. The differences between XGBoost and GBDT reflect in the following two aspects. First, the objective function of XGBoost is not just a loss function as defined in GBDT. It will add a new term $\Omega(h(x;s))$ after the loss function, which is used to reduce the complexity of the decision tree. Secondly, in GBDT, the tree just fits the first order gradient of the loss function. But XGBoost will replace the first order gradient with both first order and second order gradients when generating a new decision tree in each iteration [32].

## 5. Model Results

### 5.1. Model Performance

In this study, 75% of the data were used for the model training, while the remaining 25% were used for the model testing. Figure 3 shows the receiver operating characteristic (ROC) curves and precision–recall (P–R) curves of the models. The ROC curve reports the false positive rate and true positive rate. The false positive rate was calculated as the ratio between the number of NHRs wrongly categorized as HR and the total number of actual NHRs. The true positive rate refers to the proportion of HR correctly categorized by the model. A P–R curve reports the precision and recall, which is often used to evaluate the model performance for an imbalanced dataset. The precision is the percentage of correctly predicted HR over all of the predicted HR. The recall is calculated in the same way as the true positive rate.

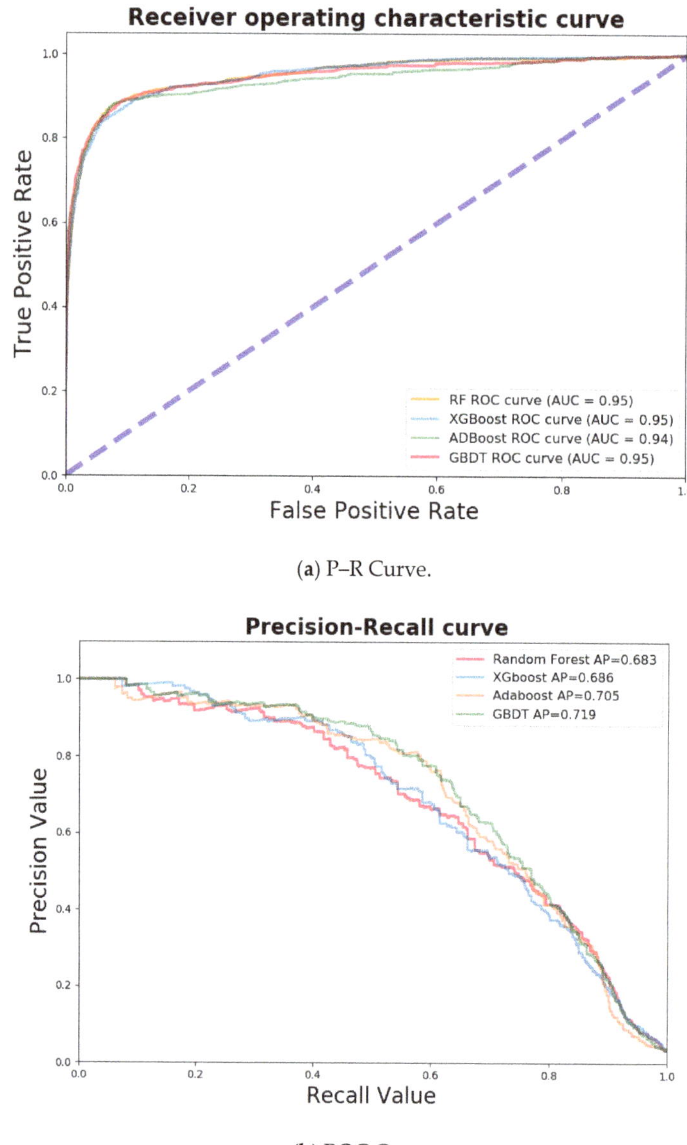


(**a**) P–R Curve.

(**b**) ROC Curve

**Figure 3.** Estimation of the model performance. ROC—receiver operating characteristic.

The four models all have a large area-under-curve (AUC), indicating that they are capable of identifying both the HR and NHR in general. According to the P–R curves, the average precision of the four machine learning models on the testing dataset were 0.683, 0.686, 0.705, and 0.719 (RF, XGBoost, AdaBoost, and GBDT, respectively). The results show that GBDT appears to be the best model, as it is able to better identify the HR of the e-bike riders.

Moreover, the thresholds can be further adjusted to obtain an even higher precision for the HR of e-bike riders, while lowering the recall. The testing dataset contains a total of 18,609 samples, in which the number of HR riders is 651, and the number of NHR riders is 17,958. The adjusted GBDT model prediction results are shown in Table 2, below.

**Table 2.** Adjusted confusion table of the gradient boost decision tree (GBDT) model. HR—high risk; NHR—non-high risk.

|  | NHR Rider | HR Rider | Threshold | Evaluation Index |
|---|---|---|---|---|
| **NHR Rider** | 17,829 | 129 | 0.959 | Recall = 0.596 |
| **HR Rider** | 263 | 388 | | Precision = 0.750 |
| **NHR Rider** | 17,682 | 276 | 0.909 | Recall = 0.70 |
| **HR Rider** | 196 | 455 | | Precision = 0.622 |

*5.2. Feature Importance*

As for the tree-structured machine learning models, the Gini index can be used to evaluate the feature importance. The higher the Gini index, the more importance a feature has. Figure 4, below, presents the importance ranking of all of the features, based on the GBDT model.

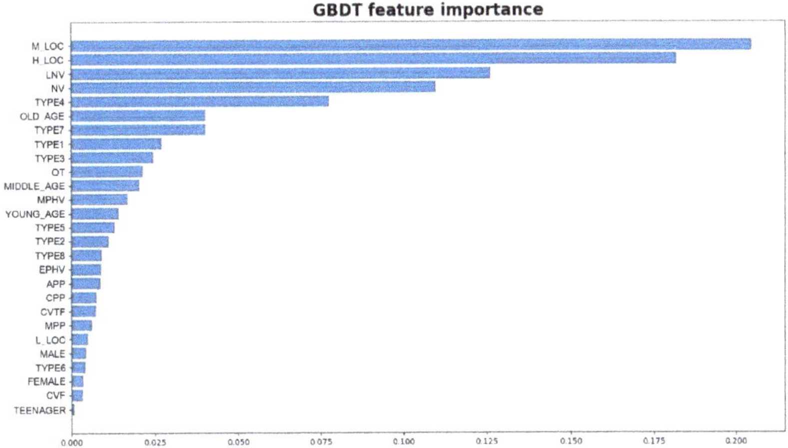


**Figure 4.** Feature importance extracted from the GBDT model.

From the results of the feature importance, it can be seen that the top three most influential features are the medium-violation location (M_LOC), high-violation location (H_LOC), and late-night violation period (LNV). It is surprising to find that the spatial and temporal characteristics of the previous violation experience have the largest effects on the crash risk of e-bike riders. A possible reason could be that those features are related to risk exposure. Violation behaviors were also found to be important, such as aggressive riding (type 4), overloading (TYPE7), not riding on the non-vehicle lane (type 1), and red-light running (type 3). On the other hand, however, violation punishment-related features were not found to be as important as other violation features. As for demographics, older drivers were also found to be an important feature.

### 5.3. Partial Dependence

To deeply examine the complex risk patterns of e-bike riders, the partial dependence was calculated for each feature [31].

$$f_{x_i}(x_i) = E_{x_{-i}}[f(x_i, x_C)] = \int f(x_i, x_C)d\mathbb{P}(x_C) \quad (3)$$

where $x_i$ is the feature $i$, for which the partial dependence function needs to be calculated, and $x_C$ is the other features used in model $f$. The partial function is estimated by calculating the averages in the training data, based on the Monte Carol method, as follows:

$$f_{x_i}(x_i) = \frac{1}{n}\sum_{i=1}^{n} f(x_i, x_{Ci}) \quad (4),$$

where $x_{Ci}$ is the actual values of the feature set $x_C$, according to the dataset, and $n$ is the number of instances in the dataset.

The major advantage of this method is to reveal complex relationships between each factor and outcome. As shown in Figure 5, for each graph, the $x$-axis indicates a feature, while the $y$-axis presents its partial dependence. It can be found that the effects of each feature tend to be non-constant across different levels. In doing so, a complex relationship between the features and outcomes can be properly identified.

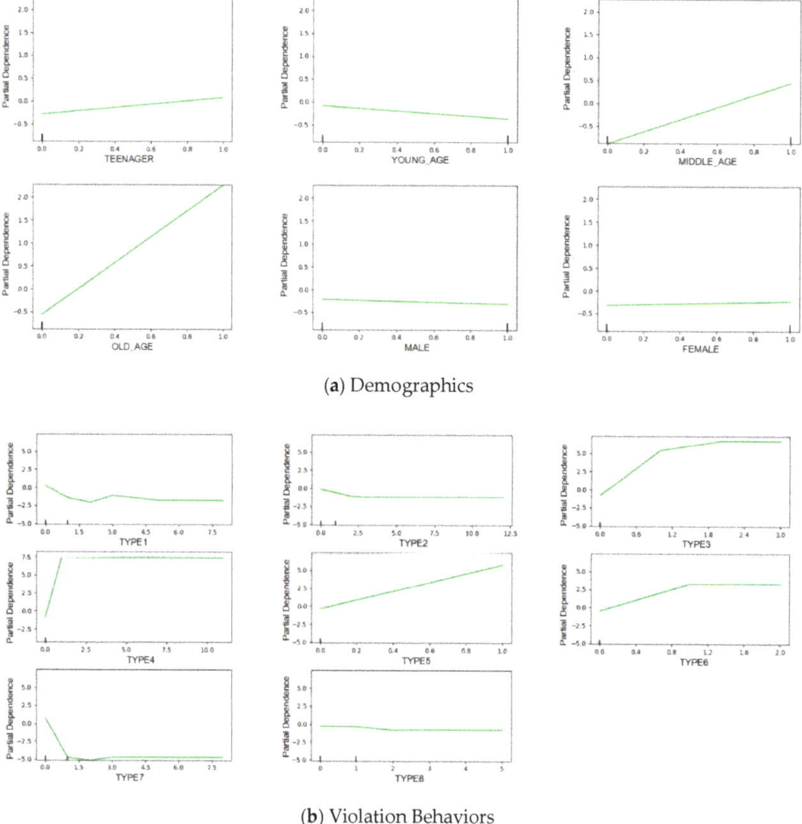

(a) Demographics

(b) Violation Behaviors

**Figure 5.** *Cont.*

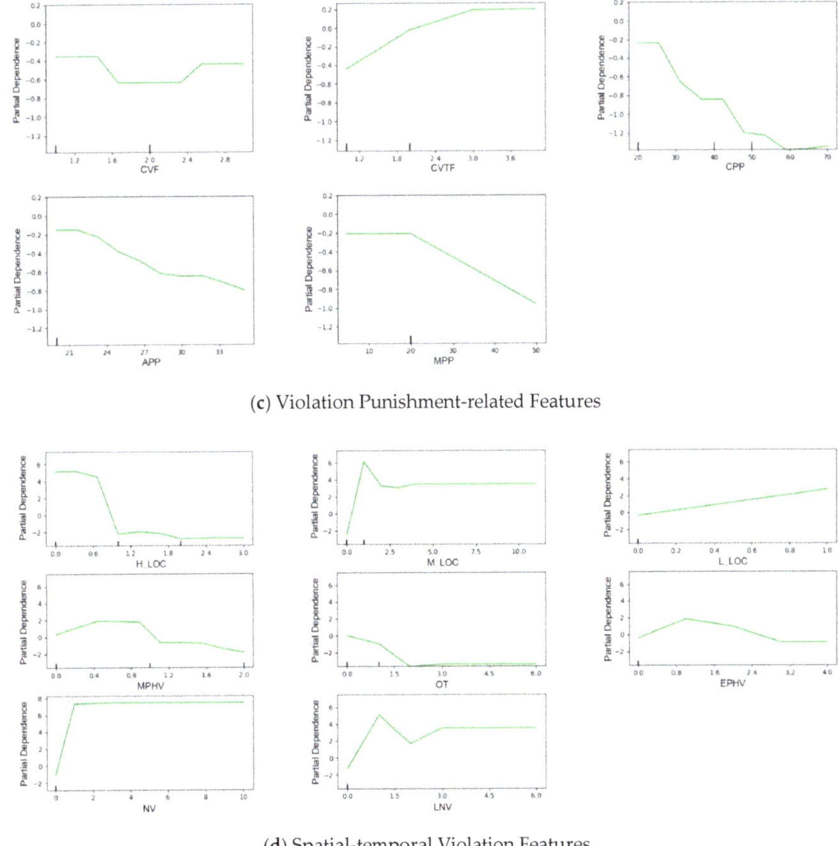

(**c**) Violation Punishment-related Features

(**d**) Spatial-temporal Violation Features

**Figure 5.** Partial dependence plots of the different features.

It appears that gender has little effect on the risk pattern of e-bike riders (Figure 5a). There is a slight trend that females are more likely to be high-risk riders. Regarding age, it has been shown that older and middle-aged e-bike riders tend to have a higher crash risk. According to previous literature, middle-aged riders are more likely to be involved in a crash [18], because their cognitive ability regarding the surrounding roadway risk is weaker than young-aged riders. Older e-bike drivers were also found to be at high risk, as a result of their diminished physical capabilities [17].

From Figure 5b, the e-bike riders with more previous violation experiences of signal violation (type 3), aggressive riding (type 4), over-speed riding (type 5), and drunk driving (type 6), are more likely to be high-risk riders. Those with more experience in lane violation (Type 1) and overloading (type 7) tend to be in the low-risk group. These findings are interesting. Sometimes e-bike riders commit lane violations (i.e., not riding on the non-vehicle lane) may not be intentional, when the non-motorized lane is occupied by other vehicles or e-bikes. E-bike riders with more overloading violation could be more skilled riders. However, these violation behaviors still deserve further investigation.

From Figure 5c, e-bike riders with many previous violation penalty fees and points tend to be NHR riders. This indicates that riders with more penalty points and fees tend to be less aggressive and more cautious, based on the current violation penalty system. Thus, the current violation penalty system can be considered to effectively alter the dangerous driving habits of HR riders.

According to Figure 5d, e-bike riders with more previous violation experience at high-frequency violation locations tend to be NHR. These violate traffic rules at middle- and low-frequency violation locations tend to be HR. At low-frequency violation locations, most e-bike riders obey traffic rules because of their awareness of traffic enforcement, dangerous roadway conditions, and so on. Thus, those who still violated the rules could be either much more aggressive or have a lack of safety awareness, exposing themselves to dangerous situations. Moreover, the violation behaviors of e-bike riders at low-frequency violation locations could be unexpected for other road users, and more likely to cause crashes. E-bike riders with more violation experiences at night (i.e., night violation (20–24) and late night violation (0–6)) are more likely to be high-risk riders. During the night, visibility could be largely decreased for drivers, because of poor light conditions. Thus, e-bike riders are more likely to be involved in a crash with motor-vehicles when they violate traffic rules.

## 6. Conclusions

In this study, the complex risk pattern of e-bike riders was examined based on a systematic machine learning framework. Three-year crash/violation records were acquired from the Kunshan traffic police department. Based on the quasi-induced exposure theory, at-fault e-bike riders were considered as a high-risk (HR) group, while other e-bike riders (including non-at-fault and non-crash-involved) were determined as a non-high-risk (NHR) group. Demographics and violation-related features were created, and four tree-structured ML models were developed to differentiate the HR group from the NHR group. The following conclusions can be drawn:

(1)   High-risk e-bike riders can be reasonably identified based on a machine learning approach. The major advantage of ML models is that they can better identify complex relationships between the features and e-bike riders' riding risk (i.e., the complex risk pattern), which are not easily captured in traditional statistical methods. Explicable methods, such as tree-based ensemble methods, are highly recommended for such tasks.

(2)   Spatial-temporal violation features were found to be important. Riders with more violation records in low- and medium-violation locations appeared to be high-risk riders. Meanwhile, those with more violation records during night were likely to be high-risk riders. Those features could be useful for identifying and characterizing high-risk e-bike riders.

(3)   Riders' violation behavior was found to be highly correlated with crash risk, as it is able to reflect riders' risky riding habits and attitudes. Violation behavior features appear to be more effective than violation punishment-related features, in terms of capturing the risk pattern of e-bike riders.

In general, the proposed method appears to be a promising tool to identify the risk pattern of e-bike riders in China. Moreover, as such, potential high-risk e-bike riders can be early identified, and certain safety interventions can be applied on this group. For example, they can be mandatorily required to attend traffic school, or can receive safety warning messages routinely. In doing so, e-bike safety in China is expected to be improved in the future.

Admittedly, this research also has some limitations. First, it would be interesting to understand if/how the selected features could support the prediction of new e-bike riders that might have some features missing. Second, it would be more convincing to compare the proposed model to other statistical/machine learning models. However, to our best knowledge, this is the first paper to identify the complex risk pattern of e-bike riders based on machine learning models. Third, the full database containing all of the registered e-bike riders can be used to validate the model. The current database only contains those with either violation or crash records. However, according to quasi-induced exposure theory [24], the current model can still be considered as a reliable tool. We recommend that future research can focus on these directions.

**Author Contributions:** Conceptualization, C.W.; Data curation, S.K and Y.S.; Methodology, C.W. and S.K.; Resources, C.W.; Supervision, C.W.; Writing—review & editing, C.W.

**Funding:** This research was supported by the National Key R&D Program of China (2018YFE0102700).

**Conflicts of Interest:** The authors declare there is no conflict of interest.

## References


1. National Bureau of Statistics of People's Republic of China. *China Statistical Yearbook*; National Bureau of Statistics of People's Republic of China: Beijing, China, 2015.
2. Guo, Y.; Li, Z.; Wu, Y.; Xu, C. Exploring unobserved heterogeneity in bicyclists' red-light running behaviors at different crossing facilities. *Accid. Anal. Prev.* **2018**, *115*, 118–127. [CrossRef] [PubMed]
3. Chen, J. Development Trend of Electric Bicycles. *For. Mach. Woodwork. Equip.* **2017**, *45*, 7–9. [CrossRef]
4. Wang, C.; Xu, C.; Xia, J.; Qian, Z.; Lu, L. A combined use of microscopic traffic simulation and extreme value methods for traffic safety evaluation. *Transp. Res. Part C Emerg. Technol.* **2018**, *90*, 281–291. [CrossRef]
5. Yao, X. The New National Standard for the Rapid Development of Electric Bicycles. *Electr. Bicycl.* **2018**, *2*, 1–5. (In Chinese)
6. Feng, Z.; Xu, Z.; Huang, D.; Jin, T.; Raghuwanshi, R.P.; Zhang, C. Electric-bicycle-related injury: A rising traffic injury burden in China. *Inj. Prev.* **2010**, *16*, 417–419. [CrossRef]
7. Brustman, R. *An Analysis of Available Bicycle and Pedestrian Accident Data: A Report to the New York Bicycling Coalition*; New York Bicycling Coalition: Albany, NY, USA, 1999.
8. Hu, F.; Lv, D.; Zhu, J.; Fang, J. Related Risk Factors for Injury Severity of E-Bike and Bicycle Crashes in Hefei. *Traffic Inj. Prev.* **2014**, *15*, 319–323. [CrossRef]
9. Haustein, S.; Møller, M. E-bike safety: Individual-level factors and incident characteristics. *J. Transp. Health* **2016**, *3*, 386–394. [CrossRef]
10. Yao, L.; Wu, C. Traffic Safety for Electric Bike Riders in China Attitudes, Risk Perception, and Aberrant Riding Behaviors. *Transp. Res. Rec.* **2012**, *24*, 49–56. [CrossRef]
11. Petzoldt, T.; Schleinitz, K.; Heilmann, S.; Gehlert, T. Traffic Conflicts and Their Contextual Factors When Riding Conventional vs. Electric Bicycles. *Transp. Res. Part F Traffic Psychol. Behav.* **2017**, *46*, 477–490. [CrossRef]
12. Du, W.; Yang, J.; Powis, B.; Zheng, X.; Ozanne-Smith, J.; Bilston, L.; Wu, M. Understanding on-road practices of electric bike riders: An observational study in a developed city of China. *Accid. Anal. Prev.* **2013**, *59*, 319–326. [CrossRef]
13. Hertach, P.; Uhr, A.; Niemann, S.; Cavegn, M. Characteristics of single-vehicle crashes with e-bikes in Switzerland. *Accid. Anal. Prev.* **2018**, *117*, 232–238. [CrossRef] [PubMed]
14. Zhou, J.; Guo, Y.; Wu, Y.; Dong, S. Assessing Factors Related to E-Bike Crash and E-Bike License Plate Use. *J. Transp. Syst. Eng. Inf. Technol.* **2017**, *17*, 229–234. [CrossRef]
15. Cherry, C.; Cervero, R. Use characteristics and mode choice behavior of electric bike users in China. *Transp. Policy* **2007**, *14*, 247–257. [CrossRef]
16. Wang, C.; Xu, C.; Dai, Y. A crash prediction method based on bivariate extreme value theory and video-based vehicle trajectory data. *Accid. Anal. Prev.* **2019**, *123*, 365–373. [CrossRef]
17. Wang, C.; Xu, C.; Xia, J.; Qian, Z. Modeling Faults among E-Bike-Related Fatal Crashes in China. *Traffic Inj. Prev.* **2017**, *18*, 175–181. [CrossRef]
18. Guo, Y.; Zhou, J.; Wu, Y.; Chen, J. Evaluation of Factors Affecting E-Bike Involved Crash and E-Bike License Plate Use in China Using a Bivariate Probit Model. *J. Adv. Transp.* **2017**, *2017*, 1–12. [CrossRef]
19. Wang, C.; Xu, C.; Xia, J.; Qian, Z. The effects of safety knowledge and psychological factors on self-reported risky driving behaviors including group violations for e-bike riders in China. *Transp. Res. Part F Traffic Psychol. Behav.* **2018**, *56*, 344–353. [CrossRef]
20. Jianan, Z.; Shuai, D.; Xinyu, Z. Characteristics of Electric Bike Accidents and Safety Enhancement Strategies. *Urban Transp. China* **2018**, *16*, 15–20. [CrossRef]
21. Shi, X.; Wang, W. Study on Accident Effects in Urban Transportation Network. *J. Highw. Transp. Res. Dev.* **2000**, *17*, 38–41.
22. Chen, J.; Li, Z.; Wang, W.; Jiang, H. Evaluating Bicycle–Vehicle Conflicts and Delays on Urban Streets with Bike Lane and on-Street Parking. *Transp. Lett.* **2018**, *10*, 1–11. [CrossRef]
23. Rose, G. E-Bikes and Urban Transportation: Emerging Issues and Unresolved Questions. *Transportation* **2012**, *39*, 81–96. [CrossRef]

24. Chandraratna, S.; Stamatiadis, N. Quasi-induced exposure method: Evaluation of not-at-fault assumption. *Accid. Anal. Prev.* **2009**, *41*, 308–313. [CrossRef] [PubMed]

25. Wang, C.; Liu, L.; Xu, C.; Lv, W. Predicting Future Driving Risk of Crash-Involved Drivers Based on a Systematic Machine Learning Framework. *Int. J. Environ. Res. Public Health* **2019**, *16*, 334. [CrossRef] [PubMed]

26. Chawla, N.V.; Bowyer, K.W.; Hall, L.O.; Kegelmeyer, W.P. SMOTE: Synthetic Minority Over-sampling Technique. *J. Artif. Intell. Res.* **2002**, *16*, 321–357. [CrossRef]

27. Elreedy, D.; Atiya, A.F. A Comprehensive Analysis of Synthetic Minority Oversampling TEchnique (SMOTE) for Handling Class Imbalance. *Inf. Sci.* **2019**, *505*, 32–64. [CrossRef]

28. Freund, Y.; Schapire, R.E. A Decisiontheoretic Generalization of On-Line Learning and an Application to Boosting. *J. Comput. Syst. Sci.* **1995**, *55*, 119–139. [CrossRef]

29. Hu, W.; Member, S.; Hu, W.; Maybank, S. AdaBoost-Based Algorithm for Network. *IEEE Trans. Syst. Man Cybern. Part B Cybern.* **2008**, *38*, 577–583.

30. Ding, C.; Cao, X.J.; Næss, P. Applying gradient boosting decision trees to examine non-linear effects of the built environment on driving distance in Oslo. *Transp. Res. Part A Policy Pract.* **2018**, *110*, 107–117. [CrossRef]

31. Friedman, J.H. Greedy Function Approximation: A Gradient Boosting Machine. *Ann. Stat.* **2001**, *29*, 1189–1232. [CrossRef]

32. Gómez-Ríos, A.; Luengo, J.; Herrera, F. A Study on the Noise Label Influence in Boosting Algorithms: AdaBoost, GBM and XGBoost. *Lect. Notes Comput. Sci.* **2017**, *10334*, 268–280.

MDPI
St. Alban-Anlage 66
4052 Basel
Switzerland
Tel. +41 61 683 77 34
Fax +41 61 302 89 18
www.mdpi.com

*Entropy* Editorial Office
E-mail: entropy@mdpi.com
www.mdpi.com/journal/entropy